LIQUID CULTURE SYSTEMS FOR *IN VITRO* PLANT PROPAGATION

Liquid Culture Systems for *in vitro* Plant Propagation

Edited by

ANNE KATHRINE HVOSLEF-EIDE

Department of Plant and Environmental Science, Agricultural University of Norway, Ås, Norway

and

WALTER PREIL

Formerly Federal Centre for Breeding Research on Cultivated Plants, Institute for Ornamental Plant Breeding, Ahrensburg, Germany

Springer

A C.I.P. Catalogue record for this book is available from the Library of Congress.

ISBN 978-90-481-6814-9 (PB)
ISBN 978-1-4020-3200-4 (e-book)

Published by Springer,
P.O. Box 17, 3300 AA Dordrecht, The Netherlands.

Sold and distributed in North, Central and South America
by Springer,
101 Philip Drive, Norwell, MA 02061, U.S.A.

In all other countries, sold and distributed
by Springer,
P.O. Box 322, 3300 AH Dordrecht, The Netherlands.

Printed on acid-free paper

Printed in the Netherlands.

Dedicated to

all Members of COST Actions 822 and 843

in 1994 - 2004

Contents

Contributing Authors

Preface

Plant cell, tissue and organ culture is the basis of plant micropropagation. In recent years, liquid culture systems based on shoot cultures or somatic embryos have become of increasing interest to commercial micropropagators for some stages of the plant propagation cycle. Various vessels have been investigated for liquid cultures, from simple devices supplying an arbitrary amount of oxygen, to complex computer-controlled bioreactors that have been especially designed for plant cell multiplication and regeneration. The reasoning behind the choice of liquid systems for micropropagation is to simplify handling and reduce labour costs, but often it gives rise to the need for costly and complicated equipment instead. In addition, liquid systems often result in bottlenecks or limiting factors greater than those encountered using gelled media. All users of plant cell, tissue or organ cultures in liquid media face similar problems, irrespective of species with which they work; namely, problems related to the initiation and development of cultures and their liquid environment; how to supply sufficient oxygen; how to maintain asepsis; and how to maintain genetic quality in spite of somaclonal, epigenetic and physiological variations. Some of these issues are more prominent in liquid culture systems, compared with those on gelled media; some may be less prominent.

This book is a result of an initiative in the scientific network of COST[1] action 843 'Quality enhancement of plant propagation through tissue culture'

1 COST is a non-commercial cooperation in the field of scientific and technical research, funded by the European Commission. More information is available at www.cost.cordis.lu/src/home.cfm.

and sponsored by the EU. The focus is on the opportunities, advantages and bottlenecks associated with liquid systems. We have built upon the expertise from previous COST actions in Europe, such as COST 87 and COST 822 that have created networks within Europe for scientists and commercial enterprises interested in developing more advanced systems for plant regeneration. Working Group 2 (WG2), "Advanced Regeneration Techniques" of COST843, realised during the first of its network meetings in Finland in 2000, that there was a need for a book dedicated to *in vitro* regeneration from liquid cultures. One thought led to another, and at the end of a creative evening, we had outlined the book you are holding in your hand! We decided to invite world-leading scientists within their fields to contribute to such a book, but firstly to be contributors to the 1st International Symposium on 'Liquid Systems for *in vitro* Mass Propagation of Plants', at Ås, near Oslo, Norway in May/June 2001.

In this book, eminent scientists from all over the world have given their perspectives from their various specialised fields. In addition, experts from industry present a commercial perspective. The various systems used for growth and developments of cultures in liquid media are presented: simple and complex bioreactors and temporary immersion systems, as well as combinations of these. The contributions comprise an essential reference for large-scale plant propagation in liquid cultures. The Scientific Committee chose the invited contributors for their valuable expertise in this specialised field, and collected these into chapters of this book; together with chapters offered by scientists from around the world.

We express our sincere appreciation to all the authors who have contributed to this book, and to our colleagues and families for their valuable support throughout this work.

Anne Kathrine (Trine) Hvoslef-Eide
Co-ordinator of WG2 in COST843

Walter Preil
Co-ordinator in COST87 and COST822

Chapter 1

General introduction: a personal reflection on the use of liquid media for *in vitro* culture

Walter Preil
Institute for Ornamental Plant Breeding, Bornkampsweg 31, D – 22926 Ahrensburg, Germany. E-mail: w.preil@bafz.de

1. Early attempts: Gottfried Haberlandt and his successors

When studying biology in the 1960s I started *in vitro* culture of pea roots at a students' course. Before that, I had never heard about Haberlandt's experiments published in 1902, and about his successors like Robbins (1922) or Kotte (1922) who grew pea root tips *in vitro* for the first time. Since students' literature searches were carried out manually in the 1960s by evaluating the relevant literature and by going back to cited previous publications, important information was often overlooked or found only by chance.

Today, electronic databases ensure a comprehensive literature overview. Publications older than thirty years which are not stored as electronic data, however, are in danger of being forgotten. This can result in reports on "new" experiments supplied for publishing "for the first time". Since reviewers and publishers sometimes do not recognize the repetitions of publications from the pre-electronic period, recent articles may present old stories being declared as new.

The translation of Haberlandt's pioneering publication from German to English by Krikorian and Berquam in 1969 led to the international acceptance of Haberlandt's role in initiating cell and tissue culture research as a new field of Plant Physiology and Applied Botany. Today, students working with plant cultures *in vitro* cannot but come across Haberlandt's name.

A.K. Hvoslef-Eide and W. Preil (eds.), Liquid Culture Systems for in vitro Plant Propagation, 1–18.

When Haberlandt (1902) started his experiments he used liquid medium for culturing isolated cells from bracts of *Lamium purpureum* in Knop's solution supplemented with 1 or 5 % (w/v) sucrose. The cells survived for more than one month. However, cell division was never observed. Similar results were achieved by Bobilioff-Preisser (1917) who cultivated mesophyll cells of *Viola lutea* and *Thunbergia alata*. The cells were kept alive on agar-gelled medium for 2-4 months. They died earlier when grown in liquid medium. No cell divisions were achieved under either of the culture conditions.

Literature on *in vitro* culture using liquid or agar-gelled media was reviewed by Fiedler (1938/1939), who summarized the unsuccessful efforts from 1902 until 1938. The review also includes reports on fruitless attempts to cultivate small tissue pieces. The recommendation based on the experiences of that period was, preferably, to use agar-gelled media. The culture conditions on agar media seemed to mimic best the natural environment, especially due to optimal oxygen supply when culturing the cells, tissues and organs on the medium surface. An exception among the numerous disappointing experiments was the successful cultivation of excised root tips of pea and corn on both agar-gelled or liquid media (Robbins 1922 a, b; Kotte 1922 a, b). A continuous growth of tomato roots tips in liquid medium was achieved by White (1934) when excising and subculturing newly-formed root meristems. This enabled more precise investigation into media compositions and other environmental factors. A considerable success in tissue culture was achieved when Nobecourt (1939), White (1939) and Gautheret (1939) reported, almost simultaneously, the indefinite growth of *Nicotiana* callus on both auxin-containing medium or on auxin-free medium when using *Nicotiana* tumour tissues which do not require growth substances.

Finally, nearly 60 years after Haberlandt's first experiments, differentiated mesophyll cells of *Macleaya* directly isolated from leaves developed in liquid media into cell clusters and calli forming organs and somatic embryos (Kohlenbach 1959, 1965, 1966, 2003). Thus the totipotency of differentiated cells predicted by Haberlandt in 1902 had been confirmed. For reviews on the early history of plant cell, tissue and organ culture see (*e.g.*). Street (1973), Gautheret (1982, 1985), Härtel (2003), Höxtermann (2003).

2. Cell suspensions for secondary metabolite production

Cell suspensions are defined as single cells or small cell aggregates in agitated liquid media. Suspensions of cultured cells capable of long-term subculturing were described first for *Nicotiana* and *Tagetes* (Muir, 1953; Muir et al., 1958) and *Daucus* (Steward and Shantz, 1956). Nickell (1956) demonstrated the feasibility of growing cell suspensions like 'micro-organisms' by cultivating highly dispersed suspensions of cells derived from hypocotyls of *Phaseolus vulgaris*. A new technique for isolating and cloning single cells by plating of filtrated suspensions on an agar layer (Bergmann, 1959, 1960) ensured the selection of somaclonal variants or mutants, to establish variant clones. Detailed overviews on culture techniques and growth patterns in cell suspension cultures were given (*e.g.*) by Rajasekhar et al. (1971), Wilson et al. (1971) and King and Street (1973). A few years later, secondary metabolite production using cell suspensions was one of the main topics of the First International Congress on Medical Plant Research "Plant Tissue Culture and its Biotechnological Application" in Munich 1976. The preface of the congress proceedings emphasized optimistically: "*The possibility of growing plant cells like microorganisms offers a basis for the development of new technologies. The production of the natural products of higher plants by means of plant cell suspension cultures seems to be possible. Consequently one can conceive the future industrial production of primary and secondary plant products by fermentation of plant cells. Prior to the industrial application of plant cell cultures, however, several premises need to be investigated. Cell culture strains have to be selected which produce the desired natural product in the highest possible yields and which simultaneously show high growth rates. Suitable fermentation technologies together with growth and production media for mass cultivation of plant cells have to be developed* " (Barz et al., 1977).

Alfermann et al. (2003) summarized the strategy outlined by Zenk et al. in 1977 as follows:

"1. *Screen plants for high accumulation of the desired natural compound(s)*
2. *Initiate callus cultures from selected high producing parent plants*
3. *Analyze these cultures for the desired product(s)*
4. *Establish cell suspension cultures from producing callus strains*
5. *Analyze the suspension cultures*
6. *Select high producing cell lines via single cell cloning using random selection based on somaclonal variations or mutagenic treatment*
7. *Ultimate objective: Selection of stable high producing cell lines*
8. *Further improvement of product yields by optimization of the culture process.*"

At that time numerous secondary products were already detected in plant tissue cultures of various plant species (Butcher, 1977). Today we have to admit that only four plant cell culture systems are or were used for large scale commercial production: Shikonin from *Lithospermum erythrorhizon*, ginsenosides from *Panax ginseng*, purpurin from *Rubia akane* and paclitaxel from *Taxus* sp. (Alfermann et al., 2003).

Two reasons may explain the relatively poor success of the theoretically convincing strategy for biotechnological production of secondary compounds. 1. Cell culture systems have to produce the metabolites more cheaply than do the conventional systems. This cannot be achieved in most cases by cell cultures. 2. Dedifferentiated cells in liquid media usually do not accumulate the desired compounds in quantities sufficient for commercialisation.

Based on the negative experience of the last decades new strategies are now being developed, like the identification and incorporation of genes for special biosynthetic steps which may alter the pattern of natural product accumulation (Alfermann et al., 2003) or the culture of highly differentiated organs, shoots or plants instead of suspended cells. Temporary immersion systems as described in this volume may solve some of the problems in accumulating the desired secondary products in future.

3. Suspension culture for plant propagation

3.1 Propagation via adventitious or axillary shoots

Enthusiastic discussions on the use of suspension culture for rapid plant propagation started in the early 1970s. Ben-Jaacov and Langhans (1972) estimated that one shoot tip of *Chrysanthemum* can produce 100,000 plantlets in less than one year when subculturing suspended organogenic callus in rotating vessels and afterwards placing the cultures on stationary medium for plantlet development. Two years later Earle and Langhans (1974 a, b) described a method using liquid Murashige and Skoog medium (1962). According to the authors that system "*could produce up to* 9×10^{14} *plantlets or 90 billion 15 cm plants within a year, a great increase over the number possible via conventional propagation*". There is no doubt that for *Chrysanthemum* such high numbers of plants can be achieved from suspension culture. Nevertheless, the described techniques are not used in Europe for vegetative propagation of *Chrysanthemum*, one of the most commercially-significant cut flower and ornamental pot plant. The reason for this unexpected situation is that conventional propagation of *Chrysanthemum* by cuttings produced in the open in southern countries and rooted in greenhouses in Northern Europe, result in cheaper plants than can be obtained from *in vitro* culture. Additionally, the risk of undesired somaclonal variants arising from proliferating cell aggregates can be avoided. However, suspension cultures were successfully used for mutagenic treatments and mutant selection in *Chrysanthemum* (Huitema et al., 1989, 1991; Preil et al., 1991).

Similarly, high numbers of plants accessible from liquid cultures of *Chrysanthemum* were also reported for other species. For *Aechmea,* Zimmer and Pieper (1975) determined propagation rates of 8 to 11-fold after four weeks subculture and calculated a theoretical production of 70 billion plants per year. In other bromeliads, when using single leaves which were separated from *in vitro*-grown plants, it was assumed to be theoretically possible to obtain more than 500 plants in a year, starting with one mother plant (Hosoki and Asahira, 1980). For carnations, Earle and Langhans (1975) reported that after four month of axillary shoot proliferation in liquid medium (assuming 60-fold increase every six weeks) 100,000 flowering plants may arise per year from one initial shoot tip. Although using a chimeral carnation cultivar, no alteration from the original type was observed, indicating that the chimeral arrangement of the petal tissue had not been disturbed by the culture procedure, i.e. only axillary shoots were initiated resulting in true-to-type clonal progeny. Propagation of chimeral cultivars *via* adventitious shoots in every case would result in rearrangements of the apical constitution and loss of the original phenotype.

Higher multiplication rates in liquid media compared to cultures on agar-gelled media were reported for many species. For example, in some *Rhododendron* cultivars, shoot production in liquid medium was 10-fold higher than with agar-gelled medium (Douglas, 1984). According to Bergervoet et al. (1989) in *Cucumis sativus* a 50 ml suspension culture might yield about fifteen hundred plants arising from adventitious buds. The growth of *Rosa chinensis* shoots cultured in liquid medium was superior to those cultured on two-phase (solid-liquid) medium or solid medium alone (Chu et al., 1993).

Despite many promising results, the use of liquid culture systems has not yet become a routine technique in the propagation of the economically most significant species. The risk of hyperhydricity (vitrification) is one of the main reasons. Further, handling of cultures on agar-gelled media is easier, in most cases for commercial laboratories. The orchid micropropagation industry is an exception to this rule. The propagators use liquid media routinely either in the initial phase of the propagation process or for proliferation of protocorm-like bodies (PLB's). Pieper and Zimmer (1976) reported that within 98 days more than 1.1 kg protocorms of *Cymbidium* were produced in liquid medium after inoculating aerated stationary vessels with 15 g PLB's. The authors calculated that 280,000 single protocorms could be obtained from one kg of protocorm biomass. A comprehensive description of micropropagation techniques including the use of liquid media was compiled for almost all orchid species of economic interest by Arditti and Ernst (1993).

3.2 Propagation via somatic embryos

Liquid media were used from the beginning of research on developmental pathways of cells leading to somatic embryos (Steward et al., 1958; Reinert, 1958). Plantlet regeneration was achieved when agar-gelled medium was inoculated with freely suspended cells or cell aggregates from *Daucus* suspension cultures (Steward et al., 1958). The similarities of cell suspension derived regenerants with zygotic embryos were soon stated (Steward, 1958). Thus the totipotency of cells and its significance for morphology and embryology had been definitely confirmed (Steward, 1968; Steward et al., 1970) as predicted by Haberlandt (1902), who summarized: " *... finally, I think that I am not making too bold a prophecy if I point to the possibility that one could probably succeed in regenerating artificial embryos from vegetative cells.*"

Tisserat et al. listed 32 families of angiosperms, 81 genera, and 132 species that have been described as producing somatic embryos as early as in 1979. Bajaj (1995) estimated that somatic embryos had been induced in more than 300 plant species belonging to a wide range of families. Examples for somatic embryogenesis achieved in 180 species of herbaceous dicots were given by Brown et al. (1995), whereas KrishnaRaj and Vasil (1995) listed 120 species of herbaceous monocots. Additionally, somatic embryogenesis has been reported for approximately 150 woody species and related hybrids (Dunstan et al. 1995).

Embryogenesis offers several advantages over other developmental pathways due to possibilities of automation of various process stages (Cervelli and Senaratna, 1995; Ibaraki und Kurata, 2001). Although substantial progress in understanding the biology of somatic embryogenesis has been achieved in the past (Merkle et al., 1995; Yeung, 1995; Nomura and Komamine, 1995; Dudits et al., 1995), insufficiencies still exist from the practical point of view, e.g. large differences in embryogenic response to cultural conditions, even in closely related genotypes, and the variation in hormonal requirement at different stages of embryo development.

Considerable progress towards commercialisation of somatic embryo production for mass propagation was achieved in conifer species (Gupta and Timmis, this volume). In *Cyclamen* a yield of some 90,000 plantlets from one litre of embryogenic suspension culture seems to be possible

(Winkelmann et al., 1998). Another study in *Cyclamen* described that from one litre of suspension culture 52,000 embryos may germinate (29 % germination rate) resulting in 33,800 plantlets *in vitro* (65 % conversion rate). In total of 27,000 young plants (80 % acclimatisation rate) could be obtained after 38 weeks culture period (Hohe et al., 2001).

Difficulties in the application of somatic embryogenesis to plant propagation of insufficiently characterised genotypes may arise from their unknown need for auxin and the duration of auxin application. In many cases, only relatively few cells from heterogeneous embryogenic cultures are able to develop into normal embryos, whereas the major part of the cell population differentiated either into callused or aberrant embryos or non-embryogenic vacuolated cell clusters. Malformed embryos regenerate sometimes in high numbers, e.g. embryos lacking cotyledons or plumules, multiheaded embryos or those with excessive root growth. Such variability of embryogenic cultures prevails, even when suspensions were cultured according to standard protocols.

Lack of reproducibility of propagation protocols is reported for many species. Variation in achieved embryo numbers is evidently often caused by loss of embryogenic competence of cultures during subcultivation. Since uniformity of embryos and reproducibility of the propagation process is one of the prerequisites of any industrial production, synchronisation of embryo development remains one of the most important problems. Whether embryogenesis becomes economically feasible for commercial mass-propagation of a wide range of cultivars, will depend on essential improvements for the embryo regeneration pathway.

Some studies have shown that capacity for embryogenesis is heritable. Two dominant genes with complementary interaction are assumed to control somatic embryogenesis in *Medicago sativa* (Wan et al., 1988; Yu and Pauls, 1993), *Cucumis sativus* (Nadolska-Orczyk and Malepszy, 1989), *Dactylis glomerata* (Tar'an and Bowley, 1997) and *Trifolium pratense* (McLean and Nowak, 1998). In *Cyclamen* a genetic control of the regeneration ability *via* somatic embryogenesis by two dominant major genes with epistatic interaction was postulated (Püschel et al., 2003). Further literature is listed by Henry et al. (1994). In future breeding programmes, selection of genotypes with high embryogenic competence might result in cultivars specifically designed for propagation *via* somatic embryos.

4. Bioreactors for plant propagation

The term "bioreactor" is not precisely defined. It is in use for large-scale vessels for plant biomass production. Bioreactors are normally connected to units controlling temperature, pH, aeration, stirring and various other devices. However, in some cases *in vitro* culture vessels that deviate from Erlenmeyer flasks, Petri dishes or culture boxes are called "bioreactors", too, when innovations and improvements of vessel design were introduced.

Bioreactors have been primarily developed for culturing microorganisms and later for plant cell suspensions to accumulate cell biomass for secondary metabolite production. For industrial cultivation of plant cells, the largest European bioreactor cascade consisting of connected vessels of 75 litres, 750 litres, 7.5 m^3, 15 m^3 and 75 m^3 volumes was established and used for suspension cultures of *Echinacea purpurea, Rauwolfia serpentina* and some other species by the Diversa Company in Ahrensburg, Germany (Rittershaus et al., 1989; Westphal, 1990). Needless to say, for plant propagation purposes bioreactors of a few litres are sufficiently large because high numbers of propagules can be obtained from small vessels.

The aim of bioreactor application is to provide optimum growth conditions by regulating chemical or physical parameters to achieve either maximum yield and high quality of the propagules, or to keep the production costs as low as possible by integration of automated facilities and simple low-cost devices. Since the mid of 1980s various publications stressed the advantages of bioreactors and discussed diverse designs or application strategies (e.g. Ammirato and Styer, 1985; Styer, 1985; Preil, 1991; Takayama, 1991; Denchev et al., 1992; Takayama and Akita, 1994, 1998; Heyerdahl et al., 1995; Ziv, 2000; Ziv, this volume).

4.1 Organogenic cultures

Plant species suitable for propagation through organogenic pathways in various types of bioreactors have been listed by Takayama (1991), Takayama and Akita (1994), Ziv (2000) and Paek et al. (this volume): for example *Anoectochilus, Anthurium,* apple, banana, *Begonia,* Boston fern, *Chrysanthemum, Colocasia, Dieffenbachia, Dioscorea,* garlic, *Gladiolus,* grape, *Hippeastrum, Lilium, Narcissus, Nerine,* Ornithogalum, *Phalaenopsis,* pineapple, *Pinellia*, poplar, potato, *Saintpaulia, Sinningia (Gloxinia), Spathiphyllum, Stevia* and strawberry. Many other species are potential candidates for scaling-up organogenic cultures. In general monocotyledonous plants, including orchids, are easier to handle in liquid systems than dicotyledonous species which tend to become more hyperhydric (vitrified) and etiolated when high plant densities accumulate in the bioreactor. An impressive example for mass propagation of shoots of *Stevia rebaudiana* using a bioreactor of 500 litre volume was given by Akita et al. (1994) and Takayama and Akita (1994).

4.2 Embryogenic cultures

In 1985 Ammirato and Styer outlined strategies for large-scale manipulation of somatic embryos in suspension culture using bioreactors. In the years following this, right up to the present day, numerous authors have repeatedly restated the general principles of bioreactor application, stressing the advantages of embryogenic pathways for plant propagation, including the production of artificial seeds (for literature on problems of artificial seeds see Redenbaugh, 1993).

No integration of bioreactors for the industrial production of somatic embryos has been achieved in any economically-important crop plant until now. Some of the difficulties in regulation of embryogenic processes have been already discussed in section 3.2. Nevertheless, in various species, bioreactor cultures can be used as a part of the propagation process, especially in scaling-up of pro-embryogenic cell masses and globular or heart-shaped embryos for further development on agar-gelled media or growth in temporary immersion systems (TIS).

Reports on embryo production in bioreactors includes a wide range of taxonomically diverse species, e.g. alfalfa (Stuart et al., 1987), sandalwood (Bapat et al., 1990), sweet potato (Bieniek et al., 1995) and *Picea sitchensis* (Ingram and Mavituna, 2000), chosen at random. Investigations on the effects of oxygen supply, carbon dioxide accumulation or medium pH in the bioreactors are of extraordinary importance for the more precise regulation of embryo development. In *Euphorbia pulcherrima* dissolved oxygen tension (pO_2) of 60 % and in *Clematis tangutica* pO_2 of 15 % resulted in high embryo yield (Preil et al., 1988; Luttmann et al., 1994). In *Cyclamen,* a significantly higher number of germinating embryos was obtained from cultures grown at 40 % pO_2 than from those grown in Erlenmeyer flasks or in bioreactors aerated with 5 %, 10 % and 20 % pO_2 (Hohe et al., 1999a). Embryo production in *Daucus* decreased significantly at 10 % dissolved oxygen (600 and 170 somatic embryos per ml at 100 % and 10 % pO_2, respectively) (Jay et al., 1992). In contrast, Shimazu and Kurata (1999) reported that the total number of carrot somatic embryos was unaffected by increasing oxygen concentrations in the range of 4-40 % pO_2. However, oxygen enrichment enhanced the number of torpedo and cotyledonary-stage embryos. These contradictory results may be due to the application of varying culture parameters, or to differently responding *Daucus* genotypes.

Carbon dioxide accumulation in bioreactor culture of *Cyclamen* was severely growth-inhibiting in comparison to CO_2-concentrations determined in Erlenmeyer flasks. By removing CO_2 from the aeration gas of the bioreactor, cell growth was as high as achieved in Erlenmeyer flasks. However, the regeneration ability of cells after being cultured in bioreactors with CO_2 accumulation, was better than those from Erlenmeyer flasks or bioreactors without CO_2-accumulation (Hohe et al., 1999b). A considerable increase of embryogenic biomass in bioreactor culture of *Clematis tangutica* was achieved when 2.5 % or 5 % (v/v) CO_2 was supplemented to the aeration gas. Supplement of 10 % (v/v) CO_2 reduced embryo yield severely (Barbon-Rodriguez, 2001). CO_2 seems to be much more involved in embryogenic processes than was expected in the past. Not surprisingly, pH of the medium also influences somatic embryogenesis. In *Daucus,* embryo production was highest when the pH of the hormone-free medium was maintained at 4.3. However, most of the embryos did not develop into plantlets. Cultures grown at pH 5.8 produced less embryos which, in contrast to pH 4.3, were able to continue development into plantlets (Jay et al., 1994). These observations indicate that control of environmental factors in bioreactors is essential to overcome difficulties in somatic embryo production and conversion of embryos into plants suitable for commerce.

5. Temporary immersion systems

A periodic immersion technique in which the plant tissue spends periods immersed in the liquid medium alternating with periods in the air was first described by Steward et al. (1952) and modified by Steward and Shantz (1956). In the so-called "Steward apparatus" or "Auxophyton", tubes are fastened with their long axes parallel to the radius of discs which are rotated at 1 rpm by a shaft held at an angle of 10-12°. The liquid medium runs from one end of the tube to the other leaving the tissue in the gas phase and *vice versa.* By similar means, Harris and Mason (1983) achieved alternate exposure and submergence of explants when tilting a flat-bottomed vessel in one direction to expose the tissue to air and in the opposite direction to submerge them in the medium. The first automated device operating according to the principle of ebb and flood was designed by Tisserat and Vandercook (1985). This soon initiated discussions on growing *in vitro* cultures which are temporarily immersed. A breakthrough was achieved when Alvard et al. (1993) introduced the RITA® vessel and Teisson and Alvard (1995) stressed temporary immersion as a new concept of plant *in vitro* cultivation using liquid media. These ideas fertilized worldwide activities in the application and testing of temporary immersion systems

(TIS), resulting in variations in vessel design, equipment and treatments such as immersion time and frequency, depending on the requirements of different plant species and cultivars.

An overview on TIS was given by Etienne and Berthouly (2002) and Berthouly and Etienne (this volume). One can only speculate why this culture technique was adopted so late as a simple, low-cost and effective propagation method, which is applicable to almost all commercially-interesting species including those listed in section 4.1. The principles of the RITA® system and of the twin-flask system of Escalona et al. (1999) represent the most convincing technical solutions combining low-cost devices and easy handling of the cultures. The practical advantages of TIS became evident when culturing e.g. somatic embryos of *Musa* spp. (Escalant et al., 1994), *Hevea brasiliensis* (Etienne et al., 1997), *Citrus* (Cabasson et al., 1997) and *Coffea arabica* (Etienne-Barry et al., 1999) as well as shoots of sugarcane (Lorenzo et al., 1998) and pineapple (Escalona et al., 1999) or potato microtubers (Jimenez et al., 1999). For details see Berthouly and Etienne (this volume).

The physiologically most important advantage of TIS is the efficient gaseous exchange between plant tissue and gas phase inside the vessel. Multiple daily air replacement by pneumatic transfer of the medium ventilates accumulated gasses like ethylene or CO_2. Additionally, uptake of nutrients and hormones over the whole explant surface ensures maximum growth. There is no doubt that TIS will play an outstanding role in future mass propagation of plants *in vitro*.

6. Concluding remarks

Liquid culture systems have significant effects on the multiplication rates and morphology of shoots, somatic embryos, microtubers or bulblets produced *in vitro*. Liquid media favour tissue organisation similar to that of aquatic plants. This, however, can complicate the acclimatisation of the propagules at the end of the propagation process. Therefore, steps have to be taken to avoid hyperhydricity (vitrification), the most serious disorder in liquid culture systems. Optimisation of aeration in bioreactors and TIS in many cases leads to propagules of excellent quality which can be acclimatized easily. Nevertheless, it must be noted that a chance contamination will quickly develop and disperse in liquid medium and is likely to lead to total loss of the culture.

Compared to agar-based systems, liquid systems are more adaptable to automation and are, therefore, suitable for the reduction of labour and costs, as media can be changed easily during scaling-up, and the cleaning of culture vessels is simplified. Some figures on reduction of production costs in TIS are discussed by Berthouly and Etienne (this volume).

In any case, the use of liquid media is only one part of the propagation scheme for various plant species, because establishing *in vitro* cultures usually starts on agar-gelled medium. After the scaling-up and multiplication phase in liquid media, gelled and solid media are used again in most cases for rooting of shoots or conversion of embryos. During this 'pre-acclimatisation' phase, tissues previously adapted to the liquid environment become 're-organised' and conditioned for the *in vivo* environment.

Temporary immersion systems will play a dominant role in future micropropagation of plants. The principles of temporary immersion of tissue cultures were already in discussion fifty years ago; they are easily understood by tissue culturists. Nevertheless, suitable equipment such as RITA® had to be invented before ideas of TIS application in mass propagation took root. The development of useable protocols and guidelines are a prerequisite for the integration of new elements in traditional *in vitro* culture systems design. The organizers of the First International Symposium on "Liquid Systems for *In Vitro* Mass Propagation of Plants" in 2002 set out to establish such guidelines.

References

Akita M, Shigeoka T, Koizumi Y & Kawamura M (1994) Mass propagation of shoots of *Stevia rebaudiana* using a large scale bioreactor. Plant Cell Rep. 13: 180-183

Alfermann AW, Petersen M & Fuss E (2003) Production of natural products by plant cell biotechnology: results, problems and perspectives. In: Laimer M & Rücker W (ed) Plant Tissue Culture 100 years since Gottlieb Haberlandt (pp. 153-166). Springer, Wien New York

Alvard D, Cote F & Teisson C (1993) Comparison of methods of liquid medium culture for banana micropropagation. Effects of temporary immersion on explants. Plant Cell, Tiss. Org. Cult. 32: 55-60

Ammirato PV & Styer DJ (1985) Strategies for large scale manipulation of somatic embryos in suspension culture. In: Zaitlin M, Day P & Hollaender A (eds) Biotechnology in Plant Science: Relevance to Agriculture in the Eigthies (pp. 161-178). Academic Press, New York

Arditti J & Ernst R (1993) Micropropagation of Orchids. John Wiley & Sons, Inc. New York

Bajaj YPS (1995) Somatic embryogenesis and its applications for crop improvement. In: Bajaj YPS (ed) Biotechnology in Agriculture and Forestry 30 – Somatic embryogenesis and synthetic seed Vol I (pp. 114-125). Springer, Berlin

Bapat VA, Fulzele DP, Heble MR & Rao PS (1990) Production of sandalwood somatic embryos in bioreactors. Current Science 59: 746-748

Barbon-Rodriguez R (2001) Efecto del dioxido de carbono sobre la embriogenesis somatica de *Coffea arabica* L. cv. Caturra rojo y *Clematis tangutica* K. Tesis, Universidad Central de las Villas, Cuba

Barz W, Reinhard E & Zenk MH (1977) Preface. In: Barz W, Reinhard E & Zenk MH (eds) Plant Tissue and Its Bio-technical Application (pp. V-VI). Springer, Berlin

Ben-Jaacov J & Langhans RW (1972) Rapid multiplication of *Chrysanthemum* plants by stem-tip proliferation. HortScience 7: 289-290

Bergervoet JHW, van der Mark F & Custers JBM (1989) Organogenesis versus embryogenesis from long-term suspension cultures of cucumber (*Cucumis sativus* L.). Plant Cell Rep. 8: 116-119

Bergmann L (1959) A new technique for isolating and cloning cells of higher plants. Nature 184: 648-649

Bergmann L (1960) Growth and division of single cells of higher plants *in vitro*. J. General Physiol. 43: 841-851

Berthouly M & Etienne H (2004) Temporary immersion system: a new concept for use liquid medium in mass propagation. (This volume pp. 161-190)

Bieniek ME, Harrell RC & Cantliffe DJ (1995) Enhancement of somatic embryogenesis of *Ipomoea batatas* in solid cultures and production of mature somatic embryos in liquid cultures for application to a bioreactor production system. Plant Cell, Tiss. Org. Cult. 41: 1-8

Bobilioff-Preisser W (1917) Beobachtungen an isolierten Palisaden- und Schwamm parenchymzellen. Beihefte Bot. Centralbl. 33: 248-274

Brown DCW, Finstad KI & Watson EM (1995) Somatic embryogenesis in herbaceous dicots. In: Thorpe TA (ed) *In vitro* embryogenesis in plants (pp. 345-416). Kluwer Academic Publishers, Dordrecht

Butcher ND (1977) Secondary products in tissue culture. In: Reinert J & Bajaj YPS (eds) Plant Cell, Tissue and Organ Culture (pp. 667-693). Springer, Berlin

Cabasson C, Alvard D, Dambier D, Ollitrault P & Teisson C (1997) Improvement of *Citrus* somatic embryo development by temporary immersion. Plant Cell. Tiss. Org. Cult. 50: 33-37

Cervelli R & Senaratna T (1995) Economic aspects of somatic embryogenesis. In: Aitken-Christie J, Kozai T & Smith MAL (eds) Automation and Environmental Control in Plant Tissue Cultures (pp. 29-64). Kluwer Academic Publishers, Dordrecht

Chu CY, Knight SL & Smith MAL (1993) Effect of liquid culture on the growth and development of miniature rose (*Rosa chinensis* Jacq. 'Minima'). Plant Cell, Tiss. Org. Cult. 32: 329-334

Denchev PD, Kuklin AI, & Scragg AH (1992) Somatic embryo production in bioreactors. J. Biotechnology 26: 99-109

Douglas GC (1984) Propagation of eight cultivars of *Rhododendron in vitro* using agar-solidified and liquid media and direct rooting of shoots *in vivo*. Scientia Horticulturae 24: 337-347

Dudits D, Györgyey J, Bögre L & Bako L (1995) Molecular biology of somatic embryogenesis. In: Thorpe TA (ed) *In vitro* Embryogenesis in Plants (pp. 267-308). Kluwer Academic Publishers, Dordrecht

Dunstan DI, Tautorus TE & Thorpe TA (1995) Somatic embryogenesis in woody plants. In: Thorpe TA (ed) *In vitro* Embryogenesis in Plants (pp. 471-538). Kluwer Academic Publishers, Dordrecht

Earle ED & Langhans RW (1974a) Propagation of *Chrysanthemum in vitro*: I. Multiple plantlets from shoot tips and the establishment of tissue cultures. J. Amer. Soc. Hort. Sci. 99: 128-131

Earle ED & Langhans RW (1974b) Propagation of *Chrysanthemum in vitro*: II. Production, growth, and flowering of plantlets from tissue culture. J. Amer. Soc. Hort.Sci. 99: 352-358

Earle ED & Langhans RW (1975) Carnation propagation from shoot tips cultured in liquid medium. HortScience 10: 608-610

Escalant JV, Teisson C & Cote F (1994) Amplified somatic embryogenesis from male flowers of triploid banana and plantain cultivars (*Musa* spp.). *In Vitro* Cell. Dev. Biol. 30: 181-186

Escalona M, Lorenzo JC, Gonzalez B, Daquinta M, Gonzalez JL, Desjardins Y & Borroto CG (1999) Pineapple (*Ananas comosus* L. Merr.) micropropagation in temporary immersion systems. Plant Cell Rep. 18: 743-748

Etienne H, Lartaud M, Michaux-Ferriere N, Carron MP, Berthouly M & Teisson C (1997) Improvement of somatic embryogenesis in *Hevea brasiliensis* (Müll. Arg.) using the temporary immersion technique. *In Vitro* Cell. Dev. Biol. 33: 81-87

Etienne H & Berthouly M (2002) Temporary immersion systems in plant micropropagation. Plant Cell, Tiss. Org. Cult. 69: 215-231

Etienne-Barry D, Bertrand B, Vasquez N & Etienne H (1999) Direct sowing of *Coffea arabica* somatic embryos mass-produced in a bioreactor and regeneration of plants. Plant Cell Rep. 19: 111-117

Fiedler H (1938/39) Die pflanzliche Gewebe- und Organkultur. Sammelreferat. Z. Bot. 33: 369-416

Gautheret R (1939) Sur la possibilité de realiser la culture indéfinie des tissus de tubercules de carotte. C.R. Acad. Sci. Paris 208: 118-120.

Gautheret RJ (1982) Plant tissue culture: the history. In: Fujiwara A (ed) Plant Tissue Culture (pp. 7-12). Proc. 5th Intl. Cong. Plant Tissue & Cell Culture, Tokyo 1982

Gautheret RJ (1985) History of plant tissue and cell culture: A personal account. In: Vasil IK (ed) Cell Culture and Somatic Cell Genetics, Vol. 2 Cell Growth, Nutrition, Cytodifferentiation and Cryopreservation (pp. 1-59). Acad. Press, Inc. Orlando, San Diego, New York

Gupta PK & Timmis R (2004) Mass propagation of conifer trees in liquid cultures. Progress towards commercialization. This volume (pp. 379-392)

Haberlandt G (1902) Culturversuche mit isolierten Pflanzenzellen. Sitzungsber. Akad. Wiss. Wien, Math.-Naturwiss. Kl. 3: 69-92; translated by Krikorian AD & Berquam DL 1969: Experiments on the culture of isolated plant cells. Bot. Rev. 35: 68-88

Härtel O (2003) Gottlieb Haberlandt (1854-1945): a portrait. In: Laimer M & Rücker W (ed) Plant Tissue Culture 100 years since Gottlieb Haberlandt (pp. 55-66). Springer, Wien New York

Harris RE & Mason EBB (1983) Two machines for *in vitro* propagation of plants in liquid media. Can. J. Plant Sci. 63: 311-316

Henry Y, Vain P & de Buyser J (1994) Genetic analysis of *in vitro* plant tissue culture responses and regeneration capacities. Euphytica 79: 45-58

Heyerdahl PH, Olsen OAS & Hvoslef-Eide AK (1995) Engineering aspects of plant propagation in bioreactors. In: Aitken-Christie J, Kozai T & Smith L (eds) Automation and Environmental Control in Plant Tissue Culture (pp. 87-123). Kluwer Academic Publishers, Dordrecht

Höxtermann E (2003) Cellular 'elementary organisms' *in vitro*: The early vision of Gottlieb Haberlandt and its realization. In: Laimer M & Rücker W (ed) Plant Tissue Culture 100 years since Gottlieb Haberlandt (pp. 67-91). Springer, Wien New York

Hohe A, Winkelmann T & Schwenkel HG (1999a) The effect of oxygen partial pressure in bioreactors on cell proliferation and subsequent differentiation of somatic embryos of *Cyclamen persicum*. Plant Cell Tiss. Org. Cult. 59: 39-45

Hohe A, Winkelmann T & Schwenkel HG (1999b) CO_2 accumulation in bioreactor suspension cultures of *Cyclamen persicum* Mill. and its effect on cell growth and regeneration of somatic embryos. Plant Cell Rep. 18: 863-867

Hohe A, Winkelmann T & Schwenkel HG (2001) Development of somatic embryos of *Cyclamen persicum* Mill. in liquid culture. Gartenbauwissenschaft 66: 219-224

Hosoki T & Asahira T (1980) *In vitro* propagation of bromeliads in liquid culture. HortScience 15: 603-604

Huitema JBM, Preil W, Gussenhoven GC & Schneidereit M (1989) Methods for the selection of low-temperature tolerant mutants of *Chrysanthemum morifolium* Ramat. by using irradiation cell suspension cultures. I. Selection of regenerants *in vivo* under suboptimal temperature conditions. Plant Breeding 102: 140-147

Huitema JBM, Preil W & de Jong J (1991) Methods for the selection of low-temperature tolerant mutants of *Chrysanthemum morifolium* Ramat. by using irradiation cell suspension cultures. III. Comparison of mutants selected with or without preselection *in vitro* at low-temperature. Plant Breeding 107: 135-140

Ibaraki Y & Kurata K (2001) Automation of somatic embryo production. Plant Cell Tiss. Org. Cult. 59: 179-199

Ingram B & Mavituna F (2000) Effect of bioreactor configuration on the growth and maturation of *Picea sitchensis* somatic embryo cultures. Plant Cell, Tiss. Org. Cult. 61: 87-96

Jay V, Genestier S & Courduroux JC (1992) Bioreactor studies on the effect of dissolved oxygen concentrations on growth and differentiation of carrot (*Daucus carota* L.) cell cultures. Plant Cell Rep. 11: 605-608

Jay V, Genestier S & Courduroux JC (1994) Bioreactor studies of the effect of medium pH on carrot (*Daucus carota* L.) somatic embryogenesis. Plant Cell, Tiss. Org. Cult. 36: 205-209

Jimenez E, Perez N, de Feria M, Barbon R, Capote A, Chavez M, Quiala E & Perez JC (1999) Improved production of potato microtubers using a temporary immersion system. Plant Cell, Tiss. Org. Cult. 59: 19-23

King PJ & Street HE (1973) Growth pattern in cell cultures. In: Street HE (ed) Plant Tissue and Cell Culture (pp. 269-337). Blackwell Sci. Publ., Oxford

Kohlenbach HW (1959) Streckungs- und Teilungswachstum isolierter Mesophyllzellen von *Macleaya cordata.* Naturwiss. 46: 116-117

Kohlenbach HW (1965) Über organisierte Bildungen aus *Macleaya cordata* Kallus. Planta 64: 37-40

Kohlenbach HW (1966) Die Entwicklungspotenzen explantierter und isolierter Dauerzellen. I. Das Streckungs- und Teilungswachstum isolierter Mesophyllzellen von *Macleaya cordata.* Z. Pflanzenphysiol. 55: 142-157

Kohlenbach HW (2003) The developmental potentials of isolated mesophyll cells and protoplasts. In: Laimer M & Rücker W (ed) Plant Tissue Culture 100 years since Gottlieb Haberlandt (pp. 99-103). Springer, Wien New York

Kotte W (1922a) Wurzelmeristem in Gewebekultur. Ber. Dtsch. Bot. Ges. 40: 269-272

Kotte W (1922b) Kulturversuche mit isolierten Wurzelspitzen. Beitr. Allg. Bot. 2: 413-434

Krikorian AD & Berquam DL (1969) Plant cell and tissue cultures: The role of Haberlandt. Bot. Rev. 35: 59-88

KrishnaRaj S & Vasil IK (1995) Somatic embryogenesis in herbaceous monocots. In: Thorpe TA (ed) *In Vitro* Embryogenesis in Plants (pp. 155-204). Kluwer Academic Publishers, Dordrecht

Lorenzo JC, Gonzalez BL, Escalona M, Teisson C, Espinosa P & Borroto C (1998) Sugarcane shoot formation in an improved temporary immersion system. Plant Cell Tiss. Org. Cult. 54: 197-200

Luttmann R, Florek P & Preil W (1994) Silicone tubing aerated bioreactors for somatic embryo production. Plant Cell, Tiss. Org. Cult. 39: 157-170

McLean NL & Nowak J (1998) Inheritance of somatic embryogenesis in read clover (*Trifolium pratense* L.). Theoretical and Applied Genetics 97: 557-562

Merkle SA, Parrott WA & Flinn BS (1995) Morphogenic aspects of somatic embryogenesis. In: Thorpe TA (ed) *In Vitro* Embryogenesis in Plants (pp. 155-204). Kluwer Academic Publishers, Dordrecht

Muir WH (1953) Cultural conditions favouring the isolation and growth of single cells from higher plants *in vitro*. Ph.D. Thesis. Univ. Wisconsins, U.S.A.

Muir WH, Hildebrandt AC & Riker AJ (1958) The preparation, isolation and growth in culture of single cells from higher plants. Am. J. Bot. 45: 589-597

Murashige T & Skoog F (1962) A revised medium for rapid growth and bio assays with tobacco tissue cultures. Physiol. Plant. 15: 473-497

Nadolska-Orczyk A & Malepszy S (1989) In vitro culture of *Cucumis sativus* L. 7. Genes controlling plant regeneration. Theoretical and Applied Genetics 78: 836-840

Nickell LG (1956) The continuous submerged cultivation of plant tissues as single cells. Proc. Natn. Acad. Sci. U.S.A. 42: 848-850

Nobécourt P (1939) Sur la pérennité et l'augmentation de volume des cultures de tissues végétaux. C.R. Soc. Biol. 130: 1270-1271

Nomura K & Komamine A (1995) Physiological and biochemical aspects of somatic embryogenesis. In: Thorpe TA (ed) *In Vitro* Embryogenesis in Plants (pp. 249-266). Kluwer Academic Publishers, Dordrecht

Paek KY, Chakrabarty D & Hahn EJ (2004)Application of bioreactor systems for large scale production of horticultural and medicinal plants. This volume (pp. 93-114)

Pieper W & Zimmer K (1976) A simple, inexpensive apparatus for *in vitro* propagation of tissues. Gartenbauwissenschaft 41: 221-224

Preil W, Florek P, Wix U & Beck A (1988) Towards mass propagation by use of bioreactors. Acta Horticulturae 226: 99-106

Preil W (1991) Application of bioreactors in plant propagation. In: Debergh PC & Zimmerman RH (eds). Micropropagation (pp. 425-445). Kluwer Academic Publishers, Dordrecht

Preil W, Huitema JBM & de Jong J (1991) Methods for the selection of low-temperature tolerant mutants of *Chrysanthemum morifolium* Ramat. by using irradiation cell suspension cultures. II. Preselection *in vitro* under low-temperature stress. Plant Breeding 107: 131-134

Püschel AK, Schwenkel HG & Winkelmann T (2003) Inheritance of the ability for regeneration via somatic embryogenesis in *Cyclamen persicum*. Plant Cell, Tiss. Org. Cult. 72: 43-51

Rajasekhar EW, Edwards M, Wilson SB & Street HE (1971) Studies on the growth in culture of plant cells. XI. The influence of shaking rate on the growth of suspension cultures. J. Exp. Bot. 22: 107-117

Redenbaugh K (1993) Synseeds. Applications of synthetic seeds to crop improvement. CRC Press Boca Raton, Ann Arbor, London, Tokyo

Reinert J (1958) Morphogenese und ihre Kontrolle an Gewebekulturen aus Carotten. Naturwissenschaften 45: 344-345

Rittershaus E, Ulrich J, Weiss A & Westphal K (1989) Large scale industrial fermentation of plant cells: Experiences in cultivation of plant cells in a fermentation cascade up to a volume of 75000 litres. BioEngineering 5: 28-34

Robbins WJ (1922a) Cultivation of excised root tips and stem tips under sterile conditions. Bot. Gaz. 73: 376-390

Robbins WJ (1922b) Effect of autolyzed yeast and peptone on growth of excised corn root tips in the dark. Bot. Gaz. 74: 59-79

Shimazu T & Kurata K (1999) Relationship between production of carrot somatic embryos and dissolved oxygen concentration in liquid culture. Plant Cell, Tiss. Org. Cult. 57: 29-38

Steward FC, Caplin S & Millar FK (1952) Investigations on growth and metabolism of plant cells. I. New techniques for the investigation of metabolism, nutrition and growth in undifferentiated cells. Ann. Bot. 16: 57-77

Steward FC & Shantz EM (1956) The chemical induction of growth in plant tissue cultures. In: Wain RL & Wightman F (eds) The Chemistry and Mode of Action of Plant Growth Substances (ed) (pp. 165-186). Butterworths Ltd., London

Steward FC, Mapes MO & Mears K (1958) Growth and organised development of cultured cells. II. Organisation in cultures grown from freely suspended cells. Amer. J. Bot. 45: 705-708

Steward FC (1958) Growth and development of cultivated cells. III. Interpretations of the growth from free cell to carrot plant. Am. J. Bot. 45: 709-713

Steward FC (1968) Totipotency of angiosperm cells. Its significance for morphology and embryology. Phytomorphology 17: 499-507

Steward FC, Ammirato PV & Mapes MO (1970) Growth and development of totipotency cells. Some problems, procedures and perspectives. Ann. Bot. 34: 761-787

Street HE (1973) Introduction. In: Street HE (ed) Plant Tissue and Cell Culture (pp. 1-10). Blackwell Sci. Publ., Oxford

Stuart DA, Strickland SG & Walker KA (1987) Bioreactor production of alfalfa somatic embryos. HortScience 22: 800-803

Styer DJ (1985) Bioreactor technology for plant propagation. In: Henke RR, Hughes KW, Constantin MJ & Hollaender A (eds) Tissue Culture in Forestry and Agriculture (pp. 117-130) Plenum Press, New York, London

Takayama S (1991) Mass propagation of plants through shake- and bioreactor-culture techniques In: Bajaj YPS (ed) Biotechnology in Agriculture and Forestry Vol 17 (pp. 495-515). Springer Verlag, Berlin

Takayama S & Akita M (1994) The types of bioreactors used for shoots and embryos. Plant Cell, Tiss. Org. Cult. 39: 147-156

Takayama S & Akita M (1998) Bioreactor techniques for large-scale culture of plant propagules. Adv. Hort. Sci. 12: 93-100

Tar'an B & Bowley SR (1997) Inheritance of somatic embryogenesis in orchardgrass. Crop Science 37: 1497-1502

Teisson C & Alvard D (1995) A new concept of plant *in vitro* cultivation liquid medium: Temporary immersion. In: Terzi M, Cella R, Falavigna A(eds) Current Issues in Plant Molecular and Cellular Biology (pp. 105-110). Kluwer Academic Publishers, Dordrecht

Tisserat B, Esan EB & Murashige T (1979) Somatic embryogenesis in angiosperms. Hortic. Rev. 1: 1-78

Tisserat B & Vandercook CE (1985) Development of an automated plant culture system. Plant Cell, Tiss. Org. Cult. 5: 107-117

Wan Y, Sorensen EL & Liang GH (1988) Genetic control of *in vitro* regeneration in alfalfa (*Medicago sativa* L.). Euphytica 39: 3-9

Westphal K (1990) Large-scale production of new biologically active compounds in plant-cell culture. In: Nijkamp HJJ, van der Plas LHW & van Aartijk J (eds), Progress in Plant Cellular and Molecular Biology (pp. 601-608). Kluwer Academic Publishers, Dordrecht

White PR (1934) Potentially unlimited growth of excised tomato root tips in a liquid medium. Plant Physiol. 9: 585-600

White PR (1939) Potentially unlimited growth of excised plant callus in an artificial medium. Am. J. Bot. 26: 59-64

Wilson SB, King PJ & Street HE (1971) Studies on the growth in culture of plant cells. XII. A versatile system for the large scale batch or continuous culture of plant cell suspensions. J. Exp. Bot. 21: 177-207

Winkelmann T, Hohe A & Schwenkel HG (1998) Establishing embryogenic suspension cultures in *Cyclamen persicum* ‚Purple Flamed'. Adv. Hort. Sci. 12: 25-30

Yeung EC (1995) Structural and development patterns in somatic embryogenesis. In: Thorpe TA (ed) *In Vitro* Embryogenesis in Plants (pp. 205-248). Kluwer Academic Publishers, Dordrecht

Yu KF & Pauls KP (1993) Identification of a RAPD marker associated with somatic embryogenesis in alfalfa. Plant Mol. Biol. 22: 269-277

Zenk MH, El-Shagi H, Arens H, Stöckigt J, Weiler EW & Deus B (1977) Formation of indole alkaloids serpentine and ajmalicine in cell suspension cultures of *Catharanthus roseus*. In: Barz W, Reinhard E & Zenk MH (eds) Plant Tissue Culture and Its Bio-technical Application (pp. 27-43). Springer, Berlin

Zimmer K & Pieper W (1975) Weitere Untersuchungen zur Kultur *in vitro* von *Aechmea*. Gartenbauwissenschaft 3: 129-132

Ziv M (2000) Bioreactor technology for plant micropropagation. Horticultural Reviews 24: 1-30

Ziv M (2004) Simple bioreactors for mass propagation of plants. This volume (pp.77-91)

I. Bioreactors

Chapter 2

Application of bioreactor design principles to plant micropropagation

Wayne R. Curtis
The Pennsylvania State University, Department of Chemical Engineering.
Address for Correspondence: 108 Fenske Laboratory, Dept. of Chemical Engineering, University Park, PA 16802-4400, USA. Phone:1-(814)863-4805, Fax:1-(814) 865-7846, E-mail: WRC2@PSU.EDU

Abstract: Principles of oxygen consumption, oxygen transport, suspension, and mixing are discussed in the context of propagating aggregates of plant tissue in liquid suspension bioreactors. Although micropropagated plants have a relatively low biological oxygen demand (BOD), the relatively large tissue size and localization of BOD in meristematic regions will typically result in oxygen mass transfer limitations in liquid culture. In contrast to the typical focus of bioreactor design on gas-liquid mass transfer, it is shown that media-solid mass transfer limitations limit oxygen available for aerobic plant tissue respiration. Approaches to improve oxygen availability through gas supplementation and bioreactor pressurization are discussed. The influence of media components on oxygen availability are also quantified for plant culture media. Experimental studies of polystyrene beads in suspension in a 30-litre air-lift and stirred bioreactors are used to illustrate design principles for circulation and mixing. Potential limitations to the use of liquid suspension culture due to plant physiological requirements are acknowledged.

Key words: oxygen transport, respiration, somatic embryogenesis, suspension culture, tissue culture

Abbreviations: see 'List of nonstandard units', 'List of variables' and 'Greek symbols'

A.K. Hvoslef-Eide and W. Preil (eds.), Liquid Culture Systems for in vitro Plant Propagation, 21–40.

1. Introduction

Bioreactors for micropropagation cover a wide range of size (0.5 – 500 l) and complexity (jelly jar to modified microbial fermenter). The intention of this review is not to focus on (or promote) any particular bioreactor configuration, but to examine bioreactor design principles as they can be applied to the proliferation of micropropagules in bioreactor systems. This general approach is intended to provide some basis for considering alternative bioreactors for a specific culture system. Although simple tissue culture vessels are bioreactors, the focus of the principles presented here is on liquid-culture bioreactors. The associated analysis for submerged-culture propagation is on the typical limitations of oxygen transfer and mixing as they apply to micropropagules such as suspended somatic embryos.

2. Micropropagule oxygen demand

Analyzing the transport of oxygen in any bioreactor system is a characterization of oxygen gradients and the thermodynamic equilibria that determine the solubility of oxygen in media and plant tissue. The rate at which oxygen is consumed is a key determinant in these gradients. It is important to keep in mind that plant tissues in culture are predominantly – if not exclusively – growing heterotrophically on sugars. Fundamentally, this is no different to the heterotrophic utilization of sugars that takes place in all living tissues. Sugar oxidation therefore becomes a primary determinant of the tissue biological oxygen demand (BOD). Not only is the plant not producing oxygen *via* photosynthesis, but the typical limitations of carbohydrate synthesis and transport are not present. As a result, heterotrophic tissue culture can be expected to become kinetically limited by its biochemical capacity to utilize nutrients. Surprisingly, the reported rates of oxygen consumption in suspensions are comparable to reports of plant respiration (Table 1). This suggests that other aspects of metabolism, such as the rate of utilization of energy derived from respiration, or transport inside the tissue might be rate-limiting for aerobic respiration.

The low solubility of oxygen in media is a fundamental limitation to oxygen use in biological systems. This can be illustrated using the average plant BOD of 10 μmol O_2 g^{-1} FW tissue. A tissue culture vessel containing 100 g FW tissue per litre would have sufficient oxygen for 8.5 hours of respiration from air, but only 15 min if it contained water saturated with oxygen from air. The consequences of oxygen solubility on micropropagule development will be discussed in more detail in subsequent sections.

Table 1: Oxygen uptake rates (OUR) of plant tissues and plant tissue culture.

Plant	OUR[(A)]
Alfalfa (*Medicago sativa*, germinating seed, 18°C)	44
Wheat (*Triticum aestivum*, seedling, 18°C)	8.8
Corn (*Zea mays*, seedling, 18°C)	6.3
Pea (*Pisum sativum*, seed, 20°C)	6.2
Rye (*Secale cereale*, seedling, 18°C)	5.0
Potato (*Solanum tuberosum*, whole plant, 19°C)	4.5
Tobacco (*Nicotiana tabacum*, whole plant, 20°C)	3.2
Plant Tissue Culture[(B)], bulk tissue	10-12
root meristem	200-300

(A) Data from compilation reference (Altman and Dittmer, 1968); μmol $O_2\ g^{-1}$ FW·hr l^{-1}.
(B) Average values from experimental measurements and quoted literature values.

3. Oxygen transfer rate

The average BOD can also be used to carry out an initial "ball park" estimate of oxygen transfer rates (OTR) needed to supply oxygen. Under balanced equilibrium conditions, total demand will be equal the oxygen transfer rate from the gas to the media ($OTR_{g\text{-}L}$):

$$OTR_{g\text{-}L} = BOD \cdot \rho_t \cdot V_t = k_La\,(C^* - C_L) \qquad \text{(Eqn. 1)}$$

Where ρ_t is the tissue density (1.04 g ml^{-1}; approximately water), V_t is the tissue volume, C^* and C_L are the equilibrium and actual medium dissolved oxygen concentrations, and k_La is a parameter used to characterize the rate of oxygen transfer from the gas phase to the medium. To provide for diffusion of oxygen into the tissue it is reasonable to take $C_L \approx C^*/2$, therefore, an estimate for k_La of 8 hr^{-1} can be calculated as the mass transfer rate needed to supply a typical plant tissue culture BOD. This initial calculation leads to a misleading conclusion, since this rate of oxygen transfer should be easy to achieve in just about any bioreactor configuration (simple shake flasks or blowing of bubbles in media would greatly exceed this level; Blanch and Clark, 1997). This apparent ease of supplying oxygen is qualitatively consistent with the observation that slow-growing plant tissues should have much lower oxygen demand than microbial cultures, none-the-less, experimentation has shown that plant tissues very often display oxygen-limited behavior. For example, the doubling of O_2 concentration in a flask headspace doubles the root extension rate and biomass accumulation rates in root culture (Asplund and Curtis, 2001).

There are several flaws in the preceding simplistic analysis of oxygen demand and oxygen transfer. An inherent assumption of the analysis is that oxygen demand is distributed evenly throughout the media. This is a reasonable assumption for microorganisms, but not for aggregates of plant tissue. Not only is the BOD not distributed throughout the bioreactor as for single cell suspensions, but the BOD is also distributed unevenly within the tissue. The oxygen demand at meristems can be ten times that of other tissues as a result of elevated metabolism and cytoplasmically dense cells (Table 1, Ramakrishnan and Curtis, 1995). Therefore, the local oxygen requirement is not dictated by the average BOD, but the BOD at the meristems. Transport of oxygen is characterized by a flux, and the flux required to support a given rate of respiration is easily calculated from the total respiration within a living tissue, divided by the surface area of the tissue.

$$Flux\left[\frac{\text{oxygen transport}}{\text{surface area}}\right]=\frac{\text{BOD}\cdot\text{FW}}{\text{A}_{\text{tissue}}} \qquad \text{(Eqn. 2)}$$

A simple calculation with yeast demonstrates the implications of working with large cell aggregates. Yeast (BOD=2000 μmol O_2 g^{-1} $FW{\cdot}hr^{-1}$; diameter=0.01 mm; FW=2.8×10^{-10} g) will have a surface flux requirement of 0.025 μmol O_2 $cm^{-2}{\cdot}hr^{-1}$. By comparison, an idealized spherical somatic embryo (BOD=300 μmol O_2 g^{-1} $FW{\cdot}hr^{-1}$; diameter=1.0 mm; FW=5.5×10^{-4} g) will have a surface flux requirement of about 5 μmol O_2 $cm^{-2}{\cdot}hr^{-1}$. This shows that even though yeast has a BOD nearly ten times greater than a plant meristem, the flux requirement at a 1 mm embryo surface could be several hundred times larger than yeast! These calculations not only shed light onto the magnitude of oxygen transfer limitations in plant tissue culture, but also the mechanistic basis of mass transfer. The k_La of equation 1 is ubiquitously used in the literature to characterize the transfer of oxygen from the gas to a surrounding liquid medium. The "k_L" is the liquid-interface mass transfer coefficient, and the "a" is the bubble interfacial area per unit volume (e.g. Singh and Curtis, 1994). Figure 1 depicts an entirely different mechanism where oxygen transfer is limited by transport at the liquid-solid interface.

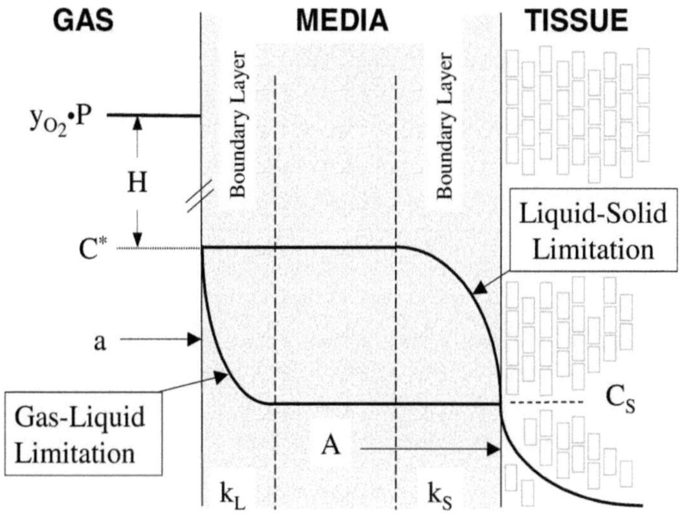

Figure 1: Schematic of gas-liquid and liquid-solid mass transfer resistance. Mass transfer is characterized by mass transfer coefficients at the gas-liquid (k_L) and liquid-solid (k_S) boundary layers. Other parameters are: (y_{O_2})=mole fraction of oxygen in the gas, (P)=gas pressure, (H)=Henry's Law coefficient describing oxygen solubility in water, (a)=interfacial area per unit bioreactor volume, (A)=surface area of tissue, (C^*)=equilibrium dissolved oxygen concentration, (C_L)=liquid dissolved oxygen concentration, (C_S)=dissolved oxygen concentration at the tissue surface.

This figure shows the two boundary layer mass transfer limitations: the gas-liquid interface adjacent to the bubbles, and the solid-liquid interface adjacent to the plant tissue. The high flux requirement at the tissue surface increases the likelihood that the limiting resistance to mass transfer will be at the plant tissue surface, and not from the gas bubbles. Under these circumstances the oxygen transport rate from the media to the tissue (OTR_{L-S}) becomes:

$$OTR_{L-S} = BOD \cdot \rho_t \cdot V_t = k_{S,required} \cdot A_{tissue} (C_L - C_S) \qquad \text{(Eqn. 3)}$$

The surface area of interest now becomes the surface area of the tissue (A_{tissue}), C_S is the dissolved oxygen at the tissue surface, and $k_{S,required}$ is the mass transfer coefficient from the liquid medium to the plant tissue surface that is required to sustain aerobic respiration. Mass transfer coefficients are determined by how fast the media flows past the plant cell aggregate surface (much like the wind-chill factor that determines the rate of heat loss from the skin surface). For a suspended object, the rate of sedimentation provides a reasonable estimate of the velocity that the plant aggregates experience (note that suspended particles will tend to move with the circulation patterns and these bulk flow velocities will not determine the velocity of the embryo relative to its surrounding media). Of particular relevance in this analysis is the suspension of somatic embryos. We have observed that somatic embryo's of oak (*Quercus rubra*) sediment at rates of 1-2 cm s^{-1} for embryo's of 1-2 mm in diameter (Singh and Curtis, 1994) which is consistent with the expected sedimentation rates for dense meristematic tissue. Once a estimate of the velocity is known, there are engineering correlations to calculate the mass transfer coefficient (k_S). For example, the correlation for a sedimenting spherical mass is:

$$k_{S,available} = \frac{D_{O_2}}{d_p}\left[2.0 + 0.6\left(\frac{\rho_{media} \cdot v_S \cdot d_p}{\mu_{media}}\right)^{\frac{1}{2}}\left(\frac{\mu_{media}}{\rho_{media} \cdot D_{O_2}}\right)^{\frac{1}{3}}\right] \quad \text{(Eqn. 4)}$$

This analysis permits a comparison of the mass transfer rates that are "required" ($k_{S,required}$) by the tissue oxygen demand, and mass transfer rates that are "available" due to the flow past the somatic embryo as it is suspended in a bioreactor ($k_{S,available}$). The "required" oxygen transfer can be calculated using equation 3 together with approximations for somatic embryo geometry and reasonable assumptions for driving force ($C_L=C^*$; $C_S=C_L/2$). Figure 2 shows the result of this comparison of tissue oxygen need *versus* oxygen availability.

Calculations are shown for both cylindrical and spherical geometry with similar trends demonstrating that exact details of geometry are not critical. Mass transfer provided in suspension is relatively insensitive to embryo size. In contrast, as the size of the tissue increases, there is a rapid increase in the required oxygen flux and $k_{s,required}$ at the surface. This means that for larger embryos (> 1 mm in size), the oxygen demand will exceed availability.

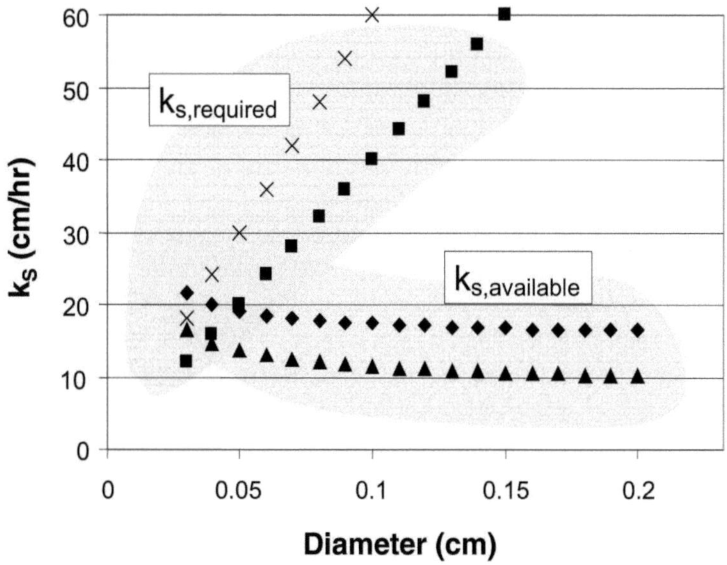

Figure 2: Comparison of oxygen transport required for aerobic respiration and availability in terms of the mass transfer coefficients calculated by equations 3 and 4.
The required mass transfer coefficient ($k_{s,required}$) based on oxygen consumption (BOD= 300 µmol O_2 cm^{-3}(tissue)·hr^{-1}) is calculated for cylindrical (×) and spherical (■) plant tissues. The available mass transfer coefficient ($k_{s,available}$) is calculated based on the sedimentation rates that would result for a cylindrical (▲) and a spherical (♦) tissue (ρ_{tissue} = 1.04 g ml^{-1}).

4. Manipulating oxygen availability

Plant tissues (and embryos in particular) can cope with reduced oxygen availability, therefore it is unclear what the physiological impact would be. None the less, recognizing the physiological changes associated with oxygen availability, it is useful to understand how availability can be controlled in bioreactors. Referring to equation 3, the only parameter that can be readily manipulated is the liquid dissolved oxygen concentration (C_L). Control of C_L is most easily discussed in terms of the limiting case where the bulk media dissolved oxygen approaches its equilibrium value (C^*) as depicted in figure 1. The equilibrium dissolved oxygen (D.O.) is dependent on environmental conditions as follows:

$$C_L = C^* = \frac{y_{O_2} \cdot P}{H} \qquad \text{(Eqn. 5)}$$

It is important to note that equilibrium dissolved oxygen (C^*) is almost never measured; instead, probes are calibrated to give a percentage readout of this complex variable. Each of the parameters of oxygen mole fraction (y_{O_2}), system pressure (P) and Henry's Law coefficient (H) present different approaches to increasing oxygen availability.

Bioreactor pressure will often enhance oxygen availability as an indirect consequence of bioreactor operation. *In situ* sterilisable bioreactors are pressure-rated autoclaves (sterilized @ ~18 p.s.i., 120°C). The typical approach of operating at a head pressure of 5-10 p.s.i. not only reduces contamination risks from leaks, but can nearly double the equilibrium dissolved oxygen (1 atmosphere = 14.7 p.s.i.). In addition, the hydrostatic pressure ($\rho_{H_2O} \cdot g \cdot depth$) can be a significant contributor to oxygen transfer driving force (1 atmosphere ≈ 33 feet of water). These observations have interesting implications for liquid culture micropropagation. The productivity of a reasonably efficient somatic embryogenic culture can generate large numbers of embryos in bioreactors smaller than 1000 l. Since scale-up would take place in tanks much smaller than traditional fermentation, the depth effect will be minimal. Use of pressurization (beyond autoclave pressures) is usually avoided due to the prohibitive cost of pressure vessels (Curtis, 1999). However, since 'large-scale' micropropagation can be implemented at a small scale, the pressure rating of the vessel (and associated steel costs) will not dominate overall equipment costs, and use of pressure could be economically feasible if it provides a substantial improvement in micropropagule development.

It is also apparent from equation 5, that an increase in the Henry's Law coefficient (H) will reduce the equilibrium dissolved oxygen pressure. Temperature has the greatest influence on H. As the temperature increases, H increases as thermal vibrations exclude oxygen, which decreases the maximum amount of oxygen that can dissolve in the media. Colder temperatures, therefore, favor oxygen solubility (15, 20, 25 and 30°C correspond to 315, 283, 258, and 236 μmol O_2 respectively in water). To take advantage of the increased oxygen solubility at lower temperatures, an organism must be able to sustain metabolism rates. This is not generally the case for terrestrial plant tissues, therefore, a reduction in temperature will usually significantly reduce the BOD. A suppression of tissue development rates is generally not desirable since it will increase the time-frame required to produce micro-propagules. This observation may, however, be useful if it is desirable to maintain aerobic respiration within a tissue by simultaneously reducing oxygen demand and increasing oxygen availability through growth at reduced temperatures. While the temperature dependence of oxygen solubility in water is widely known, the dependence of dissolved oxygen on media components receives less attention. Both the presence of inorganic salts and sugars has predictable effects on oxygen solubility. The approach to quantitatively assessing the impact of media components on solubility is more readily determined from an alternative to H that is referred to as the Bunsen coefficient. The Bunsen coefficient (α) is defined as the volume of oxygen dissolved per volume of media (calculated to a reference state of $0^{\circ}C$, 1 atmosphere pressure) when contacted with pure oxygen ($y_{O_2}=1$). The deviation of the Bunsen coefficient from pure water (α_0) can be calculated as the sum of positive and negative contributions of media components (Schumpe et al., 1981):

$$\log\left(\frac{\alpha_0}{\alpha}\right) = \sum_{i=1}^{n}\left(K_{organic}x_{organic}\right)_i + \sum_{j=1}^{m}\left(\sum_{k=1}^{w} H_k x_j z_k^2\right)_j \qquad \text{(Eqn. 6)}$$

where K_i and H_i are experimentally determined coefficients for (n) different organic components such as sugar and ions of (m) dissociated salts. These coefficients are multiplied by the respective concentrations (x_i) and also the valence squared in the case of ions. The contribution coefficients for some important plant culture media components are given in table 2A. Noting that α is roughly proportional to dissolved oxygen, and log(y) is positive for y>1, a positive value of K_i or H_i will decrease solubility. Nitrate ions (NO_3^-), therefore, reduce solubility, whereas ammonium (NH_4^+) will increase oxygen solubility. The equilibrium solubility of oxygen in several 'typical' plant culture media is presented as table 2B, demonstrating that the presence of nutrients generally suppresses oxygen solubility. This effect is relatively small (fresh MS medium has about 4% reduction), but the practice of utilizing high medium osmoticum (e.g. 5x sucrose) to 'harden' somatic embryos in preparation for transfer out of tissue culture conditions can result in a significantly reduced oxygen solubility. The Bunsen coefficient (and its counterpart the Henry's Law coefficient) provides a means of quantitatively understanding the dissolved oxygen available to the micropropagule.

The final and most obvious means of changing the equilibrium dissolved oxygen level is the alteration of the gas oxygen mole fraction. Air is composed of roughly 21% oxygen, and supplementation of plant tissue culture gas with up to 50-60 % oxygen is not usually inhibitory for plant tissues. Part of the reason for the lack of toxicity at even higher oxygen levels, is that the tissue does not actually experience these levels of oxygen since there is a drop in oxygen to the surface of the tissue due to external mass transfer, and a reduction in oxygen within the tissue due to consumption and limitations in diffusional mass transfer. Oxygen supplementation is not a panacea for increasing O_2 availability. Oxygen supplementation can be quite costly – particularly for the long culture times required for plant tissue culture. Advances in pressure swing adsorption (PSA) now provide a relatively inexpensive means of oxygen supplementation to a process of the scale under discussion for commercial micropropagation.

Another processing approach for oxygen supplementation worth noting is the separate sparging of high-purity oxygen in very small bubbles. The advantage of this approach is that the driving force for transport from these bubbles will be five times greater than ambient air, and much higher than would be accomplished if the oxygen was mixed with air sparging. The production of small bubbles is useful since the small radius of curvature will create bubbles with high internal pressure that are not prone to coalescence with the sparged air bubbles. Therefore, separate sparging and small gas bubbles avoid dilution of the driving force for oxygen transfer that results when oxygen is mixed with air. The relatively small scale of micropropagation (relative to industrial fermentation) tends to reduce the impact of costs for oxygen supplementation.

Table 2: Influence of media components on media dissolved oxygen

2A – Bunsen Coefficient contributions (Schumpe et al., 1981)	
	K_i / H_i ($l\ mol^{-1}$)
sucrose	44
glucose	8.8
NH_4^+	6.3
NO_3^-	6.2
K^+	5.0
$H_2PO_4^-$	4.5

2B – Calculated equilibrium dissolved oxygen in plant culture media (Eqn. 8)	
	O_2 @ 25°C (mg l^{-1})
Water	8.24
MS (30 g sucrose l^{-1})	7.92
DCR (20 g sucrose l^{-1})	8.04
B5 (20 g sucrose l^{-1})	8.01

MS-Murashige and Skoog, 1962; DCR-Gupta and Durzan, 1985; B5-Gamborg et al., 1968.

5. Circulation and micropropagule suspension

The relatively rapid sedimentation velocity of plant tissues results in a need for adequate circulation to provide suspension. The sedimentation rate of a nearly spherical object can be estimated from Stokes law,

$$v_{\text{sediment}} = \frac{(\rho_{tissue} - \rho_{media})g \cdot d_p^2}{18 \cdot \mu_{media}} \qquad \text{(Eqn. 7)}$$

however, the sedimentation rate is often sufficiently fast that the laminar flow assumptions for this equation are no longer valid and a more involved calculation using a (laminar-turbulent) drag coefficient is needed (Singh and Curtis, 1994). This equation, none the less, provides the important functional dependencies. The density of the media (ρ_{media}) is roughly the density of water, therefore, the more dense the tissues, the faster they will sediment. We observed the sedimentation rate of oak embryos to correspond to a tissue density of about 1.05 g ml^{-1}, which was considerably faster than unorganized plant cell aggregates (Singh and Curtis, 1994). Gravity (g) is a constant, and the effective viscosity of the media (μ_{media}) will increase when there are more cells present in the suspension, or if the suspended plant tissues are elongated (Curtis and Emery, 1993). The diameter of the suspending particle (d_p) is important for this analysis because it has a square dependence – and because plant micropropagules such as somatic embryos are extremely large by comparison to 'typical' microbial suspensions. For example, a suspension of embryos could be expected to sediment a 1 cm in a non-agitated culture flask in about 1 second. By comparison, a yeast suspension would require about 1 hour to sediment over the same depth. Preventing sedimentation of plant tissues to the bottom of a bioreactor can become a problem due to their rapid sedimentation rate and the relatively low energy inputs typically used for growth of plant tissues in bioreactors.

To study the issue of somatic embryo suspension more thoroughly, two ~ 30-l vessels were constructed. Polystyrene beads (ρ=1.05 g ml^{-1}) with a diameter of 3.2 mm (v_{sediment} = 5.1 cm s^{-1}) were used to represent somatic embryos in these suspension studies. The first 'bioreactor' used only gas sparging and had a geometrical design based on a recent low-cost bioreactor patent (Curtis, 2001) and is shown schematically in figure 3.

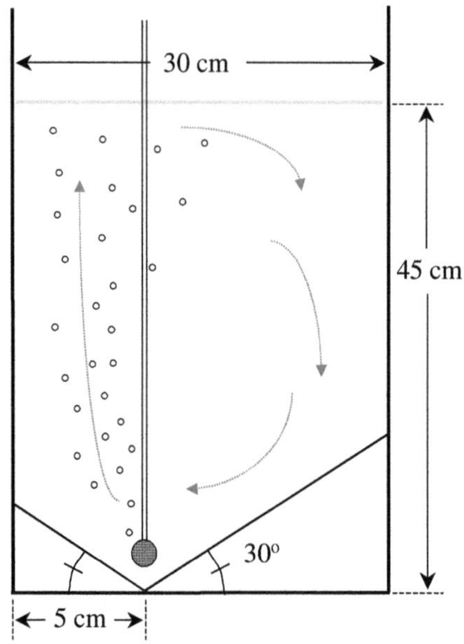

Figure 3: Schematic of the internal loop air-lift bioreactor used to study suspension and mixing intensity. The air-lift exterior is a glass tank with bottom baffling constructed of plexiglass (poly-methyl methacrylate) and sealed in place with silicone glue. Sparging was through two 0.2 μm HPLC mobile-phase spargers (Supelco, Bellfonte, PA). Off-center sparging and baffling geometry was based on previous studies to achieve high mixing with low gas flow rate.

The large glass tank had a 30° off-center baffling to facilitate circulation at minimum gas flow rates. Sparging was accomplished with dual sintered metal spargers parallel to the baffle vertex. The medium was water containing 4 g l^{-1} NaCl to provide for an ionic strength similar to plant culture medium so that bubble formation and coalescence would provide circulation patterns analogous to a plant cell culture bioreactor. A gas flow rate in this system of 1.4 l min^{-1} (or approximately 0.05 vvm; volume of gas per volume of liquid per minute) provided good circulation rates, but was inadequate to suspend the polystyrene beads. Full suspension was not achieved until approximately 0.12 vvm. This shows that suspension can be achieved at a reasonable gas flow rate with proper design. The loop-configuration for an air-lift bioreactor is particularly good for suspensions because the suspensions' requirements can be met by achieving a riser (bubble upflow region) velocity greater than the sedimentation velocity of the particles. It should be noted that simple sparging in a tank without proper design would invariably result in regions with poor circulation which would accumulate unsuspended tissue and subject it to oxygen deprivation. A qualitative observation based on other unpublished studies, is that achieving suspension in small bioreactors (1-5 l) can be more problematic than larger bioreactors (25-150 l) at comparable gassing rates. The second tank was implemented as a stirred tank using a 6-inch marine propeller and rpm tachometer. The polystyrene beads could be suspended off the bottom at 300 rpm, but uniform suspension required 350 rpm. This is a much higher rpm than typically used for plant cell suspensions at this scale, and the resulting tip speed ($\pi N \cdot D_{impeller}$) of 280 cm s^{-1} is nearly 5-times greater than our typical starting point of 50 cm s^{-1} used for plant cell suspension culture (Curtis and Singh, 1994). As will be discussed in more detail in the next section, the power level in the stirred tank is much higher than a well-designed air-lift. The chaotic flow patterns in a stirred tank will result in upward flows enhancing suspension, as well as downward flows that aid sedimentation.

6. Mixing, 'shear' and turbulence

Mixing is fundamentally different from circulation. Mixing is the dispersion of a fluid element throughout the bioreactor, while circulation is the movement of that fluid element to a different location. Mixing can be evaluated either at the scale of total bioreactor volume, or with respect to the small flow patterns that ultimately dissipate the flow energy to thermal energy. It is these small-scale 'eddies' that are often discussed in terms of fluid-dynamic damage to suspended cells. The size of the smallest flow eddies can be characterized by a turbulence length scale referred to as the Kalmogorov length scale (λ_K). This can be estimated from the rate of power dissipation per unit mass of media in the bioreactor ($\hat{P}/\rho_{media}$).

$$\lambda_K = \left(\frac{\left(\mu_{media} / \rho_{media} \right)^3}{\hat{P} / \rho_{media}} \right)^{1/4} \qquad \text{(Eqn. 8)}$$

For mechanical agitation, $\hat{P}$ can be calculated based on the impeller speed (N), impeller diameter ($D_{impeller}$) and a 'impeller power number' that depends on the impeller geometry.

$$\hat{P}_N = \frac{\rho_{media} \cdot N_P \cdot N^3 \cdot \left(D_{impeller}\right)^5}{V_{media}} \qquad \text{(Eqn. 9)}$$

The power input for gas sparging can be estimated assuming isothermal ideal gas expansion of the sparged gas over the unaerated tank height (h_0).

$$\hat{P}_g = \frac{Q_{top} P_{top}}{V_{media}} \ln\left(\frac{P_{top} + \rho_{media} \cdot g \cdot h_0}{P_{top}}\right) \qquad \text{(Eqn. 10)}$$

Note that it is necessary to specify the gas flow at the top of the tank (Q_{top}) as well as the pressure at the surface of the media (P_{top}) since both pressure and gas flow rate change throughout the depth of a tank.

The studies of polystyrene bead suspension provided a basis for examining the conditions required for growth of somatic embryos. The airlift system required a gas flow of 3.6 l min^{-1} sparging into 28 l of liquid with an unaerated liquid height of 45 cm. The resulting power dissipation rate (Eqn. 10) is approximately 0.01 Watts l^{-1} giving a Kalmogarov eddy length of 98 μm. Suspension in the agitated tank (V_{media}=38 l, N_p=0.5, N=300-350 rpm) occurred at a power level of 0.13-0.21 Watts l^{-1}. This increased energy dissipation rate results in a reduced eddy length of 45-50 μm. The significance of these calculations is depicted in figure 4.

The eddy length for energy dissipation is a fraction of the size of a typical somatic embryo (Figure 4A). This means that the plant tissue will experience large variations in local flow velocity from these turbulent eddies. This phenomenon is qualitatively the basis of fluid shear stress on suspended tissues. Figure 4B shows the analogous scenario for a yeast fermentation taking place in a high-intensity stirred tank fermentation at 2-4 Watts l^{-1}. Despite increasing power levels by 1-2 orders of magnitude, the eddy lengths are still several times larger than the diameter of yeast. As a result, yeast will move with eddies and not ‘feel’ the fluid shear experienced by larger plant tissue aggregates. It is important to keep in mind that the size of plant micropropagules alters the interpretation of “low shear” growth conditions.

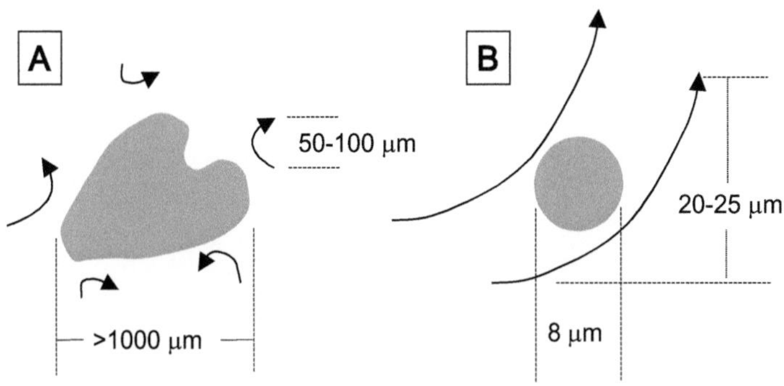

Figure 4: Schematic to depict the microscopic flow patterns calculated as the Kalmogorov eddy length (Equation 8). For somatic embryos (A) the power level is based on minimum required for suspension of 3.2 mm polystyrene beads in air-lift and stirred tank vessels; for yeast (B) the power level is a typical high-intensity stirred tank fermentation.

It is tempting to think of suspensions of cells as a homogeneous fluid, which is implied by discussion of fluid shear and stress. This is a reasonable assumption for microscopic cell suspensions, but this is not the case for suspensions of aggregates that have significant mass and differential density. It is extremely difficult rigorously to characterize flow of an inertial suspension, since this would require considerations of hydrodynamic forces on particles and the associated inertial particle acceleration. Approaches to dealing with these problems are still being developed and not within the scope of this discussion. None-the-less, a qualitative observation related to this behavior is relatively easy to understand and shown pictorially in figure 5.

This figure shows schematics of a stirred tank bioreactor with both axial (marine) and radial (Rushton) flow impellers. In the case of the radial flow impeller, a somatic embryo must dramatically change its flow direction while moving through the impeller zone. In contrast, the embryo can pass through the axial flow impeller zone without changing direction.

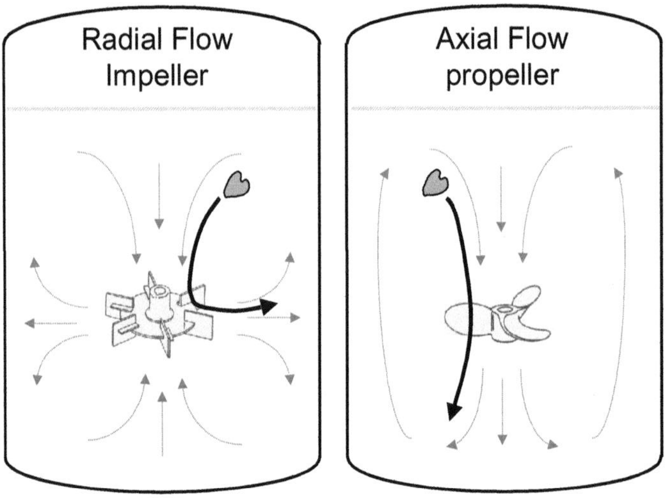

Figure 5: Implications of impeller flow patterns on 'inertial flow' and impeller impact on plant aggregates. Radial flow impellers draw flow inward from above and below with discharge towards the wall. Axial flow propellers push flow along the axis of the impeller shaft. Solid arrow represents typical embryo flow path through the impeller zone.

The more 'inertial' a particle (larger, more dense), the greater it will resist a change of flow direction. The result is that the particle will be hit by the impeller. This behavior is evident in suspension studies of polystyrene 'embryos'. During suspension with a radial flow impeller, there is a continuous sound of the beads being struck by the impeller. Impeller collisions were reduced significantly at comparable power input levels for the axial flow propeller. It is reasonable to extrapolate from these observations (and this has been experimentally confirmed) that an axial-flow impeller would perform better for aggregated plant cell suspensions. While improved performance might be attributed to 'low-shear' impeller design, the behavior has little to do with fluid shear. This concept shows that it is not necessary to analyze fully the problem to implement a design. In fact, it is possible to implement successfully bioreactors based only on empirical observation of what 'works best'. None-the-less, it is helpful to understand basic principles such as those discussed in this review to help guide the process of design, reduce 'trial-and-error' effort, and better overcome invariable problems that will arise as a system is implemented.

7. Limitations

It is worth noting that the physiological requirements for plant tissue differentiation may present some incompatibility with traditional concepts inherent in bioreactor design for submerged culture. For example, one premise of suspended liquid systems is that achieving mixing is desirable. However, if embryo development benefits from lack of homogeneity (such as the gradients inherently provided by agar-based culture), then the liquid-suspended system will never provide the desired environmental conditions for embryo development. Under these circumstances, entirely different bioreactor designs should be contemplated. Acknowledging this issue, we implemented a 300 cm^2 inclined plane perfusion bioreactor to provide for the transition from embryogenic carrot cell suspension to fully germinated seedlings by perfusing the system with auxin-free medium (unpublished). A comparable level of normal plant development did not occur in the same time frame for the carrot culture liquid suspension. The wide array of bioreactor designs that are being developed indicates that there will likely be many different 'solutions' that will depend on both the plant tissue being propagated, as well as the scale and economics associated with the particular product. For these different bioreactor configurations, the principles of mixing, oxygen transfer, and biological oxygen demand do not change, however, the details of the analysis and their application can be extremely different. Mathematics is simply a convenient way to express logic; therefore, logic (and common sense) are often the most powerful tools for bioreactor design.

List of nonstandard units

n = number of organic molecules in medium
m = number of salts in medium
w = number of dissociated ions from a dissolved salt

List of variables

a = interfacial area per unit volume (e.g. k_La), cm^{-1}
A_{tissue} = surface area of the tissue, cm^2
BOD = Biological Oxygen Demand, $\mu mol\ O_2 \bullet gFW^{-1} \bullet hr^{-1}$
C^* = Equilibrium medium dissolved oxygen content, $\mu mol\ O_2 \bullet l^{-1}$
C_L = Medium dissolved oxygen content (in bulk liquid), $\mu mol\ O_2 \bullet l^{-1}$

C_S = Medium dissolved oxygen content at tissue surface, $\mu mol\ O_2 \bullet l^{-1}$
d_p = Particle diameter, cm
D_{O_2} = Diffusion coefficient of oxygen in media, $cm^2 \bullet s^{-1}$

$D_{impeller}$ = Impeller diameter, cm
FW = Fresh Weight, g
g = Gravitational acceleration = 980 cm $\bullet\ s^{-2}$
h_0 = Height of unaerated liquid in bioreactor, cm
H = Henry's law coefficient, atm • l • ($\mu mol\ O_2)^{-1}$
H_k = Experimental coefficients for the kth dissociated medium ion in Bunsen correlation (eqn. 6), l • mol^{-1}
k_L = Mass transfer coefficient at liquid-gas interface, cm • s^{-1}
k_s = Mass transfer coefficient at solid-liquid interface, cm • s^{-1}
$k_{s,available}$= Mass transfer coefficient at solid-liquid interface that is available as a result of flow past solid, cm • s^{-1}
$k_{s,required}$ = Mass transfer coefficient at solid-liquid interface that is needed to supply oxygen demand, cm • s^{-1}
$K_{organic,i}$ =Experimental coefficients for the ith organic media component in Bunsen correlation (eqn. 6), l • mol^{-1}
N = Impeller rotational rate, s^{-1}
N_P = Impeller power number, dimensionless
OTR = Oxygen Transfer Rate, $\mu mol\ O_2 \bullet hr^{-1}$
$OTR_{g\text{-}L}$ = Oxygen Transfer Rate from the gas to liquid, $\mu mol\ O_2 \bullet hr^{-1}$
$OTR_{L\text{-}s}$ = Oxygen Transfer Rate from the liquid to tissue surface, $\mu mol\ O_2 \bullet hr^{-1}$
OUR = Oxygen Uptake Rate (specific), $\mu mol\ O_2 \bullet gFW^{-1} \bullet hr^{-1}$
P = Local system pressure, atm
P_{top} = Pressure at the top surface of liquid, atm
$\hat{P}$ = Power per unit volume, Watts • l^{-1}
$\hat{P}_N$ = Power per unit volume due to impeller mixing, Watts • l^{-1}
$\hat{P}_g$ = Power per unit volume due to gas sparging, Watts • l^{-1}
Q_{top} = volumetric gas flow rate at top (surface) of bioreactor, $cm^3 \bullet s^{-1}$
v_s, $v_{sediment}$ = Sedimentation velocity, cm • s^{-1}
V_{media} = Volume of media in bioreactor, cm^3
V_t = Tissue volume, cm^3
vvm = volume of gas sparged per volume of liquid per minute, min^{-1}
x_j = concentration of ions in liquid phase, mol • l^{-1}
$x_{organic,\ i}$ = concentration of organic components in liquid phase, mol • l^{-1}
y_{O_2} = mole fraction of oxygen in the gas phase, dimensionless
z_k = valance of dissociated ion, dimensionless

Greek symbols

α = Bunsen coefficient, volume of oxygen dissolved per volume of media @ 0^oC and 1 atm, dimensionless
α_0 = Bunsen coefficient of pure water, volume of oxygen dissolved per volume of media @ 0^oC and 1 atm, dimensionless
λ_k = Kalmogorov length scale, cm
μ_{media} = Viscosity of media, dyne • sec • cm^{-2} = g • cm^{-1} • s^{-1}
ρ_{H_2O} = Density of water, g • cm^{-3}
ρ_{media} = Media density, g • cm^{-3}
ρ_t, ρ_{tissue}= Tissue density, g • cm^{-3}

Acknowledgements

I thank Undergraduate Researchers Olatilewa O. Awe and Adam J. Pingel for implementing the polystyrene bead suspension studies in the air-lift and stirred tank bioreactors, respectively. We would like to acknowledge grant support from the National Science Foundation Grants No. BES-0003926, BES-9522033 for support of our research in bioreactor design.

References

Altman PL & Dittmer DS (1968) Metabolism, Biological Handbooks, Federation of American Societies for Experimental Biology, Bethesda Maryland

Asplund PA & Curtis WR (2001) Intrinsic oxygen use kinetics of transformed root culture, Biotechnology Progress, 17:481-489

Blanch HW & Clark DS (1997) Biochemical Engineering, Marcel Dekker, New York

Curtis WR (1999) Achieving economic feasibility for moderate-value food and flavor additives: A perspective on productivity and proposal for production technology cost reduction, In: Fu T J; Singh G; Curtis W R (eds) Plant Cell and Tissue Culture for the Production of Food Ingredients (pp. 225-236) Kluwer Academic / Plenum Publishing, New York, NY

Curtis WR (2001) Method and apparatus for aseptic growth or processing of biomass, U.S. Patent # 6,245,555, June 12

Curtis WR & Emery AH (1993) Plant cell suspension culture rheology, Biotechnology and Bioengineering, 42:520-526

Gamborg OL, Miller RA & Ojima K (1968). Nutrient requirements of suspension cultures of soybean root cells, Exp.Cell Res., 50:151-158

Gupta PK & Durzan DJ (1985) Shoot multiplication from mature trees of Douglas-fir (*Pseudotsuga menziesii*) and sugar pine (*Pinus lambertiana*), Plant Cell Reports: 177-179

Hsiao TY, Bacani FT, Carvalho EB & Curtis WR (1999) Development of a low capital investment reactor system: Application for plant cell suspension culture, Biotechnology Progress, 15(1):114-122

Murashige T & Skoog F (1962) A revised medium for rapid growth and bio assays with tobacco tissue cultures, Physiol. Plant. 15: 473-497

Ramakrishnan D & Curtis WR (1995) Elevated meristematic respiration in plant root cultures: implications to reactor design, Journal of Chemical Engineering of Japan, 28(4):491-493

Schumpe A, Quicker G & Deckwer WD (1981) Gas solubilities in microbial culture media, In: Feichter A (ed) Advances in Biochemical Engineering, Vol. 24 (pp. 1-38). Springer-Verlag, New York, NY

Singh G & Curtis WR (1994) Reactor design for plant cell suspension culture, In: Shargool P D & Ngo T T (eds) Biotechnological Applications of Plant Culture (pp. 153-184). CRC Press, Boca Raton, FL

Chapter 3

Bioreactor design for propagation of somatic embryos

Anne Kathrine Hvoslef-Eide[1], Odd Arild S. Olsen[2], Ragnhild Lyngved[3], Cristel Munster[1] & Petter H. Heyerdahl[4]

[1]Agricultural University of Norway, Dept. of Plant and Environmental Sciences, The Plant Cell Laboratory, P.O.Box 5003, N-1432 Aas, Norway. Phone +47 64 94 7830, Fax +47 64 94 7802, E-mail:trine.hvoslef-eide@ipm.nlh.no (corresponding author). [2]Fibula AS, Jutulveien 11, N-0852 Oslo, Norway. [3]Dept. of Biology, Norwegian University of Science and Technology, N-7491 Trondheim, Norway. [4]Dept. of Agricultural Engineering, P.O.Box 5065, N-1432 Aas, Norway.

Abstract: Six identical bioreactors were constructed and built at the Agricultural University of Norway to provide optimal conditions for plant cell regeneration from cells into somatic embryos ("clonal or somatic seeds"). This was made possible through cooperation in COST87 by a European network in a working group on regeneration from plant cell cultures. The bioreactor design provides gentle stirring through a slow-speed stirrer that regularly changes direction of rotation to prevent "quiet" zones in the suspension in which cells can settle and grow. In addition, the oxygen is provided, bubble–free, through thin silicone tubing loops that are hanging loose, moving with the liquid to prevent cell growth on these tubes. We used off-the-shelf components whenever possible, to reduce the costs to a minimum, which was another aim of the construction. The result was a suite of relatively inexpensive computer-controlled bioreactors that could control temperature, oxygen, pH, stirrer speed and stirrer direction. In addition, we have provided different light spectral qualities by simple means of filtering the light. Using the present software, the parameters can be set up to alter every hour during the 24 h day/night cycle. All our cultures have improved growth in the bioreactors compared to identical cultures in Erlenmeyer flasks. The cultures used are: embryogenic cultures of carrot (*Daucus carota*), Norway spruce (*Picea abies)*, birch (*Betula pendula*), cyclamen *(Cyclamen persicum)* and shoot cultures of Christmas begonia (*Begonia x cheimantha).* The paper also discusses recommendations for improvements of the current system for future revisions.

Key words: bioreactor configuration, cell culture, control, *Cyclamen persicum,* embryo culture, Erlenmeyer flasks, instrumentation, liquid culture, measurement, ornamental, regeneration, suspension, viability

*Abbreviations***:** COST- European Cooperation in the Field of Scientific and Technical Research; FDA – Fluorescein diacetate; PI – proportional + integral

A.K. Hvoslef-Eide and W. Preil (eds.), Liquid Culture Systems for in vitro Plant Propagation, 41–59.

1. Introduction

Internationally, there is considerable interest in developing vegetative propagation methods that could facilitate rapid propagation of elite material (Preil, 1991; Takayama and Akita, 1994). Considering the cost of labour in Western Europe, propagation in liquid culture, especially in bioreactors, is a possible way to produce clonal propagules at a low cost. One of the advantages of growing cell cultures in bioreactors is the opportunity to control the oxygen supply accurately in the liquid. This communication describes a self-constructed and self-built set of six identical bioreactors to provide controlled experimental conditions for cell cultures.

The designed bioreactors are closed systems or so-called batch bioreactors, which are the most frequently used s in biotechnological productions (Siebel, 1992; Glick and Pasternak, 1998). Having identical bioreactors provides the possibility to run factorial experiments to reveal interactions between the experimental parameters. The aim of the construction was to take advantage of experience from cell cultures grown in commercially-available bioreactors and to construct a bioreactor with exceptionally low shear forces, at the lowest possible price. The bioreactors on the commercial market have been designed to provide optimal growth for bacteria, yeast and, to some extent, plant cell cultures for secondary metabolite production. When aiming to grow somatic embryos for large-scale propagation, the effect of shear forces is more critical than for cell cultures for secondary metabolites. First, we designed a prototype, tested this out, improved it and then built a set of six identical "improved" bioreactors. There are two features with our bioreactors that make them more gentle to the cultured plant cells than various commercial designs; (1) the aeration system and (2) the stirring device. Both are designed to allow slow, gentle movement of the suspension, without allowing cell settlement in quiet zones in the bioreactors. So far, we have grown embryogenic cultures of carrot (*Daucus carota*) (Nissen, unpublished results), Norway spruce (*Picea abies)* (Kvaalen, 1997), birch (*Betula pendula*) (Hvoslef-Eide, 2000), cyclamen *(Cyclamen persicum)* (Hvoslef-Eide and Munster, 1997, 1998) and shoot cultures of Christmas begonia (*Begonia x cheimantha)* (Hvoslef-Eide, unpublished results) in our bioreactors.

For a more general and comprehensive description of plant cell bioreactor design than is provided in this article, see the book chapter by Heyerdahl et al. (1995).

2. Description of the bioreactor design

2.1 Vessels

To simplify the bioreactor construction and reduce the costs, it was decided to use off-the-shelf glassware, a 2000 ml vessel (Schott, Duran). It has a flat bottom, so it can stand without support in the laminar flow hood (Figure 1A). The lid was made in our workshop from stainless steel. It is sealed to the vessel rim by an o-ring and six tightening screws along the circumference (Figure 1C). Welded to, and under, the lid are the half-loops that provide the support for the oxygen supply system (Figure 1 B).

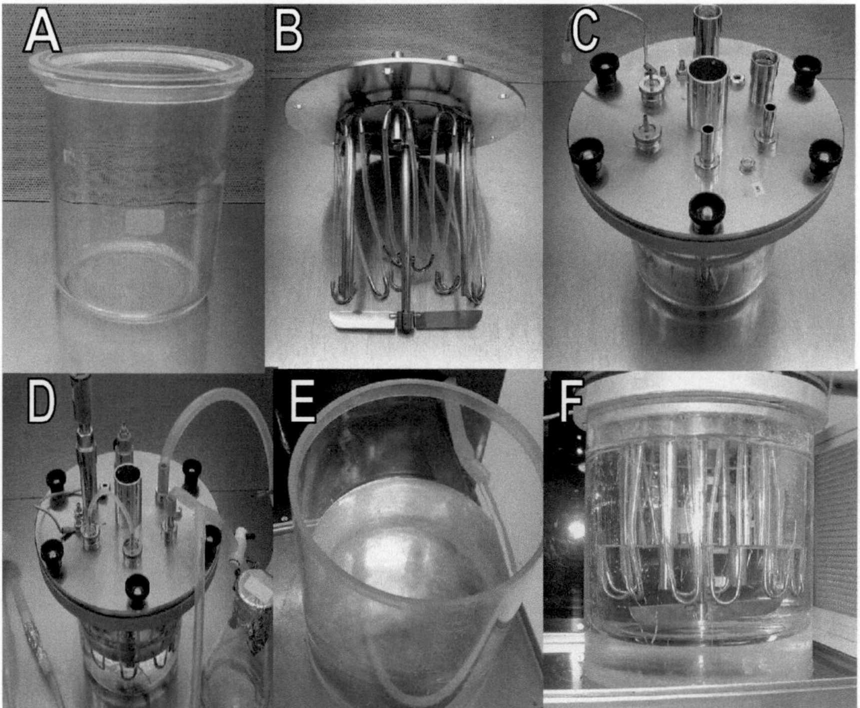

Figure 1: The assembly of the NLH bioreactors: (A) the 2 l inner glass vessel with a flat bottom, (B) the oxygen supply system with the silicone tubes joined by stainless steel hooks welded underneath the lid and the pitch-blade stirrer, (C) the glass vessel and lid assembled, (D) the vessel with the medium, oxygen electrode, pH meter, inoculation flask and sample withdrawal apparatus in place; ready to go into the autoclave, (E) the outer vessel made of acrylic supplying a water bath for temperature regulation and (F) the assembled bioreactor with light on the right, ready to be inoculated with cells.

This consists of about two metres of thin-walled silicone tubing (wall thickness 0.4 mm, tube diameter 3.8 mm, Cat. No.9.205.256, H. Jürgens GmbH & Co, Bremen) in a continuous closed circuit, except for an inlet and an outlet. The oxygen and pH electrodes, the inlet and exhaust ports for gasses, the stirrer, the inlet for medium and suspensions, the outlet for sample taking, the thermo sensor pocket, as well as the inlets for acid and alkali all go through this lid (Figure 1D). The highest risk of contamination is through the inlet for medium and cells, the sampling port(s) and the inlets for acid/alkali. To prevent this, the following measures have been taken: the acid/alkali pass through sterile filters before entering the vessel, the inlet and outlet for medium and cells are sealed, have sterile filters and are autoclaved with the medium before inoculation of cells (Figure 1D). Figure 2 shows five of the bioreactors after autoclaving, cooling off in the laminar flow hood before final assembly.

This inner sealed vessel is placed in an acrylic container, large enough to allow for free movement of water in the water-jacket thus created (Figure 1E). The water flowing through this jacket is used to control the temperature of the medium in the inner vessel. The outer vessel is fixed, but the inner vessels can be disconnected and carried to a laminar flow hood for various operations under sterile conditions, such as cell inoculation and sample withdrawal.

Figure 2: Bioreactors after autoclaving, cooling off in the laminar flow hood before final assembly in an acrylic outer container thereby providing a water-jacket for temperature control.

2.2 *Temperature measurement and control*

The temperature can be regulated between 15-35 °C and measured in the inner vessel by a commercial 100 ohm platinum temperature sensor (Termoelektro). The sensor is inserted into a stainless steel tubular pocket dipping into the medium. Platinum elements are considered to be the most stable commercial sensors available (Heyerdahl et al. 1995). The steel tube creates a sterile barrier that allows for removal of the sensor when the inner vessel is taken away to the laminar flow hood. The controlling computer continuously measures this temperature and compares it with the set point. Any deviation causes the controller to either increase or lower the temperature of the circulating water in the jacket to compensate. By this arrangement it is the medium temperature that is controlled.

Some bioreactors control only the water jacket temperature. This is less than ideal since significant temperature differences may occur between the jacket and the medium, mainly depending on how much heat is gained or lost through the lid or gained through stirring. Heat transfer through the lid is mainly dependent on the temperature difference between the medium and ambient air.

The water in the jacket is circulated by a 12 V garden fountain pump (Eheim, Type 1046, Serial nr. 90081/92011, 5 W, 5 l min^{-1}). The water passes through a small self-made coaxial tube heat exchanger and which also contains a 100 W electric heating element (Watlow Cartridge 8 x 16 mm, 1/2"NPT). The heating element is switched on and off in a 1 sec continuous cycle. By altering the on/off time-ratio, the controller can adjust the mean applied power from 0 to 100 W. Flow-controlled cold tap water flows through the heat exchanger primary side and thus cools the jacket water. The tap water valve is of a self-made pinch type and consists of a 12 V reversible-direction motor with an eccentric disk that opens or closes the tube in a continuous way. This motor is on/off controlled and runs for a predetermined time when actuated.

Thus the temperature controller can act on two actuators in parallel: heating power and cooling water flow. Our control strategy is to use the heating element mainly; the water flow valve is used only to maintain the applied heat power within a range of about 20-80%. The controller mostly operates the valve only after a temperature set point change, until new heat equilibrium is obtained.

When the inner vessel is taken away for work in the laminar flow hood, the temperature sensor is withdrawn from the pocket and placed into the water in the outer vessel. In this way the circulating water temperature is maintained at about the right temperature and ensures the quickest return to set temperature when the vessel is assembled again.

2.3 Oxygen measurement and control

The oxygen electrode (Ingold InPro 6100/220/T/N) measures the partial pressure of dissolved oxygen molecules. This is the most expensive single component of the bioreactor. The electrode must be of a type that can stand repeated autoclaving and need to be accurate and not drift too much during the course of an experiment of up to 8-10 weeks. Since the electrode membrane must be in contact with the growth medium, the electrode cannot be removed for calibration after sterilisation has taken place. The electrode is therefore calibrated *in situ* after sterilisation but before inoculation. The first calibration step is to allow laboratory air of approximately 21 % oxygen to flow through the bioreactor gas exchange tube for 24h at the set temperature. The measured value in this condition is then defined to be 100 % partial oxygen pressure (Preil, 1991). Such oxygen electrodes have an insignificant current at zero oxygen pressure. We therefore disconnect the oxygen electrode to simulate the zero point. This value thus represents the zero point error of the electrode amplifier, and this value is defined as the 0 % oxygen partial pressure point. Oxygen electrodes have linear characteristics and because two points (0 and 100%) have been defined, a linear calibration curve for the oxygen measurement of that particular experiment can be drawn. We can provide gas sparging between 0 and 150 % oxygen in the bioreactors. Zero can be provided by using pure nitrogen, while 150 % can be obtained through enrichment with pure oxygen.

Oxygen is provided bubble-free by using silicone tubing (Figure 1B) with the following specifications: inside diameter 3 mm, wall thickness 0,4 mm (H.Jürgens and Co, Bremen, Germany, No.9.205 256), similar to those used by Preil et al. (1988). This tubing allows gas exchange, but the pores are small enough to prevent bacteria and fungi going through, thereby providing a sterile barrier. We therefore do not use sterile air/oxygen/nitrogen, but laboratory air driven through the silicone tubes by an aquarium pump (Rena 301, 6 W, 600 l h^{-1}, Rena, Annecy, France) The air is enriched with pure oxygen or nitrogen as needed. This gas mixture enters through an inlet in the bioreactor lid, flows through the metal and silicone tubes and eventually is exhausted via the outlet.

The air tube consists of 20 silicone tubes, each 12 cm, connected by U-shaped stainless steel pipes welded to the lid (Figure 1B). Similar U-shaped stainless pipes are used to connect the short lengths of silicone tubing, additionally providing some 'weight' to assist the tubing to hang downwards into the medium. By this arrangement the tube loops hang loosely under the lid constantly dancing in the medium eddies. This is in contrast to designs with stationary tubes where cells might grow on the outside of the tubes.

This construction was first described at the WG2 meeting of COST 87 in Aas (Hvoslef-Eide and Heyerdahl, 1992).

The gasses cross the silicone tube walls at a rate given by the partial pressure difference across the wall, the net direction being from the high to the low partial pressure side (Luttmann et al., 1993). In addition to supplying oxygen to the medium, the tubes therefore also remove gasses produced by the cells, e.g. carbon dioxide and ethylene, which have a higher partial pressure in the medium than at the air/gas tube inlet.

Measuring the pressure, and then manipulating the mixing ratio and flow rate of air and gas through the gas exchange tube controls the oxygen partial pressure of the medium. Besides obtaining a correct oxygen concentration according to the set point, the controller must also maintain a sufficient flow through the tubes to constantly remove gasses produced by the cells.

A gas mixer designed for these bioreactors controls the flow rate of air and additional gases. The gas mixer is two independently controlled flow regulators, one for air and one for the enrichment gas. Each flow regulator consists of a solid-state gas flow sensor (Honeywell AWM 3100V), an analogue PI controller and a gas tubing pinch valve actuated by a bi-directional-controlled motor. An eccentric cam (disk) on the motor shaft, pinches the gas tubing more or less, thus effecting the valve opening and closing. The control computer sets the percent flow set point for this control loop, and the motors are actuated until the correct gas flow is achieved. The gases from the two flow controllers are then mixed in a tube prior to the bioreactor gas inlet port.

This method functions well for oxygen concentrations ranging from 50-150% when the suspension has a medium amount of cells that use oxygen. For lower oxygen concentrations, or in cases where the cell number is so small that the cells do not use enough to keep the concentrations low, we have to use nitrogen in the gas mixture to be able to maintain such low set points. On the other hand, when the cell count is high and the suspension is thickening, it is impossible to keep high set points even when supplying pure oxygen. These limitations must be monitored closely during the experiments.

2.4 pH measurement and control

We have used the pH electrodes (Ingold 405-DPAS-K8S/200 combination pH electrode and Mettler Toledo 405-DPAS-SC-K8S/200, pH 0-12, 0-130°C). There can be a considerable calibration drift in a pH electrode during autoclaving and in the course of an experiment (our unpublished results). The pH electrode is calibrated before each experiment by standard solutions at pH 4 and 7. The slope of the electrode calibration is therefore initially correct. After autoclaving, but before inoculation, a sample of the medium is withdrawn. The pH of this sample is measured with a laboratory pH meter and we adjust the bioreactor pH zero point to obtain the same reading. By adding an amount of acid and withdrawing a second sample, we could recalibrate the slope, but we do not do so because when the zero point is correct, the errors caused by slope deviations are small compared with other error sources, at least for moderate pH changes. The pH can be regulated between pH 3 and 7 in our bioreactor system.

Active regulation of the pH can be achieved by adding small amounts of acid or alkali through sterile filters into the bioreactors. We use HCl (0.1 mol) and NaOH (0.1 mol) for this purpose. Choosing the right concentration of acid or alkali is important, low enough to allow a fine control, but sufficiently concentrated to allow for pH changes without adding too much liquid and undesired Cl^- and/or Na^+ to the vessel. The controller controls the pH by comparing the measured pH to the set point and adding either acid or base accordingly. The alkali and acid actuators are independent peristaltic pumps running at constant speed when activated. The pumps are of our own design, a design that causes least wear on the tubing.

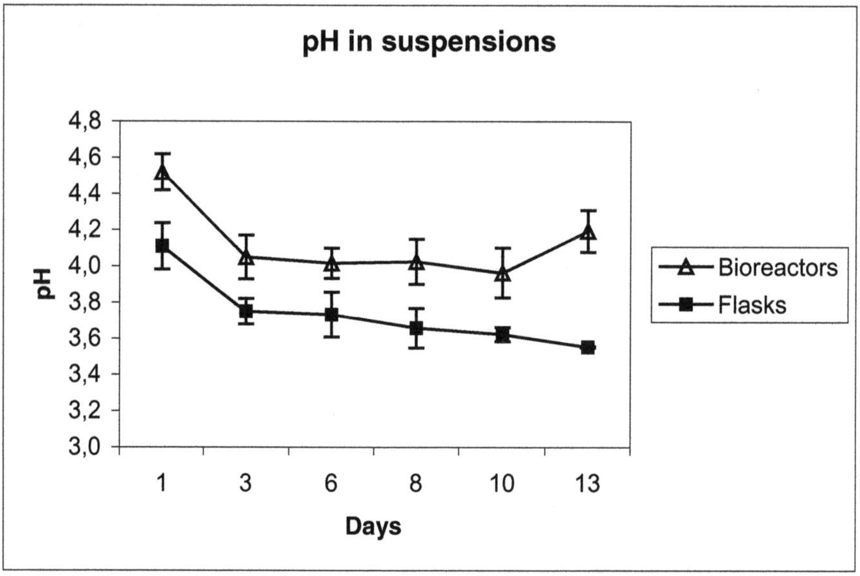

Figure 3: The temporary fall in pH after inoculation and the increase in bioreactors and in Erlenmeyer flasks of *Cyclamen persicum* Mill. cultures. Standard deviation shown on each mean measurement.

The pH can be controlled (very) precisely, but one problem then arises: acid and alkali are then more or less constantly added to maintain the set point. This produces salt that may affect the biological processes in the bioreactor. To avoid unnecessary elevation of salt concentration, our controller implements a dead band around the set point. As long as the pH is within the dead band, the controller will not activate the alkali or acid pumps. We have not used active pH regulation for our experiments in many cases, as the cells seem to be growing and developing well at the pH range provided. Therefore, we have avoided pH regulation to simplify experimental design as much as possible. But the pH of the medium is a good indicator for the condition of a cell suspension, and any infection is usually first spotted as an abnormal pH reading. Normally, with the medium we use, the pH reduces after inoculation, and then rises again (Figure 3).

2.5 *Stirring speed and stirrer direction*

The stirrer is a pitch blade stirrer made of stainless steel (Figure 1B and 1F). The shaft is held by a ball bearing pressed into a hole in the bioreactor lid. The bearing provides enough stability for the stirrer. The grease of this bearing can stand the autoclave temperature well and takes part in the sterile barrier. The motor and stirrer shaft are coupled by a slot and tap to allow for easy removal of the motor when the vessel is taken away. The rotational speed can be varied from 0 - 100 rpm. The regular changes of direction are important to obtain a good mixing at low speeds and hence reduce the shear forces. A speed as low as 30 rpm is sufficient to provide good mixing when changing the direction every 10s.

2.6 *Light quality provision*

It is possible to expose the cultures to different light qualities by using a light source providing near daylight spectrum (OFT bioLIGHTSYSTEMS, Strand Lighting, UK). This light can then be filtered through different coloured stagelight filters that will filter away parts of the light spectrum. We have used a blue filter (Strand filter No 419, primary dark blue) and a red filter (Strand filter No 406, primary red). All light treatments were at the same light quantity; 5 μmol $m^{-1}s^{-2}$ for 18 h. Measuring the light quantity inside the bioreactors and moving the light source closer or further away to provide the same quantum after filtering can obtain this. The bioreactors can be covered with aluminium foil to provide darkness.

2.7 Microprocessors, computers and software

Each bioreactor is controlled by a Personal Computer (PC). The main task of this computer is to be the user interface and for data logging. The user interface allows the operators to set parameters for the controller, specify set points, perform sensor calibration and set up other experimental properties. The set points can be specified to be constant during the experiment or to vary along a diurnal cycle, where the set points are freely specified for each of the 24 hours. Every actuator can also be manually controlled during the experiment to allow for the operators to interact actively with the process. Logged data are displayed as graphs. (Bakken, 1992; Kolstad, 1991; Myhre, 1991).

However, the PC does not take an active part in the control of the actuators; this is under the control of a dedicated microprocessor. This processor is placed in an electronic rack together with sensor amplifiers, power supply, actuator drivers and the gas mixer (Unpublished electronics and software). We developed the electronics and programs for this project. The electronics hardware is housed in a rack, one for each bioreactor. The racks are self-contained units, which make the system independent of the ever-changing technology of the PC world. The control processor communicates over a RS232 serial line with the PC or any other RS232-equipped system by a simple protocol developed for this project. This protocol and the interaction between the PC and the controller for parameter transfer; data logging etc. are described in more detail in Olsen and Heyerdahl (1994). Both the electronics and software of the control processor are substantially more simple than those of the PC and probably more stable and reliable than the PC. Since the critical online part of the system is external to the PC, the PC may malfunction without affecting the experiment.

3. Biological material and methods when starting an experiment

When starting experiments in the bioreactors, it is important to perform the experiment with the same starting material to get homogenous and comparable suspension cultures. The whole concept of six identical bioreactors is to facilitate factorial experiments to reveal differences between treatments and possible interactions. In our experience, the same genotype may vary so much from batch to batch and during a year, that experiments need to be repeated in the bioreactors both in time and space to be reliable. Generally we follow the same procedure at all times: inoculate one bioreactor and multiply the cells in culture, before inoculating all bioreactors with the same large-scale cell culture.

4. Results and Discussion

4.1 Comparing Cyclamen cell growth in Erlenmeyer flasks and in our bioreactor

We have performed comparisons using identical *Cyclamen persicum* Mill cultures simultaneously, in both Erlenmeyer flasks and our bioreactors. We started the experiment with the same inoculum for both culture systems and using the same initial cell density and the same proliferation medium (Schwenkel and Winkelmann, 1998; Winkelmann et al., 1998), following the procedures described in Hvoslef-Eide and Munster (1998; 2001). The results during the first two-week growth period in four bioreactors showed significantly better growth in bioreactors; by fresh weight, dry weight, packed cell volume (PCV) and cluster viability measured by FDA staining (Widholm, 1972) of living cells (Figure 4). The standard deviation is less in the bioreactors than in flasks, indicating a more uniform and predictable culture in bioreactors in our system. Erlenmeyer flasks are probably more variable because of the inability to control the various important environmental parameters. Preil and co-workers (Preil et al., 1988; Preil, 1991) have shown the same improved growth pattern for poinsettia cell cultures using Braun Biostat bioreactors compared with Erlenmeyer flasks when the bioreactors were depleted of excess CO_2 that may build up in a closed (recirculated gas) silicone tube system. Hohe et al. (1999a) found better growth in Erlenmeyer flasks of *Cyclamen persicum* cells compared to Applikon bioreactors (2 l) when they allowed the CO_2 to build-up in the headspace. However, in later experiments, Hohe et al. (1999b; 2001) depleted CO_2 from the circulating gas mixture and obtained improved growth in the bioreactors. This measure is not necessary in our bioreactors, since the aeration system is exhaustive and not in a closed circuit. We have investigated the gas permeability of the silicone tubes (Hvoslef-Eide and Munster, 1997) and found that the permeability could vary between tubes of different bioreactors, with build-up of CO_2 and differences in growth of cells between identical bioreactors as a consequence. We concluded that the tubing age and number of autoclave cycles to which they have been subject, could affect their permeability and now always change the tubes in all six bioreactors at the same time after this.

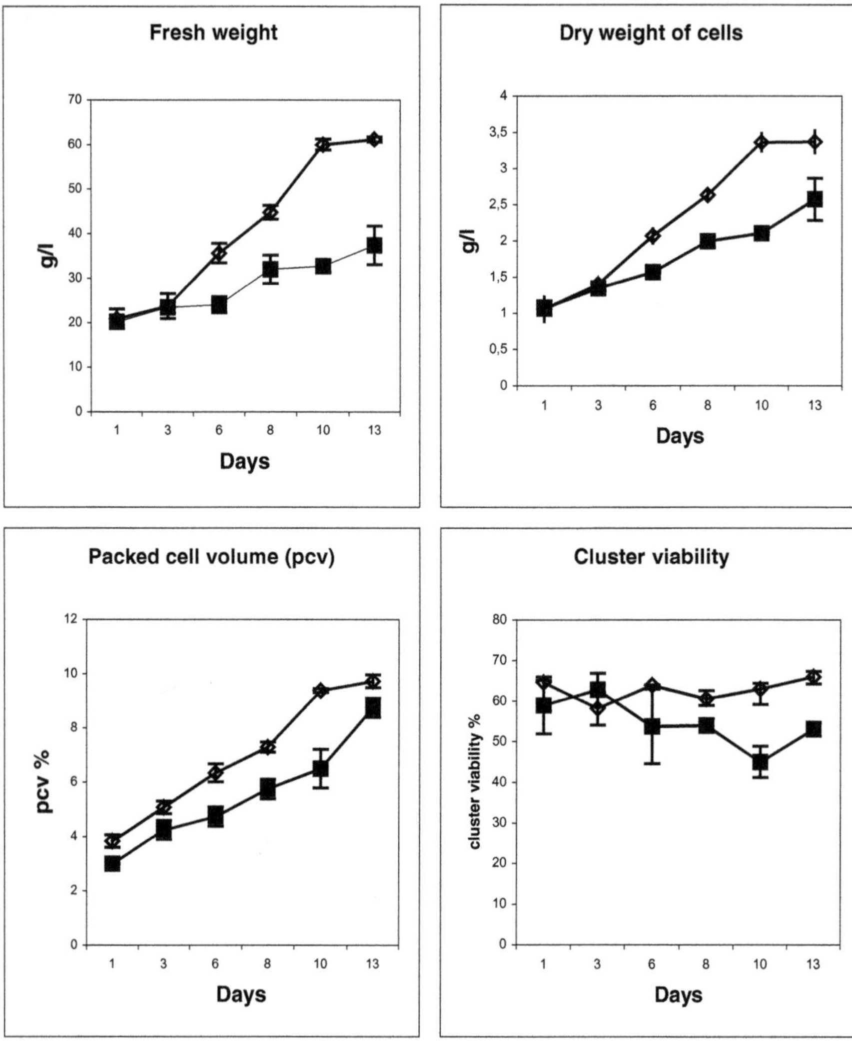

Figure 4: The difference between identical suspension cultures of *Cyclamen persicum* Mill. in bioreactors (open diamonds) and Erlenmeyer flasks (closed squares) on fresh and dry weight, packed cell volume and cluster viability during a two-week period, showing standard deviations for each measurement.

4.2 General discussion

Our system is for batch-culture, where medium is changed once each week as the cultures are diluted and cells are harvested. Such a system is not considered to be optimal with regards to maximum cell growth in such a short period of time. A continuous mode of operation, with a continuous feed of nutrients, where the physiological state of the cells during a run is more uniform, is regarded as the most efficient strategy (Glick and Pasternak, 1998). To see if more optimal nutrient conditions would give homogenous embryo formation, fed-batch systems could be investigated.

The pH values during the course of an experiment is a good indicator of the condition of the cells; at first, the pH value will always decrease because the cells will use any ammonium ions first and hence release H^+ ions into the medium. Later, the medium will be depleted of ammonium, and the cell metabolism will change to use nitrate ions and to release OH^- ions in exchange for the negatively charged nitrate. This can then be observed as a rise in pH (Figure 3). The reason the pH value begins to increase again earlier in bioreactors compared to Erlenmeyer flasks (Figure 3) is probably due to the more rapid culture growth in the bioreactors. The uptake of ammonium, compared to nitrate, leading to a reduction in pH value, is faster; the ammonium is depleted more quickly and hence the more rapid rise of pH when the cells begin to take up nitrate ions. A contamination of the bioreactor will usually first be noticeable by an immediate and large change in pH and is a good indicator of culture status.

5. Experiences and recommendations for improvements on bioreactor design

The bioreactors described have been used for ten years, and during this time experience about the advantages and limitations of the system, with regard to mechanical considerations as well as measurements, control and software has been gained using cell cultures of various plant genera and cell types.

5.1 Temperature control

The bioreactors were constructed to be able to change set point for the different parameters every hour through a 24 h cycle. The temperature control functions well when a constant temperature is required. With alternating day and night temperatures, the regulation seems to be a greater problem. Even though the available interval of temperature regulation is between 15°C and 35°C, the available power of the heater limits the temperature range in the same bioreactors to 10°C during an experiment. Because of the heat exchange from the cold water supply to the bioreactors, it is necessary to regulate the cooling water to the lowest temperature (4°C), to get the bioreactor temperature as low as 15°C. Then the heating element is not sufficiently effective to raise the bioreactor temperature to more than 25°C, without closing the water valve to that separate bioreactor. In cases where alternating temperatures are required during the day, it is recommended that systems be constructed which can use heating elements with a power of more than 100 W, and incorporate Peltier elements to cool the incoming tap water. Peltier elements are solid-state components and thus eliminate the present mechanical cooling devices that are not well suited for good control. Another approach to improvement is to use a cold and hot water mixer to provide the bioreactor with cooling water of a more appropriate temperature. The controlled mixer may either be a continuously-controlled valve or two pulse-controlled on/off valves.

Due to the cooling capacity, heating elements and heat transport, it is observed that the bioreactors may spend 2-3 hours to stabilise the temperature after a 10°C set point change. Measures must be taken to speed-up the temperature control system to provide efficient change between different day and night temperatures when required in the experimental design. Changing the control strategy can reduce the response time. In the present situation, the response is slow due to the long time taken from an instant change at the temperature control elements, and the cooling water valve and water heater, until these changes are slowly manifest as a change to the temperature of the medium. One approach often used is to control only the temperature of the water jacket (Heyerdahl et al. 1995). This removes the lag from the jacket to the medium from the control loop, thereby allowing for faster action. Unfortunately, the medium temperature may deviate from the jacket temperature as discussed above. A better approach will be to control the jacket temperature in addition to the medium temperature. This allows for a more rapid change of the jacket temperature without removing the medium temperature control. This strategy is called cascaded control: The medium controller output acts as the set point for the jacket temperature controller. Thus, when a rise in medium temperature is required the jacket temperature will increase substantially for a limited time interval and thereby transfer heat into the medium at a high rate.

To reduce temperature changes caused by varying room temperature, ideally the bioreactors should be installed in a temperature-controlled room.

5.2 Oxygen control

The oxygen concentration seems to be the first limiting factor for growth in this system. Aeration by diffusion was preferable to bubble columns and airlift bioreactors to meet the requirements of minimum oxygen stress, but the chosen strategy have some limitations. When the cell density is not too great, the oxygen concentration can be controlled from 50% to 150% by using air or oxygen enhancement. For lower oxygen concentrations or when the cell density is low we must use air mixed with nitrogen. On the other hand, when the cell density is high and the suspension is thickening, we cannot keep high set points even when supplying pure oxygen. The rate of oxygen transfer to the cells is then insufficient to support such dense suspensions. These limitations must be closely monitored during the experiments. To provide maximum growth and production in the bioreactors, an improved design for oxygen supply should be considered, e.g. to install more tubes or use thinner walled tubes to decrease diffusion resistance. The age of the silicone tubes is important with regards to how much gas exchange can occur. Having six bioreactors simultaneously running has revealed problems such as uneven gas exchange of the silicone tubes (Hvoslef-Eide and Munster, 1997). Changing the tubes routinely in all the bioreactors at the same time eliminates this variation factor.

The gas flow controllers of the gas mixer function well, but there is room for improvement. The pinch valves used cause control problems due to their high non-linearity and hysteresis. The tubes also have a limited lifetime. Malfunction of the flow sensor has also been an occasional problem. We believe this is caused by overstress of the sensor from possible pressure pulses when switching the pressure on or off. We inserted filters in series with the tubes to smooth pressure variations and also increase the flow resistance thereby letting the valves operate in a more open state where the non-linearities and hysteresis are less pronounced. Following this, we noted fewer incidents of destroyed flow sensors, because of reduced pressure over the valves, and reduced wear in the tubing.

Modern polarographic oxygen sensors are improved with regard to the AgCl-coat on the anode compared to those we used. This improvement ensures better keeping quality of the electrode, especially when high CO_2 concentrations occur in the medium.

5.3 pH control

The regulation of pH is managed by addition of acids or alkali more or less constantly. This procedure might be less than optimal for production of cells and embryos, because of the resultant high, or toxic, salt concentration in the medium. It may be better to disable the pH-control when sensitive species are grown, and just record the changing pH values. The pH-development can be explained in terms of ion exchange, rather than forcing a constant pH value in the medium during an experiment.

The pH-sensors used are sensitive to storage conditions and sterilisation, and their relative short lifespan may cause them to be a substantial part of the experimental costs. It is recommended to measure the pH of the medium in connection to sample withdrawal, due to an eventual drift in the electrode during an experiment. The measurement must be made immediately after sample withdrawal in case of high CO_2 concentrations in the sample. If the concentration of CO_2 in the sample is higher than in the air, CO_2-equilibrium will appear and pH value of the sample will rise rapidly (Cazzulino et al., 1991).

5.4 Stirring speed and stirrer direction

Agitation by one impeller, with two slightly pitched narrow blades seems to be a functional construction for propagation of plant cells and embryos. So far, minimum or no damage from shear forces is recognized. Because of the two-directional mixing (10 seconds each direction), a slow rotational speed (30 rpm) is sufficient to avoid quiet zones. Together with the silicone tubing, which provides bubble-free gas supply, the slow speed gives no foam formation, preventing the need for a foam control system.

5.5 Other recommended elements

CO_2 measurements are a good indicator of growth, and may serve as an on-line inspection. The bioreactors described do not have equipment for analysing CO_2 content. This will be the first priority when the bioreactors are to be upgraded. It would enable measuring the gas concentration in headspace or dissolved in the medium. The CO_2 sensor can also be situated outside the aseptic barrier, and so permit the measurement of CO_2 partial pressure in the exhaust from the bioreactor. Continuously connecting an infrared sensor to the lid may also perform measurements of the gas concentrations, so that samples can be withdrawn at regular intervals. Sensors for biomass, as optical density sensors, may also be used as indicators of growth. Application of image analysis technology is another alternative for measuring cell density (Cazzulino et al., 1991; Pépin et al., 1999). As well as measuring biomass concentration, these tools give information about aggregate size and distribution, pigmentation and morphology of the culture, which is a great advantage when growing embryos.

To be able to keep low oxygen concentrations, a nitrogen supply is required. If factorial experiments with both high and low oxygen concentrations are to be run, both oxygen and nitrogen are needed as enrichment gases. The best way of adding these gases, with regard to regulation and control, is to use three flow regulators, one for air, another for oxygen and another for nitrogen.

5.6 *Software*

The software used (Bakken, 1992) was developed for this specific system. The PC program can set parameters and set points in the controller non-volatile memory. In case of power failure the controllers will use these at power-up. Control is thus effective just after the power problem is fixed. The diurnal profiles are not stored in the controller, so the set point valid before the interruption will be effective until the PC is back in an active state. The diurnal cycle will also stop at the present value if the PC hardware or software crashes or freezes.

During the development of the bioreactors we saw the benefits of connecting the system to the internal and external computer networks to enable remote inspection and, possibly, control of the experiments. This would be useful during experiments conducted in cooperation with groups at other universities. This option was not implemented. Today this would be an obvious feature. Our bioreactors run unattended for substantial periods of time. To reduce the time-to-fix if problems arise, a pager could be called if states exceed alarm limits.

5.7 *Power backup*

An experiment may be severely affected by mains power failures. At our location, main power failures are rare, but real enough when they do happen. The most effective backup power is the UPS (uninterruptible power supply). This is a unit, which would be situated between the supply and the bioreactor system, which produces alternating power either from the mains power or from a back-up battery. The time the battery lasts after a power fall out depends on the power consumed by the system and the battery capacity. Most mains electricity failures will last for between about a second and a quarter of an hour. In such situations a relatively inexpensive UPS will suffice. In cases where the probability of long power failures is high, a second backup may be considered. This will in most cases be a diesel or gasoline-powered generator. Such generators take some several seconds to start and stabilize. In this case the UPS only needs battery capacity for this delay. This combined short and long term power backup may seem attractive, but beside the larger investment of the generator, compared with an UPS, the cost of maintenance must also be taken into consideration. This is due to the fact that the generator depends on its engine's automatic start. A generator without execution of proper procedures for maintenance and regular tests will provide false security.

Acknowledgements

The authors acknowledge the cooperation in COST87 and COST822, especially that of Walter Preil and his co-workers in Ahrensburg, Germany, for continuous support during the construction of the bioreactor suite. The embryogenic *Cyclamen* culture (cell line 3738-VIII) was generously shared with us through COST822 by Hans-Georg Schwenkel and his co-workers in Erfurt, Germany. The building of these bioreactors would not have been possible without the engineers in the NLH workshop; John Granly and Arne Svendsen. Tom Ringstad has been invaluable in his technical support at all times. We also acknowledge invaluable assistance from technicians in the Plant Cell Laboratory with the growth of various cultures over the years; Anne Nilsen, Jens P. Jensen, Tone I. Eliassen and Tone I. Melby. The Norwegian Research Council is thanked for financing the project of construction and building through The Biotechnology Programme in Agriculture.

References

Bakken T (1992) Adaptiv regulering av bioreaktor. M.Sc. Thesis, The Norwegian University of Science and Technology, Division of Engineering Cybernetics. In Norwegian

Cazzulino D, Pedersen H & Chin C-K (1991) Bioreactors and image analysis for scale-up and plant propagation. In: Vasil IK (ed) Scale-up and Automation in Plant Propagation (pp. 147-178). Academic Press, New York

Glick BR & Pasternak JJ (1998) Molecular Biotechnology. Principles and Applications of Recombinant DNA. 2nd Ed (pp. 399-425). ASM Press, Washington DC

Heyerdahl PH, Olsen OAS & Hvoslef-Eide AK (1995) Engineering aspects of plant propagation in bioreactors. In: Aitken-Christie J, Kozai T & Smith MAL (eds). Automation and Environmental Control in Plant Tissue Culture (pp. 87-123). Kluwer Academic Publishers, Dordrecht

Hvoslef-Eide AK & Heyerdahl PH (1992) Development and Construction of Bioreactors at NLH. COST 87, Regeneration in Suspension Culture WG, Ås/Drøbak 24-26 September 1992

Hvoslef-Eide AK & Munster C (1997) Use of bioreactors to propagate *Cyclamen persicum* by somatic embryogenesis. COST 822 Working Group 2 Meeting, Amiens, France, 4-7 September 1997

Hvoslef-Eide AK & Munster C (1998) Somatic embryogenesis in *Cyclamen persicum* Mill. through the use of bioreactors. Proc. Int. Plant. Prop. Soc. (IPPS) Vol 47: 377-382. ISSN: 0538-9143

Hvoslef-Eide AK & Munster C (2001) Light quality effects on somatic embryogenesis of *Cyclamen persicum* Mill. in bioreactors. In: Schwenkel H-G (ed) Reproduction of *Cyclamen persicum* Mill. through somatic embryogenesis using suspension culture systems (pp. 79-84). Report of Working Group 2, COST 822, European Commission, ISBN 92-894-0249-0

Hvoslef-Eide AK (2000) Bioreactor cultures of silver birch *(Betula pendula* Roth.). 4th Int. Symp. on In Vitro Culture and Horticultural Breeding, Tampere, Finland, July 2nd -July 7th 2000. ISSN: 0006-3088

Hohe A, Winkelmann T & Schwenkel HG (1999a) CO_2 accumulation in bioreactor suspension cultures of *Cyclamen persicum* Mill. and its effect on cell growth and regeneration of somatic embryos. Plant Cell Rep. 18: 863-867

Hohe A, Winkelmann T & Schwenkel HG (1999b) The effect of oxygen partial pressure in bioreactors on cell proliferation and subsequent differentiation of somatic embryos of *Cyclamen persicum.* Plant Cell, Tiss. Org. Cult. 59: 39-45

Hohe A, Winkelmann T & Schwenkel HG (2001) Development of somatic embryos of *Cyclamen persicum* Mill. in liquid culture. Gartenbauwissenschaft 66: 219-224

Kolstad T (1991) Modellering og regulering av bioreaktorer. M.Sc. Thesis, The Norwegian University of Science and Technology, Division of Engineering Cybernetics (in Norwegian)

Kvaalen H (1997) Some preliminary results with embryogenic cell suspensions of Norway spruce cultured in bioreactors. COST 822 Working Group 2, 5-8 Sept 1996, Sanremo, Italy. COST 822 Development of integrated systems for large-scale propagation of elite plants using *in vitro* techniques. Report of activities 1996. European Commission 1997, 338 pp, ISBN 92-828-2631-7

Luttmann, R, Florek, P & Preil, W (1993) Silicone-tubing aerated bioreactors for somatic embryo production. Plant Cell, Tiss. Org. Cult. 39 (2): 157-170

Murashige T & Skoog F (1962) A revised medium for rapid growth and bioassays with tobacco tissue cultures. Physiol. Plant. 15: 473-497

Myhre G (1991) Overvåking og regulering av bioreaktorer. M.Sc. Thesis, The Norwegian University of Science and Technology, Division of Engineering Cybernetics (in Norwegian)

Olsen OAS & Heyerdahl PH (1994) Symbolic access to embedded controllers. The C Users Journal, March pp 21-31

Preil W (1991) Application of bioreactors in plant propagation. In: Debergh, PC & Zimmerman R H (eds) Micropropagation: Technology and Application (pp. 425-445). Kluwer Academic Publishers, Dordrecht

Preil W, Florek P, Wix U & Beck A (1988) Towards mass propagation by use of bioreactors. Acta Hortic. 226: 99-106

Schwenkel HG & Winkelmann T (1998) Plant regeneration via somatic embryogenesis from ovules of *Cyclamen persicum* Mill. Plant Tiss. Cult. Biotechnol. 4: 28-34

Siebel MA (1992) Batch reactors. In: Jenkins R O, Leach C K & Mijnbeek G (eds) Bioreactor Design and Product Yield (pp. 103-113). Butterworth-Heinemann, Oxford

Takayama S & Akita M (1994) The types of bioreactors used for shoots and embryos. Plant Cell, Tiss. Org. Cult. 39: 147-156

Widholm JM (1972) The use of fluorescin diacetat and phenosafranine for determining viability of cultures plant cells. Stain Technol. 47: 198-194

Winkelmann T, Hohe A & Schwenkel H-G (1998) Establishing embryogenic suspension cultures in *Cyclamen persicum* 'Purple Flamed'. Adv. Hort. Sci. 12: 25-30

Chapter 4

Practical aspects of bioreactor application in mass propagation of plants

S. Takayama[1]*& M. Akita[2]
[1]*Department of Biological Science and Technology, Tokai University, 317 Nishino, Numazu, Shizuoka 410-0395, Japan.*
*(*requests of offprints: E-mail:* takayama@wing.ncc.u-tokai.ac.jp).
[2]*Department of Biotechnological Science, Kinki University, 930 Nishimitani, Uchita, Naga, Wakayama 649-6493, Japan. E-mail: akita@bio.waka.kindai.ac.jp*

Abstract: Bioreactors are an efficient tool for the production of plant propagules but, at present, their application is commercialized in only a few tissue culture companies. The present article reviews practical aspects of the use of bioreactors in the mass propagation of plants in relation to the responses of plant propagules in liquid medium, the characteristics of bioreactor culture techniques in plant propagation and discusses case studies of the use of bioreactors for several plant species including *Fragaria ananassa, Lilium* species, *Hippeastrum hybridum, Gladiolus grandiflorus, Spathiphyllum, Colocasia esculenta* and *Solanum tuberosum.* The establishment of plantlets from bioreactors and future prospects are also described.

Key words: bioreactor, liquid culture, mass propagation

Abbreviations: ABA - abscisic acid, MS - Murashige & Skoog medium (1962); FW - fresh weight; vvm - volume of gas per volume of liquid per minute

1. Introduction

The use of bioreactors in plant propagation is promising for the efficient production of propagules. As compared to conventional tissue culture techniques using solid or semi-solid medium, bioreactors require fewer culture vessels, less labor, utilities and space. At present, the use of bioreactor technology is starting to be commercialized in tissue culture nurseries, the results of laboratory-scale experiments suggest the practical applicability of the technique in plant propagation.

A.K. Hvoslef-Eide and W. Preil (eds.), Liquid Culture Systems for in vitro Plant Propagation, 61–78.

The bioreactor is a vessel widely used in the cultivation of organisms such as microbes, animal or plant cells to produce metabolites or cells (Coombs, 1986; Takayama, 1997, 2000). Bioreactor technology is also applicable to the propagation of plants. The application for plant propagation was first reported in 1981 for *Begonia* (Takayama and Misawa, 1981). The techniques have been applied to many plant species since then (Styer, 1985; Takayama et al., 1986; Wheat et al., 1986; Akita and Takayama, 1988; Levin et al., 1988; Preil et al., 1988; Hale and Young, 1991; Preil, 1991, 1995; Preil and Beck, 1991; Sondahl and Noriega, 1991; Takayama, 1991; Ziv, M., 1991, 1992; Takahashi et al., 1992; Akita and Takayama, 1993a,b; Akita et al., 1994, Akita and Takayama, 1994 a,b; Harrel et al., 1994; Luttmann et al., 1994, Takayama and Akita, 1994; Tautorus et al., 1994; Ziv et al., 1994; Heyerdahl et al., 1995; Ziv, 1995; Ziv et al., 1995; Levinet et al., 1996; Moorhouse et al., 1996; Okamoto, 1996; Ziv and Shemesh, 1996; Chatterjee et al., 1997; Akita and Ohta, 1998; Hao et al., 1998; Takayama and Akita, 1998; Escalona et al., 1999; Abdullah et al., 2000; Akita, 2000; Correll et al., 2000; Gao et al., 2000; Ingram and Mavituna, 2000; Vinocur et al., 2000; Ziv, 2000; Honda et al., 2001; Lorenzo et al., 2001; Paek et al., 2001). In plant propagation using bioreactors, the process consists of an inoculation process using small propagules with multiple shoot buds that are aseptically inoculated into the bioreactor. These are cultured for one to two months submerged in liquid medium with forced aeration leading to the production of a large number of transplantable-size propagules such as plantlets, bulblets, microcorms or microtubers. In the case of *Spathiphyllum*, about 30,000 plantlets were obtained after two months of culture in a 10-litre glass bioreactor, among them about 3,000 plantlets were transplantable into soil. The advantages of the use of bioreactor in plant propagation are as follows (Takayama and Akita, 1994).

1. Large numbers of plantlets are easily produced and scaling up is easy.
2. Since handling of cultures, such as inoculation or harvesting is easy, labor cost is saved.
3. Cultures are always in contact with the medium, facilitating uptake of nutrients and growth rate.
4. Forced aeration (oxygen supply) is performed, which improves the growth rate and final biomass achieved.
5. Cultures are moving in the bioreactor, which results in the disappearance of apical dominance and in the development of numerous shoot buds into plantlets.

In spite of these advantages, there are some shortcomings such as hyperhydricity, plantlet size variation caused by different growth stages, and microbial contamination (Takayama and Akita, 1998). In this report, we focus mainly on the fundamental and practical aspects of the use of bioreactors in plant propagation. *Spathiphyllum* plants have been used as a case study of propagation in a bioreactor and of the feasibility of reestablishment of the plants in soil.

2. Responses of plant propagules in liquid medium

In order to propagate plants in bioreactors, plantlets should be firstly cultured in liquid medium. For most terrestrial plants, growth is limited in water. In spite of this fact, growth of *in vitro*-cultured plantlets in liquid medium is stimulated in many plant species when aerated. The reasons are based on the physiological characteristics of cultures as described above (items 3, 4 and 5 above). From our experiences, many species of ferns and herbaceous angiosperms grow well, but the propagation of tree species, especially gymnosperms, is difficult in liquid medium. Many monocotyledonous plant species are suitable for propagation in liquid medium. Plantlets grown in liquid medium tend to have a hyperhydric nature (vitrification) of leaves and stems. This hyperhydric nature can be partly overcome with the use of a modified medium with lowered concentrations of salts and sugar and also by increasing the light. The morphology of propagules in the liquid medium is influenced by culture conditions such as medium components, irradiation, temperature, aeration, and differences between genera or species. Regulation of morphology and physiological status of cultures in liquid medium is important for the successful production of propagules in the bioreactor.

3. Solid culture, shake culture and bioreactor culture

In order to establish protocols for propagation of plants in the bioreactor, culture conditions should be examined precisely using solid medium and then using liquid medium in shake culture. Shake culture can be used to evaluate the suitability of a plant species for cultivation in a bioreactor. The shake culture method itself is rather easy and requires less labor than solid culture and is efficient for the production of a large number of plant propagules. However, the operation cost is still high, and a large number of culture vessels are required. Shake culture is the transition stage from solid culture to bioreactor culture. The efficiency of propagation of large number of plantlets is quite high in the bioreactor compared to solid culture or shake culture and results in a savings of labor cost (1/12.5). The number of plants produced and the manual labor required for plantlet dividing and transplant were almost the same between bioreactor and culture bottle. The efficiency of reestablishment of plants in soil is almost the same for the bioreactor and culture bottle.

Table 1: Comparison of the specifications of *Spathiphyllum* propagation in bioreactor with liquid medium and in culture bottle with solid medium

Items	Bioreactor	Culture bottle
Equipment:		
Vessel volume	20 l	500 ml
Medium volume (litre per vessel)	16.6 l (liquid)	100 ml (solid)
Number of vessels	6	1000
Inocula (number of test tubes)	96	150

4. The characteristics of bioreactor culture techniques in plant propagation

Many plant genera and species have been investigated in shake and bioreactor cultures. Cultures that grow well in liquid medium can be scaled up for bioreactor culture. This chapter will describe a case study of mass plant propagation focused mainly on the preparation of inocula, types of propagules, methods of bioreactor culture and cultivation in soil.

4.1 Preparation of 'seed cultures' (inocula)

In order to produce a large number of plantlets in a bioreactor, large numbers of propagules developing new shoots are prepared as inocula. The appropriate propagules include a) multiple shoot buds, b) regenerative tissues such as protocorm-like bodies, embryogenic or meristematic tissues, c) somatic embryos, or d) stems or shoots with a number of axillary buds. Here multiple shoot buds and stems or shoots were used as inocula because of their genetic stability. Small pieces of tissue with multiple shoot buds that have been propagated in the test tubes were the most preferable inoculum for the bioreactor. For example, in the case of *Spathiphyllum*, a piece of tissue cultured on 10 ml of agar medium was used to inoculate 1 to 2 litres of liquid medium in the bioreactor. Optimum inoculum size differs between genera or species, but usually a small inoculum size is sufficient. Multiple shoot bud cultures serially subcultured in shake flasks were used as inocula, but sometimes they only remained as multiple shoot buds without growing into plantlets even in the bioreactor. Cases like these should be considered when establishing propagation schemes.

4.2 Types of propagules produced in the bioreactor

Various types of plant propagules such as shoots, bulbs, microtubers, corms and embryos have been successfully propagated in bioreactors. Several examples are as follows:

a) Shoots: *Atropa belladonna, Begonia×hiemalis, Chrysanthemum morifolium, Dianthus caryophyllus, Fragaria ananassa, Nicotiana tabacum, Scopolia japonica, Spathiphyllum, Stevia rebaudiana, Zoysia japonica, Primula obconica.*
b) Bulbs: *Lilium* species, *Fritillaria thunbergii, Hippeastrum hybridum, Gladiolus, Hyacinthus orientalis.*
c) Corms: *Colocasia esculenta, Pinellia ternata, Caladium* sp.
d) Tubers: *Solanum tuberosum*
e) Embryos or adventitious buds: *Atropa belladonna*

The propagules produced in the bioreactor should be easily to reestablished in soil, with as little adaptation to *ex vitro* conditions as possible. Storage organs such as bulbs, corms or tubers seem to be the best choice for proliferation in bioreactors.

4.3 Configuration and operation of bioreactors: practical aspects

The bioreactors used for research are usually quite expensive, and are not practical for commercial plant propagation. In order to reduce the costs, simplicity of structure and handling, long-term maintenance of aseptic conditions, and efficient growth of propagules must be considered in designing the bioreactor. Practical bioreactors for plant propagation will provide batch cultures without sensors or controlling devices (Figure 1). This type of bioreactor is easily applicable to many plant genera and species including *Spathiphyllum* propagation as indicated in figure 2. At the end of the culture period, propagules should be of sufficient development and easily harvested, transplanted and reestablished in the soil. At present, plant genera or species with these characteristics are limited, but in the future when the regulation of the growth and physiological characteristics become highly developed, bioreactors may be more widely used in the mass propagation of plants.

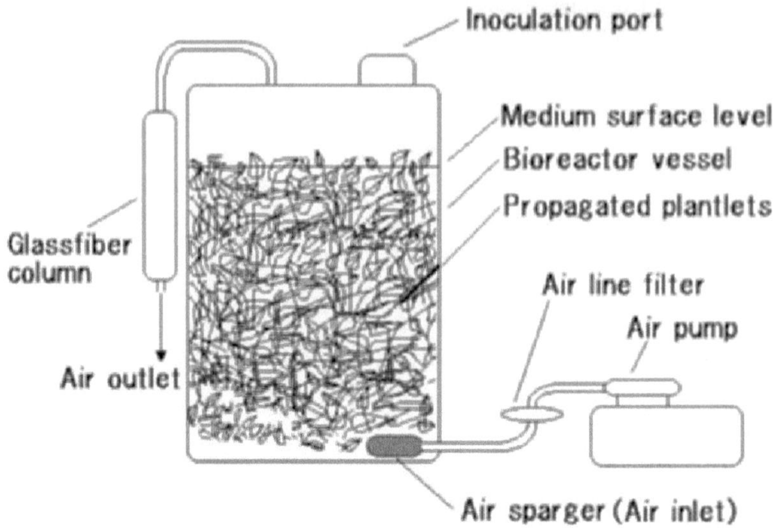

Figure 1: Simple aeration bioreactor for various plant propagation using liquid medium.

5. Case studies of the use of bioreactors in plant propagation

5.1 Strawberry (Fragaria ananassa)

Strawberry plants propagated in the field are easily infected by viruses, and so virus-free plants propagated by tissue culture are widely utilized by the growers. The propagation of strawberry is rather easy in conventional tissue culture (Boxus, 1974, 1976; Boxus et al., 1977; Damiano, 1980). The method is primarily performed to produce virus-free stock plants, but the production of large numbers of plants required for commercial cultivation in glasshouse or in the field can not be supplied trough tissue culture, so virus-free plants propagated by tissue culture are further propagated by *ex vitro* cultivation. The ultimately aim of mass propagation using bioreactors is to produce large numbers of virus-free plants to distribute directly to farmers. We reported the mass propagation of strawberry plantlets using bioreactors (Takayama et al., 1985a).

Tissue segments of multiple shoot buds induced on 10 ml of agar medium supplemented with 1 mg l^{-1} 4PU (N-(2-chloro-4-pyridyl)-N'-phenylurea) in 25 mm (diameter)×125 mm (length) test tubes were inoculated into liquid medium in a bioreactor operated at an aeration rate of 0.25 vvm under illumination of fluorescent light at 12.5 μmol $m^{-2}s^{-1}$.

The growth of plantlets in the bioreactor was stimulated by the addition of cytokinin, but they were hyperhydrated (vitrified) when harvested from the bioreactor. Therefore, the medium used for bioreactor culture contained no cytokinin or was supplemented with only a low concentration of cyotokinin.

After about one month, cultures were of transplantable size. The plantlets were taken out of the bioreactor and directly transplanted in vermiculite, covered with transparent film and cultivated in a glasshouse in shady conditions (acclimatization). The plantlets just after harvest from the bioreactor were slightly etiolated comparing to intact plants growing in the field (total content of chlorophyll was about 1/3). However, during acclimatization, chlorophyll content increased and after 2 to 4 weeks, new leaves emerged and plants were successfully acclimatized. More than 90% were successfully established in vermiculite.

5.2 Lilium species

Lilium species are important flowering plants and consist of 80 species (Willis, 1978) including over 10,000 varieties or strains. Most of strains are clonally propagated, reflecting a practical need for mass clonal propagation by tissue culture.

The process for clonal mass propagation of *Lilium* was patented and applied to bioreactor culture (Takayama and Misawa, 1982, 1983; Misawa and Takayama, 1985; Takayama et al., 1991; Takayama et al., 1996). The process consists of a) induction of multiple bulbscales by cytokinin, b) stimulation of the growth of each bulbscale in aggregated form in liquid medium, c) dissection of each bulbscale, and d) regeneration of newly formed bulblets in liquid medium. The bulblets thus obtained were easily transplanted in soil without acclimatization, and were grown to the flowering stage (Takayama et al., 1982, Takayama and Ohkawa, 1990).

The time to flowering was dependent on the species or cultivars, e.g. one year in *L. longiflorum* and *L.* × *formolongi*, or several years in *L. auratum* and *L. speciosum.* The duration also depended on bulb size. In order to shorten the time to flowering, production of larger bulblets is required. An efficient method for production of larger bulblets utilized higher medium concentrations (Takayama and Misawa, 1979), longer incubation time, and larger volume of culture medium per bulblet (Takayama et al., 1982; Takayama and Ohkawa, 1990). Culture methods satisfying these conditions were used in the bioreactor. In order to produce large numbers of bulblets in the bioreactor, large numbers of regenerative tissue segments must be inoculated. Regeneration ability of bulbscales revealed traverse gradient in one bulbscale. The most regenerative segment dissected from one bulbscale was the proximal portion while low regeneration ability was seen on the distal portion (Takayama and Misawa, 1983). In spite of these observations, cutting the bulblets randomly into pieces simplifies the process and is satisfactory. Use of such bulbscale segments as inocula resulted in the production of a large number of newly formed bulblets, and is the preferred inocula for bioreactor culture (Tsumaki and Takayama, 1992).

Although the efficiency of the production of bulblets in liquid medium (shake or bioreactor) was different between species, almost the same culture conditions were optimal within the same species or genera. However, sometimes quite different culture conditions were optimal in different strains within same species of *L. auratum* (Takayama and Okuyama, 1996).

5.3 Amaryllis (Hippeastrum hybridum)

Hippeastrum hybridum is a well known ornamental bulbous plant and is usually clonal propagated by twin scaling (Hanks, 1986; Okubo et al. 1990, Huang et al., 1990a, Stancato et al., 1995), but because of virus disorders and a low propagation rate, alternative tissue culture methods have been developed by several authors (Mii et al., 1974; Hussey et al., 1975a,b; Yanagawa et al., 1977, 1980; Fountain and Rourke, 1980; Huang et al., 1990b), but the propagation efficiency is still not adequate for production of large numbers of plants. Tissue culture of *Hippeastrum* is not difficult, but the multiplication rate is usually not high enough to produce a sufficient number of propagules for bioreactor culture. The use of cytokinin was not effective in the stimulation of the regeneration of newly formed bulblets, in fact cytokinin inhibited the growth of bulblet. In order to stimulate the regeneration and growth of *Hippeastrum hybridum in vitro*, the authors employed the noching method on the basal end of the small bulblet propagated *in vitro*. The bulblets were vertically sectioned into 4 to 8 divisions, which was easily performed and stimulated the formation of bulblets. The use of MS liquid medium containing 30 g l^{-1} sucrose in light was required. Based on this result, sectioned bulblets were cultured in the light in a 10-litre glass jar fermentor containing 5-litre of MS liquid medium. This resulted in the efficient propagation of bulblets in the bioreactor (Takayama and Yokokawa, 1996). After 4 months of cultivation in a bioreactor, 167 bulblets of 637 g FW were produced on average. When the culture period was extended to more than one year, the size of the bulblet became larger. After about one year of cultivation, bulb diameter increased to over 2 or 3 cm (Takayama and Yuasa, unpublished result).

Figure 2: Propagation of *Spathiphyllum* plants using bioreactors: (A) *Spathiphyllum* plantlets in bioreactors. (B) Close-up of plantlets growing in bioreactor. (C. and D) Harvesting the plantlets. (E) Plantlets from bioreactor. (F) Plants reestablished in Oasis growing medium after cultivated with plastic cover under irradiation.

5.4 *Gladiolus grandiflorus*

The conventional propagation method of *Gladiolus* is to use microcorms regenerated around the basal disc of the cormels. Stocks should be chosen carefully to prevent spread of diseases (Wilfret, 1980). Tissue culture techniques have enhanced the propagation efficiency (Bajaj et al., 1982; Dantu and Bhojwani, 1995; Sen and Sen, 1995; Jager et al., 1998; Park et al., 2001,). The use of bioreactors overcomes the problems encountered in conventional propagation or in tissue culture propagation on agar medium. Ziv (1990, 1991, 1992) has investigated bioreactor culture of *Gladiolus* and reported that *Gladiolus* buds proliferated profusely to clusters in bioreactor cultures supplemented with either ancymidol, paclobutrazol or uniconazol. Mechanically separated meristemoid clusters were induced to form cormlets which could either be stored or transplanted *ex vitro* without acclimatization (Ziv 1992). We also established a propagation protocol for *Gladiolus* using bioreactors in a two stage culture (Takayama et al., 1987). Shoot cultures were established in the bioreactor using MS medium with 10 g l^{-1} sucrose, and after shoots were fully grown in the bioreactor, concentrated sucrose solution was added to a final concentration of 90 g l^{-1}. By this method, microcorms from the basal position of the shoot were produced in the bioreactor. Using two stage culture in 1-litre bioreactor, the number of corms produced was about 300.

5.5 *Spathiphyllum*

Spathiphyllum is a foliage *Araceae* plant native to tropical regions and accepted as an ornamental plant because of its sweet aroma and flower shape. The propagation of *Spathiphyllum* is generally carried out by dividing, but the efficiency is not high. Therefore, tissue culture was performed (Watad et al., 1997; Ramirez-Malagon et al., 2001). Acclimatization is required for establishment of plants in soil.

We are propagating *Spathiphyllum* using bioreactors as a model experiment for direct transplant of bioreactor-cultured plants into soil under open conditions (without film cover). In our study, shoots were taken out of the bioreactor, transferred aseptically to glass bottles, and dried for one week in a glass bottle covered with paraffin-coated paper or dried gradually for several weeks using glass bottles covered with aluminum foil and paraffin-coated paper. The plants were acclimatized during desiccation. After an ambient desiccation period, plants were transferred to Oasis growing medium (Smithers-Oasis Co.), watered, with or without plastic cover, and cultivated under fluorescent light in the culture room. Almost 100% of the plants survived. Even when the plants were cultivated without plastic cover, plants could also be reestablished in Oasis growing medium, but the survival rate was lower. The maximum survival rate in open cultivation was 70%, and among them over 80% were rooted on Oasis growing medium. Treatment with ABA or mannitol (osmotic agent) also stimulated the reestablishment of the plant in Oasis growing medium in open cultivation (Takayama and Miura, Takayama and Inoue, unpublished results).

Culture conditions in the bioreactor and the subsequent treatment enhancing the acclimatization and reestablishment of tissue-cultured plantlets without protected cultivation will be important for commercial use of bioreactors.

5.6 Colocasia esculenta (Taro)

Colocasia esculenta is an important food crop, especially in tropical regions. The propagation of this plant is easily performed by division, but the commercial production of corms of this plant and transmission of virus disorders is the main problem. Tissue culture is an alternative way to propagate virus-free microcorms (Chand and Pearson, 1998; Akita and Ohta, 1996).

We have established an efficient method for propagation of microcorms of *Colocasia* using bioreactor (Takayama et al., 1989 a, b). The process of *Colocasia* propagation in a bioreactor consists of two steps.

-Step 1: Propagation of shoots in a 10 to 20-litre glass bioreactor using half-strength MS medium supplemented with 30 g l^{-1} sucrose (Takayama et al., 1989a). *Colocasia* shoots produced in this step were transplanted in soil and easily reestablished, but the manual transplantation of the shoots is laborious. To solve this problem, conditions for the production of microcorms in the bioreactor were examined. Microcorm production was quite difficult, but after serial experiments, specific conditions for the induction of microcorms were established, which were satisfactorily applied to bioreactor cultures (Takayama et al., 1989b).

-Step 2: After *Colocasia* shoots were fully grown in the bioreactor, according to step 1, the aeration rate was elevated two to five times. At the same time, the concentration of sucrose was elevated to 90 g l^{-1}. During the drying process by aeration, the cultures exhibit morphological changes to form microcorms, which developed at the base of the shoots.

Using an 8-litre bioreactor containing 6 litres of medium, 2977 microcorms were harvested after one month for step 1 and another one month for step 2. The microcorms produced were easily transplanted to soil and cultivated. After one season, a large number of well-grown corms were harvested. When storage of microcorms was required, microcorms were preserved for more than 6 months in the refrigerator.

Several *Araceae* species related to *Colocasia* such as *Anthurium andreanum, Caladium bicolor, Pinellia ternata,* and *Amorphophallus konjac* have similar *in vitro* culture characteristics.

5.7 Potato (Solanum tuberosum)

Virus-free potato microtubers are important for potato farmers because they are necessary for production of high quality potato tubers, and microtubers are easily stored, and distributed. Factors, which affect the tuberization of potatoes have been reported by many researchers (for example, Hussey and Stacey, 1981; Estrada et al., 1986) and to date, several important results have been reported (for example, Hulscher et al., 1996; Jimenez et al., 1999). Among them, we first reported the use of bioreactors for efficient production of potato microtubers (Akita and Takayama 1988, 1993a, b, 1994a, b).

We developed an efficient process that consists of two phases, including semicontinuous liquid medium surface-level control (Akita and Takayama 1994b). Virus-free shoots subcultured in test tubes were used as inocula. The process of tuber propagation in bioreactors consists of two phases: Shoot multiplication (Phase 1) and microtuber formation (Phase 2).

In Phase 1, shoots were multiplicated in MS medium containing 30 g l^{-1} sucrose under continuous irradiation. In this process, shoot cultures were grown favorably in the air space with the use of a small volume of the liquid medium. In Phase 2, liquid medium was fed to completely submerge the shoots, and at the same time, the concentration of sucrose raised to 90 g l^{-1}. After one month of cultivation in Phase 2, microtubers were produced around the medium surface. The process was further revised and was established as the semicontinuous medium surface level control method (Akita and Takayama, 1994b). Using this revised method, the microtubers were induced and developed in the entire space of the bioreactor. The transplant and cultivation conditions of microtubers produced in the bioreactor were examined and are applicable to commercial propagation (Akita and Takayama, 1993a).

Recently, a more simple system was reported by Akita and Ohta (1998). The system used a plastic bottle placed on an air-permeable sheet and potato tubers were successfully formed by using a rotary culture system without forced aeration. This type of system was scaled-up to a 10-litre bioreactor and tubers were efficiently produced (Akita, unpublished data). Moreover, it could be important for practical application of culture techniques to choose inexpensive vessels and systems for cost reduction.

6. Cultivation of bioreactor cultured plantlets *ex vitro*

The ultimate aim of the use of bioreactors for plant propagation is to produce plants for cultivating in outdoor conditions. Various types of propagules from bioreactors have been subjected to soil cultivation. The authors' experiences of the cultivation of plants in soil provided the background for the efficient establishment of plants in soil, especially in the case of *Lilium* microbulbs, *Hippeastrum* bulbs, *Solanum tuberosum* microtubers, and *Spathiphyllum* plants. The morphology of the propagules from the bioreactor, handling of the propagules in transplantation, dormancy of microtubers, requirement of the acclimatization process, growth of the shoots and roots *ex vitro*, uniformity of plant size and genetic characteristics, in relation to practical cultivation of the propagules will be published in the future (Takayama, in preparation).

7. Prospects of bioreactor technology in mass propagation

The high efficiency of plant propagation using bioreactors has been revealed, and bioreactor technology seems to be applicable to commercial propagation in the several plant species discussed herein. However, many problems still exist in scale-up and in application to other plant species. The main cause for the problems comes from the difficulty in the preparation and handling of the bioreactors, in preparation of 'seed cultures' and in finding the optimum culture conditions, which depends mainly on the types of culture, different genera or species of plants, etc. The advantage of bioreactor technology exists in the high efficiency and ease of operation . The problems have to be overcome before the general use of the technique for commercial propagation . It is important to note that the bioreactors will be a most promising system for industrial plant propagation, including process automation and robotics.

References

Abdullah MA, Ariff AB, Marziah M, Ali AM & Lajis NH (2000) Strategies to overcome foaming and wall-growth during the cultivation of *Morinda elliptica* cell suspension culture in a stirred-tank bioreactor. Plant Cell, Tissue and Organ Culture 60: 205-212

Akita M (2000) Bioreactor culture of plant organs. In: Spier R E, Griffiths B & Scragg A H (eds.), The Encyclopedia of Cell Technology (2-Volume Set)(pp.129-138). ISBN: 0-471-16123-3, John Wiley & Sons, Inc., New York

Akita M & Ohta Y (1996) Development of a system for mass propagation of *Colocasia esculenta* in large scale without forced aeration. Acta Hort. 440: 554-559

Akita M & Ohta Y (1998) A simple method for mass propagation of potato (*Solanum tuberosum* L.) using a bioreactor without forced aeration. Plant Cell Reports 18: 284-287

Akita M, Shigeoka T, Koizumi Y & Kawamura M (1994) Mass propagation of shoots of *Stevia rebaudiana* using a large scale bioreactor. Plant Cell Reports 13: 180-183.

Akita M & Takayama S (1988) Mass propagation of potato tubers using jar fermentor techniques. Acta Hortic. 230: 55-62

Akita M & Takayama S (1993a) Resting periods and the field performance of potato tubers propagated in a jar fermentor. Plant Tissue Culture Letters 10: 255-259

Akita M & Takayama S (1993b) Effects of medium surface level control on the mass propagation of potato tubers using jar fermentor culture techniques. Plant Tissue Culture Letters 10: 242-248

Akita M & Takayama S (1994a) Induction and development of potato tubers in a jar fermentor. Plant Cell, Tiss. Org. Cult. 36: 177-182

Akita M & Takayama S (1994b) Stimulation of potato (*Solanum tuberosum* L.) tuberization by semicontinuous liquid medium surface level control. Plant Cell Reports 13: 184-187

Bajaj YPS, Sidhu MMS & Gill APS (1982) Some factors affecting the *in vitro* propagation of gladiolus. Sci. Hort. 18: 269-275

Boxus PH (1974) The production of strawberry plants by *in vitro* micropropagation. J. Hort. Sci. 49: 209-210

Boxus PH (1976) Rapid production of virus-free strawberry by '*in vitro*' culture. Acta Hortic. 66: 35-38

Boxus PH, Quoirin M & Laine JM (1977) Large scale propagation of strawberry plants from tissue culture. In: Reinert J, Bajaj Y P S (eds) Applied and Fundamental Aspects of Palnt Cell, Tissue and Organ Culture (p.130-143). Springer Verlag, Heidelberg

Chand H & Pearson MN (1998) Rapid vegetative multiplication in *Colocasia esculenta* (L) Schott (taro). Plant Cell, Tissue and Organ Culture 55: 223-226

Chatterjee C, Correll MJ, Weathers PJ, Wyslouzil BE & Walcerz DB (1997) Simplified acoustic window mist bioreactor. Biotechnology Techniques 11: 155-158

Coombs J (1986) MacMillan Dictionary of Biotechnology. Macmillan Press, London, 330 pp.

Correll MJ, Wu Y, Weathers PJ (2000-2001) Controlling hyperhydration of carnations (*Dianthus caryophyllus* L.) grown in a mist reactor. Biotechnol. Bioeng. 71: 307-314

Damiano C (1980) Planning and building a tissue culture laboratory. In: Proceedings of the Conference on Nursery Production of Fruit Plants Through Tissue Culture - Applications and Feasibility (pp. 93-101). USDA

Dantu PK & Bhojwani SS (1995) *In vitro* corm formation and field evaluation of corm-derived plants of *Gladiolus*. Scientia Hortic. 61: 115-129

Escalona M, Lorenzo JC, Gonzalez B, Daquinta M, Gonzalez JL, Desjardins Y & Borroto CG (1999) Pineapple (*Ananas comosus* L. Merr) micropropagation in temporary immersion systems. Plant Cell Reports 18: 743-748

Estrada R, Tovar P & Dodds JH (1986) Induction of *in vitro* tubers in a broad range of potato genotypes. Plant Cell, Tiss. Org. Cult. 7: 3-10

Fountain WM & Rourke EN (1980) Investigations of *in vitro* culture of *Hippeastrum hybridum* cultivar apple-blossom as a method of rapid propagation. Hort. Sci. 15: 274-275

Gao WY, Fan L & Paek KY (2000) Yellow and red pigment production by cell cultures of *Carthamus tinctorius* in a bioreactor. Plant Cell, Tiss. Org. Cult. 60: 95-100

Hale A & Young RE (1991) Plant micropropagation bioreactor development. Am. Soc. Agric. Eng. Meeting.Winter 1991 (91-7542) p. 16

Hale SA & Young RE (1995) Automatic control of micropropagation medium sucrose concentration. Transactions of the ASAE 38: 1229-1235

Hanks GR (1986) The effect of temperature and duration of incubation on twin-scale propagation of Narcissus and other bulbs. Crop Res. 25: 143-152

Hao Z, Ouyang F, Geng Y, Deng X, Hu Z & Chen Z (1998) Propagation of potato tubers in a nutrient mist bioreactor, Biotechnology Techniques 12: 641-644

Harrell RC, Bieniek M, Hood CF, Munilla R & Cantliffe DJ (1994) Automated *in vitro* harvest of somatic embryos. Plant Cell, Tiss. Org. Cult. 39: 171-183

Heyerdahl PH, Olsen OAS & Hvoslef-Eide AK (1995) Engineering aspects of plant propagation in bioreactors. In: Aitken-Christie J, Kozai T, Smith L M (eds) Automation and Environmetal Control in Plant Tiss. Cult. (pp. 87-123). Kluwer Academic Publishers, Dordrecht

Honda H, Liu C & Kobayashi T (2001) Large-scale plant micropropagation, Adv. Biochem. Eng. Biotechnol. 72: 157-182

Huang CW, Okubo H & Uemoto S (1990a) Importance of two scales in propagating *Hippeastrum hybridum* by twin scaling. Sci. Hortic. 42: 141-150

Huang CW, Okubo H & Uemoto S (1990b) Comparison of bulblet formation from twin scales and single scales in *Hippeastrum hybridum* cultured *in vitro*. Sci. Hortic. 42: 151-160

Hulscher M, Krijgsheld HT & Jongedijk E (1996) Mass propagation of potato microtubers in jar fermentors. Acta Hortic. 440: 533-537

Hussey G (1975a) Propagation of *Hyacinthus* by tissue culture. Sci. Hortic. 3: 21-28

Hussey G (1975b) Totipotency in tissue explants and callus of some members of the *Liliaceae, Iridaceae* and *Amaryllidaceae*. J. Exp. Bot. 26: 253-262

Hussey G & Stacey NJ (1981) *In vitro* propagation of potato (*Solanum tuberosum* L.). Ann. Bot. 48: 787-796

Ingram B & Mavituna F (2000) Effect of bioreactor configuration on the growth and maturation of *Picea sitchensis* somatic embryo cultures. Plant Cell, Tiss.Org. Cult. 61: 87-96

Jager AK, McAlister BG & van Staden J (1998) *In vitro* culture of *Gladiolus carneus*. South African J. Bot. 6: 147

Jimenez E, Perez N, de Feria M, Barbon R, Capote A, Chavez M, Quiala E & Perez JC (1999) Improved production of potato microtubers using a temporary immersion system. Plant Cell, Tiss. Org. Cult. 59: 19-23

Levin R, Gaba V, Tal B, Hirsch S, Denola D & Vasil KI (1988) Automated plant tissue culture for mass propagation. Bio/Technol. 6: 1035-1040

Levin R, Stav R, Alper Y & Watad AA (1996) *In vitro* multiplication in liquid culture of *Syngonium* contaminated with *Bacillus* spp. and *Rathayibacter tritici*. Plant Cell, Tiss. Org. Cult. 45: 277-280

Lorenzo JC, Ojeda E, Espinosa A & Borroto C (2001) Field performance of temporary immersion bioreactor-derived sugarcane plants. *In Vitro* Cellular and Development Biology Plant 37: 803-806

Luttman R, Florek P & Preil W (1994) Silicone-tubing aerated bioreactors for somatic embryo production. Plant Cell, Tiss. Org. Cult. 39: 157-170

Mii M, Mori M & Iwase N (1974) Organ formation from the excised bulb scales of *Hippeastrum hybridum in vitro*. J. Hort. Sci. 49: 241-244

Misawa M & Takayama S (1985) A method for large-scale propagation of lilies. Japanese Patent 60-15286

Moorhouse SD, Wilson G, Hennerty J, Selby C & Mac An tSaoir S (1996) A plant cell bioreactor with medium-perfusion for control of somatic embryogenesis in liquid cell suspensions. Plant Growth Regulation 20: 53-56

Okamoto A, Kishine S, Hirosawa T & Nakazono A (1996) Effect of oxygen-enriched aeration on regeneration of rice (*Oryza sativa* L.) cell culture. Plant Cell Reports 15: 731-736

Okubo H, Huang CW & Uemoto S (1990) Role of outer scale in twin-scale propagation of *Hippeastrum hybridum* and comparison of bulblet formation from singlescales and twin scales. Acta Hortic. 266: 59-66

Paek KY, Hahn EJ & Son SH (2001) Application of bioreactors for large-scale micropropagation systems of plants, *In Vitro* Cellular and Development Biology - Plant 37: 284-292

Park IS, Choi JD, Byun MS, Goo DH & Kim KW (2001) Effects of liquid shaking culture and growth retardants on cormlet formation of *Gladiolus* `Spic and Span' *In Vitro*. J. Korean Soc. Hort. Sci. 42: 215-218

Preil W (1991) Application of bioreactors in plant proagation, - In: Debergh P C, Zimmerman R H (eds) Micropropagation Technology and Application. (pp. 425-446) Kluwer Academic Publishers, Dordrecht

Preil W & Beck A (1991) Somatic embryogenesis in bioreactor culture Acta Hortic. 289: 179-192

Preil W, Florek P, Wix U & Beck A (1988) Towards mass propagation by use of bioreactors. Acta Hortic. 226: 99-105

Ramirez-Malagon R, Borodanenko A, Barrera-Guerra JL & Ochoa-Alejo N (2001) Shoot number and shoot size as affected by growth regulators in *in vitro* cultures of *Spathiphyllum floribundum* L. Sci. Hort. 89: 227-236

Sen J & Sen S (1995) Two-step bud culture technique for a high frequency regeneration of *Gladiolus* corms. Sci. Hort. 64: 133-138

Sondahl MR & Noriega C (1991) Bioreactor systems for plant micropropagation. In: ASAE publication: Automated Agriculture for the 21st Century: Proceedings of the 1991 Symposium (pp. 497-502). 16-17 December 1991, Chicago, Illinois

Stancato GC, Mazzafera P & Magalhaes AC (1995) Dry matter partitioning during the propagation of *Hippeastum hybridum* as affected by light. Sci. Hort. 62: 81-87

Styer DJ (1985) Bioreactor technology for plant propagation, In: Henke R R, Hugher K W, Constantin M J, Hollander A (eds) Tissue Culture in Forestry and Agriculture (pp. 117-130). Penum Press

Takahashi S, Matsubara K, Yamagata H & Morimoto T (1992) Micropropagation of virus free bulblets of *Lilium longiflorum* by tank culure, 1. Development of liquid culture method and large scale propagation. Acta Hortic. 319: 83-88

Takayama S (1991) Mass propagtion of plants through shake- and bioreactor-culture techniques. In: Bajaj YPS (ed) Biotechnology in Agriculture and Forestry, Vol.17 (pp.495-515) Springer Verlag, Heidelberg

Takayama S (1997) Bioreactors for plant cell tissue and organ cultures. In: Vogel, H C & Todaro C L (eds) Fermentation and Biochemical Engineering Handbook, 2nd Edition (pp.46-70). Noyes Publications, Westwood, New Jersy, USA

Takayama S (2000) Bioreactors, Airlift. In: Spier RE, Griffiths B & Scragg AH (eds) The Encyclopedia of Cell Technology (2-Volume Set) (pp. 201-218). ISBN: 0-471-16123-3, John Wiley & Sons, Inc., New York

Takayama S & Akita M (1994) The types of bioreactors used for shoots and embryos. Plant Cell, Tiss. Org. Cult. 39: 147-156

Takayama S & Akita M (1998) Bioreactor techniques for large-scale culture of plant propagules. Adv. Hort. Sci. 12: 93-100

Takayama S, Amo T, Fukano M, Nakazawa K & Akita M (1987) Mass propagation of gladiolus by liquid culture medhods. Abst. 10th Meeting of Japanese Assoc. of Plant Tissue and Cell Cult. p. 88

Takayama S, Amo T, Fukano M & Oosawa K (1985a) Mass propagation of strawberries by jar fermentor culture. (2) Studies on the optimum conditions in a liquid medium and the establishment of mass propagation scheme using a jar fermentor. Abst. 1985 Spring Meeting of Japanese Society for Horticultural Science (pp. 210-221). Japanese Society for Horticultural Science, Tokyo

Takayama S, Arima Y & Akita M (1986) Mass propagation of plants by fermentor culture techniques. In: Abst. 6th International Congress of Plant Tissue and Cell Cuture, p. 449, International Association of Plant Tissue Culture, University of Minnesona

Takayama S, Ishio S & Oosawa K (1989a) Mass propagation of Araceous plants by jar fermentor culture techniques. (1) Mass propagation of *Colocasia esculenta* shoots. In: Abst. 1989 Spring Meeting Jap. Soc. Hort. Sci., pp. 296-297

Takayama S, Ishio S & Oosawa K (1989b) Mass propagation of Araceous plants by jar fermentor culture techniques. (1) Mass production of microcorms of *Colocasia esculenta*. In: Abst. 1989 Spring Meeting Jap. Soc. Hort. Sci., pp. 296-297

Takayama S & Misawa M (1979) Differentiation in *Lilium* bulbscales grown *in vitro*. Effect of various cultural conditions. Physiol. Plant. 46: 184-190

Takayama S & Misawa M (1981) Mass propagation of *Begonia* x *hiemalis* plantlets by shake culture. Plant Cell Physiol. 22: 461-467

Takayama S & Misawa M (1982) A scheme for mass propagation of *Lilium in vitro*. Scientia Hortic. 18: 353-362

Takayama S & Misawa M (1983) The mass propagation of *Lilium in vitro* by stimulation of multiple adventitious bulb-scale formation and by shake culture. Can. J. Bot. 61: 224-228

Takayama S, Misawa M, Takashige Y, Tsumori H & Ohkawa K (1982) Cultivation of *in vitro* propagated *Lilium* bulbs in soil. J. Amer. Soc. Hort. Sci. 107: 830-834

Takayama S & Ohkawa K (1990) Growth and Flowering of *in vitro*-propagated *Lilium auratum* bulbs in soil. Plant Tissue Culture Letters 7: 187-192

Takayama S & Okuyama K (1996) Effects of the number of inoculated bulbscales and the culture strain on differentiation and growth of *Lilium auratum* in tissue culture. J. Soc. High Technol. Agric. 8: 175-180

Takayama S, Swedlund B & Miwa Y (1991) Automated propagation of microbulbs of lilies. In: Vasil I K (ed) Cell Culture and Somatic Cell Genetics of Plants. 8: 111-131. Academic Press, New York

Takayama S & Yokokawa A (1996) Effect of abscisic acid (ABA) and light irradiation on mass propagation of *Hippeastrum hybridum* Hort. in shake- and jar fermentor-culture. J. Soc. High Technol. Agric. 8: 168-174

Tautorus TE, Lulsdorf MM, Kikcio SI & Dunstan DI 1994 Nutrient utilization during bioreactor culture, and maturation of somatic embryo cultures of *Picea mariana* and *Picea glauca-engelmannii. In Vitro* Cellular and Developmental Biology Plant 30: 58-63

Tsumaki N &Takayama S (1992) Simple mass propagation of *Lilium auratum* by shake and jar fermentor culture. J. Soc. High Technol. Agric. 3: 124-128

Vinocur B, Carmi T, Altman A & Ziv M (2000) Enhanced bud regeneration in aspen (*Populus tremula* L.) roots cultured in liquid media. Plant Cell Reports 19: 1146-1154

Watad AA, Raghothama KG & Gaba V (1997) Micropropagation of *Spathiphyllum* and *Syngonium* is facilitated by use of interfacial membrane rafts. HortSci. 32: 307

Wheat D, Bondaryk RP & Nystrom J (1986) Spin filter bioreactor technology as applied to industrial plant propagation. HortSci. 21: 819

Wilfret GJ (1980) *Gladiolus*. In: Larson R A (ed) Introduction to Floriculture (pp. 165-181). Academic Press, New York

Willis JC (1982) Dictionary of Flowering Plants and Ferns, Cambridge University Press, ISBN 052108699X

Yanagwa T & Sakanishi Y (1977) Regeneration of bulblets on *Hippeastrum* bulb segments excised from various parts of apparent bulb. J. Jap. Soc. Hort. Sci. 46: 250-260

Yanagwa T & Sakanishi Y (1980) Studies on the regeneration of excised bulb tissues of various tunicated-bulbous ornamentals. I. Regenerative capacity of the segments from different parts of bulbscales. J. Jap. Soc. Hort. Sci. 48: 495-502

Ziv M (1990) Morphogenesis of *Gladiolus* buds in bioreactors - Implication for scaled-up propagation of geophytes. In: Nijkamp J J J, Van Der Plas L H W, Artrijk J V (eds) Progress in Plant Cellular and Molecular Biology (pp. 119-124). Kluwer Academic Publishers, Dordrecht

Ziv M (1991) Morphogenic patterns of plants micropropagated in liquid medium in shaken flasks or large-scale bioreactor cultures. Israel J. Bot. 40: 145-153

Ziv M (1992) Morphogenic control of plants micropropagated in bioreactor cultures and its possible impact on acclimatization. Acta Hortic. 319: 119-124

Ziv M (1995) The control of bioreactor environment for plant propagation in liquid culture. Acta Hortic. 393: 25-38

Ziv M (2000) Bioreactor technology for plant micropropagation. Horticultural Reviews. 24: 1-30

Ziv M, Kahany S & Lilien-Kipnis H (1994) Scaled-up proliferation and regeneration of Nerine in liquid cultures: Part I. The induction and maintenance of proliferating meristematic clusters by paclobutrazol in bioreactors. Plant Cell, Tiss. Org. Cult. 39: 109-115

Ziv M, Kahany S & Lilien-Kipnis H (1995) Somatic embryos and bulblet development from bioreactor regenerated meristematic clusters of *Nerine*. Acta Hortic. 393: 203-212

Ziv M & Shemesh D (1996) Propagation and tuberization of potato bud clusters from bioreactor culture. *In Vitro* Cellular and Developmental Biology Plant 32: 31-36

Chapter 5

Simple bioreactors for mass propagation of plants

Meira Ziv
Department of Agricultural Botany, The Hebrew University of Jerusalem, P.O. Box 12, Rehovot 76100, ISRAEL.
Tel.: +972-8-9489915; Fax:+972-8-9467763; E-mail: meira@agri.huji.ac.il

Abstract: Bioreactors provide a rapid and efficient plant propagation system for many agricultural and forestry species, utilizing liquid media to avoid intensive manual handling. Large-scale liquid cultures have been used for micropropagation through organogenesis or somatic embryogenesis pathways. Various types of bioreactors with gas-sparged mixing are suitable for the production of clusters of buds, meristems or protocorms. A simple glass bubble-column bioreactor for the proliferation of ornamental and vegetable crop species resulted in biomass increase of 3 to 6-fold in 3-4 weeks. An internal loop bioreactor was used for asparagus, celery and cucumber embryogenic cultures. However, as the biomass increased, the mixing and circulation were not optimal and growth was reduced. A disposable pre-sterilized plastic bioreactor (2 to 5-litre volume) was used for the proliferation of meristematic clusters of several ornamental, vegetable and woody plant species. The plastic bioreactor induced minimal shearing and foaming, resulting in an increase in biomass as compared to the glass bubble-column bioreactor.

A major issue related to the use of liquid media in bioreactors is hyperhydricity, that is, morphogenic malformation. Liquid cultures impose stress signals that are expressed in developmental aberrations. Submerged tissues exhibit oxidative stress, with elevated concentrations of reactive oxygen species associated with a change in anti-oxidant enzyme activity. These changes affect the anatomy and physiology of the plants and their survival. Malformation was controlled by adding growth retardants to decrease rapid proliferation. Growth retardants ancymidol or paclobutrazol reduced water uptake during cell proliferation, decreased vacuolation and intercellular spaces, shortened the stems and inhibited leaf expansion, inducing the formation of clusters. Using a two-stage bioreactor process, the medium was changed in the second stage to a medium lacking growth retardants to induce development of the meristematic clusters into buds or somatic embryos. Cluster biomass increased 10 to 15-fold during a period of 25-30 days depending on the species. Potato bud clusters cultured in 1.5 litres of medium in a 2-litre capacity bioreactor, increased during 10-30 days. Poplar *in vitro* roots regenerated buds in the presence of thidiazuron (TDZ); the biomass increased 12-fold in 30 days. Bioreactor-regenerated clusters were separated with a manual cutter, producing small propagule units that formed shoots and initiated roots. Clusters of buds or meristematic nodules with reduced shoots, as well as arrested leaf growth, had less distortion and were optimal for automated cutting and dispensing. In tuber-, bulb-

A.K. Hvoslef-Eide and W. Preil (eds.), Liquid Culture Systems for in vitro Plant Propagation, 79–93.

and corm-producing plants, growth retardants and elevated sucrose concentrations in the media were found to enhance storage organ formation, providing a better propagule for transplanting or storage.
Bioreactor-cultures have several advantages compared with agar-based cultures, with a better control of the contact of the plant tissue with the culture medium, and optimal nutrient and growth regulator supply, as well as aeration and medium circulation, the filtration of the medium and the scaling-up of the cultures. Micropropagation in bioreactors for optimal plant production will depend on a better understanding of plant responses to signals from the microenvironment and on specific culture manipulation to control the morphogenesis of plants in liquid cultures.

Key words: airlift bioreactors, clusters, growth retardants, hyperhydricity, oxidative-stress

Abbreviations: ANC - ancymidol; APX - ascorbate peroxidase; BA - benzyl aminopurine; CAT - catalase; DW - dry weight; FW - fresh weight; 2iP-2 - isopentenyladenine; PAC - paclobutrazol; ROS - reactive oxygen species; TDZ - thidiazuron

1. Introduction

Micropropagation is currently applied to a large number of agricultural and forestry species, but is still costly due to intensive manual manipulation throughout the various culture phases. Automation and scaled-up liquid cultures for *in vitro* plant propagation are essential to overcome some of the limitations imposed by labour–intensive methods and high production costs. Progress in tissue culture automation depends on the use of liquid cultures in bioreactors that can provide rapid proliferation, mechanized tissue cutting and separation, and automated dispensing (Ziv, 1995a; Aitken-Christie et al., 1995).

Liquid cultures have been used for plant culture in both agitated vessels and in bioreactors, for somatic embryogenesis (Styer, 1985; Nadel et al., 1990; Preil, 1991, Scragg, 1992) and as a system for plant propagation through the organogenic pathway (Ziv, 1992; Takayama, 1991). The propagation aspects of several plant species in bioreactors and some of the problems associated with the operation of bioreactors were reviewed by Ziv (1995b; 2000) and Takayama and Akita (1998).

Propagation in bioreactors through the organogenic pathway has been achieved in banana, Boston fern, *Spathiphyllum*, strawberry, potato, poplar, coffee, *Gladiolus*, lilies, pineapple, orchids, *Ornithogalum, Nerine, Narcissus* and *Cyclamen* (Ziv 2000).

In various types of bioreactors, the mechanical or gas-sparged mixing provided stirring and aeration (Scragg, 1992). Large-scale cultivation of plant cells, embryos or organs has made use of airlift or bubble-column bioreactors instead of stirred-tank bioreactors, due to their low shear properties. The main advantages of airlift bioreactors is their relatively simple construction, the lack of regions of high shear stress, reasonably high mass and heat transfer and reasonably high yields at low input rates (Denchev et al., 1992). However, bioreactor configuration must be determined according to mixing and aeration requirements of specific plants or tissues, as well as for minimization of shear stress.

In general, bioreactor-culture offers many advantages, including better control of the culture conditions; optimal supply of nutrients and growth regulators; renewal of the culture atmosphere; changing the medium during the culture period according to the developmental stage; filtration of the medium for exudates; contamination control; and production of clusters of buds or somatic embryos for automated handling of the propagules.

The use of liquid cultures in bioreactor for plant propagation imposes several problems such as leakage of endogenous growth factors, the need for an initial high concentration of the inoculum, hyperhydricity and malformation, foam development, shearing and oxidative stress (Ziv, 2000). The major disadvantage encountered is the problem of shoot malformation in liquid media (Ziv, 2000). Attempts to control hyperhydric deformities have focused on better aeration and intermittent plant submergence in the medium, using temporary immersion bioreactors (Teisson et al., 1996; Escalona et al., 1999). Levin (personal communication) used two disposable plastic bioreactors (Life Reactor, Osmotek) as an ebb and flow system for intermittent submergence in micropropagated *Spathiphyllum.* Growth retardants (Ziv, 1992; 2000) have been used to reduce leaf expansion, thus minimising shoot malformation.

The present paper reviews the use of simple bioreactors for micropropagation and describes the advantages and limitation of bubble-column and airlift bioreactors used for plant proliferation.

2. Morphogenesis and growth in bioreactors

At present, commercial propagation of most plants is carried out through the organogenic pathway in agar-gelled cultures, even though the protocols are long and costly. The commercial use of bioreactors for unipolar structures such as protocorms, buds or shoots is limited to a small number of plants, mainly due to the hyperhydricity problem in the leaves and shoots. Controlling shoot growth and providing culture conditions that reduced abnormal leaves and enhanced formation of bud or meristematic clusters in potato, *Gladiolus* and *Ornithogalum dubium,* by use of growth retardants, resulted in a high proliferation rate, and limited hyperhydricity (Levin et al.,

1997; Ziv and Hadar, 1991; Takayama, 1991; Takayama and Akita, 1998; Ilan et al., 1995; Ziv et al., 1998; Ziv, 2000).

The development of spherical meristematic or bud clusters in liquid cultures provided a highly proliferating and rapidly growing system. The clusters were amenable to the control of the medium components, to mechanical separation, and to automated inoculation as an efficient delivery system to the final stage for plant growth (Levin et al., 1997; Ziv et al., 1998). In most species, cluster formation appears to be associated with the continuous submergence, circulation and agitation of the plant biomass in the medium, as well as with a balanced ratio of growth-promoting and growth-retarding regulators. The formation of condensed organized structures, in which the shoots are reduced to buds or meristematic tissue in liquid media, has been reported for several plant species. The clusters were made-up of densely-packed meristematic cells, actively dividing and forming new meristemoids on the outer surfaces. The meristemoids surrounded loosely-packed cells in the centre and exhibited some vascularisation as was shown in liquid cultured poplar clusters (McCown et al., 1988). In banana, on the other hand, the clusters were made-up of condensed buds surrounding a central core with a cavity (Ziv et al., 1998). Protocorm-like clusters were also induced in liquid-cultured *Gladiolus* buds (Ziv, 1990) and in several species of the complex *Brodiaea* (Ilan et al., 1995) by addition of PAC or ANC (gibberellin biosynthesis inhibitors) to the medium. In *Phalaenopsis*, protocorm-like bodies were proliferated in bioreactors and regenerated on agar-based medium (Young et al., 2000). In potato and banana (Figure 1), bud clusters were induced by a balanced ratio between kinetin and ancymidol in the liquid medium (Ziv et al., 1998). In woody species, McCown et al. (1988) and Aitkens et al. (1995) described nodules in poplar in liquid medium and in radiata-pine in agar-gelled medium, respectively, induced by a balanced ratio of growth regulators. Levin et al. (1997) worked with several ornamental species and described a several-fold increase of an organogenic biomass consisting of clusters, which proliferated in bioreactors. These were separated mechanically prior to dispensing to agar-gelled cultures for further growth. The production of clusters in *Philodendron* cultured in liquid media required the presence of BA and an inductive treatment for 24-48 h with ancymidol. The short inductive treatment annulled a carry-over dwarfing effect of ancymidol on leaf and shoot development after transplanting of the clusters to agar-gelled medium for further plant growth (Ziv and Ariel, 1991). Growth regulators, promoting or retarding substances when supplied in a specific balanced ratio, apparently act as morphogenic signals and control the development of the spherical meristematic or bud clusters.

Figure 1: A cluster of banana buds from a bubble-glass-bioreactor (9.75 µmol ancymidol, 22.2 µmol BA and 11.4 µmol IAA).

2.1 *Types of bioreactors*

Various types of simple bioreactors with mechanical or gas-sparged mixing were used in plant and cell cultures to provide stirring, circulation and aeration (Takayama, 1991; Scragg, 1992; Takayama and Akita, 1998, Ziv 2000). Mechanically-stirred bioreactors depend on impellers, including a helical ribbon impeller (Archambault, 1994), magnetic stirrers or vibrating perforated plates. Aeration, mixing and circulation in bubble-column or airlift bioreactors, is provided by air entering the vessel from the side or from the basal opening through a sparger: As the air bubbles rise they lift the plant biomass and provide the required oxygen (Merchuk, 1990, Styer, 1985; Preil, 1991; Cazzulino et al., 1991).

It has been shown that mixing by gas sparging in bubble-column or airlift bioreactors lacking impellers or blades, was far less damaging for clusters than mechanical stirring, since shearing stress was minimized (Ziv and Hadar, 1991; Ilan et al., 1995; Ziv and Shemesh, 1996). The use of plastic disposable airlift bioreactors for bud or meristematic clusters was found to increase proliferation and reduce shearing stress (Ziv et al. 1998).

The main advantage of airlift bioreactors is their relatively simple construction, the lack of regions with high shearing potential, reasonably high gas and nutrient transfer rates and relatively high yields at low input rates (Denchev et al., 1992). A bubble-free oxygen supply bioreactor with silicone tubing was found suitable for embryogenic cell suspensions and provided foam-free cultures (Luttman et al., 1994). For hairy root cultures an acoustic mist bioreactor was found to increase root biomass significantly (Chatterjee et al., 1997). Simple glass, polycarbonate cylinder (Figure 2), or plastic bioreactors were used for propagation of ornamental, vegetable and woody plant species. These included the foliage plants – Boston fern, *Philodendron* and *Spathiphyllum*; the ornamentals – *Gladiolus*, lilies, *Ornithogalum dubium*, *Narcissus, Nerine*, *Brodiaea;*, and the vegetable crops – cucumber, asparagus and potato. Poplar root cultures regenerating buds, and banana were also propagated in bioreactors. In banana, the type of culture and the volume of the medium affected biomass growth (Table 1). Increasing the concentration of ancymidol decreased biomass growth, increased bud or meristem number per cluster and decreased leaf elongation and expansion (Table 2).

Table 1: The effect of culture type and medium volume on banana bud cluster proliferation in the presence of 2.5 ppm ancymidol after 30 days

Culture vessel capacity (ml)	Medium (volume ml)	FW (g)		Growth value [(1)]	No. buds per cluster
		Initial	Final		
Erlenmeyer 250	50	6.5 ± 0.7	17.5 ± 1.8	1.46	5.6 ± 0.6
Erlenmeyer 500	100	10.8 ± 1.2	57.2 ± 3.1	4.29	6.7 ± 0.8
Glass bioreactor 1000	500	17.6 ± 2.8	123.5 ± 2.1	6.01	12.0 ± 1.8
Disposable bioreactor 2500	1250	21.66 ± 2.8	269.4 ± 4.8	11.4	15.1 ± 2.1

[(1)]Growth value: $(FW_{final}-FW_{initial})/FW_{initial}$.

Table 2: The effect of ancymidol on banana cluster proliferation in disposable plastic bioreactors after 26 days culture

Ancymidol (μmol)	Δ Growth [(1)] (FW g)	Buds or meristems per cluster	Leaf tissue (% FW)
0	172 ± 16	12 ± 0.8	89
3.95	165 ± 12.2	18 ± 2.1	86
9.75	111 ± 9.7	26 ± 3.1	58
19.50	80 ± 5.3	48 ± 3.9	17

[(1)]Initial inoculum 30g l^{-1} medium

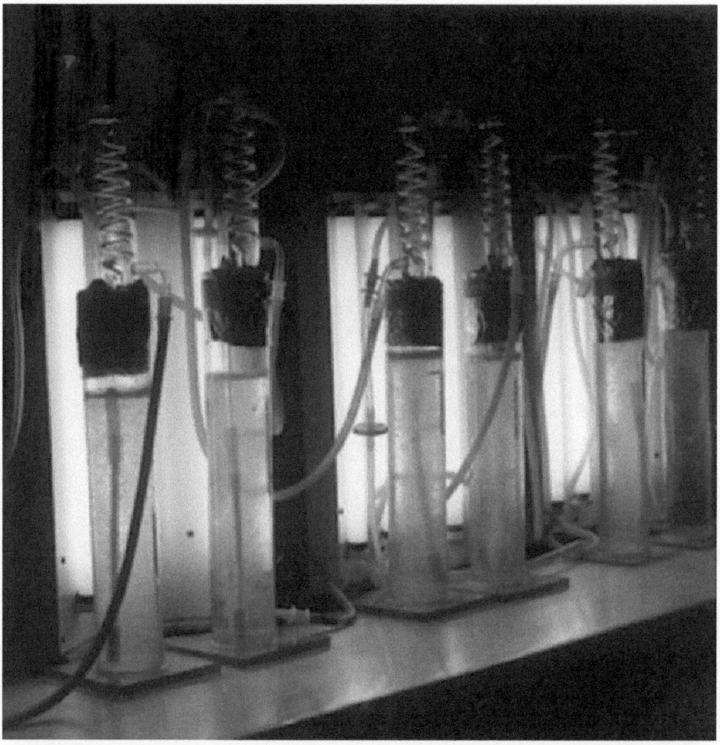

Figure 2: A polycarbonate internal-loop bioreactor used for asparagus embryogenic clusters.

Temporary immersion - or ebb and flow bioreactors - were described as a better aeration system through the periodic immersion and exposure to the gaseous phase (Etienne et al., 1997). Temporary immersion was used for banana, rubber and coffee (Teisson and Alvard, 1995; Etienne et al., 1997; Teisson et al., 1996), for tea (Akula et al., 2000) and on a much larger scale for pineapple and sugarcane (Lorenzo et al., 1998; Escalona et al., 1999). Temporary immersion was found to enhance potato microtuber growth in a rotating bioreactor (Yu et al., 2000). The periodic immersion cycle adapted to each species was found to decrease hyperhydricity, but not eliminate it altogether.

2.2 *The gaseous atmosphere*

The culture vessel gas composition is influenced by the volume of the vessel, the volume of the medium and the ventilation. In bioreactors, control of the gaseous phase depends on the gas flow and can be easily manipulated to provide the required concentrations of O_2, CO_2 and C_2H_4. In airlift or bubble-column bioreactors, the air supplied is used for both mixing and aeration. The importance of aeration and the gaseous phase were shown in potatoes cultured in airlift bioreactors. Induction of tubers was inhibited under continuously submerged conditions. Microtubers developed only after the shoots elongated and reached the gaseous phase. A two-phase culture, substituting the growth medium with a tuber-induction (9% (w/v) sucrose) medium, enhanced tuber formation from shoots which developed above the medium and were exposed to the gaseous phase (Akita and Takayama, 1994).

Oxygen concentrations in liquid cultures depend on the presence of O_2 in the gas phase above and in the air bubbles inside the medium, as well as in the dissolved O_2 in the medium. Air is sparged through a sparger located at the base of the bioreactor. Oxygen requirements may vary from one species to another, and concentration of O_2 in liquid cultures in bioreactors can be regulated by agitation or stirring and through aeration, gas flow and air bubble size.

Increasing O_2 concentrations, in the sparging air, from 21% to 80% (v/v), in bioreactor cultures of Boston fern clusters, enhanced growth values (final FW-initial FW/ initial FW) from 0.61 to 0.92 (Ziv and Hadar, 1991). Reducing O_2 concentrations to 10% (v/v) affected cell differentiation in bioreactor cultures of carrot embryogenic tissue. Under these conditions embryo production was severely inhibited (Jay et al., 1992).

There are reports that high aeration rates, rather than excessive oxygen concentrations, inhibit growth and that reduced growth could be due to depletion of CO_2 or to the removal of various culture volatiles, including CO_2 (Hegarty et al., 1986). CO_2 enrichment in an illuminated bioreactor culture of *Brodiaea* clusters did not affect biomass growth. Increasing CO_2 from 0.3% to 1% (v/v) gave a similar growth value, 4.2 and 5.3, respectively, under the two CO_2 concentrations and 135 μmol $m^{-1}s^{-1}$ light intensity (Ilan et al., 1995). In *Cyclamen persicum* high CO_2 concentrations correlated with increased production of pro-embryogenic masses (Hvoslef-Eide and Munster, 1998).

2.3 *Mineral nutrients consumption*

In bioreactors, in which either humidified air or condensers are used to prevent dehydration, the concentration of the nutrients in the medium is affected mainly by absorption rate and by cell lysis (Archambault et al., 1995). Differentiation and proliferation of micropropagated fern, *Gladiolus*, and *Nerine* nodular clusters in bioreactors was better on ½-strength, rather than on the full-strength, MS minerals (Ziv, unpublished). This was also true for *Lilium* bulblets differentiating on bulb scales, which were cultured in bioreactors (Takayama, 1991). A drop in pH to 4.5 and lower values and the subsequent increase to pH 5.5 were attributed to the initial utilization of ammonium and to the uptake of nitrate at a later stage.

The composition of minerals, monitored during the culture of *Brodiaea* in liquid media changed, with phosphate, ammonia, nitrate and potassium depleting faster and prior to the depletion of Ca^{++} and Mg^{++} (Ilan et al., 1995). In general, biomass growth is limited by the availability of phosphate, nitrogen and carbohydrates, and to a lesser extent by the availability of calcium, magnesium and other ions.

2.4 *Carbohydrate supply and utilization*

Sucrose, and to a lesser extent glucose, fructose, or sorbitol, are the most commonly used carbohydrates *in vitro*. In general, sucrose is removed rather rapidly from the medium and after 10-15 days the sucrose can be completely depleted or reduced to 5-10 g l^{-1} from an initial concentration of 30 g l^{-1} in both agar-gelled and liquid cultures. At the same time glucose and fructose that appear in the medium due to sucrose hydrolysis increases in the presence of invertase in the culture medium, and can reach concentrations of 5-10 g l^{-1}. In embryonic suspension cultures of celery, the addition of mannitol reduced cell lysis and enhanced somatic embryogenesis. When 40 g l^{-1} mannitol was added, a higher number of embryos was produced and the frequency of singulated normal embryos was increased (Nadel et al., 1990).

The biomass of Boston fern meristematic clusters in a bubble-column bioreactor was increased with the increase in sucrose concentrations from 7.5 g l^{-1} to 30 g l^{-1}, while higher concentrations caused a decrease in cluster growth. Elevated sucrose concentrations in the medium caused a decrease in the clusters size and leaf chlorophyll content (Ziv and Hadar, 1991). *Gladiolus* clusters cultured in the presence of growth retardants had a higher concentration of starch - 845 as compared to 585 mg g^{-1} DW in the control (Ziv, 1992). Potato microtubers grew at a faster rate in a rotating bioreactor when the medium was replaced frequently and the number of tubers >1 g increased four-fold when 8% (w/v) sucrose was used (Yu et al., 2000).

2.5 *Growth regulator effects*

The availability of growth regulators in bioreactor cultures can be more effective in controlling the proliferation and regeneration potential than in agar cultures, due to the direct contact of plant cells and aggregates with the medium.

In embryogenic cultures of *Nerine*, auxin and cytokinins were used to induce proembryogenic clusters. Embryogenic expression was achieved, however, only after a short exposure to 2-iso-pentenyladenine (2iP) and further subculture to a growth regulator-free medium (Lilien-Kipnis et al., 1994). The addition of paclobutrazol with an elevated concentration of phosphate enhanced bulblet growth in *Nerine* (Vishnevetsky et al., 2000). Excised aspen roots cultivated in liquid medium in bioreactors in the presence of thidiazuron (TDZ) regenerated buds on the entire root surface, as compared to BA-treated root explants, in which the buds regenerated only in close proximity to the site of initiating lateral roots. Roots cultured in the presence of ancymidol developed clusters that were separated mechanically and developed normal shoots on a rooting medium (Vinocur et al., 2000).

Since one of the major problems in liquid cultured plants is malformation of shoots and hyperhydricity, the induction of meristematic or bud clusters with arrested leaf growth (McCown et al., 1988; Ziv, 1991; Ziv and Shemesh, 1996) was one of the solutions to reduce hyperhydricity. The use of relatively high cytokinin concentrations or growth retardants, which inhibit gibberellin biosynthesis, was the most effective method to reduce shoot and leaf growth and to promote the formation of meristematic clusters (Ziv, 1990). In *Spathiphyllum,* addition of PAC decreased biomass growth (Figure 3). The removal of PAC after subculture, enhanced growth in cultures treated with the highest concentrations of growth retardant (Figure 4). However, even after the removal of PAC on subculture, shoot development was inhibited in particular in cultures that were treated with high concentrations of the retardant (Figure 5).

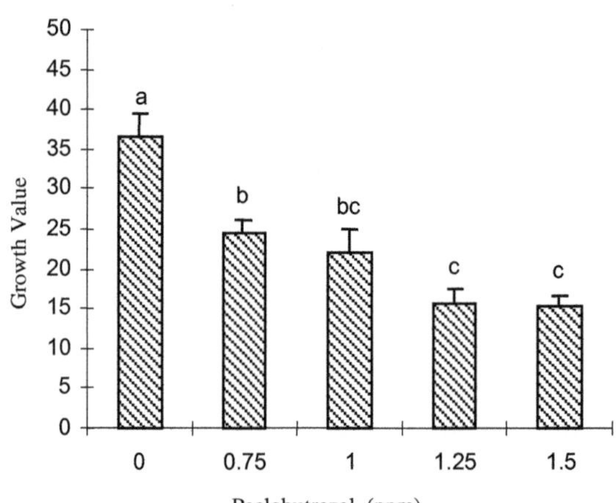

Figure 3: The effect of paclobutrazol on the growth of *Spathiphyllum* in liquid culture.

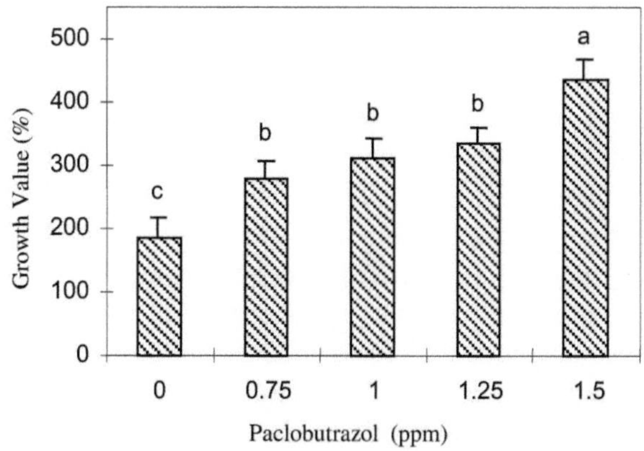

Figure 4: Clusters growth of *Spathiphyllum* after subculture to a medium lacking paclobutrazol following a pretreatment with the growth retardant.

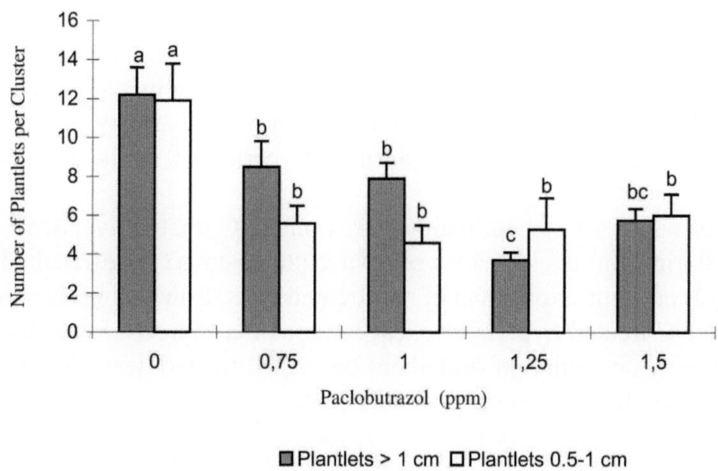

Figure 5: Shoot and bud growth of *Spathiphyllum* clusters in paclobutrazol – free medium after a pre-treatment with the growth retardant.

2.6 *Shearing and oxidative stress*

In many plants cultivated in bioreactors, the continuous aeration, mixing and circulation cause shearing damage and cell wall breakdown. The cell debris adheres to the vessel and causes foaming that prevents adequate liquid circulation and oxygen supply. Foaming was reduced when ½-strength MS medium minerals were used (Ziv, 1995b) and by lowering the concentration of calcium in the medium (Takayama et al., 1991). Disposable plastic bioreactors with a volume of 2 and 5 litre capacity used for organogenic micropropagation were found to provide good circulation with reduced cell shearing, cell damage and foaming. Meristematic and bud clusters are less shear-sensitive than large, vacuolated cells (Ziv et al., 1998).

The excessive accumulation of water in plant tissue (the most characteristic symptom of hyperhydricity) can result in oxygen depletion in the cells, induce oxidative stress, production of reactive oxygen species (ROS) and cause injury to the plant tissue. Several developmental processes in tissue cultured plants can be affected by ROS leading to recalcitrance and loss of morphogenetic competence (Benson, 2000). ROS such as H_2O_2 and hydroxyl free-radicals react instantaneously with almost any substrates (Levine, 1999).

In *Narcissus* liquid cultures, antioxidant enzyme activities were found to correlate with hyperhydric shoots and leaf section explant malformation (Chen and Ziv, 2001). In ancymidol (ANC) treated hyperhydric *Narcissus* shoots, ascorbate peroxidase (APX) and catalase (CAT) activities were significantly greater than in their non-treated, non-hyperhydric counterparts. In the ANC-treated hyperhydric *Narcissus* leaf sections, APX and CAT activities were significantly less than in the non-hyperhydric ones cultured in medium lacking ANC, especially during the period of meristematic centre formation.

The formation of meristematic centres on ANC-treated hyperhydric leaf sections during the 3rd and 4th weeks in culture, could have resulted from lower H_2O_2 concentrations than in non-treated ones. Lower concentrations of H_2O_2 in ANC-treated hyperhydric leaf sections may cause a lower hydroxyl free–radical accumulation and therefore ANC-treated leaf sections may have the ability to develop meristematic centres.

3. Conclusion

The understanding of the signals and mechanisms that control organogenesis and somatic embryogenesis in liquid media will greatly advance the use of bioreactors for commercial micropropagation. The immediate microenvironment, both chemical and physical factors, is the major factor involved in the control of normal morphogenesis of plants. The physiological status of the plant tissue will affect cell interaction with the environmental signals and will determine developmental events, proliferation and growth. Further basic and applied research can provide the information necessary for an efficient and economic use of bioreactors for plant propagation.

References

Aitken-Christie J, Kozai T & Takayama S (1995) Automation in plant tissue culture. General introduction and overview. In: Aitken Christie J, Kozai T & Smith MAL (eds) Automation and Environment Control in Plant Tissue Culture (pp. 1-18). Kluwer Acad. Publ., Dordrecht

Akita M & Takayama S (1994) Induction and development of potato tubers in a jar fermentor. Plant Cell Tissue Organ Culture 36: 177-182

Akula A, Becker D & Bateson M (2000) High-yielding repetitive somatic embryogenesis and plant recovery in a selected tea clone, "TRI-2025", by temporary immersion. Plant Cell Rep. 19: 1140-1145

Ammirato PV & Styer DJ (1985) Strategies for large scale manipulation of somatic embryos in suspension culture. In: Zaitlin M, Day P & Hollaender A (eds) Biotechnology in Plant Science: Relevance to Agriculture in the Eighties (pp. 161-178). Academic Press, New York

Archambault J, Lavoie L, Williams RD & Chavarie C (1995) Nutritional aspects of *Daucus carota* somatic embryo cultures performed in bioreactors. In: Terzi M, Cella R & Flavinga A (eds) Current Issues in Plant Molecular and Cellular Biology (pp. 681-687). Kluwer Acad. Publ., Dordrecht

Attree SM, Pomery MK & Fowke LC (1994) Production of vigorous, desiccation tolerant white spruce [*Picea glauca* (Moench.) Voss.] synthetic seeds in a bioreactor. Plant Cell Rep. 13: 601-606

Benson E (2000) Do free radicals have role in plant tissue culture recalcitrance? Special symposium: *in vitro* plant recalcitrance. *In Vitro* Cell Bio. 36: 163-170

Cazzulino D, Pederson H & Chin CK (1991) Bioreactors and image analysis for scale-up and plant propagation. In: Vasil IK (ed) Cell Culture and Somatic Cell Genetics of Plants (pp. 147-177). Academic Press, San Diego, CA

Cervelli R & Senaratna T (1995) Economic analysis of automated embryogenesis. In: Aitken-Christie J, Kozai T & Smith MAL (eds) Automation and Environmental Control in Plant Tissue Culture (pp. 29-64). Kluwer Acad. Publ., Dordrecht

Chatterjee C, Corell MJ, Weathers PJ, Wyslouzil BE & Wacers DB (1997) Simplified acoustic window mist bioreactor. Biotechnology Techniques 11: 159-158

Chen J & Ziv M (2001) The effect of ancymidol on hyperhydricity, regeneration, starch and antioxidant enzymatic activities in liquid-cultured Narcissus. Plant Cell Rep. 20(1): 22-27

Denchev PD, Kuklin AI & Scragg AH (1992) Somatic embryo production in bioreactors. J. Biotech. 26: 99-109

Escalona M, Lorenzo JC, Gonzales BL, Daquinta M, Borroto CG, Gozales JI & Desjardine Y (1999) Pineapple (*Ananas comosus* L. Merr.) micropropagation in temporary immersion system. Plant Cell Rep. 18: 743-748

Etienne H, Lartaud M, Michaux-Ferrière N, Carron MP, Berthouly M & Teisson C (1997) Improvement of somatic embryogenesis in *Hevea brasiliensis* (Müll. Arg.) using the temporary immersion technique. *In Vitro* Cell. Dev. Biol. Plant 33: 81-87

Hegarty PK, Smart NJ, Scragg H & Fowler MW (1986) The aeration of *Catharanthus roseus* L.G. Don suspension cultures in airlift bioreactors: the inhibitory effect at high aeration rates on culture growth. J. Expt. Bot. 37: 1911-1920

Hvoslef-Eide AK & Munster C (1998) Somatic embryogenesis of *Cyclamen persicum* Mill. In bioreactors. Comb. Proc. Int. Plant Prop. 47: 377-382

Ilan A, Ziv M & Halevy AA (1995) *In vitro* propagation in liquid culture and acclimatization of *Brodiaea.* Scientia Hort. 63: 101-112

Jay V, Genestier S & Courduroux JC (1992) Bioreactor studies on the effect of dissolved oxygen concentrations on growth and differentiation of carrot (*Daucus carota* L.) cell cultures. Plant Cell Rep. 11: 605-608

Levin R, Stav R, Alper Y & Watad AA (1997) A technique for repeated non-axenic subculture of plant tissues in a bioreactor on liquid medium containing sucrose. Plant Tissue Culture Biotech. 3: 41-45

Levine A (1999) Oxidative stress as a regulator of environmental responses in plants. In: Lerner HR (ed) Plant Responses to Environmental Stresses: From Phytohormones to Genome Reorganization (pp. 247-264). Marcel Dekker Inc NY

Lilien-Kipnis H, Azizbekova N & Ziv M (1994) Scaled-up proliferation and regeneration of *Nerine* in liquid cultures. Part II. Ontogeny of somatic embryos and bulblet regeneration. Plant Cell, Tiss. Org. Cult. 39: 117-123

Lorenzo JC, Gonzalez BL, Escalona M, Teisson C, Espinosa P & Borroto C (1998) Sugarcane shoot formation in an improved temporary immersion system. Plant Cell, Tissue Organ Culture 54: 197-200

Luttman R, Florek P & Preil W (1994) Silicone-tubing aerated bioreactors for somatic embryo production. Plant Cell, Tiss. Org. Cult. 39: 157-170

McCown BH, Zeldin EL, Pinkalla HA & Dedolph RR (1988) Nodule culture: a developmental pathway with high potential for regeneration, automated propagation and plant metabolite production of woody plants. In: Hanover MW & Keathley DE (eds) Genetic Manipulation of Woody Plants (pp. 146-166). Plenum Press, New York

Merchuk JC (1990) Why use air-lift bioreactors? Trends in Biotechnology. 8: 66-71

Murashige T & Skoog F (1962) Revised medium for rapid growth and bioassay with tobacco tissue cultures. Physiol. Plant. 15: 473-497

Nadel BL, Altman A & Ziv M (1990) Regulation of large scale embryogenesis in celery. Acta Hort. 280: 75-82

Preil W (1991) Application of bioreactors in plant propagation. In: Debergh PC and Zimmerman RH (eds) Micropropagation: Technology and Application (pp. 425-455). Kluwer Acad. Publ., Dordrecht

Sakamoto Y, Onishi N & Hirosawa T (1995) Delivery systems for tissue culture by encapsulation. In: Aitken-Christie J, Kozai T & Smith MAL (eds) Automation and Environmental Control in Plant Tissue Culture (pp. 215-244). Kluwer Acad. Publ., Dordrecht

Scragg AH (1992) Large-scale plant cell culture: methods, applications and products. Current Opinion Biotech. 3: 105-109

Stuart DA, Strickland SG & Walker KA (1987) Bioreactor production of alfalfa somatic embryos. Hort. Sci. 22: 800-803

Styer DJ (1985) Bioreactor technology for plant propagation. In: Henke RR, Hughes KW, Constantin MJ & Hollaender A (eds) Tissue Culture of Forestry and Agriculture (pp. 117-130). Plenum Press, New York

Takayama S (1991) Mass propagation of plants through shake and bioreactor culture techniques. In: Bajaj Y P S (ed) Biotechnology in Agriculture and Forestry: High-Tech and Micropropagation, Vol. 17 (pp. 1-46). Springer-Verlag, Berlin

Takayama S & Akita M (1998) Bioreactor techniques for large-scale culture of plant propagules. Adv. Hort Sci, 12: 93-100

Tautorus TE & Dunstan DI (1995) Scale-up of embryogenic plant suspension cultures in bioreactors. In: Jain M, Gupta PK & Newton RJ (eds) Somatic Embryogenesis in Woody Plants, Vol. 1 (pp. 265-269). Kluwer Acad. Publ., Dordrecht

Teisson C & Alvard D (1995) A new concept of plant *in vitro* cultivation liquid medium: temporary immersion. In: Terzi M et al. (eds) Current Issues in Plant Molecular and Cellular Biology (pp. 105-110). Kluwer Acad. Publ., Dordrecht

Teisson C, Alvard D, Berthouly B, Cote F, Escalant JV, Etienne H & Lartaud M (1996) Simple apparatus to perform plant tissue culture by temporary immersion. Acta Hort. 40: 521-526

Vishnevetsky J, Zamski E & Ziv M (2000) Carbohydrate metabolism in *Nerine sarniensis* bulbs developing in liquid culture. Physiol. Plant. 108(4): 361-369

Young PS, Murthy HN & Yoeup PK (2000) Mass multiplication of protocorm-like using bioreactor system and subsequent plant regeneration in *Phalaenopsis*. Plant Cell, Tiss. Org. Cult. 63: 67-72

Yu WC, Joyce PJ, Cameron DC & McCown BH (2000) Sucrose utilization during potato microtuber growth in bioreactors. Plant Cell Rep. 19: 407-413

Ziv M (1990) Morphogenesis of gladiolus buds in bioreactors - implication for scaled-up propagation of geophytes. In: Nijkamp HJJ, van der Plas LHW & van Aatrijk J (eds) Progress in Plant Cellular and Molecular Biology (pp. 119-124). Kluwer Acad. Publ., Dordrecht

Ziv M (1991) Vitrification: Morphological and physiological disorders of *in vitro* plants. In: Debergh PG & Zimmerman RH (eds) Micropropagation: Technology and Application (pp. 45-69). Kluwer Acad. Publ., Dordrecht

Ziv M (1992) The use of growth retardants for the regulation and acclimatization of *in vitro* plants. In: Karssen CM, Van Loon LC, Vreugdenhil D (eds) Progress in Plant Growth Regulation (pp. 809-817). Kluwer Acad. Publ., Dordrecht

Ziv M (1995a) *In vitro* acclimatization. In: Aitken-Christie J, Kozai T & Smith MAL (eds) Automation and Environmental Control in Plant Tissue Culture (pp. 493-516). Kluwer Acad. Publ., Dordrecht

Ziv M (1995b) The control of bioreactor environment for plant propagation in liquid culture. Acta Hort. 319: 25-38

Ziv M (2000) Bioreactor technology for plant micropropagation. Horticultural Reviews 24: 1-30

Ziv M & Ariel T (1991) Bud proliferation and plant regeneration in liquid-cultured philodendron treated with ancymidol and paclobutrazol. J. Plant Growth Regul. 10: 53-57

Ziv M & Hadar A (1991) Morphogenic pattern of *Nephrolepsis exaltata* Bostoniensis in agar-gelled or liquid culture. Implication for mass propagation. Israel. J. Bot. 40: 7-16

Ziv M & Shemesh D (1996) Propagation and tuberiation of potato bud clusters from bioreactor culture. *In Vitro* Cell. Dev. Biol. Plant 32: 31-36

Ziv M, Ronen G & Raviv M (1998) Proliferation of meristematic clusters in disposable pre-sterilized plastic bioreactors for large-scale micropropagation of plants. *In Vitro* Cell Dev. Biol. Plant 34: 152-158

Chapter 6

Application of bioreactor systems for large scale production of horticultural and medicinal plants

K. Y. Paek, Debasis Chakrabarty & E. J. Hahn
Research Center for the Development of Advanced Horticultural Technology, Chungbuk National University, Cheongju, South Korea

Abstract: Automation of micropropagation *via* organogenesis or somatic embryogenesis in a bioreactor has been advanced as a possible way of reducing costs. Micropropagation by conventional techniques is typically a labour-intensive means of clonal propagation. The paper describes lower cost and less labour-intensive clonal propagation through the use of modified air-lift, bubble column, bioreactors (a balloon-type bubble bioreactor), together with temporary immersion systems for the propagation of shoots, bud-clusters and somatic embryos. Propagation of *Anoectochilus*, apple, *Chrysanthemum*, garlic, ginseng, grape, *Lilium*, *Phalaenopsis* and potato is described. In this chapter, features of bioreactors and bioreactor process design specifically for automated mass propagation of several plant crops are described, and recent research aimed at maximizing automation of the bioreactor production process is highlighted.

Key words: *Anoectochilus,* apple, automated masspropagation, *Chrysanthemum*, dissolved oxygen, garlic, ginseng, grape, *Lilium*, micropropagation, mixing, nutrients, pH, *Phalaenopsis*, potato, secondary metabolites, siberian ginseng, somatic embryogenesis

Abbreviations: BTBB–balloon type bubble bioreactor; DO–dissolved oxygen; IEDC–induced embryogenic determined cells; PLB–protocorm-like body; PPF–photosynthetic photon flux; STR–stirred tank reactor

1. Introduction

Large-scale plant production through cell tissue and embryo cultures using bioreactors is promising for industrial plant propagation. Bioreactors are usually described in a biochemical context as self-contained, sterile environments which capitalize on liquid nutrient or liquid/air inflow and outflow systems, designed for intensive culture and affording maximal

A.K. Hvoslef-Eide and W. Preil (eds.), Liquid Culture Systems for in vitro Plant Propagation, 95–116.

opportunity for monitoring and control over microenvironmental conditions (agitation, aeration, temperature, dissolved oxygen, pH, etc.). Three main classes of culture system in bioreactors can be distinguished: 1) those producing biomass (cells or organogenic or embryogenic propagules, shoots or roots as the final product), 2) those producing metabolites and enzymes and 3) those used for biotransformation of exogenously added metabolites (which may be precursors in a metabolic pathway).

The use of plant cell cultures is focused on the production of valuable natural products such as pharmaceuticals, flavours and fragrances and fine chemicals. More than 20,000 different chemicals are produced from plants, and about 1,600 new plant chemicals are discovered each year (Sajc et al., 2000). Much of the research in this field has been done by private industry. Various problems associated with low cell productivity, slow growth, genetic instability, and an inability to maintain photoautotrophic growth has limited the application of plant cell cultures (Bourgaud et al., 2001). In spite of potential advantages of the production of secondary metabolites in plant cell cultures, only shikonin, ginsenosides and berberine have been produced on a large scale, and all three process plants are located in Japan (Bourgaud et al., 2001).

Secondary metabolites are currently obtained commercially by extraction from whole plants or tissue. Large-scale plant tissue culture is an attractive alternative to the traditional method of plantation or plant cell culture. It offers various advantages including controlled supply of biochemicals independent of plant availability (cultivation season, pests and politics), well defined production systems which results in higher yields and more consistent quality of the product and also it overcomes the drawback of plant cell culture systems.

Automation of micropropagation in bioreactors has been advanced by several authors as a possible way of reducing costs of micropropagation (Preil, 1991; Sharma, 1992; Takayama and Akita, 1994; Christie et al., 1995; Leathers et al., 1995; Son et al., 1999; Ibaraki and Kurata, 2001; Chakrabarty and Paek, 2001; Paek et al., 2001). Bioreactors containing liquid media are used for large-scale growth of various tissues. The use of bioreactor for micropropagation was first reported in 1981 for *Begonia* propagation (Takayama and Misawa, 1981). Since then it has proved applicable to many species and plant organs including shoots, bulbs, microtubers, corms and somatic embryos (Paek et al., 2001).

In this chapter, features of bioreactors and bioreactor process design specifically for automated mass propagation of different horticultural and medicinal plants are described, and recent research aimed at maximizing automation of the bioreactor production process is highlighted.

2. Bioreactor design

Bioreactors devoted to mass propagation includes systems for cultivating cells, tissues, somatic embryos or organogenic propagules (e.g. bulblets, corms, nodules, microtubers, and shoot clusters) in liquid suspensions. Until the mid 1970s, traditional microbial technology provided the main source of knowledge and equipment for the cultivation processes, almost exclusively in the form of stirred tank reactors (STR) with flat blade turbines for agitation. Today, a relatively large number and variety of reactor systems are available, allowing a rational selection of an appropriate reactor for a given process. Still, most of the standard equipment designed for microbial cultivation does not meet the special requirements for cultivation of fragile plant cells or cell aggregates.

Takayama and Akita (1994), Heyerdahl (1995), Walker (1995), Lee (1997), Sajc et al. (2000), Honda et al. (2001), Paek et al. (2001), Paek and Chakrabarty (2003) reviewed different reactor configurations for plant cell suspensions, plant tissue and organ cultures. The relative advantages and selection criteria for various reactor configurations were discussed for specific process applications.

Those bioreactors are fundamentally classified by agitation methods and vessel construction into: a) mechanically agitated (stirred tank bioreactor, rotating drum bioreactor, spin filter bioreactor), b) pneumatically agitated and non-agitated bioreactors (simple aeration bioreactor, bubble column bioreactor, air-lift bioreactor, balloon type bubble bioreactor-BTBB).

Numerous modifications of the conventional STR with bubble aeration have been developed that have a variety of impeller designs (Honda et al., 2001). Notwithstanding, the STR presents several limitations such as high power consumption, high shear forces and problems with sealing and stability of rotating shafts in tall bioreactors. Air-lift bioreactors combine high loading of 'solid' particles and good mass transfer, which are inherent for three-phase fluidized beds. Air bubbles, using internal or external recirculation loops, generate efficient mixing in the liquid phase. The main advantages of air-lift bioreactors are low shear forces, low energy requirements, and simple design. Rotary drum reactors have significantly higher surface area to volume ratios than other reactor types. As a consequence, mass transfer is achieved with comparably less power consumption, according to Danckwert's surface renewal theory (Dankwert, 1951). These features are favorable for bioprocesses utilizing shear-sensitive tissues, as well as for photo bioreactors (Sajc et al., 2000).

In a bubble column bioreactor the bubbles create less shear forces, so that it is useful for plant organ cultures especially for propagation of various species through tissue culture of shoots, bulbs, corms and tubers. The disadvantages of air-lift and bubble column bioreactors are foaming induced by large volumes of air, and growth of cells in the head space. The phenomenon of foaming and cell growth on the wall of the vessel is due to the diameter of the vessel and the top of the vessel being the same (Paek et al., 2001). To overcome this problem we designed a bioreactor that has a larger top-section diameter and/or a balloon type bubble bioreactor. The layout of BTBB is shown in figure 1. By using a concentric tube for cell lifting at the riser part of the vessel base, foaming was much reduced. This bioreactor was found to be reliable for cell, tissue and organ culture. Pilot scale BTBBs of 300, 500 and 1000 litres have been utilized for the production of biomass of various valuable plant species (Paek et al., 2001).

Recently, we developed a novel type ebb and flood bioreactor system (a periodic immersion system) for the mass propagation of several plant species. The lay out of the ebb and flood bioreactor system is shown in figure 2. The principal equipment in an ebb and flood bioreactor is the same as that in the BTBB. However in order to avoid the complete submersion of explants in the liquid medium, a supporting net was used to hold the plant material. In this system, medium is pumped from a storage tank into the culture vessel. A series of channels helps to supply nutrient solution evenly to the plant material, resulting in uniform growth. The medium remains in the vessel for a few minutes, after which it drained back to the storage tank for re-use. The drainage process is controlled by a solenoid valve at intervals of between 4 to 8 hours, depending on plant species and explant type.

3. Physico-chemical parameters

The design and operation of a bioreactor are mainly determined by biological needs and engineering requirements, which often include a number of factors: efficient oxygen transfer and mixing, low shear and hydrodynamic forces, effective control of the physico-chemical environment and ease of scale–up.

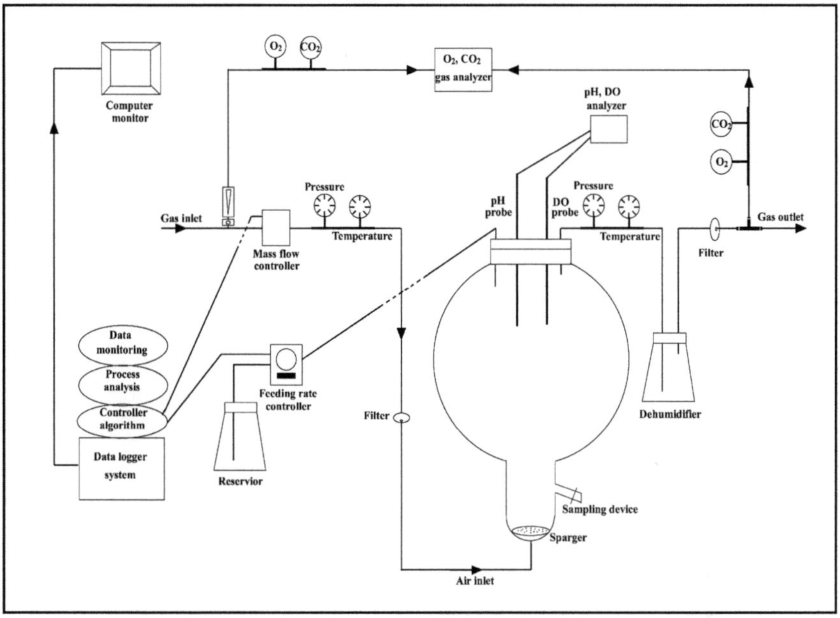

Figure 1: Configuration of a balloon-type bubble bioreactor system.

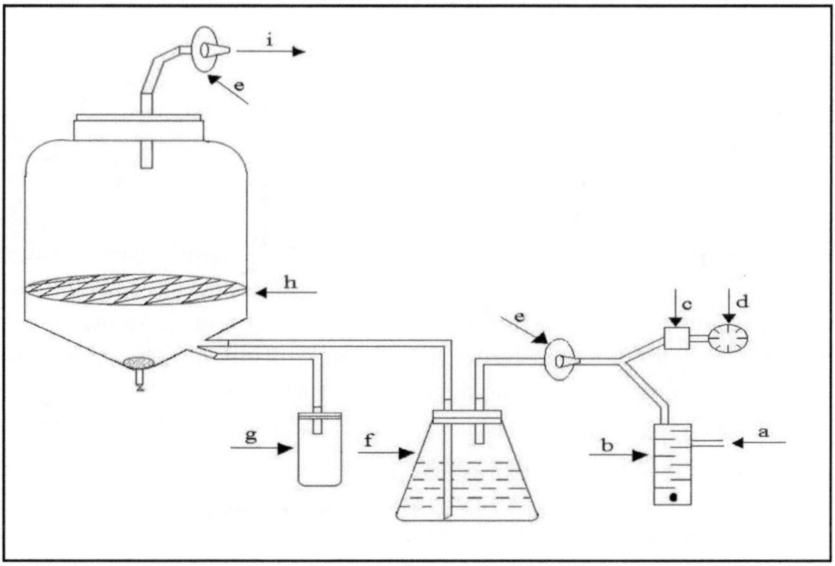

Figure 2: Layout of an ebb and flood bioreactor system (a) Air inlet; (b) Air flow meter; (c) Timer; (d) Solenoid valve; (e) Membrane filter; (f) Medium reservoir; (g) Sampling port; (h) Supporter (net); (i) Air outlet.

3.1 Dissolved oxygen

One of the main functions of the bioreactor is to promote the mass transfer of oxygen from the gaseous to the liquid phase. Since oxygen is only sparingly soluble in water (0.25 mmol l^{-1} at 25°C. 1 atm., 21% (v/v) O_2 in the air), it is necessary to drive the diffusion of oxygen into the aqueous phase to meet the demand of actively growing tissues or cells (Leathers et al., 1995). This is accomplished by modifying operational parameters such as aeration rate, agitation speed, impeller design, gas mixing and bioreactor configuration. Culture mixing is also important because dissolved oxygen (DO) must be transported rapidly to the culture tissues or cells. In general, it is essential that the dissolved oxygen concentration remains above the critical DO_2 level at all the times for optimal cell growth (Leather et al., 1995; Sajc et al., 2000). The critical dissolved oxygen concentration, DO_{2crit}, can be described as the dissolved oxygen concentration above which no further increase in specific oxygen uptake rate can be measured. At DO_2 levels below the DO_{2crit}, cells have reduced energy (ATP) levels, which may have direct effects on cellular metabolism and morphology (Leathers et al., 1995). Practically, this is important because the DO_{2crit} value is used for designing appropriate bioreactor operating systems to ensure that an oxygen-limiting condition does not suppress the metabolic activity of the culture. Therefore, supplying adequate amounts of oxygen (above the DO_{2crit}) is a major concern in bioreactor scale-up.

3.2 Mixing

The other key parameter is mixing, which is necessary to distribute equally cells or tissues, and nutrients throughout the liquid phase (Leathers et al., 1995, Sajc et al., 2000; Honda et al., 2001). Mixing is normally carried out by sparging, mechanical agitation or a combination of these two, but the magnitude of hydrodynamic forces associated with mixing should be small enough not to cause cell or tissue damage, but sufficient to stimulate selected cell functions. However, there has been little quantitative work on the effect of hydrodynamic forces on plant tissue engineering. Previous studies have focused mostly on the kinetics of cell growth and product formation, and the effect of hydrodynamic conditions on the structure and composition of plant tissue is not well understood.

3.3 pH

Changes in pH during culture have been reported by several authors (Dussert et al., 1995; Hilton and Wilson, 1995; Yu, 2000; Lian et al., 2002b). These changes appeared to be related to the balance between ammonium in the medium - as shown by several authors (Hilton and Wilson, 1995; Escalona et al., 1999; Lian et al., 2002b). Clear inhibitory effects of a culture at pH 5.0 on embryogenesis were found by Lazzeri et al. (1987). Precise recording of fluctuations in parameters, like pH in computer controlled bioreactor cultures, will improve the repeatability of complex biological process.

3.4 Nutrients

Nutrient availability is a major chemical factor involved in scaling up. For large-scale culture in a bioreactor several aspects play an important role. Periodic measurement of the individual nutrients at different times provides information regarding nutrient uptake, biomass and metabolite production in bioreactors. We investigated detailed analysis of the dynamics of various nutrient compounds during *Lilium* bulblet growth in BTBB and it was found that ammonium, nitrate and phosphate became exhausted from the medium. After 16 weeks of culture considerable amounts of K^+, Mg^{2+}, Ca^{2+}, Na^+ and Cl^- were still present in the medium, but the limiting growth factor was sugar, rather than the main nutrient (Lian et al., 2002b). Similar investigations were carried out with *Begonia*, rice suspension culture, carrot somatic embryo, ginseng adventitious roots and potato microtuber growth during bioreactor culture (Törmälä et al., 1987; Schmitz and Lörz, 1990; Archambault et al., 1995; Yu, 2000; Yu et al., 2000). In spite of these results, there is still a need for detailed investigation on hormonal interactions and the dynamics of various nutritional compounds. Offline analyses of changes in nutrient and hormone concentrations during bioreactor culture will present new possibilities for the better manipulation of embryogenesis and organogenesis.

4. Bioreactors in automated masspropagation

For production of cells, somatic embryos or organogenic propagules (e.g. bulblets, corms, nodules, microtubers or shoot clusters), bioreactor culture is one of the most promising ways for scaling up the system and there have been several reports on large-scale propagation of horticultural and medicinal plants by using bioreactors. However, inconsistencies in optimizing bioreactor types and culture parameters have been reported. Although the main source of these inconsistencies may be due to species-to species variations, careful consideration is needed in interpreting these results. So, once the culture conditions have been established in a small-scale bioreactor, cultures can be easily scaled up to large-scale (500-1000 litre bioreactors).

4.1 Secondary metabolites

The production of secondary metabolites using plant cells has been the subject to extended research. In 1959 the first report on the large-scale cultivation of plant cells appeared (Tulecke and Nickell, 1959). In the last few years, much success has been achieved in the field of plant cell fermentation and scaling up. Plant cells now can be cultivated in volumes up to 75,000 litre (Rittershaus et al., 1989) (for reviews see Hisihimoto and Azechi, 1988; Dornenburg and Knori, 1995; Bourgaud et al., 2001). Among hundreds of secondary plant products that have been investigated with undifferentiated cell cultures, shikonin, ginsenosides and berberine are presently produced on a large scale and indeed these are the most successful stories of an industrial scale-up linking plant cell culture with bioreactor technology.

Although undifferentiated cell cultures mainly have been studied, a large interest has been shown in hairy root and other organ cultures. Hairy roots, once established, can be grown in a medium with low inoculum with a high growth rate. Several bioreactor designs have been reported for hairy root cultures taking into consideration their complicated morphology and shear sensitivity (Giri and Narasu, 2000). The main problem associated with hairy root cultures in bioreactors is the restriction of nutrient oxygen delivery to the central mass of tissue results in a pocket of senescent tissue. Due to branching, the roots form an interlocked matrix that exhibits a resistance to flow. The ability to exploit hairy root culture as a source of bioactive compounds depends on development of bioreactor system where several physical and chemical parameters must be taken into consideration.

4.2 Micropropagation

Automation of organogenesis in a bioreactor has been advanced as a possible way of reducing costs of micropropagation (Takayama and Akita, 1994; Leathers et al., 1995; Chakrabarty and Paek, 2002; Paek et al., 2001). Organogenic plant progagules are cultivated intensively in bioreactors for the end result of producing transplants for mass production. Intensive cultivation of such structures as potato microtubers and bulblets of lily is another strategy for producing progagules, which can be handled for direct planting in the field. Micropropagation by axillary shoot proliferation is typically a labour-intensive means of producing elite clones, but recently the adaptation of air-lift, bubble column, BTBB, ebb and flood and temporary immersion bioreactors for propagation of shoots and bud-clusters has provided a workable means for scale up. Some of the most advanced plant tissue culture work that has been progressed to research-scale bioreactors is based on production of crop species such as *Stevia rebaudiana*, *Begonia*, *Chrysanthemum*, apple, grape, pineapple, garlic and *Phalaenopsis*.

4.3 Somatic embryogenesis

Somatic embryogenesis also offers a potential system for large-scale plant propagation in automated bioreactors. Conventional micropropagation requires intensive labour which often limits its commercial viability and application. Somatic embryos could be easier to handle since they are relatively small and uniform in size, and they do not require cutting into segments and individual implanting onto media during proliferation. In addition, somatic embryos have the potential for long-term storage through cryopreservation or desiccation, which facilitates flexibility in scheduling production and transportation and therefore fits large-scale production. The production of somatic embryos in bioreactors has been reported for a number of species (for reviews see Denchev et al., 1992; Cervelli and Senaratna, 1995; Moorhouse et al., 1996; Timmis, 1998; Ibaraki and Kurate, 2001; Paek and Chakrabarty, 2003), but many improvements are needed for the practical automatic somatic embryo production systems that can cope with synchronization of the somatic embryo development, identifying the occasional embryo abnormality during culture, and overcoming the difficulties in embling acclimatization.

5. System examples

5.1 Ginseng

Ginseng, *Panax ginseng* of the *Araliaceae* family, is one of the most valuable oriental herbs. It (usually, the dried root) has been used as a healing drug and health tonic in countries such as China, Japan and Korea since ancient times (Tang and Eisenbrand, 1992). In recent years, ginseng has been used increasingly as a health tonic, in the form of a variety of commercial health products including ginseng capsules, soups, drinks and cosmetics, which are distributed around the world. Until now, ginseng has been reported to contain saponins, antioxidants, peptides, polysaccharides, fatty acids, alcohols and vitamins (Huang, 1993). The saponins, known as ginsenosides, are widely believed to be the major bioactive compounds of ginseng.

Generally, the ginseng roots on the market are from farms. Field cultivation is a time-consuming and labour-intensive process, so the use of the plant cell and tissue culture process has been investigated as an alternative for the more efficient production of ginseng.

Ginseng tissue culture was first documented in 1964 (Luo et al., 1964); since then numerous studies have been reported (as reviewed by Wu and Zhong, 1999). The large-scale suspension culture of ginseng cells was first reported by Yasuda et al. (1972). An industrial scale ginseng cell culture process was initiated in the 1980s at the Nitto Denko Corporation (Ibaraki, Osaka, Japan) using 2,000 and 20,000-litre STR bioreactors to produce ginseng biomass. Two types of bioreactors are commonly used for these plant cell suspension cultures: STR and air-lift types. Studies on ginseng cells in STRs suggested that the agitator design and the agitation rate are major factors affecting cell growth and saponin production (Furuya et al., 1984). Up to now, the industrial application of ginseng cell culture has found only a few commercial applications worldwide. The reasons are probably due to the slow growth of ginseng cells and the higher water content of cultured cells compared with field-grown plants. Transformed root (hairy root) cultures offer a promising alternative method that can partially avoid these problems (Yosikawa and Furuya, 1987; Yu, 2000), but hairy roots usually produce opine-like substances which are lethal to mammalian cells. Therefore, we started work on ginseng adventitious root culture, which provides an efficient means of biomass production due to their fast growth and stable metabolite production. A series of experiments was conducted to establish efficient ginseng adventitious root growth and ginsenoside production in liquid media (Yu et al., 2001a; Son and Paek, 2001) and subsequently we established a pilot-scale culture of multiple adventitious roots induced from callus using a BTBB bioreactor system (Figure 3 a, d and e) (Yu et al., 2000a;Yu et al., 2001a, b; Choi et al., 2001). In the bioreactor, roots were tangled and formed ball-like structures. Root interiors became brown and their saponin content decreased sharply compared to actively growing roots. Cutting root cultures during the culture period appears to be a necessary procedure to promote root growth and prevent deterioration. Therefore, multiple adventitious roots growing in bioreactors were sliced by a blade connected to the top-driven motor. We measured a 150-fold growth rate increase obtained at day 56 of culture when roots were cut during the culture in a 500-litre BTBB: the total saponin content in harvested adventitious roots reached approximately 1% of dry weight, which corresponds to half of the content in field-grown ginseng. Later, we were able to increase the total saponin contents (up to 4-5%) by using elicitors such as methyl jasmonate (Yu et al., 2000b). An industrial-scale ginseng adventitious root culture has been initiated by CBN Biotech, Cheongju, Korea, in 500 to 1000-litre BTBBs to produce ginseng biomass using the above-mentioned protocol.

Ginseng was formerly a wild plant found only in a few isolated areas in Korea and northwestern China. Nowadays, wild ginseng (mountain ginseng) is rarely available. Therefore, we initiated further work to induce and culture adventitious roots of mountain ginseng through the same process (Lian et al., 2002a) as in the case of ginseng and the commercial application of this mountain ginseng is now under trial.

5.2 Siberian ginseng

Siberian ginseng (*Eleutherococcus senticosus*) is an endangered medicinal woody plant species. Conventional propagation is difficult because of the long-term stratification required to induce both maturation and germination of the zygotic embryos (Isoda and Shoji, 1994). The low frequency rooting of cuttings hinders propagation. Plant regeneration of Siberian ginseng through direct somatic embryogenesis (Choi et al., 1999a; Gui et al., 1991) and indirect embryogenic callus and cell suspension culture (Choi et al., 1999b) has been reported.

Paek et al. (2001) reported the production of Siberian ginseng somatic embryos using a BTBB. Induced embryogenic-determined cells (IEDC) were transferred to a 1000 ml Erlenmeyer flask containing 500 ml culture medium and cultured for 3 weeks. Then the cultured cells were transferred into a 20-litre BTBB. Viable plantlets were regenerated from mature embryos upon transfer of the embryos from the bioreactor to soil.

Kim and Kim (2001) reported the efficient mass production of Siberian ginseng somatic embryos in bioreactors where the somatic embryos at the torpedo stage were transferred to 5 to 10-litre air-lift bioreactors and cultured for 10-15 days: the somatic embryos developed into emblings (Figure 3 b).

Recently we developed a protocol for large-scale production of Siberian ginseng somatic embryos in a 500-litre BTBB (Figures 3 c and f). By inoculating 3.5 kg IEDC, 60 kg mature somatic embryos (5.7 kg dry weight) were harvested from a 500-litre BTBB (unpublished). This protocol is being applied on a large scale in Korea for the commercial production of secondary metabolites from mature somatic embryos of Siberian ginseng (Microplants Co., Ltd., Daejon, South Korea and CBN Biotech, Chungbuk National University, Cheongju, South Korea). Similarly, using the same protocol, more than 500,000 somatic embryos of thornless *Aralia elata,* at different developmental stages, were harvested from a 10-litre BTBB after 6 weeks of culture.

5.3 Phalaenopsis

Phalaenopsis is an important ornamental orchid, which is difficult to propagate vegetatively. There is a number of tissue culture reports of its propagation, but few use bioreactors. We have now reported the mass multiplication of *Phalaenopsis* in bioreactors (Figures 4 a and b) (Park et al., 2000). Continuous immersion culture (air-lift column and air-lift-balloon bioreactor), and temporary immersion cultures (with or without a charcoal filter attached) were used for the culture of protocorm-like bodies (PLB) sections. In all four cases, 2 litres modified Hyponex medium (Kano, 1965; 1 g l^{-1} of '6.5N-4.5P-19K' + 1 g l^{-1} of '20N-20P-20K' + 1% (w/v) potato homogenate) was used and 20 g of inoculum (~1000 PLB explants) was inoculated into the medium. For the temporary immersion bioreactors, PLB sections were placed on a plastic net installed in the vessel. The system was programmed to immerse the PLB sections in the medium for 5 min in every 125-minute period *via* a timer and solenoid valve. In continuous-immersion culture, PLB sections were submerged in liquid medium during the entire culture period. A temporary immersion culture with charcoal medium-filter attached was found to be most suitable for PLB culture and about 18,000 PLBs were harvested from 20 g of inoculum. These PLBs were regenerated into plantlets and transplanted to pots containing peat moss and perlite (1:1) (Figure 4c).

5.4 Anoectochilus

A simple protocol is described for the *in vitro* mass propagation of *Anoectochilus formosanus,* an endangered orchid, using an automated low cost bioreactor system. Comparative studies between culture on gelled media and in a bioreactor (balloon type bubble bioreactor-BTBB), using both nodal and shoot-tip explants, demonstrated that shoot multiplication was most efficient in BTBB culture when using nodal explants. Shoots grown on a TDZ-containing medium grew slowly and had small leaves. To overcome this problem, shoots were transferred to a medium without TDZ; comparative studies between cultures on gelled medium and in bioreactors (continuous immersion with air supply, continuous immersion without air supply and temporary immersion using ebb and flood) demonstrated that plantlet growth was greatest in a continuous-immersion bioreactor with an air supply. Regenerated shoots with well-developed roots were acclimatized and then grown in a greenhouse (Ket et al., 2003).

5.5 *Apple*

Many woody plant species are sensitive to the liquid medium environment in a detrimental way. Hyperhydricity frequently occurs with tissues grown in or on liquid media as a result of contact with the liquid and other micro environmental parameters present at that time (Christie et al. 1995). Recently we developed a novel type ebb and flood bioreactor system for the mass propagation of apple rootstock M9 EMLA. Although the multiplication rate was highest in immersion culture (5-litre BTBB), a large number of hyperhydric plantlets was produced. With the ebb and flood system, hyperhydricity was reduced as compared to the immersion system. In an attempt to completely eliminate the hyperhydricity, we supplied compressed air inside the bioreactor chamber to reduce the humidity. This approach significantly reduced the hyperhydricity during the bioreactor culture of apple plantlets (Chakrabarty et al., 2003) (Figure 5a). Plantlets regenerated during bioreactor culture were transferred to hydroponic culture for *ex vitro* rooting and acclimatization (Figure 5b).

5.6 *Chrysanthemum*

We investigated the effects of environmental factors (PPF, air temperature, air volume, medium composition, inoculation density and types of medium supply) on the growth and quality of *Chrysanthemum* plants in bioreactors (Kim, 2001). Optimum culture environments for bioreactor culture (10-litre air-lift column type with raft) were: a NH_4^+:NO_3^- ratio of 20:40, 25°C air temperature, 100 mmol m^{-2} s^{-1} PPF, 0.1 vvm air volume and an inoculation density of 40 to 60 nodal cuttings per bioreactor culture (Figure 5c). Supplementation of the culture with sugar-free medium after 8 weeks of culture resulted in higher growth rates as compared to supplementation with sugar-containing medium.

5.7 *Garlic*

Garlic is vegetatively propagated, but its low propagation rate as well as long time to produce a sufficient number of seed bulbs for practical cultivation led to the development of an *in vitro* protocol for micropropagation. However, the rate of multiplication and growth of microbulbs during *in vitro* culture are not sufficiently high to be of practical use. In order to achieve the efficient and automated production of garlic bulblets, bioreactors have been tested to verify their value for large-scale micropropagation (Kim, 2002) (Figure 5d). Comparative studies between gelled media culture and BTBB (immersion, ebb and flood and immersion culture with a net to avoid the complete immersion of plant materials) indicated that shoot multiplication was most efficient in immersion culture. Microbulbs cultured under the ebb and flood system also showed a high rate of bulbing and the highest number of ideal and 'competent' bulblets (<0.2g) was achieved when bulblets were cultured in a ebb and flood system for 12 weeks with 2 medium flushes per day.

5.8 Grape

Nodal cuttings of the grape rootstock clone 5BB were grown in BTBB by three different culture methods: ebb and flood, raft culture with a supplementary air supply and raft culture without a supplementary air supply (Shim, 2002). Plantlet growth was greatest in raft culture with the supplementary air supply, showing maximum fresh weight, dry weight and shoot length (Figure 5e). The originating morphological position of the nodal cuttings used in the experiments influenced the growth and survival rate of grape plantlets cultured in a bioreactor system and nodal cuttings from upper portion were found to be the best explant source.

5.9 Lilium

Lilium is an important floricultural crop for bulb and cut flower production. Numerous studies have been reported on regeneration of bulblets from excised lily bulb scales (Robb, 1957; Hackett, 1969; Allen, 1974; Anderson, 1977; Novak and Petru, 1981; Takayama and Misawa, 1983; Varshney et al., 2000) and it is now the current commercial method used for vegetative propagation of lilies. In our previous study, we reported a two-stage culture process of *Lilium* micropropagation (bulblet formation and then their development) in 5 to 20 litre batches of non-stirred bioreactors within 60 days (Lim et al., 1998; Kim, 1999, Kim et al., 2001). The fed-batch culture system using a bioreactor was used for the mass production of *Lilium* bulblets by Seon et al., 2000. Recently we developed a more efficient two-stage bioreactor culture of *Lilium* bulblets using BTBB (Lian, 2001; Lian et al., 2002b). In the first stage bulblets were induced from chopped bulb scales using an ebb and flood bioreactor. Although the percentage of bulblet formation was lower in ebb and flood system as compared to gel-based culture, nevertheless we have harvested a large number of bulblets from each batch culture. This bulblet culture in bioreactors (ebb and flood) will reduce the labour manipulations required for gel-based culture and facilitate scaling-up of bulblet production. The second stage is to promote the growth using continuous immersion bioreactors. Bulblets of *Lilium* Oriental Hybrid 'Casablanca' grew at a faster rate when the medium was exchanged with new medium frequently in a BTBB (immersion type) (Lian et al., 2003) (Figure 5f). Uptake of sugar and other minerals indicate that high sucrose concentration are necessary for optimal bulblet growth. Although high sucrose concentrations could be maintained by the exchange method, the sucrose supplied was rapidly hydrolyzed into glucose and fructose when medium was replaced with new medium every 2, 6 and 12 weeks of the bioreactor culture. Mineral absorption also displayed variation, both in quantity and selectivity of the organic nutrients supplied. During the growth of bulblets, fast exhaustion of NH_4^+, NO_3^-, SO_4^{2-} and $H_2PO_4^-$ occurred, whereas consumption of K^+, Mg^{2+}, Ca^{2+}, Na^+ and Cl^- was slow. There was also a rapid reduction in pH of the medium following the addition of, or exchange with, fresh medium during the bulblet growth.

5.10 Potato

We also applied bioreactor culture for shoot production and subsequently microtuber growth of potato (Piao et al., 2003). Nodal cuttings were transferred to a 20-litre BTBB or a 10-litre column-type bioreactor equipped with an ebb and flood system or continuous immersion system for medium circulation and growth for 4 weeks at 25°C under a PPF of 100 mmol m^{-2} s^{-1} with a 16-h photoperiod. The cultures were then maintained for another 8 weeks at 25°C in darkness for microtuber formation. The analysis of tuber classification according to size showed that addition of BAP in the culture medium influenced the formation of microtubers larger than 1.1 g. It has also been observed that there is a strong influence of medium renewal on individual microtuber growth during bioreactor culture of potato.

Figure 3: (a) Bioreactor culture (5-litre sized BTBB) of adventitious roots of ginseng (*Panax ginseng* C.A. Meyer); (b) Mass propagation of Sibirian ginseng somatic embryos in 5-litre BTBB from embryogenic cells; (c) Pilot scale BTBB (500-litre); (d) Pilot scale column type bioreactor (1000-litre); (e) Harvesting of adventitious roots from 500-litre BTBB eight weeks after inoculation; (f) Harvesting of Sibirian ginseng somatic embryos from 500-litre BTBB after 30 days of culture

Figure 4: (a) *Phalaenopsis* PLB proliferation in the bioreactor (1-litre column-type) after 5 weeks; (b) Shoot regeneration and plantlet growth from PLB in the bioreactor; (c) Acclimatized plantlets.

6. Conclusion

The use of bioreactors has led to the development of a technology suitable for large-scale plant propagation; currently, various plant species are propagated in bioreactors for biomass production as well as large-scale micropropagation. However, many plant species are sensitive to liquid medium in a detrimental way. Hyperhydricity frequently occurs with tissues grown in or on liquid media and transplanting the shoots in the soil is not easy because most shoots are etiolated and succulent and easily damaged by handling or environmental stress. For bioreactor culture, research aimed at improvement of the physical and chemical environments - such as an increased number of air exchanges, increased PPF and CO_2 content - is necessary for the better practical use of this technique. According to Vasil (1994) 'The most difficult and intractable problems in the use of bioreactors for large-scale plant propagation are in the biology and not in the engineering'.

Figure 5: (a) Apple plantlets in a BTBB after 40 days of culture; (b) Large scale production of apple plantlets after using bioreactor system; (c) Mass propagation of *Chrysanthemum* plantlets in a column type bioreactor; (d) Multiple shoot formation of garlic in a 5-litre BTBB, inset: microbulb formation; (e) Growth and developments of grape shoots in a 5-litre BTBB after 40 days of culture; (f) Bulblet growth of *Lilium* oriental hybrid 'Casablanca' in a 5-litre BTBB.

Acknowledgements

This work was supported by Korea Science and Engineering Foundation (KOSEF) through Research Center for the Development of Advanced Horticultural Technology at Chungbuk National University, Cheongju, 361-763, Korea.

References

Allen TC (1974) Control of viruses in lilies. In: *Lilies* and other *Liliaceae*. Royal Horticultural Society

Anderson WC (1977) Rapid propagation of *Lilium*, cv. Red Carpet. *In Vitro* 13: 145

Archambault J, Lvoie L, Villiams RD & Chavarie C (1995) Nutritional aspects of Daucus carota somatic embryo cultures performed in bioreactors. In: Terzi M (ed) Current Issues in Plant Molecular and Cellular Biology (pp. 681-687). Kluwer Academic Publishers, Dordrecht

Bourgaud F, Gravot A, Milesi S & Gonteir E (2001) Production of plant secondary metabolites: a historical perspective. Plant Sci. 161: 839-851

Cervelli R & Senaratna T (1995) Economic aspects of somatic embryogenesis. In: Christie JA, Kozai T, Smith ML (eds) Automation and Environmental Control in Plant Tissue Culture (pp. 29-64). Kluwer Academic Publishers, Dordrecht

Chakrabarty D & Paek KY (2001) Recent advances in *in vitro* manipulation and propagation of medicinal plants. In: Proceedings of KSMCS International Symposium, 2001 (pp. 26-61). The Korean Society of Medicinal Crop

Chakrabarty D, Hahn EJ, Yoon YS & Paek KY (2003) Micropropagation of apple root stock 'M9 EMLA' using bioreactor. J. Hortic Sci.& Biotech. 78: 605-609

Choi SM, Son SH, Yun SR, Kwon OW, Seon JH & Paek KY (2000) Pilot-scale culture of adventitious roots of ginseng in a bioreactor system. Plant Cell, Tiss. Org. Cult. 62: 187-193

Choi YE, Yang DC, Yoon ES & Choi KT (1999a) High-efficiency plant production via direct somatic single embryogenesis from preplasmolysed cotyledons of *Panax ginseng* and possible dormancy of somatic embryos. Plant Cell Rep. 18: 493-499

Choi YE, Yang DC & Yoon ES (1999b) Rapid propagation of *Eleuterococcus senticosus* via direct somatic embryogenesis from explants of seedlings. Plant Cell, Tiss. Org. Cult. 58: 93-97

Christie JA, Kozai T & Takayama S (1995) Automation in plant tissue culture. General introduction and overview. In: Christie JA, Kozai T, Smith M L (eds) Automation and Environmental Control in Plant Tissue Culture (pp. 1-18). Kluwer Academic Publishers, Dordrecht

Danckwerts PV (1951) Significance of liquid-film coefficients in gas-absorption. Indust. Eng. Chem. 43: 1460-1467

Denchev PD, Kuklin AI & Scragg AH (1992) Somatic embryo production in bioreactors. J. Biotech. 26: 99-109

Dornenburg H & Knor D (1995) Strategies for the improvement of secondary metabolite production in plant cell culture. Enzyme Micro. Tech. 17: 674-684

Dussert S, Verdeil JL, Rival A, Noirot M & Morel JB (1995) Nutrient uptake and growth of *in vitro* coconut (*Cocos nucifera* L.) calluses. Plant Science 106: 185-193

Escalona M, Lorenzo JC, Gonzalez B, Daquinta M, Gonzalez JL, Desjardins Y & Borroto CG (1999) Pineapple (*Ananas comosus* L. Merr) micropropagation in temporary immersion systems. Plant Cell Rep. 18: 743-748

Furuya T, Yoshikawa, T Orihara Y & Oda H (1984) Studies of the culture conditions for *Panax ginseng* cells in jar fermentors. J. Nat. Prod. 47: 70–75

Giri A & Narasu ML (2000) Transgenic hairy roots: recent trends and applications. Biotechnol. Adv. 18: 1-22

Gui Y, Guo Z, Ke S & Skirvin RH (1991) Somatic embryogenesis and plant regeneration in *Acanthopanax senticosus*. Plant Cell Rep. 9: 514-516

Hackett WP (1969) Aseptic multiplication of lily bulblets from bulb scales. Proc. Int. Plant Propag. Soc. 19: 105-108

Heyerdahl PH, Olsen OAS & Eide KH (1995) Engineering aspects of plant propagation in bioreactors. In: Christie JA, Kozai T, Smith ML (eds) Automation and Environmental Control in Plant Tissue Culture (pp. 87-123). Kluwer Academic Publishers, Dordrecht

Hilton MG & Wilson PDG (1995) Growth and the uptake of sucrose and mineral ions by transformed root cultures of *Datura stramonium*, *Datura candida×aurea wrightii*, *Hyoscyamus muticus* and *Atropa belladonna*. Planta Medica, 61: 345-350

Hishimoto T & Azechi S (1988) Bioreactor for large scale culture of plant cells. In: Bajaj YPS (ed) Biotechnology in Agriculture and Forestry, Vol. 4 (pp. 209-220). Springer-Verlag, Berlin

Honda H, Liu C & Kobayashi T (2001) "Large scale plant propagation". In: Scheper T (ed) Advances in Biochemical Engineering/Biotechnology, Vol. 72 (pp. 157-182). Springer-Verlag Berlin, Heidelberg

Huang KC (1993) The Pharmacology of Chinese Herbs (pp. 21-45). CRC press, Boca Raton, FL

Ibaraki Y & Kurata K (2001) Automation of somatic embryo production. Plant Cell, Tiss. Org. Cult. 65: 179-199

Isoda S & Shoji J (1994) Studies on the cultivation of *Eleutherococcus senticosus* Maxim. II On the germination & raising of seedlings. Nat. Med. 48: 75-81

Ket N, Chakrabarty D, Hahn EJ & Paek KY (2003) Micropropagation of an endangered jewel orchid (*Anoectochilus formosanus*) using bioreactor system. (communicated)

Kim EK (2002) Propagation of multiple shoots and microbulbs using bioreactor system in garlic. M. S. thesis. Chungbuk National University, Korea

Kim JW & Kim HS (2001) Mass production of Siberian ginseng (*Eleutherococcus senticosus*) somatic embryos by cell culturing. Oriental Pharmacy Exp. Med. 1: 34-38

Kim SJ (2001) Effect of environmental conditions on growth and quality of chrysanthemum plantlets in bioreactor culture. M. S. thesis. Chungbuk National University, Korea

Kim YS (1999) Development of large scale process for mass production of lily bulblets by bioreactor system. M. S. thesis. Chungbuk National University, Korea

Kim YS, Hahn EJ & Paek KY (2001) A large scale production of *Lilium* bulblets through bioreactor culture. Acta Hort. 560: 383-386.

Lazzeri PA, Hildebrand DE & Collins GB (1987) Soybean somatic embryogenesis: Effects of nutritional, physical and chemical factors. Plant Cell, Tiss. Org. Cult. 10: 209-220

Leathers RR, Smith MAL & Christie AJ (1995) Automation of the bioreactor process for mass propagation and secondary metabolism. In: Christie JA, Kozai T, Smith ML (eds) Automation and Environmental Control in Plant Tissue Culture (pp. 187-214). Kluwer Academic Publishers, Dordrecht

Lee YH (1997) Design of bioreactor system for small and pilot scale cultivation of plant cells. M. Sc thesis. Chungbuk National University, Korea

Lian ML (2001) Several factors affecting bulblet production in vitro and establishment of mass production system of *Lilium* Hybrid using bioreactor. Ph D thesis. Chungbuk National University, Korea

Lian ML, Chakrabarty D & Paek KY (2002a) Effect of plant growth regulators and medium composition on cell growth and saponin production during cell suspension culture of mountain ginseng (P. ginseng C. A. Meyer). J. Plant Biol. 45: 201-206

Lian ML, Chakrabarty D & Paek KY (2002b) Growth and the uptake of sucrose and mineral ions by *Lilium* bulblets during bioreactor culture. J. Hort. Sci. Biotech. 77: 253-257

Lian ML, Chakrabarty D & Paek KY (2003) Bulblet formation from bulbscale segments of *Lilium* Oriental Hybrid 'Casablanca' using bioreactor system. Biologia Plantarum 46: 199-203

Lim S, Seon JH, Paek KY & Son SH (1998) Development of pilot scale process for mass production of *Lilium* bulblets *in vitro*. In: Drew R. A. (ed) Proc. Int. Symp. Biotechnology Tropical and Subtropical Species. Acta Hort.: 237-241

Luo SW, Huang WH, Gao GY & Yu RC (1964) Tissue culture of *Panax ginseng*. Plant Physiol. Commun. 2: 26-38 (in Chinese)

Moorhouse SD, Wilson G, Hennerty MJ, Selby C & Antsaoir SMA (1996) Plant cell bioreactor with medium perfusion for control of somatic embryogenesis in liquid cell suspensions. Plant Growth Regul. 20: 53-56

Novak FJ & Petru E (1981) Tissue culture propagation of *Lilium* hybrids. Scientia Horticulturae. 14: 191-199

Paek KY & Chakrabarty D (2003) Micropropagation of woody plants using bioreactor. In: Jain S. M. & Ishii K. (eds) Micropropagation of Woody Trees and Fruits (pp. 756-756). Kluwer Academic Publishers, Dordrecht

Paek KY, Hahn EJ & Son SH (2001) Application of bioreactors of large scale micropropagation systems of plants. *In vitro* Cell. Dev. Biol.-Plant. 37: 149-157

Park SY, Murthy HN & Paek KY (2000) Mass multiplication of protocorm like bodies using bioreactor system and subsequent plant regeneration in *Phalaenopsis*. Plant Cell, Tiss. Org. Cult. 63: 67-72

Piao XC, Chakrabarty D & Paek KY (2003) A simple method for mass production of potato microtubers using a bioreactor system. Current Science. 84: 1129-1132

Preil W (1991) Application of bioreactors in plant propagation In: Debergh PC, Zimmerman RH (eds) Micropropagation (pp. 425-445). Kluwer Academic Publishers, Dordrecht

Rittershaus E, Ulrich J, Weiss A & Westphal K (1989) Large scale industrial fermentation of plant cells: Experiences in cultivation of plant cells in a fermentation cascade up to a volume of 75000 litres. BioEngineering 5: 28-34

Robb SM (1957) The culture of excised tissue from bulb scale of *Lilium speciosum* Thunb. J. Exp. Bot. 8: 348-352

Sajc L, Grubisic D & Novakovic GV (2000) Bioreactors for plant engineering: An out for further research. Biochem. Eng. J. 4: 89-99

Schmitz U & Lörz H (1990) Nutrient uptake in suspension culture of Gramineae. II. Suspension cultures of rice (Oryza sativa L). Plant Science. 66: 95-111

Sharma NK (1992) Automation in plant tissue culture: problems and prospects. Curr. Sci. 7: 507-512

Shim SW (2002) Environmental conditions affecting *in vitro* growth and acclimatization of '5BB' grape rootstock. Ph. D thesis Chungbuk National University, Korea

Son SH, Choi SM, Kwon SR, Lee YH & Paek KY (1999) Large scale culture of plant cell and tissue by bioreactor system. J. Plant Biotech. 1: 1-8

Son SH & Paek KY (2001) Large scale production of medicinal plant species: the application of bioreactors for production of ginseng roots. In: Saxena P K (ed) Development of Plant-based Medicines. Conservation, Efficacy and Safety (pp. 139-150). Kluwer Academic Publisher, Dordrecht

Takayama S & Misawa MA (1983) Scheme for mass propagation of *Lilium in vitro*. Physiologia Plantarum. 48: 121-125

Takayama S & Akita M (1994) The types of bioreactors used for shoots and embryos. Plant Cell, Tiss. Org. Cult. 39: 147-156

Takayama S & Misawa M (1981) Mass propagation of *Begonia hiemalis* plantlet by shake culture. Plant Cell Physiol. 22: 461-467

Tang W & Eisenbrand G (1992) Panax ginseng C. A. Mayer. Chinese Drugs of Plant Origin (pp. 710-737). Springer-Verlag, Berlin

Timmis R (1998) Bioprocessing for tree production in the forest industry: conifer somatic embryogenesis. Biotech. Progress. 14: 156-166

Törmälä T, Jokinen K, Majamemi A & Tuominen U (1987) Utilization of media components by begonia shoots in submerged culture. In: Neijssel OM., Van der Meer RR, Luyben KCAM (eds). Proc. 4th Europe Cong. Biotech. 2 (pp. 391-394). Elsevier Science, Amsterdam

Tulecke W & Nickell LG (1959) Production of large amounts of plant tissue by submerged culture. Science, 130: 863-864

Varshney A, Dhawan V & Srivastava PS (2000) A protocol for *in vitro* mass propagation of Asiatic hybrids of lily through liquid stationary culture. *In vitro* Cellular and Development Biology-Plant. 36: 383-391

Vasil IK (1994) Automation of plant propagation. Plant Cell, Tiss. Org. Cult. 62: 187-193

Walker PN (1995) System analysis and engineering. In: Christie JA, Kozai T, Smith ML (eds) Automation and Environmental Control in Plant Tissue Culture (pp. 65-85). Kluwer Academic Publishers, Dordrecht

Wu J & Zhong JJ (1999) Production of ginseng and its bioactive components in plant cell culture: Current technological and applied aspects. J. Biotechnol. 68: 89-99

Yasuda S, Satoh K, Isshii T & Furuya T (1972) Studies on the cultural conditions of plant cell suspension culture. In: Terui G. (ed). Fermentation Technology Today (pp. 697-703). Soc. Fermen. Tech. Osaka, Japan

Yoshikawa T & Furuya T (1987) Saponin production by cultures of *Panax ginseng* transformed with *Agrobacterium rhizogenes*. Plant Cell Rep. 6: 449-453

Yu KW, Gao WY, Hahn EJ & Paek KY (2001a) Effects of micro elements and nitrogen source on adventitious root growth and ginsenoside production in ginseng (*Panax ginseng* C. A. Meyer). J. Plant Biol. 44: 179-184

Yu KW, Gao WY, Hahn EJ & Paek KY (2001b) Effects of NH_4^+:NO_3^- ratio and ionic strength on adventitious root growth and ginsenoside production in bioreactor culture of *Panax ginseng* C. A. Meyer. Acta Hort. 560: 259-261

Yu KW, Hahn EJ & Paek KY (2000a) Production of adventitious ginseng roots using bioreactors. Kor. J. Plant Tissue Cult. 27: 309-315

Yu KW, Gao WY, Son SH & Paek KY (2000b) Improvement of ginsenoside production by jasmonic acid and some other elicitors in hairy root culture of ginseng (*Panax ginseng* C. A. Meyer). *In Vitro* Cell. Dev. Biol. 36: 424-428

Yu WC, Joyce PJ, Cameron DC & McCown BH (2000) Sucrose utilization during potato microtuber growth in bioreactors. Plant Cell Rep. 19: 407-413

Chapter 7

Membranes to reduce adherence of somatic embryos to the cell lift impeller of a bioreactor

Seppo Sorvari*, Riitta Mäkeläinen, Katriina Ahanen & Otto Toldi[1]
MTT Horticulture, Toivonlinnantie 518, FIN-21500 Piikkiö, Finland.
**requests for offprints; Fax: +358-2-4772299; E-mail: seppo.sorvari@mtt.fi*
[1]Agricultural Biotechnology Center, P.O. Box 411, H-2101 Gödöllö, Hungary

Abstract: Membranes less attractive to embryos were tested as a replacement for nylon screens to prevent adherence of somatic embryos, cell clusters and cells to different sites of the bioreactor, a feature considered undesirable in plant cell suspension cultures. The results showed that the loss of embryogenic cell-mass could be halved by using silicone or track membranes. For aeration purposes, these membranes are as satisfactory as nylon screens conventionally used in cell lift impellers.

Key words: artificial seed, dialysis membrane, microseed, nylon screen, silicone membrane, somatic embryogenesis, somatic seed, track membrane

1. Introduction

Somatic embryogenesis is a highly effective cloning method whereby large numbers of embryos can be produced in a minimum space. The idea of somatic embryogenesis dates back to the 1950's when Reinert (1958, 1959) and Steward and co-workers (1958) reported the differentiation of bipolar structures resembling sexually-produced embryos in carrot cell cultures. One of Reinert's primary observations was that withdrawal of auxin from the nutrient medium triggered the formation of somatic embryos. This observation frequently has been made since that time (Kamada and Harada, 1979; Sung and Okimoto, 1981, 1976; Nomura and Komamine, 1985; Sorvari et al., 1997), and the technique is still successfully applied.

To compete with micropropagation, the production of somatic embryos should be fast and capable of producing millions of new plants daily. However, before it can be as effective as is true seed propagation, and

A.K. Hvoslef-Eide and W. Preil (eds.), Liquid Culture Systems for in vitro Plant Propagation, 117–125.

economically viable, numerous basic problems need still to be resolved. Above all, both the differentiation of somatic embryos must be a reliable process that can be widely applied to most commercial important plant species, and the genetic stability should be comparable to that of micropropagated plants. The quality of the embryos and the storage techniques need to be satisfactory to ensure that the germination rate will be about the same as for true seeds.

Use of somatic embryos as a competitive cloning method requires that the whole process, from the initiation of suspension culture to the sowing-ready microseeds, should be automated. Bioreactors are of key importance in the development of automated processes for the production of somatic embryos. A key part of the bioreactor vessel is its mixing system. Mixing is needed to oxygenate the suspension culture; also, some physical movement is necessary to keep the developing somatic embryos apart and separate, which would otherwise adhere together. Several types of bioreactors are available commercially, from those with simple marine blade mixing systems, to those with bubble column, air lift loop, liquid impelled loop and cell lift impelled systems (Scragg, 1993). The cell lift impelled bioreactor (CLIB), with or without a double screen, was primarily developed for highly fragile animal cell cultures, nevertheless it seems to be reasonably well suited for plant cell cultures. The CLIB is a satisfactory alternative to the conventional stirred tank reactor (CSTR), which is not well suited for shear-sensitive animal and plant cells and somatic embryos.

CLIB takes advantage of what is known as the cell-lift principle. In the CLIB with double screen, oxygenation of the nutrient medium is carried out by sparging sterile oxygen or air into the cell-free cavity. Even a slow rotation of the impeller discharge ports located in the upper part of the cell lift impeller causes a centrifugal force sufficient to create a weak differential pressure between the lower and upper parts of the impeller. Through this action, cells or embryos are lifted into the central draft tube and then expelled *via* the three discharge ports into a continuous recirculation loop. Shear forces are low in the CLIB and even relatively large somatic embryos can be cultured without major damage. However, even though the double screen is made either of very fine stainless steel or nylon mesh, plant cells and embryos tend to adhere to the finest meshes causing problems with aeration of the medium and the blocking of tubes and probes within the bioreactor. Even the smallest fault in mesh structure allows the cells to pass into the aeration cavity between the double screens causing further clogging of the mesh from inside the cavity.

The aim of this research was to study the possibility of replacing the mesh by membranes to which somatic cells and embryos do not easily adhere, whilst allowing adequate aeration of the growing cell mass.

2. Material and methods

2.1 Initiation of bioreactor culture

The basic protocol for carrot somatic embryo production and maintenance was as follows. The primary explants consisted of hypocotyl sections of seedlings of the domestic carrot (*Daucus carota* L.) var. Duke. Suspension cultures were initiated by placing hypocotyl sections of sterile germinated seedlings, each about 0.5-1 cm long into 100 ml Erlenmeyer flasks. Each flask contained 20 ml of modified MS (Murashige and Skoog, 1962) liquid nutrient medium supplemented with 1 mg l^{-1} 2,4-D: 15 hypocotyl sections were added per flask. The initial cultures were incubated on a gyratory shaker (100 rpm) for 4 weeks with a 16-h day (40-50 μmol $s^{-1}m^{-2}$) at 24/22±1°C day/night temperature. The first subcultures were prepared by sieving the suspension through a stainless steel mesh of pore size 355 μm. The cells passing through the mesh were collected by 5 min centrifugation (100 g), and about 0.5-0.7 ml of the packed cell volume (PCV) was resuspended in 20 ml of fresh MS medium supplemented with 1.0 mg l^{-1} 2,4-D. For maintenance of subsequent subcultures, the same procedure was repeated every 2 weeks.

For the initiation of somatic embryogenesis, the 8-day-old subcultured suspensions were successively sieved through a series of nets of 355, 200, 100, 45 μm mesh, and finally through a net of 27 μm mesh. The fractions remaining on the net of 27 μm mesh were washed 4 times with plant growth regulator (PGR)-free MS solution. The cells were then suspended in PGR-free MS medium to the density of $3x10^4$ cells ml^{-1}, and 20 ml were transferred to a new 100-ml Erlenmeyer flask. For the differentiation of cells to embryos the cultures were kept in the dark on a gyratory shaker (100 rpm) for 2 weeks at 25/23±1°C in a 16h daily photoperiod.

The inoculation volume of packed cells in the bioreactor cultures was 3.0 ml or 2.6 ml of the 45-100 μm fraction per 3.5 l or 3 l of growth medium, respectively. The bioreactor used in this study was the New Brunswick Celligen Plus™ with the CellLift® mixing and aeration system. The revolution speed of CellLift® was, at the beginning of the process 25 rpm, and 35-45 rpm at the end.

2.2 Membranes used in the bioreactor

The membranes used in place of the nylon screen as a cover for the CellLift® were a dialysis membrane, a silicone membrane and a nuclear track membrane. The nylon screen was the original screen from the New Brunswick, developed especially for the CellLift® impeller. The dialysis membrane was a Spectra/Por® 1 membrane with molecular weight cut-off (MWCO) of 6000-8000Da from Spectrum Medical Industries, Inc. The silicone membrane was acquired from a local industrial plastics reseller; it was 400 μm thick, and the thinnest one available. Nuclear track membrane was obtained from Nerox Filter Oy (Tampere, Finland). Nuclear track membranes are produced by the physico-chemical treatment of autoclavable poly(ethylene terephthalate) films exposed to heavy ion beams. Film thickness is 10 μm, pore size from 0.05 to 2.0 μm and pore density from 10^5 to $3x10^9$ pores per cm^2 (Apel et al., 1992).

2.3 Experimental design

The cell lift impeller has two sites for the aeration screens (Figure 1). One site is on the outer rim and the other is located around the central draft tube. Because factory-made membrane tubes were not available, they were made by hand in the laboratory. The seams were sealed with silicone glue and the self-made sleeves were tightened to the cell lift impeller with O-rings. Depending on the space between the two membranes of the cell lift impeller, the initial volume of the bioreactor was either 3 l or 3.5 l. If the inter-membrane space was kept air filled, the initiation volume was 3 l, otherwise it was 3.5 l. The bioreactor was autoclaved in the Boxer 200/110 autoclave. The total autoclaving time including cooling was 5.5 hours, of which the effective autoclaving at 121°C was 15 min. Because the autoclaving of the bioreactor was a long process, sucrose, meso-inositol, casein hydrolysate and the oxygen probe were autoclaved separately at 121°C for twenty minutes. After the autoclaving process the nutrient components were added in a laminar flow hood and the oxygen probe was assembled into place. After cooling to room temperature in the laminar flow hood, the cell suspension was added to the bioreactor vessel in the volumes as described above. The changes in D.O. (Dissolved Oxygen %) were recorded daily and the changes in pH every second day. No attempts were made to regulate the pH were made, but from about 7th to 10th day additional pure oxygen was given to prevent too low D.O. concentrations.

The cell lift impeller was equipped with membranes and nylon screen in 5 different combinations, as follows. In 'D/N', the outer rim was dialysis membrane and the inner nylon screen. In 'N/N', both the outer and inner rims were nylon screens. In 'S/N', the outer rim was silicone membrane and inner nylon screen. In 'S/S', both the outer and inner rims were silicone screens. In 'T/T', both the outer and inner rims were track membranes. In S/S and T/T the cavity between the membranes was filled with air, whereas in other cases the cavity was filled with nutrient medium.

Each experiment was run twice, each for 2 weeks. At the end of the experiment, non-adhering embryogenic cell mass was collected separately from the material adhering to the different sites of the bioreactor vessel and the cell lift impeller. For the assay of the dry weight, the collected samples were dried at 60°C overnight and weighed.

Figure 1: A cell lift impeller with the screen of the outer rim removed to allow a view into the double screen cavity. Growing carrot somatic embryos completely cover the nylon screen of the cell lift draft tube.

3. Results and discussion

Growth of plant cell and somatic embryo cultures in a bioreactor is not without its problems, because the proembryogenic cells tend to form relatively large clusters and somatic embryos have a tendency to adhere for surfaces, and to the tubes and probes that extend into the growing medium. The nylon screen that is commonly used on the cell lift impeller is highly susceptible to the blocking by carrot somatic embryos (Figure 1). One solution could be to use direct air bubbling of the nutrient medium as a mixing system. However, there are indications that direct bubbling disturbs the growth of cells in a suspension culture (Hegarty et al., 1986). Alternatively, it has been shown that a bubble-free silicone tubing system has sufficient capacity to supply oxygen adequate for the growth of *Euphorbia pulcherrima* cell suspension (Luttman et al., 1994). We tested different membranes analogous to the silicone tubing, as replacement for the nylon screen. The most significant result was, that in the experiments S/S and T/T (Figure 2), the loss of embryogenic material was reduced significantly. Whereas the loss of embryogenic material with double nylon screen was 25%, with S/S it was only 10% and with T/T it was 11%. Although the adherence of embryos was reduced markedly, many sites still remained for accumulation of cells and embryos. Experiments D/N and S/N showed that the screen in the central draft tube was one of the main locations for somatic embryo attachment. Evidently, because of centrifugal force, adherence is more pronounced in the central draft tube than on the outer rim of the cell lift impeller, and the central draft tube should therefore have a very smooth surface.

In general, the oxygen (Figure 3) and pH (Figure 4) values followed the same pattern in all experiments. If the pH is not regulated artificially, it reduces for the first 6-8 days and then increases back to the approximate initial value. When air is used, none of the aeration systems can provide sufficient enough oxygen in the second half of the culture period, therefore pure oxygen must be added to the aeration system. For this, simultaneously the circulation speed of the cell lift impeller must be gradually increased up to 35-40 rpm.

Dialysis, silicone and track membranes are materials to which plant cells and embryos do not easily adhere. Theoretically the large mesh size of nylon screen allows greater aeration of the nutrient medium, but in practice the cells, cell clusters and developing embryos quickly block the screen and prevent effective aeration (Figure 3).

For the preparation of different membranes, the dialysis membrane, nylon screen and track membrane are inelastic and especially dialysis membrane and track membrane in particular are difficult to glue into the form of a sleeve. Therefore new membranes had to be prepared for every experiment. Dialysis membranes are available in sleeve form, but not in the cell lift impeller-fitting diameters required. Silicone membrane is the easiest to use because it is elastic. After the experiments described above, a method was developed in the laboratory to make seamless silicone tubes that fit exactly to the draft tube and the outer rim of the cell lift impeller.

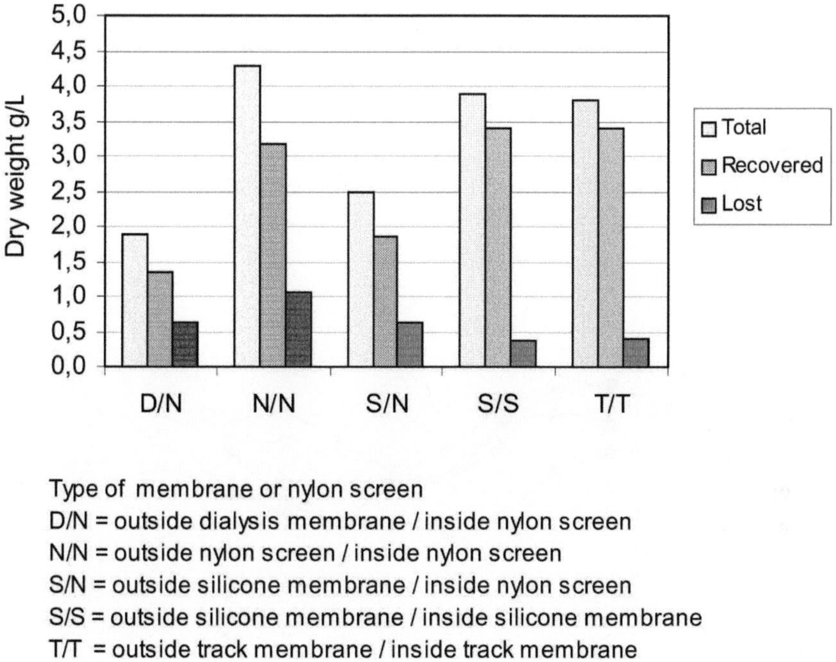

Figure 2: Influence of dialysis membrane, nylon screen, silicone membrane and track membrane on the recovery of carrot somatic cells after culturing of two weeks in a bioreactor equipped with cell lift impeller.

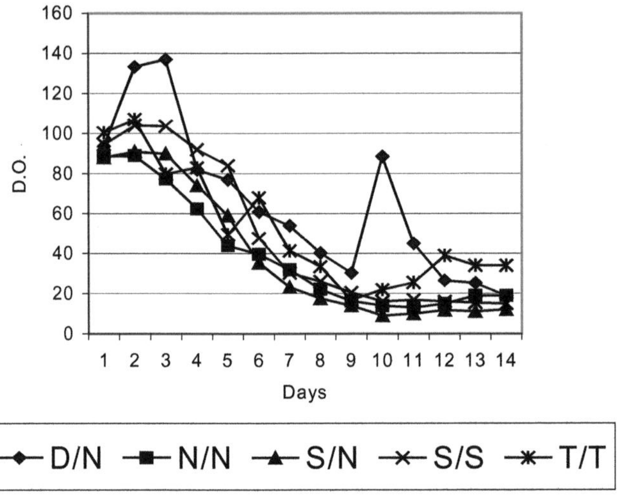

Figure 3: Changes of D.O. (Dissolved Oxygen %) in the bioreactor vessel during the culturing of carrot somatic embryos for two weeks. Abbreviations: see figure 2.

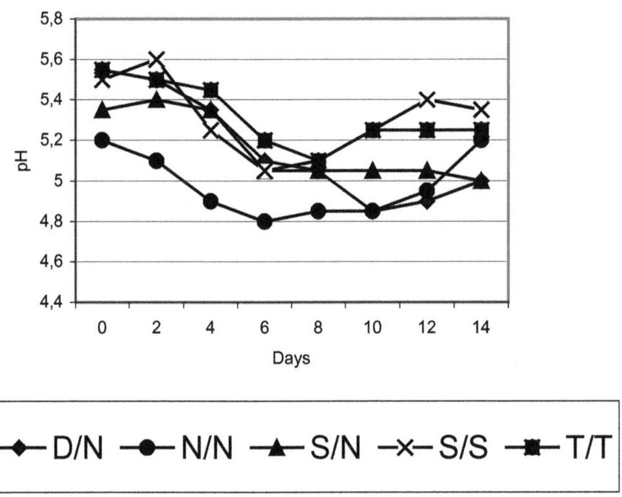

Figure 4: Changes of pH in the bioreactor vessel during the culturing of carrot somatic embryos for two weeks. Abbreviations: see figure 2.

4. Conclusion

Bioreactors are of key importance in the development of automated technology for plant cloning either *via* exposed somatic embryos or *via* microseeds. Plant cells require different conditions from animal or microbial cells and are particularly prone to adhere to bioreactor sites. In this study we have shown that, through choice of suitable material - for example, for the aeration system - the loss of embryogenic material can be reduced substantially.

References

Apel P, Didyk A, Zhitariuk NI, Larionova IE, Mamonova TI, Orelovich OL, Samoilova LI & Yanini IV (1995) Properties of track membranes of different structure characteristics. Scientific Instrument Engineering 5: 50-56

Hegarty PK, Smart NJ, Scragg AH, & Fowler MW (1986) The aeration of *Catharantus roseus* L.G. Don suspension cultures in airlift bioreactors: The inhibitory effect at high aeration rates on culture growth. J. Exp. Bot. 37: 1911-1920

Kamada H & Harada H (1979) Studies on the organogenesis in carrot tissue cultures. I. Effects of growth regulators on somatic embryogenesis and root formation. Z. Pflanzenphysiol. 91: 255-266

Luttman R, Florek P & Preil W (1994) Silicone-tubing aerated bioreactors for somatic embryo production. Plant Cell, Tissue and Organ Culture 39: 157-170

Murashige T & Skoog F (1962) A revised medium for rapid growth and bioassays with tobacco tissue cultures. Physiologia Plantarum 15: 473-497

Nomura K & Komamine A (1985) Identification and isolation of single cells, that produce somatic embryos at a high frequency in carrot suspension culture. Plant Physiol. 79: 988-991

Reinert J (1958) Morphogenese und ihre Kontrolle an Gewebekulturen aus Carotten. Naturwissenschaften 45: 244-245

Reinert J (1959) Ueber die Kontrolle der Morphogenese und die Induktion von Adventivembryonen an Gewebekulturen aus Karotten. Planta 53: 318-333

Scragg AH (1993) Commercial and technical perspectives. In: Hunter C (ed) *In Vitro* Cultivation of Plant Cells (pp 151-178). Butterworth-Heinemann, Oxford

Sorvari S, Toldi O, Ahanen K, Viinamäki T, Hakonen T & Tahvonen R (1997) Using polysaccharides and galactomannans as gelling agents in capsule formation of artificial seeds. J. Am. Soc. Hortic. Sci. 122: 878-883

Steward FC, Mapes MO & Smith J (1958) Growth and organized development of cultured cells. I. Growth and division of freely suspended cells. American Journal of Botany 45: 693-703

Sung Z (1976) Turbidimetric determination of plant cell culture growth and its application. Plant physiol. 57: 460-462

Sung Z & Okimoto R (1981) Embryonic proteins in somatic embryos of carrot. Proc.Natl.Acad.Sci.U.S.A. 78: 3683-3687

Chapter 8

Cost-effective mass cloning of plants in liquid media using a novel growtek bioreactor

Satyahari Dey
Biotechnology Department, Indian Institute of Technology, Kharagpur, India-721302.
Fax: 91-3222-755303; E-mail: satyahari01@yahoo.com

Abstract: A low-cost Growtek bioreactor has been designed, patented and commercialised. It has unique features such as a floating and rotating explant-holder with perforated explant support and a side tube for medium changing, culture feeding and for content monitoring. The bioreactor can be operated both in static and agitated modes. Extensive performance studies have been conducted using representatives of trees (*Santalum album*), commercial ornamentals (*Dendranthema grandiflora*), monocotyledonous horticultural species (*Ananas comosus*), tuber crops (*Solanum tuberosum*) and a medicinal plant (*Catharanthus roseus*). In comparison to propagation in agar-gelled media as well as in liquid media using other culture vessels, this bioreactor exhibited 1.2 – 23.3 times shoot production, minimised root injuries by 32 – 48 %, reduced contamination by 12 – 18 % and reduced incubation time by 16- 42 %. Thousands of *Ananas comosus* plantlets raised in this bioreactor have been field tested. Additionally, it was found to be effective for hairy root culture of *C. roseus*.

Key words: *Chrysanthemum, Catharanthus,* cost-effectiveness, Growtek bioreactor, liquid medium, mass cloning, pineapple, potato, *Santalum*

Abbreviations: BAP- 6-benzylaminopurine; GA- gibberellic acid; GI- growth rate; IAA- indole-3-acetic acid; IBA- indole-3-butyric acid; MS- Murashige and Skoog' s (1962) medium; NAA- naphthalene acetic acid

1. Introduction

The industrial production of tissue cultured plants has largely been dominated by herbaceous ornamental species and a few vegetable, fruit or plantation crops (banana, oil palm etc.). The success with woody and semi-woody plants has been rare (Smith, 1997; Gupta et al., 1993). The high cost

A.K. Hvoslef-Eide and W. Preil (eds.), Liquid Culture Systems for in vitro Plant Propagation, 127–141.

of production (~ US$ 0.10 – 0.15 per unit) owing to the labour-intensive nature (labour cost may be 50-85 % of production cost), prejudicing economic viability, was the single most important reason that discouraged *in vitro* industrial propagation of many species (Vasil, 1994; Goldstein, 1999). Plant tissue culture was practised initially with agar-gelled media. It was soon realized that agar was one of the costliest ingredients in the medium, though not a nutrient. Many gelling and non-gelling matrices were tested in order to achieve cost-effectiveness, by substituting agar (Sorvari, 1986; Henderson and Kinnersley, 1988; Bhattacharya et al., 1994). Subsequently, the use of liquid media, scale-up in bioreactors (Preil, 1991; Takayama, 1991; Das et al., 1999) and induction of automated production were some of the alternatives explored for the minimization of cost of production through improvement in propagation efficiencies (Tisserat, 1991; Smith and Spomer, 1995; Hvoslef-Eide and Melby, 2000; Dey, 2001). The prospects for temporary immersion have also been discussed (Etienne et al., 1997; Jimenez et al., 1999). The other aspects of cost minimisation are the use of low-cost culture vessels, prevention of contamination, improved quality of plantlets and their enhanced field survival. The successful adaptations of these alternatives may also enhance the scope for commercial exploitation of somatic embryogenesis, plant secondary metabolite production (Curtis and Emery, 1993; Hunter and Kilby, 1999) and heterologously-expressed healthcare products of human origin (Doran, 2000; Meyer et al., 2002). The recent attempts at the production of such new generation products as plantibodies (Peeters et al., 2001; Stoger et al., 2002) are indicators of further need for developing the most cost-effective bio-processes based on plant cell and tissue culture in liquid media. The use of liquid media in these cases will offer benefits of increased nutrient uptake, greater availability of dissolved oxygen, easier dispensing, automated scale up and process control, periodic sampling and more productivity.

Our laboratory has been working for more than a decade on cost-minimisation aspects through novel bio-process (Indian patent application No. 197/Cal/2001), product (Bhattacharya et al., 1994) and equipment development. This article describes the performance of the novel Growtek bioreactor for mass cloning of several commercially-important plants.

Figure 1: (A): Growtek bioreactor with floating explant-holder (a) main vessel with side tube (b) inside of the lid (c) shown separately; (B): pineapple shoot cluster production in glass jars and Growtek; (C): pineapple shoot cluster propagation in Life Guard and Growtek.

2. Materials and methods

2.1 Culture vessels

Growtek bioreactor (Indian patent No. 183604/2000), Life Guard culture box (107 X 107 X 96 mm h; Sigma Cat. No. C8062) with a Life Raft membrane raft (Sigma M7413), Magenta GA-7 (77 X 77 X 97 mm h; Sigma V8505, Phytacon (140 X 140 mm h; Sigma), Erlenmeyer flasks (250 ml) and locally available glass jars with metallic threaded caps (76 X 114 mm h) were used. Embryogenic calli of *Santalum album* were raised in borosilicate culture tubes without a rim (32 X 200 mm h).

The Growtek bioreactor (Figure 1 A) has unique features including a floating, rotating, non-absorbing explant-holder with perforated (a) explant-support matrix; a side–tube with silicon rubber septum for changing media and online monitoring of the medium environment (pH, dissolved oxygen, temperature etc.), a lid with a central and downwardly projected slope (c) for minimising condensate accumulation inside; and a polycarbonate body (b) with perfect transparency that will be satisfactory for up to 80 autoclave cycles.

Growtek was used in both static and agitated modes (at 100 rpm on a rotary shaker) for pineapple and *Chrysanthemum* shoot multiplications.

2.2 Plants and culture conditions

Extensive performance studies have been conducted using representative species of trees (sandalwood, *Santalum album* L. IITBT 08), commercial ornamentals (*Chrysanthemum*: *Dendranthema grandiflora* Tzvelev, cv. Birbal Sahni), monocotyledonous horticultural species (pineapple, *Ananas comosus* L. Merr., cv. Queen), tuber crops (potato, *Solanum tuberosum* cv. Kufri Jyoti), and a medicinal plant (Madagaskar periwinkle, *Catharanthus roseus* L. G. Don., var. pink).

The following optimised phytohormone doses were used in MS media (Murashige and Skoog, 1962) with 3 % (w/v) sucrose. Sandalwood: 4.44 µmol BAP, 1.14 µmol IAA, 0.58 µmol GA; *Chrysanthemum*: 0.88 µmol BAP, 0.57 µmol IAA; pineapple: 26.6 µmol BAP, 1.07 µmol NAA; potato: 4.44 µmol BAP, 1.14 µmol IAA; periwinkle hairy root culture: hormone-free.

The quantity of inoculum used for each vessel was as follows. Sandalwood: 1 g (fresh weight) embryogenic callus mass for embryo maturation and 400 somatic embryos for a root injury study; pineapple: 5

shoot clusters, each of about 500 mg; periwinkle hairy roots: 500 mg fresh weight; potato and *Chrysanthemum*: 5 nodal segments.

All cultures have been maintained and mass produced during several years in the optimised media. In order to retain their satisfactory performance over a reasonable period of time, a regular rotation of subculture between gelled and liquid media was followed, as in the following schedules.

Multiple shooting in pineapple, *Chrysanthemum* and potato: 2 subcultures in gelled medium followed by 4 in liquid medium for mass-production; somatic embryogenesis in sandalwood: 2 in gelled and 8 in liquid; periwinkle hairy roots: 2 in gelled and 10 in liquid. Each subculture was of 3-4 weeks duration. Performances reported here are for liquid subculture stages, in comparison to growth in gelled media as mentioned in data tables.

The other physical conditions were as follows. Media (50 ml in each vessel) were autoclaved at 104 kPa (121 ^{0} C) for 15 min. The pH of the medium was adjusted to 5.7 $\pm$ 0.1 before autoclaving. Incubation was in culture racks maintained at 25 $\pm$ 2 ^{0} C., 60-70 % relative humidity and at 16 h day length (47 µmol m^{-2} s^{-1}) provided by timer-controlled cool white fluorescent lamps. Periwinkle hairy roots were kept dark. Unless otherwise mentioned, the agar gel concentration for media for all purposes was 0.7 % (w/v) and for some somatic embryo germination it was 1.0 % (w/v).

For pineapple shoot production and sandalwood embryogenesis in liquid media, pH was always maintained in the range 4.8 – 5.8 and 5.2 – 5.8 respectively (by monitoring and control through the side tube) for consistent performance of the cultures. Change of medium between rooting and *in vitro* hardening was done through the side tube.

2.3 *Growth rate (GI)*

The growth rate was calculated on a dry weight basis from the following relationship-

$$GI = \frac{\text{Final biomass weight} - \text{initial biomass weight}}{\text{Initial biomass weight}}$$

The biomass was lyophilised before measuring dry weight.

2.4 *Root injury index*

Root injury was calculated during uprooting of plantlets/somatic seedlings from agar-gelled media (somatic embryos were 10 – 15 mm long; and rooted shoots had 4 – 6 roots). Even if the plant had only one damaged root, it was classified as 'damaged'. To determine whether or not a root was injured, observations were made using a dissecting microscope.

$$\text{Root injury \%} = \frac{\text{Number of plantlets with injured roots x 100}}{\text{Total number of plantlets}}$$

2.5 *Rooting, hardening and field cultivation of pineapple*

Growtek was extensively used for pineapple mass propagation that led to field trials. Rooting of pineapple shoots was achieved in 10 days (nearly 100%) in half-strength MS macro-element solution (with full strength microelement solution and 2% (w/v) sucrose) supplemented with 9.8 μmol IBA. Rooted pineapple plantlets (6-8 cm long) were hardened first *in vitro* in a Growtek for one week (in photoautotrophic mode, in quarter-strength MS macroelements and full-strength microelement solutions without sucrose and other organic supplements) and then in a greenhouse (Figure 2, C) for one month before field transfer. Plantlets were transferred to perforated polythene pots filled with sand and vermicompost (1:1). The potted plantlets were kept in sunlit greenhouse (relative humidity 75-85 %) racks, shaded partially by agronet. In the first week, 75% sunlight shading was achieved, followed by 50% shading and no shading in the 2nd and subsequent weeks, respectively. Field cultivation (Figure 2 D) was conducted in the Science & Technology Entrepreneurship Park of IIT-Kharagpur (www.stepiitkgp.com). Field survival was nearly 100%: it eventually resulted in normal fruiting.

2.6 *Cost analysis for culture vessels*

For Life Guard and temporary immersion systems (TIS) prices are taken from the product list of Sigma Chemical Co., USA and for Growtek from Tarsons Products, India (2002). The relative cost factor is calculated on the basis of the Growtek price.

Table 1: Efficiencies of different culture vessels for multiple shoot production of pineapple and *Chrysanthemum* in liquid media

Vessel	Number of shoots produced in 4 weeks			
	Pineapple		*Chrysanthemum*	
	Static	Agitated	Static	Agitated
Glass jar or Erlenmeyer flask (gelled medium)	14.5 ± 1.7	-	27.8 ± 1.7	-
Life Guard	39.1 ± 3.2	-	29.2 ± 2.3	-
Growtek	310.0 ± 6.2 (21.4)	338.1 ± 7.3 (23.3)	33.4 ± 2.1 (1.2)	46.3 ± 2.2 (1.7)

Data are means ± S.E. for 15 replicates for each vessel type. Figures in parentheses indicate fold of multiplication in comparison to glass jar.

Table 2: Performances of different culture vessels for biomass production during hairy root culture of *Catharanthus roseus* and multiple shoot culture of *Solanum tuberosum*

Vessel	Medium	Days to reach GI 2.0	
		Hairy root	Multiple shoot
Glass jar/ Erlenmeyer flask	Gelled	20.2 ± 1.1 (142)	28.3 ± 1.5 (137)
Life Guard	Liquid	16.5 ± 1.4 (116)	24.8 ± 1.4 (120)
Growtek	Liquid	14.2 ± 0.6	20.7 ± 0.8

Data are means ± S.E. for 15 replicates each. Figures in parentheses indicate prolonged incubation time (%; basis Growtek) at the mean values for respective vessel.

Table 3: Peripheral fungal contamination in culture for different vessel types

Vessel	Medium	Contamination % *
Phytacon	Gelled	20.1 ± 1.8 a
Life Guard	Liquid	19.7 ± 1.4 a
Magenta	Gelled	19.5 ± 1.1 a
Glass jar	Gelled	16.4 ± 0.6 b
Growtek	Liquid	3.5 ± 0.3

Data are means ± S.E. for 25 replicates for each vessel type (5 each for pineapple, potato and *Chrysanthemum* multiple shoot cultures; sandalwood somatic seedlings and *C. roseus* hairy root cultures).

* Values for other vessel types are significantly different from that of Growtek (at 0.01 level; t-test). Means followed by different letters indicate significant difference at 0.01 level (Anova and F-test) for other vessels.

3. Results

3.1 Comparison of shoot multiplication in different culture vessels

Pineapple and *Chrysanthemum* multiple shoot production was compared during 4-week periods in jars, Erlenmeyer flasks, Life Guard and Growtek devices. The data presented in table 1 clearly show the better performance in Growtek compared with other culture vessels. Pineapple responded well in liquid culture in Growtek. In static conditions a 21.4-fold increase in shoot production was observed in comparison to production in agar-gelled medium in jars, this was further enhanced when agitated (Table 1) on a rotary shaker. The health of plantlets grown in Growtek was much better than those raised in jars or Life Guard (Figure 1 B and C).

3.2 Comparison of biomass production for periwinkle hairy roots and potato multiple shoots

Figure 2 B and table 2 show the suitability of Growtek for hairy root culture. Table 2 presents data showing more biomass production in Growtek. Plants in both Life Guard and jars took longer times (116 – 120 % and 137 – 142 % respectively) to exhibit a growth rate 2.0. Faster growth was observed in Growtek, both for hairy root and for multiple shoot production. Somatic embryos grown in Growtek (Figure 2 A) also exhibited higher embryonic biomass in comparison to those from agar-gelled medium in glass jars or in Magenta vessels (Table 4).

3.3 Air-borne fungal contamination in culture

Air-borne fungal spores may contaminate cultures during incubation due to drainage of condensed water vapour from the seal of caps down the inside wall of the vessels. Such contaminants normally colonized along the periphery of the medium surface. Data presented in table 3 show that such contamination occurs to the extent of about 20%. The lowest incidence was for Growtek (3.5%), followed by glass jars (16.4%) and other vessels with push-fit types of caps (the maximum contamination; ~20%).

Figure 2: (A): Somatic embryogenesis of sandalwood in Growtek (top view);
(B): Hairy root culture of periwinkle in Growtek (top view);
(C): Pineapple plantlets hardened in greenhouse and ready for field transfer;
(D): Tissue-cultured pineapple at the fruiting stage in the field.

3.4 Root injury of plantlets and somatic seedlings before transfer for hardening

Both potato and *Chrysanthemum* plantlets and sandalwood somatic embryos suffered from significant root injuries (about 48%) while being uprooted from agar-gelled media. Root injury was absent in the case of plantlets and somatic seedlings raised in Growtek (Table 4).

3.5 Cost of different commercially available culture vessels

Growtek appears to be the lowest cost among the three sets of culture apparatus available on the market (Table 5). Life Guard and TIS are 1.20 were 21.84 times more expensive.

Table 4: Comparison of root injury in gelled and liquid media before transfer for hardening

Vessel	Medium	Root injury (%)		
		Potato plantlet	*Chrysanthemum* plantlet	Sandalwood somatic embryo
Glass jar/ Phytacon/ Magenta	Gelled (0.7 % agar)	44.9 ± 2.6	35.2 ± 3.2	33.6 ± 1.5 (20.1 ± 3.4)
	Gelled (1.0 % agar)	-	-	44.4 ± 3.7 (35.6 ± 3.5)
Growtek	Liquid	0.0	0.0	0.0 (248.2 ± 12.2)

Data represent means ± S.E. for 25 replicates for each plant type. Figures in parenthesis are number of normal healthy cotyledonary somatic embryos produced in each condition.

Table 5. Comparison of costs of culture vessels designed for liquid media

Vessel	Approx. price ( )	Relative cost factor (Basis Growtek)
Life Guard with membrane raft (Sigma C8062 & M7413)	7.00	1.20
Temporary Immersion System (Sigma)	126.70	21.84
Growtek	5.80	

4. Discussion

The performance of the Growtek bioreactor presented in this article is important in view of the concern expressed by many researchers about minimisation of production costs for *in vitro* mass propagation and secondary metabolite production (Dey, 2001; Sutton and Polonenko, 1999; Vasil, 1994; Goldstein, 1999; Zobayed et al., 2001; Bhattacharya et al., 1990). The response of *Chrysanthemum* (Table 1) under static condition is broadly similar in all culture vessels with gelled medium and the difference is not significant between gelled and liquid media using Life Guard. Agitated liquid medium in Growtek has however, resulted in enhanced growth (1.7 times vs. 1.2 times; in comparison to gelled medium). A production increase of 23.3-fold was obtained for pineapple grown in Growtek in the agitated mode. The explant holder, being circular and floating, offers the unique advantage of agitating cultures in liquid media, when using a rotary shaker. The number of shoots obtained per Growtek (338 ± 7.3) is probably the maximum. Escalona et al (1999) attempted automated scale-up in a bioreactor and obtained 192 'competent' plants per litre of medium in a temporary immersion system. The cultivars (Smooth Cayenne) and culture conditions were however different. Other reports concerning pineapple micropropagation (Lakshmi Sita et al., 1974; Zepeda and Sagawa, 1981) were not targeted for scale up and cost reduction. Using this protocol one million pineapple plantlets can be raised in 8 weeks using 3000 Growtek and 50 m^2 space. The 100% field survival of pineapple plantlets is better than reported earlier (Escalona et al., 1999; Soneji et al., 2002). The differences in the response of *Chrysanthemum* and pineapple may be explained by the interaction between the tissue and the support matrix in the processes of nutrient uptake. The better performance of pineapple is likely to be due to greater adherence to the explants holder, as reported for other plants (Facchini and Di Cosmo, 1991; Bhattacharya et al., 1994). We believe that pineapple shoots were able to draw a few ions more rapidly when grown in Growtek (data not presented here) because of reduced water-stress due to the specially-designed floating explants-holder. Better growth of pineapple in the Life Guard and Growtek bioreactors compared with glass jars may be attributed to the elimination of the influence of impurities (normally present in the low-cost tissue-culture grade agar used in this study), as well as reduced diffusion barriers (Debergh, 1983). The healthier shoots in Growtek (Figure 1 C) are due, possibly, to a more suitable gas /vapour phase inside the culture vessel in comparison to inside the Life Guard. The latter and the similar other vessels have a flat ceiling which accumulate (Kavanagh et al., 1991) more condensate (which also partly blocks the air passage because water droplet accumulation at the junction of body and cap) leads to restricted gas exchange including more ethylene accumulation. About 35% light reduction through lid was also observed. The central downward slope in Growtek prevented this problem. Such accumulation in flat lids also caused higher fungal contamination as explained later in this section.

Cultures in both glass jars and Life Guard bioreactors took longer times than Growtek (116-142%; Table 2) in reaching a GI 2.0 for hairy root and multiple shoots, indicating a better physico-chemical microenvironment inside Growtek. Hairy root cultures in suspended agitated conditions in conventional bioreactors create rheological problems and more root injury (Curtis and Emery, 1993). Fragmentation of hairy roots in air-lift or stirred tank bioreactors reduce productivity, and even the metabolite profile during secondary metabolites production (Takayama, personal communication). The shear stress management for high-value product formation in hairy roots requires considerable attention (Curtis and Emery, 1993; Sharp and Doran, 2001). The possibility of using Growtek in both static and agitated modes simulates features of both gelled and liquid medium systems. This unique combination can be fruitfully utilized for laboratory-scale studies on secondary metabolite biosynthesis in hairy roots and for better somatic embryogenesis (Table 4).

The literature clearly records the problems encountered with culture contamination from bacteria (Maes et al., 1998; Cassells, 1991). Culture contamination from air-borne fungal spores is an especially serious problem in tropical and semi-tropical climate (particularly during the rainy season). This problem occurs during incubation on illuminated, but uncooled, shelves owing to drainage of along the edge of the medium along the inner wall. The effectiveness of five different vessels (Table 3) shows the order: Growtek >glass jar> Magenta/Life Guard/ Phytacon.

It is obvious that more contamination occurred in vessels with push-fit type lids (providing straight air passage between the inside and outside) having flat lids. The condensate trickles down periodically sucking in contaminated air. The better result in Growtek compared with glass jars is due to a coarser thread area in the former (Figure 1 B). Higher depth of thread rim in Growtek favours proper exchange, thereby better shoot and root health.

The perforated explant-holder (Figure 1 A, a) permits the free access of nutrient media to tissue surfaces, without sinking the latter, but perforations are small enough to prevent root entry. This surface growth of roots helps the easy and speedy transfer of rooted plantlets without injury (Table 4). This in turn offers a greater greenhouse and field survival rate for plantlets (Figure 2 C, D). It has been observed earlier that healthy and uninjured roots lead to successful field survival of plantlets (Gangopadhyay et al., 2002;

Bhattacharya et al., 1994). Root health is seriously affected by ethylene accumulation in vessels with inadequate gas exchange (Zobayed et al., 2001; de Klerk, 2001). The near 100% field survival of pineapple plantlets may be correlated with both healthy, uninjured roots.

It is clear, therefore, that Growtek is an effective bioreactor for many aspects of propagation of plant cells and tissues. It is also cheaper than other commonly used apparatus meant for the use of liquid media (Table 5). Apart from the cost, other culture vessels may require additional expenses for operation (e.g., wetting agent in Life Guard; air delivery system in TIS). The superiority of any of these available bioreactors will be dependent on eventual cost-effectiveness in mass cloning and convenience of operation without compromising the quality of the plantlets.

In conclusion, this article reports the usefulness of Growtek in terms of enhanced multiplication rates, reduced bioreactor costs, saving in incubation time, the minimisation of contamination and plantlet transfer without root injury. Experiments continue to be conducted in our institute to use Growtek for *in vitro* molecular pharming, production of secondary metabolites, bioremediation, solid-state fungal cultivation and aseptic seed germination.

Acknowledgements

The author acknowledges the useful experimental assistance provided by Mr. P. Saha, Mr. S. De and Mr. A. Pradhan. Help from STEP, IIT-Kharagpur, in the field trial is thankfully recorded.

References

Bhattacharya PS, Dey S, Das N & Bhattacharyya BC (1990) Rapid mass propagation of *Chrysanthemum morifolium* by callus derived from stem and leaf explants. Plant Cell Rep. 9: 439-443

Bhattacharya P, Dey S & Bhattacharyya BC (1994) Use of low-cost gelling agents and support matrices for industrial scale plant tissue culture. Plant Cell Tiss. Org. Cult. 37: 15-23

Cassels AC (1991) Problems in tissue culture: contamination. In: PC Debergh and RH Zimmermann (eds) Micropropagation Technology and Application (pp. 31-44). Kluwer Academic Publishers, Dordrecht

Curtis WR & Emery AH (1993) Plant cell suspension culture rheology. Biotechnol. Bioengng. 42: 520-526

Das Susobhan, Das Surajit, Pal S, Mujib A, Sahoo SS, Dey S, Ponde NR & Das Gupta S (1999) A novel process for rapid mass propagation of *Santalum album* L. in liquid media and bioreactor. In: Giberti G et al. (eds) Agricultural Production, Post Harvest Techniques, Biotechnology (proc, WOCMAP2) Acta Hort. 522: 281-286

Debergh P (1983) Effects of agar brand and concentration on the tissue culture medium. Physiol. Plant. 59: 270-276

Dey S (2001) Mass cloning of *Santalum album* L. through somatic embryogenesis: scale up in bioreactor. Sandalwood Research Newsletter 13: 1-3

Doran PM (2000) Foreign protein production in plant tissue cultures. Current Opinion in Biotechnology 11: 199-204

Escalona M, Lorenzo JC, Gonzalez B, Daquinta M, Gonzalez JL, Desjardins Y & Borroto CG (1999) Pineapple (*Ananas comosus* L. Merr.) micropropagation in temporary immersion systems. Plant Cell Rep. 18: 743-748

Etienne H, Lartaud M, Michaux-Ferriere N, Carron MP, Berthouly M & Teisson C (1997) Improvement of somatic embryogenesis in *Hevea Brasiliensis* (Mull. Arg.) using the temporary immersion technique. *In Vitro* Cell. Dev. Biol. 33: 81-87

Facchini PJ & Di Cosmo F (1991) Plant cell bioreactor for production of protoberberine alkaloids from immobilized *Thalictrum rugosum* cultures. Biotechnol. Bioengineering 37: 397-403

Gangopadhyay G, Das S, Mitra SK, Poddar R, Modak BK & Mukherjee KK (2002) Enhanced rate of multiplication and rooting through the use of coir in aseptic liquid culture media. Plant Cell, Tiss. Org. Cult. 68: 301-310

Goldstein WE (1999) Economic considerations for food ingredients produced by plant cell and tissue culture. In: Fu et al. (eds) Plant Cell and Tissue Culture for the Production of Food Ingredients (pp. 195-213). Kluwer Academic/Plenum Publishers, New York

Gupta PK, Pullman G, Timmis R, Kreitinger M, Carlson WC, Grob J & Welty E (1993) Forestry in 21st century: the biotechnology of somatic embryogenesis. Biotechnology 11: 454-459

Henderson WE & Kinnersley AM (1988) Corn starch as an alternative gelling agent for plant tissue culture. Plant Cell, Tiss. Org. Cult. 15: 17-22

Hunter CS & Kilby NJ (1999) Betalains: their accumulation and release *in vitro*. In: Robert D Hall (ed.) Plant Cell Culture Protocols. Methods in Molecular Biology, Vol. 111 (pp. 403-410), Humana Press Wageningen

Hvoslef-Eide AK & Melby TI (2000) Summary of results from bioreactor experiments with *Cyclamen* propagation via somatic embryogenesis. COST 843 Working Group 2 Meeting, Advanced Propagation Techniques (pp. 18-20). Tampere, Finland, July 7-10

Jimenez E, Perez N, de Feria M, Barbon R, Capote A, Chavez M & Quiala E (1999) Improved production of potato microtuber in a temporary immersion system. Plant Cell, Tiss. Org. Cult. 59: 19-23

Kavanagh K, Drew AP & Maynard C (1991) The effect of the culture vessel on micropropagation. In: Bajaj YPS (ed) Biotechnology in Agriculture and Forestry, High-Tech and Micropropagation- I, Vol. 17 (pp. 202-211), Springer-Verlag, Berlin, Heidelberg

Klerk Geert-Jan de (2001) Rooting of microcuttings: theory and practice. *In Vitro* Cell. Dev. Biol. (Plant) 37(3): p. 19 (Part-II)

Lakshmi Sita G, Singh R & Iyer CPA (1974) Plantlets through shoot tip cultures in pineapple. Curr. Sci. 43: 724-725

Maes M, Crepel C, Werbrouck S & Debergh P (1998) Perspectives for a DNA–based detection of bacterial contamination in micropropagated plant tissue. Plant Tissue Culture and Biotechnology 4: 49-56

Meyer JE, Pepin MF & Smith MAL (2002) Anthocyanin production from *Vaccinium pahalae*: limitations of physical microenvironment. Journal of Biotechnology 93: 45-57

Murashige T & Skoog F (1962) A revised medium for rapid growth and bioassays with tobacco tissue cultures. Physiol. Plant. 15: 473-497

Peeters K, Wilde CD, Jaeger GD, Angenon G & Depicker A (2001) Production of antibodies and antibody fragments in plants. Vaccine 19: 2756-2761

Preil W (1991) Application of bioreactors in plant propagation. In: Debergh PC & Zimmermann RH (eds) Micopropagation: Technology and Application (pp. 425-445). Kluwer Academic Publishers, Dordrecht

Sharp JM & Doran PM (2001) Strategies for enhancing monoclonal antibody accummulation in plant cell and organ cultues. Biotechnol. Prog., 17: 979-992

Smith DR (1997) The role of *in vitro* methods in pine plantation establishment: the lesson from New Zealand. Plant Tissue Culture and Biotechnology 3: 63-73

Smith MAL & Spomer A (1995) Vessels, gels, liquid media and support systems. In: Aitken-Christie J, Kozai T & Smith L (eds) Automation and Environmental Control in Plant Tissue Culture (pp. 371-404). Kluwer Academic Publishers, Dordrecht

Soneji JR, Rao PS & Mhatre M (2002) Germination of synthetic seeds of pineapple (*Ananas comosus* L. Merr.)Plant Cell Rep. 20: 891-894

Sorvari S (1986) Comparison of anther cultures of barley cultivars in barley–starch and agar gelatinized media. Ann. Agric. Fenn. 25: 249-254

Stoger E, Sack M, Fischer R & Christou P (2002) Plantibodies: applications, advantages and bottlenecks. Current Opinion in Biotechnology. 13: 161-166

Sutton BCS & Polonendo (1999) Commercialisation of plant somatic embryogenesis. In: Jain SM, Gupta PK & Newton RJ (eds) Somatic Embryogenesis in Woody Plants (Forestry Sciences), Vol. 4 (pp. 263-291). Kluwer Academic Publishers, Dordrecht/Boston/London

Takayama S (1991) Mass propagation of plants through shake–and bioreactor–culture techniques. In: Bajaj YPS (ed) Biotechnology in Agriculture and Forestry, High-Tech and Micropropagation- I, Vol. 17 (pp. 495-515). Springer-Verlag, Berlin, Heidelberg

Tisserat B (1991) Automated systems. In: Y P S Bajaj (ed) Biotechnology in Agriculture Forestry, High-Tech and Micropropagation, Vol. 17 (pp. 420-431). Springer-Verlag, Berlin, Heidelberg,

Vasil IK (1994) Automated plant propagation. Plant Cell, Tiss. Org. Cult. 39: 105-108

Zepeda C & Sagawa V (1981) *In vitro* propagation of pineapple. Hort. Sci. 16: 495-497

Zobayed SMA, Afreen F & Kozai T (2001) Physiology of *Eucalyptus* plantlets grown photo-autotrophically in a scaled–up vessel. *In Vitro* Cell. Dev. Biol (Plant) 37: 807-813

Chapter 9

Multiplication of *Chrysanthemum* shoots in bioreactors as affected by culture method and inoculation density of single node stems

Eun-Joo Hahn & Kee-Yoeup Paek*

Research Center for the Development of Advanced Horticultural Technology, Chungbuk National University, Cheongju, Chungbuk 361-763, Korea.

**(Corresponding author) Phone: +82 43 266 3245; Fax: +82 43 272 5369;*

E-mail: ejhahn@chungbuk.ac.kr

Abstract: Single node cuttings (1 cm in length) of *Chrysanthemum* were cultured on gelled and liquid media to compare shoot multiplication efficiency. Liquid culture resulted in greater fresh weight, dry weight, shoot length and leaf area compared to gelled culture. Shoots from liquid culture grew vigorously without hyperhydricity, showing 100% *ex vitro* survival. To determine optimal inoculation density of single nodes in a bioreactor, different numbers of single nodes (20 or 40 or 60 or 80) were placed into a 10-litre column-type bioreactor. Shoot length was greatest at the 80-node inoculation, with the least number of branches, indicating the best inoculation density tested for shoot multiplication in bioreactors. In the final experiment, single-node cuttings in bioreactors were treated with three different culture systems: ebb and flood, deep flow technique (DFT) culture and immersion. Results indicated that the DFT culture led to the greatest fresh weight, shoot length and leaf area, followed by the ebb and flood culture, while the immersion culture suppressed shoot multiplication due to the lack of oxygen and the high water potential. Our results suggested the possibility of large-scale production of *Chrysanthemum* shoots in bioreactors.

Key words: deep flow technique, ebb and flood culture, immersion culture, gelled culture

Abbreviations: DFT – deep flow technique; DW – dry weight; FW – fresh weight; PPF - photosynthetic photon flux; vvm – volume of gas per volume of liquid per minute

A.K. Hvoslef-Eide and W. Preil (eds.), Liquid Culture Systems for in vitro Plant Propagation, 143–153.

1. Introduction

Chrysanthemum is one of the most popular cut flowers, cultivated in the largest area in South Korea (Hahn, 1998). Plant multiplication has mostly been *via* cuttings, but this method has a low multiplication rate and often the plants are of low quality. Because cuttings are obtained repeatedly from mother plants, they may be subjected to any virus infection and degeneration, thereby increasing production costs (Hahn et al., 1998; Kim, 2001).

These problems have been solved by applying micropropagation methods, which are routinely applied to the clonal propagation of a variety of horticultural plants including *Chrysanthemum* (Ben-Jaacov and Langhans, 1972; Earle and Langhans, 1974a, 1974b; Pierik, 1988). Once disease-free shoots are obtained through shoot tip cultures, a large number of nodal cuttings can be obtained within a short period, thus showing the advantages of high multiplication rate and plant uniformity compared with conventional vegetative propagation *via* cuttings (Chu, 1992; Debergh and Read, 1991).

Although there have been advances in micropropagation systems through improvement of the microenvironments (Kozai et al., 1992), there are still problems to be solved. In general, micropropagation has been done through conventional gelled culture systems using small-scale culture vessels where sucrose concentrations and relative humidity are extremely high, and CO_2 concentrations and photosynthetic photon fluxes (PPF) are low. As a result, conventional micropropagation requires is costly and has a relatively low multiplication efficiency (Bi et al., 1997).

To increase multiplication efficiency, micropropagation should be scaled-up. Bioreactor systems have been introduced for mass propagation of horticultural plants (Levin and Vasil, 1989; Takahashi et al., 1992) and have proved their potential for large-scale micropropagation. They are highly effective for mass production of valuable plants because so many propagules can be obtained at a time. There are reports of mass propagation of some horticultural plants using bioreactor systems, for example for *Phalaenopsis* (Park et al., 2000), oriental lily (Lian et al., 2003), garlic (Kim et al., 2003) and potato (Piao et al., 2003). However bioreactor systems are still mostly used for cell and root cultures for secondary metabolite production (Paek et al., 2001; Rittershaus et al., 1989).

To establish large-scale micropropagation of *Chrysanthemum* through a bioreactor system, we first compared multiplication rate of *Chrysanthemum* shoots between gelled and liquid cultures. Based on those results, shoots were propagated in bioreactors using variations of culture method and inoculation density to determine the optimal conditions.

2. Materials and methods

2.1 *Plant material*

Axillary buds of *Chrysanthemum* (*Dendranthema grandiflorum* Kitam 'Cheonsu') were washed with running tap water and surface-sterilized in a 1.0% sodium hypochlorite solution for 30 seconds followed by rinsing 2-3 times with sterilized distilled water. Shoot tips, with two leaf primordia, were excised from the axillary buds and placed in 100 ml Erlenmeyer flasks containing 20 ml MS (Murashige and Skoog, 1962) gelled media supplemented with 0.5 mg l^{-1} benzyladenine and 30 g l^{-1} sucrose to induce adventitious shoots. The pH of the medium was adjusted to 5.8 before autoclaving. After 5 weeks of culture, the adventitious shoots were cut into 1.5 cm-length single nodes and propagated in polypropylene growth vessels (107 × 107 × 97 mm, Osmotek, Israel) containing 50 ml MS basal media supplemented with 30 g l^{-1} sucrose. Cultures were maintained at 25/18°C (day and night) with 70 μmol m^{-2} s^{-1} PPF during the 16-h photoperiod.

2.2 *Gelled and liquid cultures*

Ten single node explants were placed into a 900 ml square-type glass vessel containing 200 ml gelled or liquid MS medium supplemented with 30 g l^{-1} sucrose. To gel the medium, 2.4 g l^{-1} gelrite was used. In liquid culture, a plastic net was placed inside the culture vessel to support the explants. Gas-permeable microporous filters (Mill-Seal, Millipore, Tokyo; pore size 0.5 μm) were attached on top of the culture vessels and the air exchange rate in the culture vessel was controlled to 0.1 vvm. Cultures were maintained for 5 weeks as above. Fresh weight, dry weight, shoot length, stem diameter, number of leaves, leaf area, water potential and chlorophyll content of leaves were recorded after 5 weeks. Leaf area and chlorophyll content of fully-developed leaves were measured with a leaf area meter (Skye Co, UK) and a chlorophyll meter (SPAD-502, Minolta, Japan). Leaf water potential was recorded for the fourth–youngest, fully-developed leaves using a water potential measuring instrument (Tru Psi, Decagon, USA). After measurements, shoots were cut into 1.5 cm-length single nodes, transplanted to plug trays (2.7 × 2.7 × 4 cm; 128 cells per tray) filled with sand and grown for 4 weeks in greenhouse conditions to investigate the number of plants that survived.

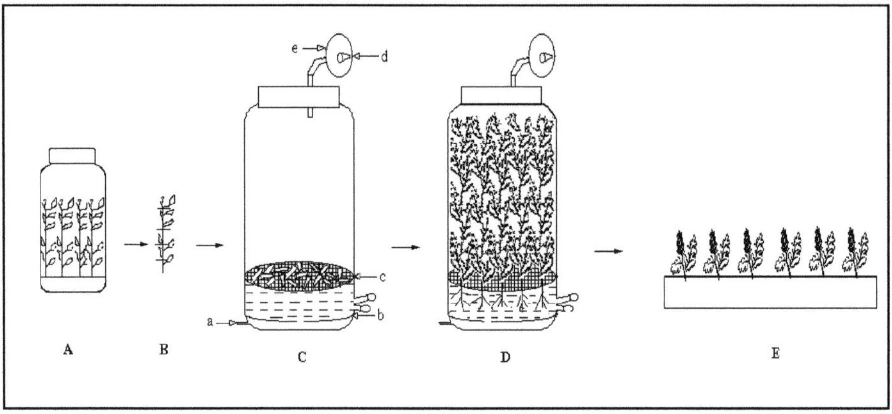

Figure 1: Schematic diagram of multiplication of *Chrysanthemum* shoots in bioreactors:
A: Shoots induced from meristem culture; B: Single node stems (1.5 cm in length);
C: Bioreactor culture of single node stems (a: Air inlet, b: Sparger, c: Plastic net to support explants, d: Air outlet, e: Membrane filter); D: Shoot multiplication in a bioreactor;
E: *Ex vitro* rooting of single node cuttings.

2.3 *Bioreactor culture: effect of the number of single nodes inoculated into a bioreactor*

Plant material, culture medium, air temperature, PPF, the number of air exchanges and CO_2 concentration inside the vessels were maintained the same as those in liquid culture (above). Groups of single node cuttings (20, 40, 60, or 80) were placed into 10-litre column-type bioreactors containing 4 litres MS liquid medium supplemented with 30 g l^{-1} sucrose. Cultures were maintained for 12 weeks, followed by measurements of fresh weight, plant height, stem diameter, number of branches and leaves. The process of bioreactor culture of *Chrysanthemum* shoots is described in figure 1.

2.4 *Effect of culture method*

Five-litre column-type bioreactors were used for the experiment. Forty single nodes were placed into a bioreactor and cultured for 10 weeks by three different ways: ebb and flood culture, deep flow technique (DFT) culture and immersion culture. For the ebb and flood culture and the DFT culture, a plastic net was placed into the bioreactor to suspend the explants and the medium was supplied to the underneath of them. In the ebb and flood culture, 1.5 litre MS liquid medium containing 30 g l^{-1} sucrose was supplied for 30 min at 2-hour intervals for the first 2 weeks and at 1-hour intervals from the third week until the end of culture. The medium was subirrigated and circulated using a timer and a solenoid valve (Seki Co., Seoul, Korea). In DFT culture, the medium was continuously supplied to the underneath of the explants. In the immersion culture, the explants were placed on the base of the bioreactor (without a net) and cultured in the medium. CO_2 concentration inside the bioreactor was recorded every two weeks during culture period using: analysis was by gas chromatography (HP 6890, Hewlett Packard, Wilmington, USA). Measurements were made after 10 weeks, of fresh weight, shoot length, leaf number, leaf area, chlorophyll content, and water potential.

3. Results and discussion

3.1 Gelled and liquid culture

There was a significant difference in growth of *Chrysanthemum* plantlets grown on gelled and liquid media. Total fresh weight (2292 mg per plantlet) and total dry weight (185 mg per plantlet) in liquid culture were more than double those in gelled culture (929 mg per plantlet FW and 72 mg per plantlet DW) (Table 1). Liquid culture also resulted in greater shoot length and leaf area compared with gelled culture. Number of leaves, chlorophyll content and stem diameter were also greater in liquid culture. Furthermore, shoots from liquid cultures grew vigorously without hyperhydricity, showing similar leaf water potentials to those in gelled culture. Survival, *ex vitro,* was 100% regardless of whether the culture medium was gelled or liquid (Table 1).

The absorption of nutrients from the medium was accelerated when the nutrients were supplied in liquid medium (Takayama and Akita, 1994); growth was greater compared with that in gelled medium. Avila et al. (1998) also reported leaf areas that were double and shoots with greater length nodes for potato cultures in liquid media, as we found in our results. Hyperhydricity is generally known to occur more frequently in liquid media due to the high water potential of leaves (Paek and Han, 1989). In our results, however, leaf water potential in liquid and gelled media was not different, with slightly lower values of water potential and no hyperhydricity (data not shown) in liquid cultures. This result indicated that the air supply (0.1 vvm) decreased the relative humidity inside the culture vessel (Hahn and Paek, 2001) and the leaf water potential, preventing hyperhydricity. There have been similar results for *Rehmannia glutinosa* and several horticultural crops (Cui et al., 2000; Kubota et al., 1997). These results suggest that bioreactor culture may be possible for the large-scale shoot multiplication of *Chrysanthemum*.

Table 1: Growth of *Chrysanthemum* plantlets after 5 weeks of gelled and liquid culture

		Gelled culture	Liquid culture
Fresh weight (mg per plantlet)[y]	Shoot	844 ± 21	1986 ± 226
	Root	86 ± 14	306 ± 36
Dry weight (mg per plantlet)[y]	Shoot	66 ± 4	160 ± 14
	Root	6 ± 1	26 ± 2
Shoot length (cm)		4.8 ± 0.1	8.3 ± 0.4
No. leaves/plantlet		11.2± 0.2	13.4 ± 0.5
Leaf area (cm^2 per plantlet)		20.4 ± 1.5	49.9 ± 1.3
Chlorophyll content[z]		40.1 ± 1.4	42.3 ± 4.5
Stem diameter (mm)		2.1 ± 0.02	2.5 ± 0.02
Water potential (MPa)		-2.40 ± 0.5	- 2.84 ± 0.7
Ex vitro survival (%)[x]		100	100

[z]SPAD value.
[y]Each value represents mean±standard error of 3 replicate vessels each with 5 plantlets recorded after five weeks of culture.
[x]100 single nodes produced *in vitro* per number of single nodes that grow to 3 cm-length transplants × 100.

Table 2: Effects, after 12 weeks growth in liquid medium, of the number of single nodes inoculated into a 10-litre bioreactor on their fresh weight, shoot length, stem diameter, number of branches and leaves and *ex vitro* survival

Number of single nodes inoculated	Fresh weight (g)[x]	Stem length (cm)	Stem diameter (mm)	No. branches per plantlet	No. leaves per plantlet	*Ex vitro* survival (%)[y]
20	7.84 ± 1.1	23.4 ± 2.3	2.5 ± 0.05	8.33 ± 1.5	59.2 ± 6.8	100
40	9.67 ± 1.7	26.7 ± 1.5	2.4 ± 0.11	6.61 ± 0.8	63.8 ± 5.2	100
60	9.54 ± 2.1	26.9 ± 2.4	2.2 ± 0.09	6.58 ± 1.0	62.7 ± 6.4	100
80	8.44 ± 0.9	28.3 ± 2.0	2.2 ± 0.07	4.52 ± 0.9	54.1 ± 4.1	100

[x]Each value represents mean±standard error of 2 replicate vessels each with 10 plantlets recorded after 12 weeks of culture.
[y]200 single nodes produced *in vitro* per number of single nodes that grow to 3 cm-length transplants × 100.

Table 3: Effects of culture system on fresh weight, stem length, stem diameter, leaf number, leaf area and leaf water potential of *Chrysanthemum* plantlets after 10 weeks of bioreactor culture

Liquid culture system	Fresh weight (g per plantlet)	Stem length (cm)	Stem diameter (mm)	No. leaves per plantlet	Leaf area (cm^2 per plantlet)	Water potential (MPa)
Ebb and flood	1.73 b[z]	12.8 b	1.9 b	17.5 a	36.9 b	- 3.29
DFT	2.63 a	15.8 a	2.2 a	17.0 a	45.3 a	- 2.91
Immersion	1.54 b	11.9 b	1.9 b	14.0 b	34.5 b	- 3.76

[z]Mean separation within columns by Duncan's multiplication range test, 5% level.

3.2 *Bioreactor culture: the effect of the number of single node inoculum*

Shoot multiplication in early stage was enhanced by high inoculation density (80 single-node cuttings), but became similar among 40, 60, and 80-node inoculations as the cultures proceeded (data not shown). Shoot lengths increased with increasing numbers of single nodes, but stem diameters and the numbers of branches were reduced (Table 2). Shoot length was greatest with 80 nodes per inoculation, but there was least branching (branches are not considered to be shoots and require much longer time for acclimatization, with less plantlet survival due to shoot fragility), indicating that the largest number of single nodes for rooting would be obtained from this treatment. Twenty-node inoculation resulted in least fresh weight and shoot length but with greater numbers of branches, which resulted in smaller number of single nodes for rooting, compared to other treatments (Table 2).

Niu and Kozai (1997) reported the decrease of photosynthesis and shoot growth with higher numbers of explants when potatoes were cultured *in vitro*, in contrast to our results. This could be explained by the difference in the volume of the culture vessels between the experiments. Under ventilation, CO_2 concentration inside a culture vessel gets higher with the increase of the vessel volume. (Paek et al., 2001; Li et al., 2001). Our result suggested that a high inoculation density did not inhibit shoot multiplication if a large volume of culture vessel was used. In this regard, a 80-node inoculation was shown to be effective in obtaining large numbers of single-node cuttings, which survived 100% after transplanting.

3.3 *The effect of culture method*

DFT culture led to greatest fresh weight, shoot length and leaf area but the differences were minor compared to plant development in the ebb and flood culture. On the other hand, growth was inhibited in immersion culture (Table 3). The negative effect of immersion culture on plantlet growth was mainly due to the lack of oxygen and an high water potential, because the explants were immersed into the medium for the entire culture period (Nguyen and Kozai, 1998; Ziv, 1991). As a result, immersion culture suppressed photosynthetic activity and increased the number of vitrified leaves (data not shown). Difference in plantlet growth among the three culture systems was associated with CO_2 concentrations inside the culture vessels during the light periods. CO_2 concentration in each bioreactor was approximately 1500 μmol mol^{-1} at the initial stage of culture. CO_2 concentration in DFT culture decreased by 400 μmol mol^{-1} after 2 weeks of culture; in ebb and flood culture CO_2 concentration decreased to 750 μmol mol^{-1}, after which remained constant up to week ten. However, CO_2 concentrations >1400 μmol mol^{-1}, was recorded after 2 weeks in immersion culture, then the CO_2 concentration gradually decreased (Figure 2). The results indicated that the explants in DFT and ebb and flood culture systems started to photosynthesize from their initial stages, but photosynthesis in immersion culture was severely suppressed during the early stage of culture. Photosynthesis of plantlets *in vitro* and CO_2 concentration inside the culture vessel are closely related each other, as reported elsewhere (Fujiwara et al., 1987; Kubota et al., 1997; Kim et al., 2003).

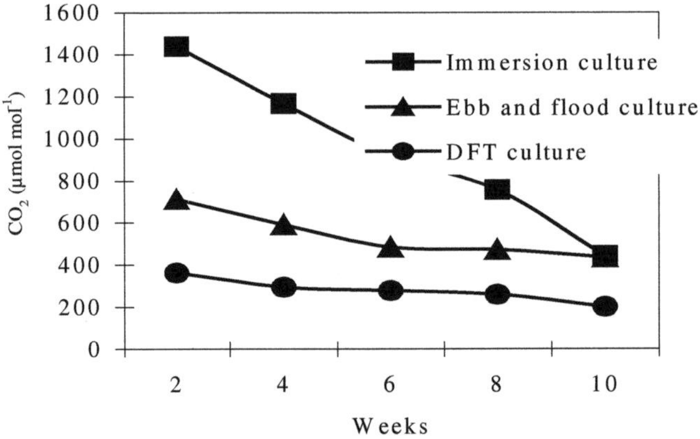

Figure 2: Changes of CO_2 concentration inside 5-litre column-type bioreactors during multiplication of *Chrysanthemum* shoots.

4. Conclusion

Multiplication of horticultural plants through bioreactor systems is, initially, costly, but long-term, it can increase multiplication efficiency remarkably by applying large-scale culture vessels and optimizing the physical and chemical environments. Since plantlets in bioreactors are grown in optimized culture conditions, proliferation rates are much greater than those in conventional gelled cultures. In addition, 100% survival *ex vitro* can be achieved with faster growth after transplantation. Our results suggested the possibility of large-scale production of *Chrysanthemum* shoots through bioreactor systems. Further studies are needed to optimise culture conditions, such as medium composition, rate and frequency of air exchanges and CO_2 enrichment to maximize multiplication efficiency.

Acknowledgment

This work was supported in part by the grant from the Research Center for the Development of Advanced Horticultural Technology funded by the Korea Science and Engineering Foundation and the Ministry of Agriculture and Forestry.

References

Avila A, Pereyra SM & Arguello JA (1998) Nitrogen concentration and proportion of NH_4^+-N affect potato cultivar reaponse in solid and liquid media. Hort. Science 33: 336-338

Ben-Jaacov J, Langhans RW (1972) Rapid multiplication of *Chrysanthemum* plants by stem-tip proliferation. HortScience 7: 289-290

Bi J, Ou F, Wang Y, Hao Z, & Liu C (1997) Bioreactors for plant micropropagation. 4th Asia-Pacific Biochemical Engineering Conference. Vol. 2 (pp. 687-690)

Chu IYE (1992) Perspective of micropropagation industry. In: Kubota K & Kozai T (ed) Transplant Production Systems (pp. 137-150). Kluwer Academic Publishers, Dordrecht

Cui YC, Hahn EJ, Kozai T & Paek KY (2000) Number of air exchanges, sucrose concentration, photosynthetic photon flux, and differences in photoperiod and dark period temperatures affect growth of *Rehmannia glutinosa* plantlets *in vitro*. Plant Cell, Tiss. Org. Cult. 62: 219-226

Debergh PC & Read PE (1991) Micropropagation. In: Debergh PC & Zimmerman RH (eds) Micropropagation, Technology and Application (pp. 1-13). Kluwer Academic Publishers, Dordrecht

Earle ED & Langhans RW (1974a) Propagation of *Chrysanthemum in vitro*: I. Multiple plantlets from shoot tips and the establishment of tissue cultures. J. Amer. Soc. Hort. Sci. 99: 128-131

Earle ED & Langhans RW (1974b) Propagation of *Chrysanthemum in vitro*: ɪɪ. Production, growth, and flowering of plantlets from tissue cultures. J. Amer. Soc. Hort. Sci. 99: 352-358

Fujiwara K, Kozai T & Watanabe I (1987) Fundamental studies of environment in plant tissue culture vessels. (3) Measurements of carbon dioxide gas concentration in closed vessels containing tissue cultured plantlets and estimates of net photosynthetic rates of the plantlets. J. Agr. Meteorol. 43: 21-30

Hahn EJ (1998) The establishment of microponic system in a plant factory by controlling environments and acclimatization method for mass production of *in vitro* produced *Chrysanthemum* plantlets. Ph.D Diss. Univ. of Seoul, Seoul, Korea

Hahn EJ, Bae JH & Lee YB (1998) Growth and leaf surface characteristics of *Chrysanthemum* plantlets in micropropagation and microponic system. J. Kor. Soc. Hort. Sci. 39: 838-842

Hahn EJ & Paek KY (2001) High photosynthetic photon flux and high CO_2 concentration under increased number of air exchanges promote growth and photosynthesis of four kinds of orchid plantlets *in vitro*. *In Vitro* Cell. Dev. Biol. 37: 678-682

Kozai T, Fujiwara K, Hayashi M & Aitken CJ (1992) The *in vitro* environment and its control in micropropagation. In: Kurata K and Kozai T (ed) Transplant production systems (pp. 247-282). Kluwer Academic Publishers, Dordrecht

Kim EK, Hahn EJ, Murthy HN & Paek KY (2003) High frequency of shoot multiplication and bulblet formation of garlic in liquid cultures. Plant Cell, Tiss. Org. Cult. 73: 231-236

Kubota C, Fujiwara K, Kitaya Y & Kozai T (1997) Recent advances in environment control in micropropagation. In: Goto E (ed) Plant Production in Closed Ecosystems (pp. 153-169). Kluwer Academic Publishers, Dordrecht

Levin R & Vasil IK (1989) An integrated and automated tissue culture system for mass propagation of plants. *In Vitro* Cell. Dev. Biol. 25: 21-27

Li RW, Murthy HN, Kim SK & Paek YP (2001) CO_2-enrichment and photosynthetic photon flux affect the growth of *in vitro*-cultured apple plantlets. J. Plant Biology 44: 87-91

Lian ML, Chakrabarty D & Paek KY (2003) Growth of *Lilium* oriental hybrid 'Casablanca' bulblet using bioreactor culture. Sci. Hort. 97: 41-48

Murashige T & Skoog F (1962) A revised medium for rapid growth and bioassays with tobacco tissue culture. Physiol. Plant. 15: 473-497

Nguyen QT & Kozai T (1998) Environmental effects on the growth of plantlets in micropropagation. Environ. Cont. Biol. 36: 59-75

Niu G & Kozai T (1997) Simulation of the growth of potato plantlets cultured photoautotrophically *in vitro*. Trans. ASAE. 40: 255-260

Paek KY, Hahn EJ & Son SH (2001) Application of bioreactors large-scale micropropagation systems of plants. *In Vitro* Cell. Dev. Biol. 37: 149-157

Paek KY & Han BH (1989) Physiological, biochemical and morphological characteristics of vitrified shoot regenerated *in vitro*. J. Kor. Soc. Plant Tiss. Cult. 18: 151-162

Park SY, Murthy HN & Paek KY (2000) Mass multiplication of protocorm-like bodies (PLBs) using bioreactor system and subsequent plant regeneration in *Phalaenopsis*. Plant Cell, Tiss. Org. Cult. 63: 67-72

Piao XC, Charkrabarty D, Hahn EJ & Paek KY (2003) A simple method for mass production of potato microtubers using a bioreactor system. Current Sci. 8: 1129-1232

Pierik RLM (1988) *In vitro* culture of higher plants as a tool in the propagation of horticultural crops. Acta Hort. 226: 25-40

Rittershaus E, Ulrich J, Weiss A & Westphal K (1989) Large scale industrial fermentation of plant cells: Experiences in cultivation of plant cells in a fermentation cascade up to a volume of 75,000L. BioEngineering. 5: 28-34

Takahashi S, Matsubara H, Yamagata H & Morimoto T (1992) Micropropagation of virus free bulblets of *Lilium longiflorum* by tank culture. 1. Development of liquid culture method and large-scale propagation. Acta Hort. 319: 83-88

Takayama S & Akita M (1994) The types of bioreactors used for shoots and embryos. Plant Cell, Tiss. Org. Cult. 39: 147-156

Ziv M (1991) Vitrification: Morphological and physiological disorders of *in vitro* plantlets. In: Debergh PC & Zimmerman RH (eds) Micropropagation, Technology and Application (pp. 45-69). Kluwer Academic Publishers, Dordrecht

Chapter 10

Control of growth and differentiation of bioreactor cultures of *Physcomitrella* by environmental parameters

Annette Hohe[1] & Ralf Reski[2]
[1]*Department of Plant Propagation, Institute of Vegetable and Ornamental Crops, Kuehnhaeuser Strasse 101, 99189 Kuehnhausen, Germany. E-mail: hohe@erfurt.igzev.de.*
[2]*Plant Biotechnology, Freiburg University, Schänzlestraße 1, 79104 Freiburg, Germany. E-mail: ralf.reski@biologie.uni-freiburg.de*

Abstract: The effect of physical and chemical environmental parameters on growth and differentiation of suspension cultures of the moss *Physcomitrella patens* in bioreactors was investigated. By supplementation of the aeration gas with 2 % (v/v) CO_2 as well as by continuous illumination, growth of this photoautotrophic growing batch culture was markedly enhanced, resulting in a doubling time of 1.2 d. The growth rate of semi continuously-growing bioreactor cultures was not affected by controlling the pH of the culture medium with set points at 4.5 or 7.0. However, growth of the culture at pH 7.0 resulted in increased caulonema development, thus showing a distinct effect on moss differentiation. The impact on research and plant biotechnological applications of the potential to control moss growth and differentiation by environmental parameters is discussed.

Key words: CO_2, light, moss, pH, photoautotrophic culture, suspension culture

Abbreviations: vvm - volume of gas per volume of liquid per minute

1. Introduction

During the last decade, the moss *Physcomitrella patens* has emerged as a model system for basic and applied plant sciences due to several characteristics which make it unique among lower and higher plants (Reski, 1998a). It is the only plant known to date showing high rates of homologous recombination in its nuclear DNA, allowing reverse genetics by gene targeting (Reski, 1998b; Reski, 1999; Schaefer, 2001). Compared to higher plants, genetic and phenotypic analysis is more straightforward due to the haploid status of the moss gametophyte. Thus, several functional genomics

A.K. Hvoslef-Eide and W. Preil (eds.), Liquid Culture Systems for in vitro Plant Propagation, 155–161.

projects have been started in order to analyse gene functions on a large scale (Nishiyama et al., 2001; Egener et al., 2002). Additionally, *Physcomitrella* is an ideal organism for physiological analyses because of its clearly separated and well-defined developmental stages (e.g. Bhatla et al., 2002). Moreover, it is used also in applied plant biotechnology as a production system for heterologous proteins (Reutter and Reski, 1996).

Thus, bioreactor cultures of *Physcomitrella* protonema suspensions have been established for several reasons: i) protonema suspensions are used for protoplast isolation for large scale transformation (Hohe et al., 2001; Hohe and Reski, 2002). ii) Bioreactor cultures represent a 'contained environment' for the production of heterologous compounds (Reutter and Reski, 1996). iii) Bioreactor culture offers a unique possibility for studying the effect of environmental conditions on growth and development (Hohe et al., 2002). This, in turn, is the basis for physiological studies as well as for the optimisation of culture conditions for the production of heterologous compounds.

In this paper we report on the effect of physical and chemical growth conditions, i.e. gas atmosphere, light intensity and pH, on the growth and development of bioreactor cultures of *Physcomitrella* protonema.

2. Materials and methods

Protonema suspension cultures of *Physcomitrella patens* were grown in a modified Knop medium containing 4.24 mmol $Ca(NO_3)_2$, 1.02 mmol KCl, 1.84 mmol KH_2PO_4, 3.36 mmol $MgSO_4$ and 0.045 mmol $FeSO_4$, pH 5.8 (Reski and Abel, 1985). For batch cultures in bioreactors the medium was additionally supplemented with 2.5 mmol ammonium tartrate.

Bioreactor cultures were done in glass vessels with a working volume of either 5 or 10 litres, respectively (Applikon, Schiedam, The Netherlands). Cultures were aerated with 0.3 vvm air or air supplemented with 2 % (v/v) CO_2 and stirred with a marine impeller running with 400 (10-litre vessel) or 500 rpm (5-litre vessel). Light intensity was 190 (10-litre vessel) or 120 µmol $s^{-1}m^{-2}$ (5-litre vessel) provided by fluorescent tubes (Philips TLD 25) either continuously or with a light/dark periodicity of 16/8 hours. Control of pH was by automatic titration of 0.5 N KOH or HCl (ADI 1030 control device, Applikon, Schiedam, The Netherlands).

Growth was determined by measuring the dry weight of the cultures; two 50 ml samples were taken per day and the cell material dried to constant weight at 105°C for 2 hours.

Culture was either done as a batch culture starting with approximately 30 mg l^{-1} dry weight, or as semi continuous culture, i.e. in a daily rhythm a certain amount of suspension was harvested and replaced by fresh medium in order to maintain a certain density of the suspension (determined by measurement of the dry weight). Here, the average dilution rate was 0.18 - 0.2 d^{-1}.

All experiments were executed in parallel bioreactor runs using identical inoculum and repeated at least twice.

3. Results

The effect of three environmental parameters on bioreactor cultures of *Physcomitrella* was investigated for two purposes - optimisation of culture growth, as well as analysing the effect of these parameters on moss development.

In 10-litre batch cultures the effect of light and CO_2 on the photoautotrophically growing moss cultures was studied. Cultures grown under standard conditions with 16 h light per day and aerated with air grew with a doubling time (t_d) of 2.3 d (Figure 1). By using continuous illumination t_d was reduced to 1.7 d. Supplementation of the aeration gas with 2 % (v/v) CO_2 resulted in a reduction of t_d to 1.3 d.

Using both, continuous illumination and CO_2-enriched aeration gas, the doubling time was reduced to 1.2 d (Figure 1): the growth rate was doubled compared to the control. The pH of the culture medium decreased throughout the culture period, probably due to ammonium uptake (Figure 1). This pH decrease was especially pronounced in cultures showing a high growth rate, thus reflecting higher nitrogen uptake of the more vigorously growing cultures.

Growing suspension cultures without pH-control in ammonium-free medium resulted in an average pH of 5.8 throughout a semi continuous bioreactor culture in 5-litre vessels (Figure 2). Control of the pH with set points of 7.0 and 4.5, respectively, had a pronounced effect on the growth rate during exponential growth (day 0 – 3) before the start of the semi continuous culture mode (Figure 2). Here the doubling time was 3 d for the culture without pH control and 2.8 d at pH 4.5, but reduced to 1.7 d at pH 7.0. However, the pH did not affect culture growth during the semi continuous culture mode: here the dilution rate was between 0.18 and 0.2 d^{-1} for all cultures corresponding to a doubling time of 3.9 to 3.5 d. In contrast, pH control markedly affected protonema differentiation: At pH 4.5 the cultures predominantly developed chloronema, whereas at pH 7.0 caulonema development was increased.

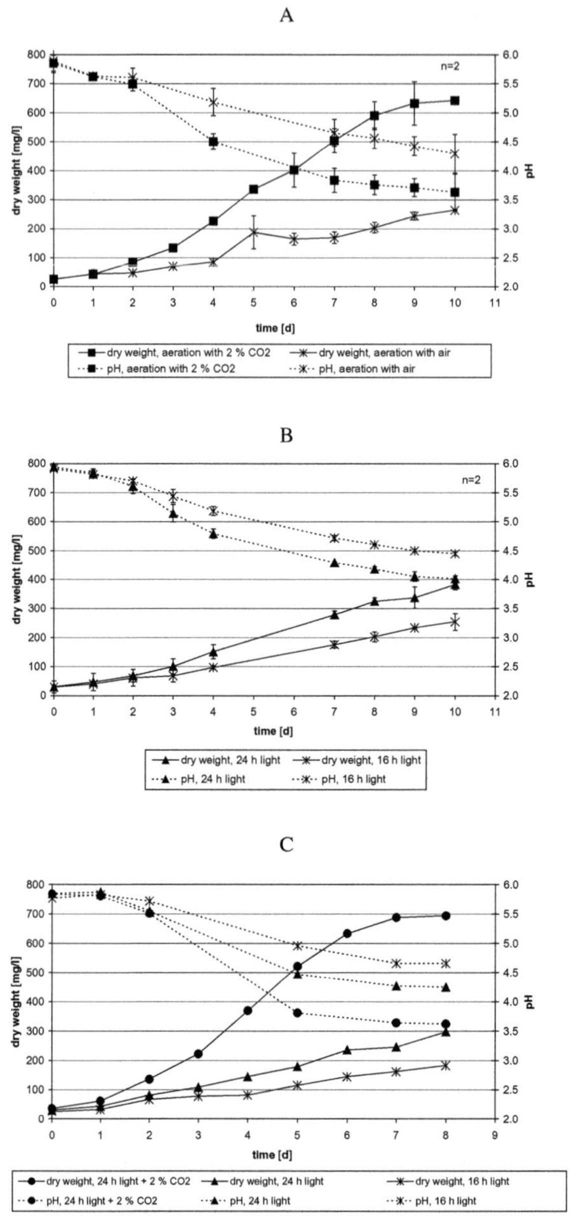

Figure 1: Effect of CO_2 supplementation of the aeration gas and continuous illumination on growth of photoautotrophically growing *Physcomitrella* in bioreactor batch cultures in ammonium-supplemented medium. A: comparison of aeration with air and air supplemented with 2 % (v/v) CO_2. B: comparison of growth in photoperiods of 16 and 24 h. C: growth curve obtained in culture aerated with CO_2-enriched air and in continuous illumination. Growth curves in one figure were obtained by running parallel bioreactor cultures using identical inoculum.

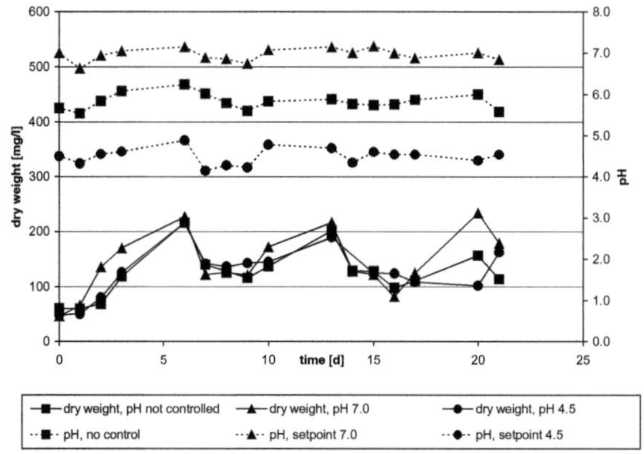

Figure 2: Semi continuous bioreactor cultures of *Physcomitrella* in ammonium-free medium growing at different pH values, (without pH control the pH value averaged pH 5.8 compared to growth in bioreactors with pH set points at pH 7.0 and 4.5).

4. Discussion

Physcomitrella is a very fast and stably growing plant suspension culture that can be grown in standard stirred tank glass bioreactors without difficulty. Moreover, the growth rate of the photoautotrophically growing culture was markedly increased by aeration with CO_2-enriched air and continuous illumination. Here the culture realised a doubling time of 1.3 d which is extraordinarily fast growing compared to suspension cultures of higher plants that usually grow with doubling times of 2-3 d (Scragg, 1995).

Controlling the pH value of the suspension culture resulted in a distinct effect on moss development with a pH of 7.0 promoting caulonema development. Transition from the chloronema to the caulonema stage was shown to be auxin dependent (Reski, 1998a). Moreover, more than 90 % of the total auxin of a *Physcomitrella* suspension culture accumulates in the medium (Reutter et al., 1998), so that the pH of the culture medium might well affect culture differentiation by interfering with auxin metabolism and uptake as happens in higher plants (Minocha, 1987).

Our results show a tight control of growth and differentiation of *Physcomitrella* by environmental conditions which opens a wide range of possibilities for further research and application in plant biotechnology. Optimisation of the growth rate can be used for high yields in the production of heterologous proteins by transgenic *Physcomitrella* cultures. In this respect, the semi continuous culture in our experiments is an important step towards continuous cultures. Since the growth rate during the semi continuous culture mode was not affected by the pH of the culture medium, the pH could be adjusted according to the requirements for the production of specific compounds, which is especially important if they are excreted and harvested from the medium.

On the other hand, environmental conditions can be used to control, precisely, moss differentiation, which offers unique possibilities for research in plant physiology. This is of special interest since chloronema and caulonema not only represent different developmental stages of moss protonema, but react differently to externally applied growth regulators (Reski and Abel, 1985) and represent cells arrested in different phases of the cell cycle (Reski, 1998a). Thus, the possibility of controlling differentiation by environmental parameters also presents new opportunities for studying hormone physiology and the plant cell cycle.

Acknowledgement

This work was performed in a joined project of Freiburg University and BASF Plant Science GmbH.

References

Bhatla SC, Kiessling J & Reski R (2002) Observation of polarity induction by cytochemical localization of phenylalkylamine-binding receptors in regenerating protoplasts of the moss *Physcomitrella patens*. Protoplasma 219: 99-105

Egener T, Granado J, Guitton MC, Hohe A, Holtorf H, Lucht JM, Rensing S, Schlink K, Schulte J, Schween G, Zimmermann S, Duwenig E, Rak B & Reski R (2002) High frequency of phenotypic deviations in *Physcomitrella patens* plants transformed with a gene-disruption library. BMC Plant Biology 2: 6

Hohe A & Reski R (2002) Optimisation of a bioreactor culture of the moss *Physcomitrella patens* for mass production of protoplasts. Plant Sci. 163: 69-74

Hohe A, Decker EL, Gorr G, Schween G & Reski R (2002) Tight control of growth and cell differentiation in photoautotrophically growing moss (*Physcomitrella patens*) bioreactor cultures. Plant Cell Rep. 20: 1135-1140

Hohe A, Schween G & Reski R (2001) Establishment of a semicontinuous bioreactor culture of *Physcomitrella patens* for mass production of protoplasts. Acta Hort. 560: 425-428

Minocha SC (1987) pH of the medium and the growth of cells in culture. In: Bonga JM & Durzan DJ (eds) Cell and Tissue Culture in Forestry Vol. 1 (pp. 125-141). Martinus Nijhoff Publishers, Dordrecht, Boston, Lancaster

Nishiyama T, Hiwatashi Y, Sakakibara K, Kato M & Hasebe M (2000) Tagged mutagenesis and gene-trap in the moss, *Physcomitrella patens* by shuttle mutagenesis. DNA Res. 7: 9-17

Reski R & Abel WO (1985) Induction of budding on chloronemata and caulonemata of the moss, *Physcomitrella patens*, using isopentenyladenine. Planta 165: 354-358

Reski R (1998a) Development, genetics and molecular biology of mosses. Bot. Acta 111: 1-15

Reski R (1998b) *Physcomitrella* and *Arabidopsis*: The David and Goliath of reverse genetics. Trends Plant Sci. 3: 209-210

Reski R (1999) Molecular genetics of *Physcomitrella*. Planta 208: 301-309

Reutter K & Reski R (1996) Production of a heterologous protein in bioreactor cultures of fully differentiated moss plants. Plant Tiss. Cult. Biotechnol. 2: 142-147

Reutter K, Atzorn R, Hadeler B, Schmülling T & Reski R (1998) Expression of the bacterial ipt gene in *Physcomitrella* rescues mutations in budding and plastid division. Planta 206: 196-203

Schaefer DG (2001) Gene targeting in *Physcomitrella patens*. Curr. Opin. Plant Biol. 4: 143-150

Scragg AH (1995) The problems associated with high biomass levels in plant cell suspensions. Plant Cell Tiss. Org. Cult. 43: 163-170

II. Temporary Immersion Systems

Chapter 11

Temporary immersion system: a new concept for use liquid medium in mass propagation

M. Berthouly[1] & H. Etienne[2]

[1].*Centre de Coopération Internationale en Recherche Agronomique pour le Développement – Amis (CIRAD-AMIS), CIRAD, TA 40/03, Avenue Agropolis, 34398 Montpellier Cedex 5, France. Tel (33) 4 67 61 71 07, Fax : (33) 4 67 61 56 05. E-mail : berthouly@cirad.fr*

[2].*Centre de Coopération Internationale en Recherche Agronomique pour le Développement-Cultures Pérennes (CIRAD-CP), CIRAD, TA 80 / PS3, Boulevard de la Lironde, 34398 Montpellier Cedex 5, France. Tel: (33) 4 67 61 71 35; Fax: (33) 467 61 71 20*
E-mail: herve.etienne@cirad.fr

Abstract: Mass propagation of plants by tissue culture is labour intensive and costly. Gelling agents have many drawbacks: they are not inert medium components and do not enable easy automation for commercial mass propagation. So liquid culture systems are considered to have advantages, e.g. culture conditions are much more uniform, media can be changed easily. The use of liquid medium for *in vitro* culture has many advantages and has been the subject of many studies over many years. It has also frequently been considered an ideal technique for mass production as it reduces manual labor and facilitates changing the medium composition. Techniques and culture vessels of varying complexity have been developed as a result of studies.

The major disadvantage of a liquid medium is hyperhydricity, which is a severe physiological disorder. So we considered that to compensate for this problem it would be necessary to expose the plant to the liquid medium intermittently rather than continuously. For this the bioreactors previously developed are not suitable as they are mainly adapted to bacterial culture and do not take into account the specific requirements of plant cells and tissues, such as sensitivity to shear forces, mechanical damages or foam formation in bubble aerated bioreactors.

So temporary immersion systems for plant micropropagation have been described and grouped into 4 categories according to operation: i) tilting and rocker machines, ii) complete immersion of plant material and renewal of nutrient medium, iii) partial immersion and a liquid nutrient renewal mechanism, iiii) complete immersion by pneumatic driven transfer of liquid medium and without nutrient medium renewal. The positive effects of temporary immersion on micropropagation are indicated for shoot proliferation and microcuttings, microtuberization and somatic embryogenesis. Immersion time, i.e. duration or frequency, is the most critical parameter for system efficiency. Optimizing the volume of nutrient medium and the volume of container also substancially improves efficiency, especially for shoot proliferation. Temporary immersion also generally improves plant tissue quality. It results in

A.K. Hvoslef-Eide and W. Preil (eds.), Liquid Culture Systems for in vitro Plant Propagation, 165–195.

increased shoot vigour and quantity of morphologically normal somatic embryos. Hyperhydricity, which seriously affects cultures in liquid medium, is eliminated with these culture systems or controlled by adjusting the immersion times.
Plant material propagated by temporary immersion performs better during the acclimatization phase than material obtained on semi-solid or liquid media. Successful regeneration of *Solanum tuberosum* microtubers and *Coffea arabica* somatic embryos produced in temporary immersion bioreactors after direct sowing on soil has been demonstrated. As was predicted, when using liquid medium for micropropagation, several investigations have confirmed large gains in efficiency from temporary immersion. The parameters most involved in reducing production costs are, firstly a large reduction in labour, followed by a reduction in shelving area requirement and the number of containers used, along with better biological yields.
Scaling up embryogenesis and shoot proliferation procedures involving temporary immersion systems are now taking place, in order to commercialize this process. To improve this system as well in research as in commercial production, CIRAD has developed a new simple and specific apparatus for plant tissue culture using temporary immersion in liquid medium.

Key words: acclimatization, bioreactor, hyperhydricity, organogenesis, shoot proliferation, somatic embryogenesis, temporary immersion

1. Introduction

1.1 Current limitations of micropropagation

Current techniques for micropropagation require a large number of small containers, gelified media and aseptic conditions, and then there remains a complicated and costly production technology. Plant micropropagation involves periodic transfers of plant material to fresh media, after subcultures of 4 to 6 weeks, due to exhaustion of the nutrients in the medium and also because of continuous tissue growth and proliferation, which is rapidly limited by the size of the culture container (Debergh et al., 1992). Agar products are not inert and complicate automation. High production costs generally limit the commercial use of micropropagation to markets with a very high unit value, such as ornamentals, foliage plants and selected fruit crops (Sluis and Walker, 1985; Simonton et al., 1991). Labour generally accounts for 40 to 60% of production costs. Cutting and planting represent the most expensive part of the micropropagation process (Chu, 1995). Although tissue handling is the major part of the work and the most technical, there is also the cleaning, filling and handling of a large number of containers (Maene and Debergh, 1985). Other major costs come from losses occurring during acclimatization in greenhouses and stem and root hyperhydricity (Reuther, 1985). It has been concluded for various species that extensive expansion of micropropagation would only take place if new technologies became available to automate procedures, and if acclimatization protocols were improved (Kitto, 1997).

1.2 Advantages of liquid media for plant micropropagation

Using liquid media in micropropagation processes is considered to be the ideal solution for reducing plantlet production costs and for considering automation (Debergh, 1988; Aitken-Christie, 1991). Indeed, liquid culture systems provide much more uniform culturing conditions, the media can easily be renewed without changing the container, sterilization is possible by ultrafiltration and container cleaning after a culture period is much easier. In addition, with liquid culture media, much larger containers can be used and more of the container volume can be used, whereas agar media necessitate flat culturing. Transfer times can be reduced since explants are no longer positioned, but in many cases merely placed in contact with the liquid medium.

Moreover, plant tissues from numerous species have performed better when cultured in liquid medium rather than on an agar medium. For instance, a larger number of shoots were produced in *Prunus persica* L. Batsch (Hammerschlag, 1982). More somatic embryos were produced in soft-red winter wheat (Jones and Petolino, 1988) and *Gossypium hirsutum* (Gawel and Robacker, 1990). The development of scaled-up liquid bioreactor cultures was considered from the outset for both embryogenic and organogenic regeneration pathways (Ziv, 1995). However, somatic embryogenesis appears to be the least labour-intensive, hence the most appropriate for an automated system in liquid medium, as stated by Aitken-Christie and Jones (1987) for organogenesis, a prerequisite for automation is to obtain a culture system in which shoots or somatic embryos can be produced in the same container for a long period without transfer, thereby enabling regular or total harvesting of shoots or somatic embryos for acclimatization and plant conversion, respectively.

1.3 Micropropagation systems with liquid medium

Bioreactors developed in the past are not suitable for micropropagation as they are mainly adapted to bacterial culture and do not take into account the specific requirements of plant cells, such as sensitivity to shear forces, mechanical damage or foam formation in bubble aerated bioreactors (Teisson et al., 1999). The advantages of *in vitro* culture in a liquid medium are therefore often counterbalanced by technical problems such as asphyxia, hyperhydricity, shear forces and the need for complex equipment. In order to avoid such problems, other procedures have been developed that include culture supports such as paper bridges, cellulose blocks or sponges (Etienne et al., 1991; Smith and Spomer, 1995; Wataad et al., 1997), a raft to support plants over stationary liquid (Connor and Meredith, 1984; Hamilton et al., 1985), adding liquid medium to established cultures on agar (Maene and Debergh, 1985) and mist bioreactors (Weathers and Giles, 1988; Tisserat et al., 1993). A series of temporary immersion techniques has also been proposed for micropropagation, based on a principle similar to that of mist bioreactors, preferring temporary contact between the plants and the liquid medium rather than permanent contact.

2. Different temporary immersion culture systems

In 1983, Harris and Mason described work on tilting machines designed to achieve temporary immersion, in order to combine aeration and liquid medium culture. At the time, they pointed out that Stewart et al. had noticed as early as 1952 that carrot root explants did not grow quickly when immersed in a liquid medium, and they deduced that it was due to a lack of oxygen. They designed an apparatus known as an "auxophyton", which turned the culture containers on a wheel, exposing the explants alternately to the air, or immersing them in the liquid. After 20 days' culture, the carrot explants weighed 2.6 times more than those cultured on an agar medium.

Since the work by Harris and Mason, a wide range of semi-automatic systems using the temporary immersion principle have been developed (Table 1). All these systems respect the conditions mentioned by Teisson et al. (1999): avoid continuous immersion, which adversely affects growth and morphogenesis, provide adequate oxygen transfer, sufficient mixing and limit shear levels, enable sequential medium changes and automation, reduce the risk of contamination, be as cheap as possible. The proposed systems differ in container size, the type of culture support, the existence of a computerized immersion control or a simple timer, through use either of a peristaltic pump, or an air pump, or mechanical motion of the container to displace the liquid, through recycling or not of the medium, and lastly through separation or incorporation of the medium tank with the culture container. These systems are also easier to use than conventional bioreactors, and longer subcultures are possible in most of them. The originality of temporary immersion culture systems is to enable partial or total contact of programmable duration between the explant and the liquid medium. The systems that have been designed can be divided into the following four categories (diagrams can be found in figures 1 A-E):

2.1 Systems with tilting or rocker machines

Two machines were described by Harris and Mason (1983). The tilting machine inclines Erlenmeyers flasks 30 degrees in opposite directions; it has a capacity of 400 50-ml Erlenmeyer flasks or 320 125-ml flasks. The Rocker machine rolls 70 910-ml wide-mouth jars lying on their sides, or tilts 120 455-ml wide-mouth jars standing upright 30-40 degrees every 30 sec. These machines do not include replenishment of the liquid culture medium.

2.2 Systems with complete immersion and a liquid medium renewal mechanism

Tisserat and Vandercook (1985) developed a large elevated culture chamber that was periodically drained and then refilled with fresh medium in a sterile environment. The automated plant culture system (APCS, Figure 1A) consists of silicone tubing, 2 impeller pumps, 2 glass medium reservoir bottles, a 3-way stainless steel valve, a plant culture chamber, and an interface module containing relay boards. This system provides a long-term method for *in vitro* plant culture.

A

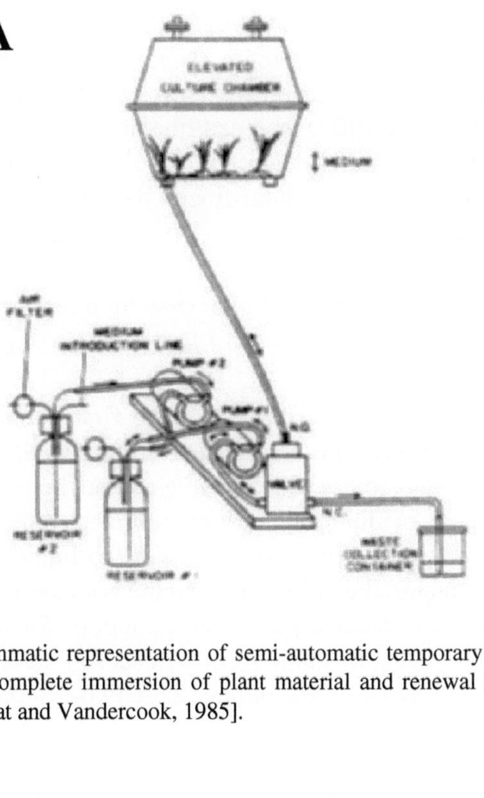

Figure 1 A: Diagrammatic representation of semi-automatic temporary immersion systems: APCS system with complete immersion of plant material and renewal of the liquid culture medium [from Tisserat and Vandercook, 1985].

2.3 Systems with partial immersion and a liquid medium renewal mechanism

The plant explant is always positioned on a culture support (agar medium, propylene screen, cellulose plugs). Liquid culture medium is frequently supplied, then withdrawn into a drain-off recipient vessel, so as to imbibe the support, stabilize the composition of the culture medium and extend the duration of subcultures, whilst avoiding or postponing the need to change the medium. Only the base of the plant material is partially immersed. Two models have been published:

- Aitken-Christie and Jones (1987) and Aitken-Christie and Davies (1988) proposed a semi–automatic process in large polycarbonate containers measuring 250 x 390 x 120mm (Figure 1B). In their system, *Pinus* shoots were grown on an agar medium, with automatic addition and withdrawal of liquid medium by peristaltic pumps on a periodic basis. The liquid from the fresh medium recipient vessel came into contact with the explants for 4 to 6 hours, using a vacuum suction system, then went to the drain-off recipient vessel. This system follows on from the work by Maene and Debergh (1987), who had shown the positive effects of a adding liquid nutrient medium or auxins in the final *in vitro* stages.
- Simonton et al. (1991): their system featured a computer-controlled pumping apparatus that intermittently supplied liquid medium to plants cultured in 7-litre vessels (Figure 1C). Plant material rested on a perforated polypropylene screen that was attached to the inside of the vessel. Control capabilities included medium introduction and depth regulation within four individual culture vessels, medium cycling on an assigned schedule, schedule adjustment during a culture period, and medium replacement.

2.4 Systems with complete immersion by pneumatic driven transfer of liquid medium and without medium replenishment

Different systems were described after the first publication by Alvard et al. (1993). These include the most recent temporary immersion systems. They are simple and easy to use. They enable contact between all parts of the explant and the liquid medium, along with complete renewal of the culture atmosphere by forced ventilation, which drives the liquid towards the plant material. The plant material can be placed in the container in bulk, removing the need to position the plant material on a support. Such systems include pneumatic transfer of the medium from a reservoir tank to the container holding the plants. To avoid excess tubing, these two compartments are

B

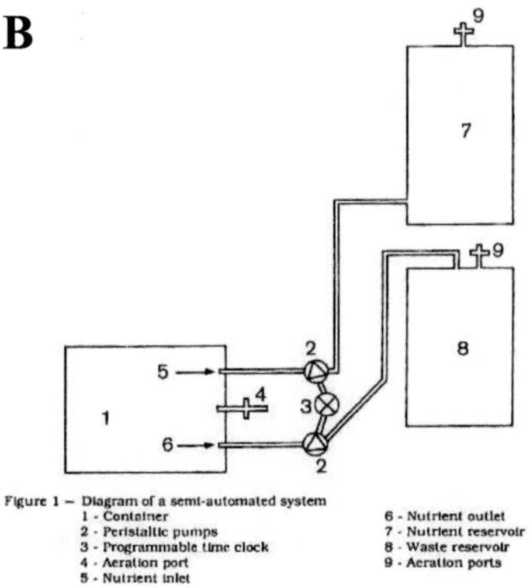

C

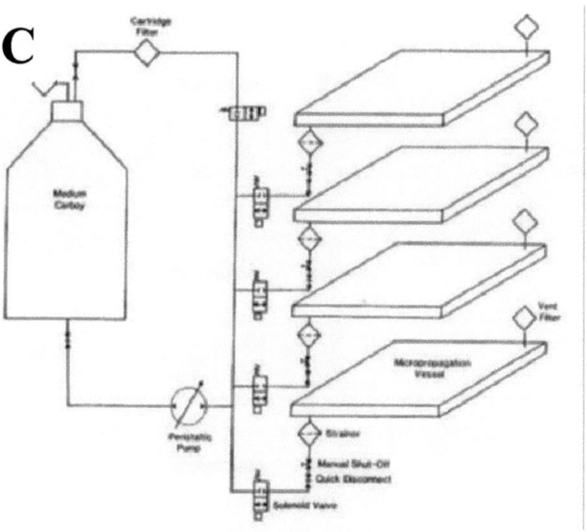

Figure 1 B and C: Diagrammatic representation of semi-automatic temporary immersion systems with partial immersion and with a liquid-nutrient renewal process.
B) from Aitken-Christie and Davies, 1988; C) from Simonton et al., 1991.

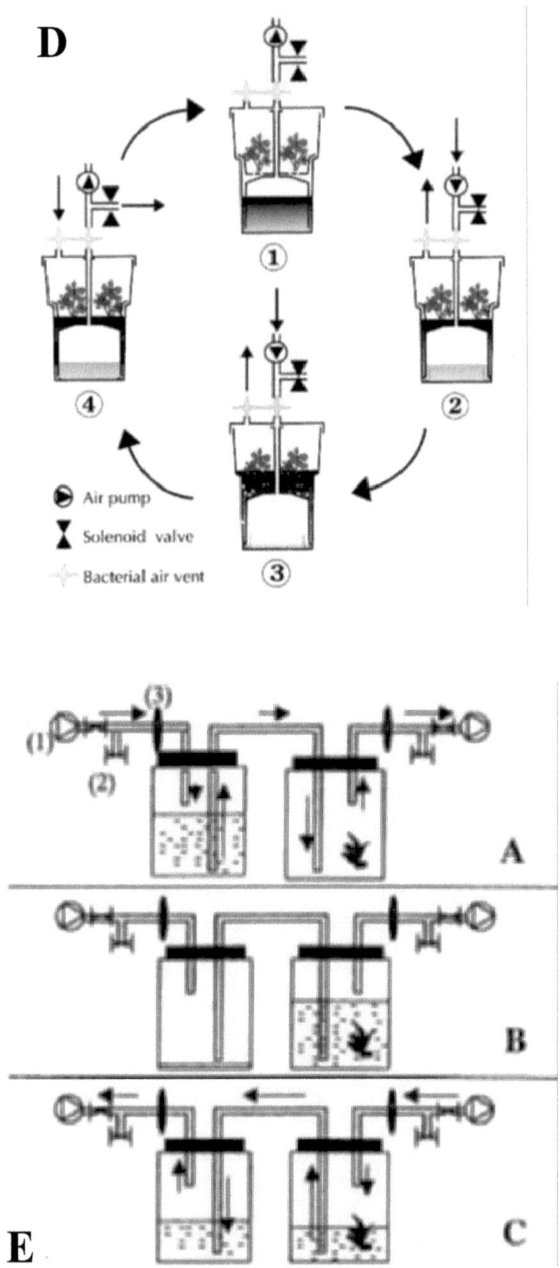

Figure 1 D and E: Diagrammatic representation of semi-automatic temporary immersion systems with complete immersion of plant material by pneumatic driven transfer of liquid medium and without medium renewal. D) RITA® system from Alvard et al., 1993;
E) BIT® twin flasks system from Escalona et al., 1999.

preferably part of the same vessel. Pressure is applied through a solenoid valve by a compressor connected to a programmable plug. This application determines the time and duration of floodings. As these systems do not include a fresh medium tank, the culture medium has to be changed after 4 to 6 weeks. However, replacement is rapid and there is no need to transfer the plant material. Two variants of this system have been developed and are currently on the market: the Recipient for Automated Temporary Immersion system (RITA®) and the twin flasks system (BIT®).

- The RITA® system (Figures 1D and 2B; Teisson and Alvard, 1995). The 1-litre vessel comprises two compartments, an upper one with the plant material and a lower one with the medium. The pressure applied in the lower compartment pushes the medium into the upper one. Plants are immersed as long as pressure is applied. During the immersion period, air is bubbled through the medium, gently stirring the plants and renewing the headspace atmosphere inside the culture vessel, with the pressure escaping through outlets on the top of the apparatus. The RITA® system is mainly intended for mass propagation by somatic embryogenesis.

- The twin flasks system (BIT®) (Figures 1E and 2D; Escalona et al., 1998). The easiest way to carry out pneumatically driven temporary immersion is to connect two glass or plastic flasks - from 250 ml to 10 litres - by tubing, and apply alternative pressure to push the medium into the respective recipient vessels. The RITA® vessel can easily be adapted to this configuration. In 1994, Akita and Takayama proposed a similar system known as the 'system for semi-continuous medium surface level control culture', adapted to potato tuberization. That system incorporates a forced aeration in the culture vessel, which is not found in the BIT® system.

3. Effect of temporary immersion on biological yield from different micropropagation processes

When envisaging commercial use of automated temporary immersion systems, it is important to take comprehensive measurements of growth, production and quality on the cultured material, and compare them to data obtained on material produced with the conventional culture systems.

3.1 Shoot proliferation and microcuttings

A great deal of work has proved that temporary immersion stimulates shoot proliferation. In 1987, Aitken-Christie and Jones showed better shoot growth could be obtained for radiata pine with a method involving replenishment of a liquid nutrient medium than with monthly transfers on an agar medium. This system enabled continuous shoot growth and monthly harvests for 18 months, without transferring the plant material. Shoots obtained with partial and temporary immersion in the nutrient medium were longer and of better quality than those obtained on agar media. A complete and convincing demonstration of the efficacy of temporary immersion was carried out for the proliferation of banana meristems. Alvard et al. (1993) showed that applying liquid medium strongly influenced the development and proliferation rate for micropropagated banana explants. When four liquid medium culture methods were compared to the conventional method on agar medium, they obtained the following results after culturing for 20 days: i) shoots placed in a simple liquid medium or on a cellulose support barely proliferated or not at all; ii) shoots on an agar medium, those subjected to partial immersion and those in a bubble-aerated medium had multiplication rates of 2.2 to 3.1 and, iii) the highest proliferation rate (>5) was found for explants subjected to temporary immersion in the medium. These authors obtained their results using a RITA® bioreactor, with 20 min immersions every 2 h. A Cuban team obtained similar results with banana, using the twin flasks system (Teisson et al., 1999).

Serviceberry shoots (*Amelanchier* x *grandiflora* Rehd. 'Princess Diana') were cultured in the temporary immersion system described by Simonton et al. (1991) and compared to those obtained either on agar medium or in liquid medium in baby food jars, and on agar medium or in non-cycling liquid medium in a 7-litre vessel (Krueger et al., 1991). A combination of liquid medium, a 7-litre recipient vessel and intermittent contact with the liquid medium gave significantly higher proliferation rates than in any other combination tested. Compared with conventional treatment (agar medium in baby food jars), cultures that grew in intermittent contact with the culture medium gave higher values for the number of shoots (x 2.6), shoot weight (x 2.1), shoot length (x 1.2) and culture weight (x 2.2).

With sugarcane, Lorenzo et al. (1998) also showed with the Twin Flasks system that temporary immersion clearly stimulated shoot formation and length. The multiplication rate (23.9 shoots per 30 days) increased by 6 compared with the standard protocol (3.96 shoots per 30 days; Jiménez et al., 1995). Similar results were obtained with another 3 genotypes. Likewise, Escalona et al. (1999) used the same bioreactor for pineapple meristem propagation to show that temporary immersion stimulated the multiplication rate, along with the fresh and dry weights, after culturing for 42 days (Figures 2E, F). The multiplication rates increased by 300% and 400% respectively, compared to those obtained using systems with liquid or solid supports.

Orchids and cow tree (*Mitragyna*) grown in the APCS temporary immersion system developed by Tisserat and Vandercook (1985) grew more quickly than on an agar medium. Based on fresh weight and volume measurements, they revealed a four-fold increase for these two parameters in the case of orchid after culturing for 270 days, and an increase of 1.8 in the case of cow tree after 45 days. The same tests on aster shoot cultures did not reveal any difference in plant growth and development. Nevertheless, the aster plants regenerated by temporary immersion were much more vigorous when planted in the nursery, thereby revealing a better physiological condition.

In the case of coffee (*C. arabica* and *C. canephora*), multiplication by microcuttings in a semi-solid medium is of limited value, due to the slow growth of orthotropic shoots. The multiplication rate is around 6 or 7 every 3 months (Söndhal et al., 1989). When a RITA® temporary immersion system was used, that result was obtained in just 5 to 6 weeks (Berthouly et al., 1995). Cyclically immersed *Vitis vinifera* L., in side-to-side tipping culture vessels using tilting machines with alternate exposure and submergence intervals of 30 s, or longer, produced seven times as many shoots in 90 days as explants maintained on agar media (Harris and Stevenson, 1982; Harris and Mason, 1983). Similar increases in shoot numbers have been obtained with *Arctostaphylos uva ursi* (L.), *Amelanchier alnifolia* Nutt., *Nicotiana tabacum* 'Xanthi-nc' and *Fuchsia hybrida* 'Swingtime' (Stevenson and Harris, 1980).

3.2 Microtuberization

Plant growth and tuberization of potato (*Solanum tuberosum* L.) was clearly stimulated by temporary immersion in a twin flasks system (Akita and Takayama, 1994). The number of tubers formed, i. e. approximatively 500 to 960 tubers for a culture, was much better than obtained earlier (approx. 220 tubers for a culture (Akita and Takayama, 1993)). Total tuber weight and homogeneity also increased. On the other hand, there was no tuber formation under complete or continuous immersion conditions. Teisson and Alvard (1999) confirmed the efficiency of temporary immersion for potato microtuberization, working with a double RITA® system based on the twin flasks method. Three microtubers were obtained per single node 10 weeks after inoculation. Fifty percent of the microtubers exeeded 0.5 g and sprouted, still in a temporary immersion system. This system was very quick and efficient as three to four shoots started to develop from one single tuber. Similar results were obtained with 3 different cultivars (Teisson and Alvard, 1999) and the process seemed to be easily improvable with an ultimate target of direct transfer to the field.

3.3 Somatic embryogenesis

3.3.1 Proliferation of embryogenic callus

Temporary immersion culture systems have proved more successful in achieving embryogenic tissue proliferation than conventional systems using an agar medium or suspensions in Erlenmeyer flasks. Tisserat and Vandercook (1985) quantified the growth of carrot and date palm callus in a totally automated culture system called APCS, in which immersions of 5 to 10 min were applied every two hours. Compared to cultures on an agar medium, there was a growth increase of 1.9-fold for carrot and 4-fold for date palm. Likewise, an improvement in callus quality was noted for the 2 species and a large quantity of somatic embryos and plants was obtained for carrot. Similarly, for different *Coffea arabica* genotypes tested, embryogenic callus growth was greater with temporary immersion than in a stirred liquid medium in an Erlenmeyer flask (Berthouly et al., 1995).

3.3.2 Embryo development

The production and quality of somatic embryos has been improved for various species by temporary immersion culture. In *Citrus deliciosa*, Cabasson et al. (1997) compared the efficacy of various culture systems on somatic embryo development. Somatic embryos derived from suspension cultures were plated on semi-solid medium, maintained in suspension culture or temporarily immersed. About 60% of somatic embryos plated on gelified medium developed to the cotyledonary stage, but were hyperhydric. Continuous growth in a suspension culture at 100 rpm hindered cotyledon and protoderm formation, and somatic embryos were unable to develop beyond the globular stage. Temporary immersion in a RITA® bioreactor promoted somatic embryo development, i.e. 66% of the somatic embryos produced were cotyledonary, and were morphologically similar to nucellar embryos. Escalant et al. (1994) showed with several banana and plantain cultivars that after two months of temporary immersion culture in a RITA® bioreactor, three times more embryos were produced than on an agar medium (1,375 embryos vs. 450). Temporary immersion promoted adventive embryogenesis from the epidermal cells of primary embryos. After 6 months, the initial number of somatic embryos in the one-litre bioreactor had multiplied by 40, i.e. 6,000 embryos. On the other hand, after 2 months' incubation on an agar medium, the embryos had changed into a compact, white callus. High germination rates (60 to 70%) were obtained by transferring part of the embryos produced with temporary immersion onto agar medium.

With rubber tree (*Hevea brasiliensis*), somatic embryo production on an agar medium gives low and not particularly reproducible yields of embryos of poor morphological quality (Etienne et al., 1993). Transferring embryogenic callus to a RITA® type temporary immersion container greatly improved the quantity of somatic embryos produced and their quality, and enabled routine production (Etienne et al., 1997b). Somatic embryo production in a liquid medium was three to four times greater than on a semi-solid medium, i.e. 400 embryos per g callus fresh weight. Temporary immersion also reduced the proportion of abnormal embryos by half, with an increase in the germination rate. During germination, temporary immersion considerably increased root development (+60%) and epicotyl emergence (+35%). According to Teisson et al. (1999), around 150 *Hevea* embryos at the cotyledonary stage were harvested 4 to 8 weeks after transferring embryogenic callus to the bioreactor. However, in order to regenerate fully developed plants, the germinated material had to be transferred to a semi–solid medium. So far, plant conversion has seemed to require plantlets to be in an upright position, which was impossible to achieve in a RITA® vessel, as their position changes with every flooding.

With *Coffea sp*., mass somatic embryo production in Erlenmeyer flasks and in bioreactors has been well mastered for some time for *C. canephora*, for which the production of several hundred thousand somatic embryos per gram of inoculum has been reported (for review, Berthouly and Etienne, 1999). *Coffea arabica* is a more recalcitrant species. With a view to large-scale dissemination of improved F_1 hybrids of *C. arabica* in Central America, reproducible production was successfully achieved with around twenty clones using a RITA® type temporary immersion bioreactor (Etienne et al., 1997a). Depending on the genotypes, yields ranging from 15,000 to 50,000 somatic embryos per gram of embryogenic suspension were recorded. Such yields are around twice of those obtained in Erlenmeyers flasks under optimum conditions. However, the most spectacular effect has been the degree of improvement seen in the quality of coffee somatic embryos produced by temporary immersion (Figure 2A). Whilst the proportion of normal torpedo type embryos is around 30% in a bioreactor or in Erlenmeyer flasks (Zamarripa et al., 1991; Noriega and Söndahl, 1993), it is usually over 90% with temporary immersion. This improvement in quality is reflected in higher plant conversion rates on an agar medium (80 to 90%) and especially in the successful regeneration of plants under *ex vitro* conditions after direct sowing on horticultural substrate of somatic embryos produced in a temporary immersion bioreactor (Etienne-Barry et al., 1999; Figure 2C). Embryo to plant conversion rates of 70-80% are routinely achieved under such conditions for all genotypes.

3.3.3 *Synchronization of embryo production*

In *Citrus*, temporary immersion also improved the synchrony of regeneration by suppressing secondary embryogenesis at the onset of germination, contrary to the results obtained in conventional cultures (Cabasson et al., 1997). In rubber, temporary immersion also increased synchrony during the development and germination phases when compared to cultures on a gelified medium (Etienne et al., 1997b). In coffee, when compared to cultures on a gelified medium, the synchronization of development and germination for embryos grown in temporary immersion is notable (Etienne-Barry et al., 1999). It is encouraged by high culture densities (1,500 to 3,000 embryos per 1-litre bioreactor), for which 66% of the embryos are at the same germination stage (Figure 2A).

In banana and plantain (Escalant et al., 1994), unlike *Citrus*, temporary immersion promoted a process of embryo proliferation by secondary somatic embryogenesis. Under the culture conditions used, embryos formed continually in cascades, each forming 4 to 5 new embryos, which separated and continued the phenomenon. The somatic embryos formed either at the base or from a few epidermal cells of the primary embryos. With the temporary immersion system, it was possible for banana and coffee somatic embryos to divide the population easily at any time, by transferring part of it to other bioreactors.

4. Culture parameters affecting the efficacy of temporary immersion systems

The main reason for the efficacy of temporary immersion systems is probably that they combine ventilation of the plant tissues, and intermitent contact between the main part or the entire surface of the explants and the liquid medium. These two characteristics are not usually combined in other liquid culture procedures.

4.1 Immersion time

In culture systems with temporary tissue immersion, it is clear that the immersion time is very important, since it governs nutrient uptake and expression of hyperhydricity. The immersion times used for different work vary considerably (Table 1). This is probably due to the large variety of species, micropropagation processes and temporary immersion systems used. Long immersion times (1 h every 6 h) prove to be efficient for potato tuberization, whereas very short immersion times (1 min every 12 h) stimulate somatic embryo production most in coffee and rubber (Etienne et al., 1997a,b). Likewise, very frequent immersions (30 sec every 30 sec) can prove to be highly efficient in tilting machines for grapevine shoot propagation (Harris and Mason, 1983).

Krueger et al. (1991) showed the importance of immersion frequencies for the proliferation of serviceberry shoots. Hyperhydricity was observed with immersions for 5 min every 30 min, but was not seen with immersions for 5 min every 60 min. On the other hand, the first combination is better for the number of shoots obtained. In order to combine the advantages of both combinations, the authors recommended using the first combination during an initial proliferation cycle, then the second combination to maintain shoot quality. They also revealed the existence of a period of adaptation, when switching to the lowest frequencies. Stress resulted in partial desiccation of the shoots, but the material recovered later. With radiata pine shoots grown on agar medium, liquid nutrient replenishment at a frequency of twice a week more effectively stimulated growth (FW x 1.2) and shoot quality (x 1.5) than frequencies of once every 2 or 4 weeks (Aitken-Christie and Jones, 1987).

Berthouly et al. (1995) showed with coffee microcuttings that the immersion time substantially affected the multiplication rate, estimated by the number of micronodes produced after 6 weeks. Indeed, immersion times of 1, 5 and 15 min applied every 6 h gave multiplication rates of 3.5, 5.4 and 8.4 respectively in their experiment. In addition, the optimum immersion time varied depending on the coffee species used. For instance, it was 15 min every 6 h for *C. arabica* microcuttings and only 1 min every 6 h for *C. canephora* microcuttings.

In coffee, modifying the immersion time greatly affects somatic embryo production. According to Berthouly et al. (1995), 15 minutes' immersion every 6 hours led to successive development and germination of embryos, whereas for an identical culture medium, immersions of 1 minute every 24 hours halted embryo development and stimulated the production of secondary embryos.

We recently observed that increasing the frequency for short immersions (1 min) stimulated somatic embryo formation and quality in *C. arabica*. Thus, average yields of 480, 2,090 and 3,100 embryos were obtained per 1-litre bioreactor, with 60, 79 and 85% torpedo type embryos, for daily frequencies of 1, 2 and 6 immersions, respectively. Hyperhydricity was not observed with such immersion conditions. On the other hand, increasing immersion times by 5 min or more led to a considerable reduction in somatic embryo production and in their quality, becoming all the more critical as the immersion frequencies increased. For example, 15 min immersions applied 2 or 6 times per day led to hyperhydrated embryo frequencies of 64 and 90%, respectively. It is likely that each culture stage requires adaptation of the immersion length and frequency to obtain optimum results.

4.2 Volume of liquid medium

It is particularly important to optimize the liquid medium volume when using temporary immersion systems without medium renewal, such as the twin flasks and RITA® systems or tilting and rocker machines. Lorenzo et al. (1998) found an optimum volume of medium per explant for sugarcane shoot proliferation in the BIT® twin flasks system. An increase in multiplication rate from 8.3 shoots per 30 days to 23.9 shoots per 30 days was obtained by multiplying the volume of standard medium by ten from 5.0 to 50.0 ml per explant. However, the volume of medium used did not affect the length of the shoots formed. Higher volumes proved to be less efficient. According to the authors, the explanation can be found in the secretion of chemical molecules that stimulate shoot formation, which would seem to be diluted when large volumes of medium are used. Using the same temporary immersion system, Escalona et al. (1999) similarly demonstrated with pineapple that an optimum medium volume exists for shoot proliferation, which was estimated to be 200 ml per explant for that species. In this case, larger volumes also led to a drop in the proliferation rate.

4.3 Volume of the culture container

For all temporary immersion systems, the volume of the container, hence the head space, is much larger than in the containers used for conventional procedures. Moreover, containers ranging in size from 1 to 20 litres can usualy be adapted to the system. Krueger et al. (1991) demonstrated that the large size of their culture container (7 l) had a positive effect on micropropagation efficiency for serviceberry, notably by avoiding culture overcrowding and encouraging shoot elongation, when compared to those obtained in 140 ml baby food jars. Monette (1983) had already shown for

Vitis vinifera L. that longer shoots were obtained in larger containers. Grapevine explant growth was so strong in liquid medium using tilting machines, that it soon became necessary to use larger containers such as 910 ml square wide-mouth Mason jars, as opposed to 125 ml flasks to avoid crowding of the cultures, benefiting from a larger aperture to remove material, and larger volumes of medium to prevent early deficiencies in certain constituents of the medium (Harris and Mason, 1983). Using larger containers means that larger volumes of media can be used, which can have a positive effect on plant material proliferation and growth.

4.4 Oxygenation and forced ventilation

Work by Alvard et al. (1993) on banana meristem propagation clearly showed that a lack of oxygen in the liquid culture medium was a major limiting factor for small explant growth. In their experiment, the absence of liquid medium stirring led to explant asphyxia. Bubble aeration of the explant or medium encouraged growth but partial immersion of the explant did not provide sufficient oxygen. Temporary immersion clearly proved to be the most effective culture system. However, for other systems, the positive effect of aeration has not been proved. Aitken-Christie and Jones (1987) showed that aeration alone did not stimulate shoot growth in radiata pine and was not a contributing factor to the increased growth found with nutrient replenishment. As early as 1952, Stewart et al. attributed better shoot growth to an improved oxygen supply through alternating immersion. Yet, the pliofilm seal on their recipient vessel had no negative effect on growth after culturing for two months.

In systems using pneumatic propulsion of the nutrient medium, this type of operation was found to cause forced ventilation leading to complete renewal of the culture atmosphere on each immersion. According to Teisson and Alvard (1995), gas exchanges in such a system primarily occur during immersion and are caused indirectly by movement of the liquid and directly by the air pump under the most frequently used culture conditions that correspond to complete renewal of the culture atmosphere after 5 minutes of immersion for a 1-litre bioreactor. Such forced ventilation with air containing the gas concentrations and relative humidity of the culture room probably has positive effects (Krueger et al., 1991). The relative humidity resulting from forced ventilation may stimulate transpiration in the plants, which will then be more effectively adapted to *ex vitro* conditions (Wardle et al., 1983).

Table 1: List of the main publications describing micropropagation processes using temporary immersion culture. For each publication, the temporary immersion system used, the gain in biological yield compared with processes conventionally used, existence or not of hyperhydricity and acclimatization success rate are indicated

Family – Species (Common Name)	**Morphogenetic Pathway**	**Temporary Immersion System**	**Immersion times used ***
Vitis vinifera (grape)	Shoot proliferation	Tilting and rocker machines	30 sec every 30 sec
Potinera sp. (orchid)	Shoot proliferation	Automated plant culture system (APCS)	5-10 min every 12 h
Callistephus hortensis (aster)	Shoot proliferation	Automated plant culture system (APCS)	5-10 min every 12 h
Pheonix dactylifera (date palm) *Daucus carota* (carrot)	Embryogenic callus proliferation	Automated plant culture system (APCS)	5-10 min every 12 h
Mitragyna inermis (cow tree)	Shoot proliferation	Automated plant culture system (APCS)	5-10 min every 12 h
Pinus radiata D. Don (radiata pine)	Shoot hedge proliferation	Liquid nutrient medium on agar with liquid replishment system	4 to 6 h every 3 days
Amelanchier x *grandiflora* 'Princess Diana'	Shoot meristem proliferation	Programmable micro-propagation apparatus using cycled medium (Simonton et al., 1991)	5 min every 30 or 60 min
Musa acuminata (banana)	Shoot meristem proliferation	1-litre bioreactor with 2 compartments (RITA®)	20 min every 2 h

* The best immersion times are presented when several frequency/duration combinations were tested

Increase in efficacy (x fold) compared with agar	Hyperhydricity	Subsequent successful acclimatization	Authors
Shoot No. x 7, shoot length increased, rooting faster and more efficient	Not determined	Not determined	Harris and Mason, 1983
Shoot FW x 4	No	Not determined	Tisserat and Vandercook, 1985
Shoot biomass (FW) x 1	No	Yes (shoots and flowers larger after)	Tisserat and Vandercook, 1985
Date palm: callus FW x 3.2, better quality Carrot: callus FW x 1.9, better quality and plant regeneration	No	Not determined	Tisserat and Vandercook, 1985
Shoot FW x 1.8	No	Not determined	Tisserat and Vandercook, 1985
Shoot FW x 1.2, and shoot health x 2	Yes but reduced compared to agar medium	Yes	Aitken-Christie and Jones, 1987
Shoot No. x 2.6, shoot weight x 2.1, shoot length x 1.2	No with 5 min every 1 h; Yes with 5 min every 30 min	Not determined	Krueger et al., 1991
Shoot No. x 2.5	No	Not determined	Alvard et al., 1993

Table 1: continued

Family – Species (Common Name)	**Morphogenetic Pathway**	**Temporary Immersion System**	**Immersion times used ***
Solanum tuberosum L. (potato)	Tuberization	10-litre twin jar fermentors	60 min every 6 h
Triploid banana and plantain (*Musa* spp.)	Somatic embryogenesis	1-litre bioreactor with 2 compartments (RITA®)	1 min every 6 h
Coffea arabica and *Coffea canephora*	Microcutting	1-litre bioreactor with 2 compartments (RITA®)	15 min every 6 h (*C. arabica*); 1 min every 6 h (*C. canephora*)
Hevea brasiliensis (rubber tree)	Somatic embryogenesis	1-litre bioreactor with 2 compartments (RITA®)	1 min every 12 h (embryo development); 15 min every 6 h (germination)
Coffea arabica (coffee)	Somatic embryogenesis	1-litre bioreactor with 2 compartments (RITA®)	1 min every 12 h
Citrus deliciosa	Somatic embryogenesis	1-litre bioreactor with 2 compartments (RITA®)	1 min every 4 h
Saccharum sp. (sugarcane)	Shoot meristem proliferation	10-litre twin flasks (BIT®)	2 min every 9 h
Ananas comosus (pineapple)	Shoot meristem proliferation	10-litre twin flasks (BIT®)	2 min every 3 h
Coffea arabica (coffee)	Somatic embryogenesis	1-litre bioreactor with 2 compartments (RITA®)	1 min every 12 h (embryo development); 5 min every 12 h (germination)

* The best immersion times are presented when several frequency/duration combinations were tested

Increase in efficacy (x fold) compared with agar/liquid	**Hyperhydricity**	**Subsequent successful acclimatization**	**Authors**
Tuber No. x 3-4 (/fermenter)	No	Direct planting without acclimatization	Akita and Takayama, 1994
Embryo No. x 3 (/agar)	No	No	Escalant et al., 1994
Shoot No. x 2 (/agar)	No, at the immersion times used; Yes, for longer times	Not determined	Berthouly et al., 1995
Embryo No. x 4 (/agar); 85% cotyledonary embryos (vs 26% agar)	No	No	Etienne et al., 1997
Embryo No. x 2 (/agar) 90% torpedo embryos (vs 30 % agar)	No	No	Etienne et al., 1997
66% well formed cotyledonary embryos (vs 0% in suspension and 60% hyperhydric cotyledonary embryos on solid medium)	No (but systematically observed with agar medium)	No	Cabasson et al., 1997
Shoot No. x 4.0 to 6.0 (/agar)	No	Yes, efficient	Lorenzo et al., 1998
Shoot No. x 3.0 (/liquid) and x 4.0 (/agar)	No	Yes, efficient	Escalona et al., 1999
90% well formed torpedo embryos (vs 30% in liquid medium) 75% embryo-to-plant conversion in *ex vitro* conditions	No	Yes, efficient by direct sowing of germinated somatic embryos	Etienne-Barry et al., 1999

Figure 2: A) Aspect of coffee (*Coffea arabica*) germinated somatic embryo production in a RITA® type temporary immersion bioreactor
B) View of a culture room equiped with RITA® bioreactors
C) Direct sowing on horticultural substrate of coffee somatic embryos mass produced in a RITA® bioreactor
D) BIT® type twin flasks system
E) Pineapple explants proliferating in a twin flasks system
F) Pineapple stems elongating in a BIT® bioreactor

5. Effect of temporary immersion on plant quality

5.1 Morphological characteristics of plant material produced in a bioreactor

Pineapple leaves borne on temporarily immersed proliferating shoots were smaller than those produced in a liquid medium (Escalona et al., 1999). Clusters of shoots produced during the proliferation of axillary buds of shoots in a bioreactor were almost spherical and the shoots all formed around a central region. Some of the shoots, which were too small for *ex vitro* rooting and acclimatization, required an elongation phase in the same bioreactor. Cow tree (*Mitragyna inermis*) shoots formed in temporary immersion were considerably longer and had more foliage than those produced on an agar medium (Tisserat and Vandercook, 1985). Serviceberry (*Amelanchier* x *grandiflora* 'Princess Diana') shoots propagated by temporary immersion were also heavier and longer than those obtained on an agar medium (Krueger et al., 1991). Shoots of grape and *Amelanchier alnifolia* produced in a liquid medium on tilting or rocker machines were longer and rooted better and more quickly than those produced on agar media (Harris and Mason, 1983). The positive effect of temporary immersion on shoot elongation probably resulted from the larger volume of the container (Monette, 1983; Krueger, 1991).

Temporary immersion had a highly positive effect on the development of *Citrus* somatic embryos, leading to the development of cotyledons and protoderm, which does not occur in suspensions (Cabasson et al., 1997). The embryos obtained were morphologically identical to nucellar embryos. Likewise, in *Coffea*, *Hevea* and *Musa*, somatic embryos produced in a RITA® type bioreactor revealed a morphological quality that was generally better than that obtained on a solid medium or in Erlenmeyer flasks (Etienne et al., 1993, 1997; Escalant et al., 1994). In *Hevea,* the proportion of morphologically abnormal somatic embryos was reduced by half with temporary immersion, compared to those produced on agar medium. Development problems due to continuous stirring in suspensions or in conventional bioreactors are also a source of shear forces that can wound the plant material, can also substantially reduced in temporary immersion systems. Stirring, which have negative effects on polar distribution of growth regulators, which is essential during early ontogenetic stages (Liu et al., 1993). Effective germination in a liquid medium was successfully obtained for the first time with coffee using temporary immersion (Etienne-Barry et al., 1999). Densities exceeding 1,600 embryos per 1-litre bioreactor had a positive effect on germinated embryo morphology by stimulating elongation of the embryonic axis (+4-5 mm), an increase in fresh weight

(+100%) and a reduction in cotyledon area. These three morphological changes were positively correlated with the rate of conversion into plants after sowing on horticultural substrate, and to plantlet growth rates.

The quality of tissues grown in a temporary immersion system can change during subcultures, revealing a period of adaptation to this system. The percentage of normal waxy (abundant tubular epicuticular wax) radiata pine shoots harvested monthly increased significantly over the culture period from 41% at the first harvest to 93% at the eighth harvest, and remained high at 97% from the ninth to twelfth harvests (Aitken-Christie and Jones, 1987). We also observed a similar adaptation phenomenon to changes in the immersion times in coffee somatic embryo cultures.

5.2 *Hyperhydricity*

Micropropagation in liquid culture media increases nutrient uptake and promotes growth, but hyperhydricity is frequently seen. Continuous contact of plant tissues with the liquid medium, be it total or partial, is the source of hyperhydricity (Debergh et al., 1981; Ziv et al., 1983; Hussey, 1986). It is characterized by different degrees of morphological and physiological disorders including a glassy, waterlogged-tissue appearance, disordered growth in the shoot system, and more specifically in the leaves (Ziv, 1995). It was also shown that increasing aeration in the culture recipient (Hussey, 1986) and intermittent contact between the plant material and the liquid culture medium (Aitken-Christie and Jones, 1987) could reduce hyperhydricity. These two characteristics are combined in most temporary immersion systems.

In a work on banana micropropagation, Alvard et al. (1993) did not report any hyperhydricity symptoms in shoots grown in a temporary immersion system, whereas stems immersed in a liquid medium with continuous bubble–aeration revealed hyperhydricity in the outer leaf sheaths. Serviceberry shoots propagated in a liquid medium revealed severe hyperhydricity symptoms, such as highly translucent, curled and thickened leaves and stems (Krueger et al., 1991). Temporary immersions of 5 minutes every hour were found to prevent hyperhydricity. However, these authors noted that more frequent immersions (5 minutes every 30 seconds) resulted in this physiological problem.

In the long term, the proportion of wet shoots (no tubular epicuticular wax, small amounts of globular epicuticular wax) in radiata pine was significantly smaller in a temporary immersion culture system with liquid medium replenishment, than on agar medium, and fell from 59% at the first harvest to 7% at the eighth (Aitken-Christie and Jones, 1987). According to these authors, this change was probably due to the fact that monthly harvests of axillary shoots were formed closer to the upper part of the container, hence far from the agar and the surface of the liquid, where humidity was higher and where the liquid remained on the surface of the needles.

Coffea arabica microcuttings revealed little or no hyperhydricity with immersion times of 15 min every 6 hours, but longer immersion times caused it to occur (Berthouly et al., 1995). For such immersion times, *Coffea canephora* cuttings were glassy. In this species, which is more susceptible to this problem, immersion times had to be reduced to 1 min every 6 hours.

Somatic embryos of *Citrus* grown in a temporary immersion system did not reveal any hyperhydricity, unlike those obtained in suspensions or even more so, those grown on an agar medium (Cabasson et al., 1997). Rubber tree somatic embryos produced in a temporary immersion system did not reveal any hyperhydricity symptoms (Etienne et al., 1997b). However, a glassy appearance was found in germinating material placed in a plant conversion medium, but that was probably due to the excessive immersion times used (15 min immersion every 6 hours). Just as was seen with coffee, hyperhydricity risks are greater in the later phases of somatic embryogenesis (i.e. germination and conversion into plants). Short immersion times need to be used to avoid this problem. We found that vitrification of *C. arabica* somatic embryos was primarily caused by the immersion times, but not by the frequency if short immersions (1 min) were used. Several 15 min immersions per day led to embryo populations that were mostly glassy. Using temporary immersion therefore, generally reduces hyperhydricity problems, making it possible to control them by adjusting immersion frequencies and times.

5.3 Acclimatization of material produced in a temporary immersion system

Losses during the acclimatization stage can severely handicap conventional micropropagation procedures. Indeed, the physiological condition and hyperhydricity of plants grown on an agar medium or in a liquid medium make this a problematic stage. Temporary immersion systems have resulted in more successful acclimatization. *Ex vitro* acclimatization and rooting of pineapple shoots (Escalona et al., 1999) and sugarcane shoots (Lorenzo et al., 1998) produced in a temporary immersion system under semi-industrial conditions has proved efficient and routine. In pineapple, the survival rate increases linearly with shoot size. Shoots measuring over 6 cm can be grown directly in the greenhouse, with 90 to 100% successful acclimatization (Escalona et al., 1999). Smaller shoots require a longer culture period prior in the bioreactor to acclimatization. Similarly, coffee microcuttings (*C. arabica* and *C. canephora*) obtained by temporary immersion can be acclimatized directly after rooting induction in the same bioreactor (Berthouly et al., 1995). After planting, the shoots and flowers of aster plants obtained from shoot tips grown in a temporary immersion system were larger than those from an agar medium (Tisserat and Vandercook, 1985). Shoots of radiata pine harvested from a nutrient replenishment system were very effectively rooted and no decline in rooting ability was found during successive harvests for 18 months (Aitken-Christie and Jones, 1987). Given the larger proportion of waxy shoots compared to wet shoots, which root poorly and have low survival rates (Aitken-Christie et al., 1985), when compared to a conventional system on agar, using a temporary immersion system provided an unexpected bonus during the acclimatization stage. Shoots of grape and *Amelanchier alnifolia* produced in a liquid medium on tilting machines also rooted better and more quickly than those produced on agar media (Harris and Mason, 1983).

Potato tubers propagated in a twin flask system could be stored under room conditions and directly transplanted to soil without any acclimatization (Akita and Takayama, 1994). Lastly, in *Coffea arabica*, temporary immersion has enabled mass production of germinated somatic embryos capable of successfully regenerating into plants (70-80% embryo-to-plant conversion) after direct sowing on horticultural substrate (Etienne-Barry et al., 1999).

6. Impact of temporary immersion culture on production costs

The few investigations carried out so far confirm the spectacular increase in efficiency that could be expected from using a liquid medium culture procedure for micropropagation. For the proliferation of sugarcane shoots, Lorenzo et al. (1998) calculated that using temporary immersion reduced costs by 46% compared with the standard procedure on an agar medium. The cost of a shoot produced by temporary immersion was only 0.00104 USD as opposed to 0.00191 USD for a shoot produced on agar medium. This gain primarily resulted from a drastic reduction in direct work (from 0.00032 to 0.00004 USD per shoot produced) and in the space required in culture chambers (from 0.00017 to 0.00009 USD per shoot produced).

Using temporary immersion to propagate pineapple shoots resulted in a 100-fold increase in the number of shoots during the four-month period following inital culture of the crown buds (Escalona et al., 1999). This protocol reduced production costs per pineapple plant by 20% when compared to the conventional method in liquid medium. According to the authors, the key parameters for this success were: a reduction in the number of containers and in material handling, eliminating of cutting and explantation, elimination of *in vitro* rooting, and a reduction in contamination levels.

Long-term maintenance of radiata pine shoot hedges by nutrient replenishment led to a reduction in manpower requirements, facilitated the move towards automation and reduced the cost of micropropagated trees (Aitken-Christie and Jones, 1987). This was the first report on a method of culturing shoots as hedges for a period of up to 18 months without manual subculturing. The shoots were harvested in 600 ml glass jars at a rate of 672 shoots per hour and approximately 1,100 shoots were produced per square metre of agar surface per month. Based on that harvesting rate, this system is around 7 times cheaper than the normal method of radiata pine shoot subculturing on agar medium as described by Smith (1985). Culturing in liquid medium on tilting and rocker machines, as described by Harris and Mason (1983), led to lower costs by reducing the volume of medium by 50% and of agar by 90% or more, in addition to gains in shoot yields and shoot size, and better rooting.

In somatic embryogenesis processes, the late culture phases are the most expensive, due to the amount of handling required. For instance, with coffee and banana, for which mass production procedures have been established, temporary immersion bioreactors are used either for the production and germination of somatic embryos as for coffee (Etienne-Barry et al., 1999), or just for germination as for banana (Teisson et al., 1999). The production and germination of coffee somatic embryos in a temporary immersion bioreactor, combined with direct sowing of germinated embryos under *ex vitro* conditions reduced handling times to 13% and shelving area requirements to 6.3% of the values obtained with conventional acclimatization of plants developed on agar media (Etienne-Barry et al., 1999). Moreover, the time spent *in vitro* was reduced by 3 months.

Scaling-up experiments with temporary immersion systems are currently under way, with a view to commercial production. For some years, the French research organization CIRAD has been using mass propagation by somatic embryogenesis with the RITA® system to disseminate selected F_1 hybrids of *C. arabica* in Central America in cooperation with CATIE and PROMECAFE institutions and in Tanzania, in cooperation with the ARTI of Lyamungu. A Cuban team at the Centro de Biopletas in Ciego de Avila is also preparing for commercial propagation of sugarcane and pineapple using meristem proliferation techniques with the BIT® twin flasks system.

7. Conclusions

A literature review of temporary immersion reveals that this original culture system has raised considerable interest mainly since the beginning of the 90'. Many authors have indicated that temporary immersion has positive effects at all stages of the shoot proliferation and somatic embryogenesis processes. Plant cell and tissue growth and proliferation rates are generally better than those obtained on agar media or in bioreactors. Regenerated plantlets and somatic embryos are of better quality, i.e. they reveal greater similarities with *ex vitro* plants and zygotic embryos, respectively. Better results have also been obtained during acclimatization. Temporary immersion combines the advantages of solid culture media (maximum gas exchanges) and liquid media (increased nutrient uptake). Explant immersion times and frequencies are probably the parameters requiring the greatest attention when developing a micropropagation procedure. Their optimization, probably towards lower values, results in higher biological yields through greater control of morphogenetic response, but also through the control of hyperhydricity. This is a major advantage over traditional bioreactors.

Little research has been carried out on the effects of temporary immersion on the physiology of plant material. Better knowledge of such an impact is currently needed to optimize culture conditions in these simplified bioreactors, including data on culture density which is a determining factor but has been superficially examined until now, immersion times for each culture stage, duration of subcultures, and the chemical composition of media. The beneficial effect of temporary immersion systems in reducing the action of toxic substances or growth inhibitors exuded by the cultured material, either through a rinsing or dilution effect has not been assesed either. It is likely that this effect is all the greater in that used medium is eliminated or in contact with the plant tissue for only a limited period.

In addition to its positive effects on multiplication and plant material quality, temporary immersion is also an opportunity to benefit from the economic advantages of using liquid media in micropropagation. The semi-automatic systems proposed have gradually become simpler. The simplicity and low cost of recently developed bioreactors are compatible with their use for large-scale propagation. The significant reduction in the production cost of vitroplants brought about by temporary immersion means that economically viable use of somatic embryogenesis, microtuberization and shoot proliferation can be envisaged for several economically important species.

References

Aitken-Christe J & Davies HE (1988) Development of a semi-automated micropropagation system. Acta Hort. 230: 81-87

Aitken-Christie J, Jones C & Bond S (1985) Wet and waxy shoots in radiata pine micropropagation. Acta Hortic. 166: 93-100

Aitken-Christie J & Jones C (1987) Towards automation: radiata pine shoot hedges *in vitro*. Plant Cell, Tiss. Org. Cult. 8: 185-196

Akita M & Takayama S (1994) Stimulation of potato (*Solanum tuberosum* L.) tuberization by semicontinuous liquid medium surface level control. Plant Cell Rep. 13: 184-187

Alvard D, Côte F & Teisson C (1993) Comparison of methods of liquid medium culture for banana micropropagation. Effects of temporary immersion of explants. Plant Cell, Tiss. Org. Cult. 32: 55-60

Berthouly M, Dufour M, Alvard D, Carasco C, Alemano L & Teisson C (1995) Coffee micropropagation in a liquid medium using the temporary immersion technique. In: ASIC Publishers (eds) 16th International Scientific Colloquium on Coffee, Kyoto, Japon (pp. 514-519). Vevey, Switzerland

Berthouly M & Etienne H (1999) Somatic embryogenesis of coffee. In: Jain SM, Gupta PK & Newton RJ (eds) Somatic Embryogenesis in Woody Plants, Vol. 5 (pp. 259-288). Kluwer Academic Publishers, London

Cabasson C, Alvard D, Dambier D, Ollitrault P & Teisson C (1997) Improvement of *Citrus* somatic embryo development by temporary immersion. Plant Cell,Tiss. Org. Cult. 50: 33-37

Chu I (1995) Economic analysis of automated micropropagation. In: Aitken-Christie J, Kozai T, Smith MAL (eds) Automation and Environmental Control in Plant Tissue Culture (pp. 19-27). Kluwer Academic Publischers, Dordrecht

Connor AJ & Meredith CP (1984) An improved polyurethane support system for monitoring growth in plant cell cultures. Plant Cell, Tiss. Org. Cult. 3: 59-68

Debergh P (1988) Improving mass propagation of *in vitro* plantlets. In: Kozai T (ed) Horticulture in High Technology Era (pp. 45-57). International Symposium on High Technology in Protected Cultivation, Tokyo

Debergh P, Harbaooui Y & Lemeur R (1981) Mass propagation of globe artichoke (*Cynara scolymus*) Evaluation of different hypotheses to overcome vitrification with special reference to water potential. Physiol. Plant. 53: 181-187

Escalant J-V, Teisson C & Côte F (1994) Amplified somatic embryogenesis from male flowers of triploid banana and plantain cultivars (*Musa* spp.). *In Vitro* Cell. Dev. Biol. 30: 181-186

Escalona M, Lorenzo JC, González B, Daquinta M, Fundora Z, Borroto CG, Espinosa D, Arias E & Aspiolea ME (1998) New system for *in vitro* propagation of pineapple (*Ananas comosus* (L.) Merr). Pineapple News 5: 5-7

Escalona M, Lorenzo JC, González B, Daquinta M, González JL, Desjardins Y & Borroto CG (1999) Pineapple (*Ananas comosus* L. Merr) micropropagation in temporary immersion systems. Plant Cell Rep. 18: 743-748

Etienne H, Berger A & Carron MP (1991) Water status of callus from *Hevea brasiliensis* during induction of somatic embryogenesis. Physiol. Plant. 82: 213-218

Etienne H, Bertrand B, Anthony F, Côte F & Berthouly M (1997a) L'embryogenèse somatique: un outil pour l'amélioration génétique du caféier. In: ASIC Publishers (eds) 17th International Scientific Colloquiun on Coffee, Nairobi, Kenya (pp. 457-465). Vevey, Switzerland

Etienne H, Lartaud M, Michaux-Ferrière N, Carron MP, Berthouly M & Teisson C (1997b) Improvement of somatic embryogenesis in *Hevea brasiliensis* (Müll. Arg.) using the temporary immersion technique. *In Vitro* Cell. Dev. Biol.-Plant. 33: 81-87

Etienne-Barry D, Bertrand B, Vásquez N & Etienne H (1999) Direct sowing of *Coffea arabica* somatic embryos mass-produced in a bioreactor and regeneration of plants. Plant Cell Rep. 19: 111-117

Gawel NJ & Robacker CD (1990) Somatic embryogenesis in two *Gossypium hirsutum* genotypes on semi-solid versus liquid proliferation media. Plant Cell, Tiss. Org. Cult. 23: 201-204

Hamilton R, Pederson H & Chin CK (1985) Plant tissue culture on membrane rafts. BioTechniques. March/April, p. 96

Hammerschlag F (1982) Factors affecting establishment and growth of peach shoots *in vitro*. HortScience 17: 85-86

Harris RE & Mason EB (1983) Two machines for *in vitro* propagation of plants in liquid media. Can. J. Plant Sci. 63: 311-316

Harris RE & Stevenson JH (1982) *In vitro* propagation of *Vitis*. Vitis 21: 22-32

Hussey G (1986) Problems and prospects in the *in vitro* propagation of herbaceous plants. In: Withers LA & Alderson PG (eds) Plant Tissue Culture and its Agricultural Applications (pp. 69-84). Butterworths, Boston

Jiménez E, Pérez J, Gil V, Herrera J, García Y & Alonso E (1995) Sistema para la propagación de la caña de azucar. In: Estrade M, Riego E, Limonta E, Tellez P & Fuente J (eds) Avances en Biotecnología Moderna, Vol 3 (pp. 11-20). Elfos Scientiae, Cuba

Jones AM & Petolino JF (1988) Effects of support medium on embryo and plant production from cultured anthers of soft-red winter wheat. Plant Cell, Tiss. Org. Cult. 12: 243-261

Kitto SL (1997) Commercial micropropagation. Hort. Sci. 32: 1012-1014

Krueger S, Robacker C & Simonton W (1991) Culture of *Amelanchier x grandiflora* in a programmable micropropagation apparatus. Plant Cell, Tiss. Org. Cult. 27: 219-226

Liu CM, Xu Z-H & Chua N-H (1993) Auxin polar transport is essential for the establishment of bilateral symmetry during early plant embryogenesis. The Plant Cell 5: 621-630

Lorenzo JC, Gonzalez BL, Escalona M, Teisson C, Espinosa P & Borroto C (1998) Sugarcane shoot formation in an improved temporary immersion system. Plant Cell, Tiss. Org. Cult. 54: 197-200

Maene L & Debergh P (1985) Liquid medium additions to established tissue cultures to improve elongation and rooting *in vivo*. Plant Cell, Tiss. Org. Cult. 5: 23-33

Monette PL (1983) Influence of size of culture vessel on *in vitro* proliferation of grape in a liquid medium. Plant Cell Tiss. Org. Cult. 2: 327-332

Noriega C & Söndahl MR (1993) Arabica coffee micropropagation through somatic embryogenesis via bioreactors. In: ASIC publishers (eds) 15th International Scientific Colloquium on Coffee, Montpellier, France (pp. 73-81). Vevey, Switzerland

Reuther G (1985) Principles and application of the micropropagation of ornemental plants. In: Schäfer-Menuhr A (ed) *In Vitro* Techniques: Propagation and Long-Term Storage (pp. 1-14). Martinus Nijhoff, Dordrecht

Simonton W & Robacker C (1988) Alternative system for micropropagation. Amer. Soc. Agr. Engineers, Technical Paper N° 88-1028

Simonton W, Robacker C & Krueger S (1991) A programmable micropropagation apparatus using cycled medium. Plant Cell, Tiss. Org. Cult. 27: 211-218

Sluis CJ & Walker KA (1985) Commercialization of plant tissue culture propagation. Intl. Assoc. Plant Tiss. Cult. Newsl. 47: 2-12

Smith DR (1985) Pinus radiata. In: Bajaj YPS (ed) Trees. Biotechnology in Agriculture and Forestry, Vol 1 (pp. 274-291). Springer Verlag, Berlin

Smith MAL & Spomer LA (1995) Vessels, gels, liquid media and support systems. In: Aitken-Christie J, Kozai T & Smith MAL (eds) Automation and Environmental Control in plant tissue culture (pp. 371-405). Kluwer Academic Publishers, Dordrecht

Söndhal SP, Nakamicra F, Medince Filho MP, Carvalho A, Fazuoli LC & Caslo WM (1989) Coffee Handbook of Plant Cell Culture, Vol III, Chap. 21

Stevenson JH & Harris RE (1980) *In vitro* plantlet formation from shoot tip explants of *Fuchsia hybrida* cv. Swingtime. Can. J. Bot. 58: 2190-2192

Teisson C & Alvard D (1995) A new concept of plant *in vitro* cultivation liquid medium: temporary immersion. In: Terzi M et al. (eds) Current Issues in Plant Molecular and Cellular Biology (pp. 105-110). Kluwer Academic Publishers, Dordrecht

Teisson C & Alvard D (1999) *In vitro* production of potato microtubers in liquid medium using temporay immersion. Potato research 42: 499-504

Teisson C, Alvard D, Lartaud M, Etienne H, Berthouly M, Escalona M & Lorenzo JC (1999) Temporary immersion for plant tissue culture. In: Plant Biotechnology and *In vitro* Biology in the 21st Century, Proceedings of the IXth International Congress of Plant Tissue and Cell Culture, Section H: Novel micropropagation methods (pp. 629-632). Jerusalem

Tisserat B, Jones D & Galletta PD (1993) Construction and use of an inexpensive *in vitro* ultrasonic misting system. Hort.Technology 3: 75-79

Tisserat B & Vandercook CE (1985) Development of an automated plant culture system. Plant Cell, Tiss. Org. Cult. 5: 107-117

Wardle K, Dobbs EB & Short KC (1983) *In vitro* acclimatization of aseptically cultured plantlets to humidity. J. Amer. Soc. Hort. Sci. 108: 386-389

Wataad A, Raghothana KG, Kochba M, Nissim A & Caba V (1997) Micropropagation of Spathiphyllum and Syngonium is facilited by use of interfacial membrane rafts. HortScience 32: 307-308

Weathers PJ, Cheetham RD & Giles KL (1988) Dramatic increases in shoot number and lengths for *Musa, Cordyline* and *Nephrolepsis* using nutrient mists. Acta Hort. 230: 39-44

Weathers PJ & Giles KL (1988) Regeneration of plants using nutrient mist culture. *In vitro* Cellular and Development Biology 24: 727-732

Zamarripa A, Ducos JP, Tessereau H, Bollon H, Eskes AB & Petiard V (1991) Developpement d'un procédé de multiplication de masse en masse du caféier par embryogenèse somatique en milieu liquide. In: ASIC Publishers (eds) 14th International Scientific Colloquium on Coffee, San Francisco, US (pp. 392-402). Vevey, Switzerland

Ziv M, Meir G & Halevy AH (1983) Factors influencing the production of hardened glaucous carnation plantlets *in vitro*. Plant Cell, Tiss. Org. Cult. 2: 55-65

Chapter 12

Mass propagation of tropical crops in temporary immersion systems

Elio Jiménez González
Institute of Plant Biotechnology. Central University of Las Villas. Carretera a Camajuaní Km 5 ½. Santa Clara, Villa Clara. Cuba. E-mail: ejimenez@ibp.uclv.edu.cu

Abstract: Temporary immersion systems (TIS) have been described for in vitro multiplication of a wide range of tropical crops. Laboratory protocols are available for shoot multiplication, somatic embryo and microtuber production. Seven species are now commercially propagated by this culture technique (*Ananas comosus, Coffea arabica, Cymbopogon citratus, Musa* sp., *Phalaenopsis, Saccharum* sp., *Solanum tuberosum*) with different regeneration pathways and a variety of TIS designs. Beside the development of methods for producing somatic embryos in TIS, shoot multiplication protocols are the most applied from the commercial point of view. RITA® proved to be a suitable tool for research and laboratory scale, but for commercial application larger vessels are frequently used. Most important tropical species are commercially propagated in TIS using twin flasks ranging from 5-10 litres.
Production strategies for plant propagation in TIS, either by organogenesis or somatic embryogenesis are discussed and examples are given to illustrate the different possibilities for TIS integration in propagation of *Musa sp.* and *Solanum tuberosum* for shoot multiplication and microtuber production. For sugarcane (*Saccharum* sp.) somatic embryo production in bioreactors, embryo germination in TIS and field performance of regenerated plants are described.

Key words: automation, liquid medium, shoot multiplication, somatic embryos, microtubers

Abbreviations: ANC - ancymidol; ARTI - Agricultural Research and Training Institute (Lyamungu, Tanzania); 6-BAP – 6-benzylaminopurine; CATIE – Centro Agronómico Tropical de Investigación Enseñanza (Cali, Colombia); CIAT - Centro Internacional de Agricultura Tropical; CIRAD - Centre de Coopération Internationale en Recherche Agronomique pour le Développement (Montpellier, France); CLAYUCA – Consorcio Latinoamericano y del Caribe de Apoyo a la Investigación y Desarrollo de la Yuca; PBZ - paclobutrazol; DO2 – dissolved oxygen; INIVIT – Instituto Nacional de Investigaciones en Viandas Tropicales; PROMECAFE – Programa Cooperativo Regional para el Desarrollo Tecnológico y la Modernización de la Caficultura de Centroamérica

A.K. Hvoslef-Eide and W. Preil (eds.), Liquid Culture Systems for in vitro Plant Propagation, 197–211.

1. Introduction

Commercial *in vitro* propagation is currently applied to several tropical plants. Among food crops bananas, plantains, sugarcane and pineapples are the most successful examples. However, *in vitro* propagation methods are available for a large number of other tropical plants, which are normally used on a laboratory scale for research purposes, germplasm preservation or propagation of selected plants in research or academic institutions.

The present limitation to the wider commercial application of micropropagation in tropical crops in developing countries is the high cost of the plants when using the traditional micropropagation methods, e.g. bananas and plantains (0.15 – 0.30 USD), sugarcane (0.12 – 0.20 USD), pineapple (0.10 – 0.25 USD) (Pérez et al, 1998).

Labour costs in Latin America are lower compared to developed countries in North America, Europe or Asia, but still remain as the major part of propagation costs. Manual labor for *in vitro* propagated plants in Latin America represents from 45 – 60 % of total costs (Pérez et al., 1998).

Partial automation of *in vitro* propagation and especially temporary immersion systems (TIS) may help to reduce propagation costs (Etienne and Berthouly, 2002). TIS can increase the efficiency of propagation processes by reducing the cost as a result of labour reduction, savings in shelf space, high biological yields, improved plant quality or low investment and maintenance costs. For pineapple shoot multiplication the application of TIS resulted in 20 % cost reduction compared to cultures in conventional liquid medium (Escalona et al., 1999) and sugarcane propagation in TIS saved 46 % of the costs in comparison with the standard procedure in semisolid medium (Lorenzo et al., 1998).

2. TIS from laboratory to commercial application

TIS have been described for *in vitro* multiplication of a wide range of tropical crops such as *Ananas comosus, Camellia sinensis, Citrus deliciosa, Coffea s.p, Colocasia sp., Eucalyptus sp., Hevea brasiliensis, Manihot esculenta, Musa sp., Psidium guajava, Saccharum sp., Solanum tuberosum* (Table 1). Laboratory protocols have been developed by using different regeneration systems: Shoot multiplication, somatic embryos (either multiplication and/or germination) and induction of storage organs (microtubers).

Table 1: Examples of tropical plants propagated in TIS on a laboratory scale

Plant species	**Regeneration system**	**TIS**	**Reference**
Ananas comosus	Shoots	twin flasks	Escalona et al., 1998
Camellia sinensis	Somatic embryos including germination	RITA ®	Akula et al., 2001
Citrus deliciosa	Somatic embryos including germination	RITA ®	Cabasson et al., 1997
Coffea arabica and Coffea canephora	Shoots	RITA ®	Berthouly et al., 1995
Colocasia sp.	Shoots	10-l twin flasks	Medero eta al., 2001
Hevea brasiliensis	Somatic embryos including germination	RITA ®	Etienne et al., 1997
Eucalyptus	Shoots	RITA ®	Mc Alister et al., 2002
Manihot esculenta	Shoots	10-l twin flasks	Medero et al., 2001
Musa sp.	Shoots	Nalgene filtration unit	Alvard et al., 1993
	Somatic embryos including germination	RITA ®	Escalant et al., 1994
	Somatic embryos including germination	10-l twin flasks	Gómez et al., 2002
Psidium guajava	Somatic embryo germination	RITA ®	Vilches et. al., 2002
Saccharum sp.	Shoots	twin flasks	Lorenzo et al., 1999
	Somatic embryo germination	1-l twin flasks	Jimenez et al., 2002
Solanum tuberosum	Microtubers	10-l twin flasks	Jimenez et al., 1999
	Microtubers	RITA ®	Teisson & Alvard, 1999

Several system designs and culture vessels are used: modified Nalgene filtration units as described by Alvard et al. (1993), RITA® system (Teisson and Alvard, 1995) and the twin flask system, which uses a pair of bottles connected by silicone tubes, one as medium container and the other for plant growth (Escalona et al., 1999; Jiménez et al., 1999). In the latter case also different vessels have been used, ranging from small glass vessels (250 ml) up to 5 or 10-litre vessels, either of glass or polycarbonate. For a detailed description of the different immersion culture systems see the review by Etienne and Berthouly (2002) and Berthouly and Etienne (this volume pp. 161-190).

A successful example of the practical application of TIS on a laboratory scale is the protocol developed at the Instituto Nacional de Investigaciones en Viandas Tropicales (INIVIT, Cuba) for *Manihot esculenta* shoot multiplication. The aim is to speed up the selection process in genetic improvement programs as well as for the distribution of elite seed. The technology was transferred to CIAT (Colombia) in the frame of a regional collaboration consortium for research on *Manihot* (CLAYUCA). It is currently used for germplasm distribution among participating countries and it is planned to expand it to all members in the future (Mederos et al., 2001). *Colocasia* sp. plants are also produced at INIVIT on a laboratory scale by shoot multiplication in 10-litre TIS. The objective is to provide elite plants for the national seed production program and for the growers. Seven species are currently propagated on a commercial scale (Table 2).

Table 2: Examples of tropical plants propagated in temporary immersion systems (TIS) on a commercial scale

Plant species	**Regeneration system**	**TIS**	**Institution**
Ananas comosus	Shoots	10-l twin flasks	Centro de Bioplantas, Cuba
Coffea arabica	Somatic embryos incl. germination	RITA ®	CIRAD-CATIE-PROMECAFE, Costa Rica, and CIRAD-ARTI, Tanzania
Cymbopogon citratus	Shoots	5-l twin flasks	BioPlanta GmbH, Germany, and Institute of Plant Biotechnology, Cuba
Musa sp.	Shoots	10-l twin flasks	Institute of Plant Biotechnology, Cuba
Phalaenopsis	Shoots	5-l twin flasks	Institute for Ornamental Plant Breeding, Ahrensburg, Germany
Saccharum sp.	Shoots	10-l twin flasks	Centro de Bioplantas, INICA, Cuba
Solanum tuberosum	Microtubers	10-l twin flasks	Institute of Plant Biotechnology, Cuba

A scale up and production facility was established at Centro de Bioplantas (Ciego de Avila, Cuba) for *Ananas* shoot multiplication in 10-litre twin flask TIS. Shoots multiplied in TIS are used as starting material in order to speed up the propagation process in standard semisolid medium and also for the production of transplants in a two stage system which involves multiplication and elongation in TIS (Escalona et al., 1999). The same group also developed a protocol for sugarcane shoot multiplication (Lorenzo et al., 1999), this technology was transferred to a commercial facility at the Instituto Nacional de Investigaciones de la Caña de Azúcar (INICA) which is used for the multiplication of elite varieties and basic seed production.

CIRAD (Montpellier, France) has been using mass propagation by somatic embryogenesis with the RITA® system to disseminate clones of selected F_1 *Coffea arabica* hybrids in Central America (in cooperation with CATIE and PROMECAFE) and in Tanzania (in cooperation with ARTI of Lyamungu) (Etienne and Berthouly, 2002).

BioPlanta GmbH (Leipzig, Germany) in collaboration with IBP (Cuba) developed a scale up and production facility for biomass (shoots) and secondary metabolite production in *Cymbopogon citratus, Lavandula offcinialis, Hypericum perforatum* and *Fabiana imbricata* using TIS 5-litre twin flasks (Hohe et al., 2002).

The Institute for Ornamental Plant Breeding (Ahrensburg, Germany) established a procedure for *Phalaenopsis* propagation in 5-litre twin flasks, which comprises a first step of adventitious shoot multiplication followed by rooting in TIS (Hempfling and Preil, 2002).

At the Institute of Plant Biotechnology (Santa Clara, Cuba) a production facility with 10-litre twin flask TIS was built (Figure 1). Potato microtubers are produced according to the protocol described by Jiménez et al. (1999). Tubers are directly transplanted to greenhouse or field for elite seed production (Pérez et al., 2000). A second application is banana propagation by shoot multiplication in medium containing growth retardants in 10-litre TIS and subsequent elongation and rooting in TIS using a hormone free medium (Albany et al., 2002).

From the above mentioned examples some conclusions may be drawn:

- Regeneration system: Beside the development of methods for producing somatic embryos in TIS, shoot multiplication still remains the main alternative for tropical crop propagation.
- TIS design: RITA® is a suitable tool for research and lab scale, however for commercial application larger vessels are frequently used. Most important tropical species are commercially propagated in twin flasks TIS ranging from 5 to 10 litres.

Figure 1: Production facility at the Institute of Plant Biotechnology, Cuba (left). Potato microtuber production (up right), and banana propagation in 10-litre twin flask TIS (low right).

3. Production strategies for plant propagation in TIS

For TIS integration into commercial propagation using organogenic systems, establishment of standard *in vitro* culture procedure is advisable which includes: donor plant selection, culture initiation and shoot multiplication. Afterwards TIS can be used for further shoot proliferation and multiplication. Shoots produced can subsequently be rooted in semisolid or static liquid medium or elongated and rooted in TIS by a simple medium exchange in the same culture vessel. The last option is the most attractive from the commercial point of view due to savings in man power by avoiding manipulation and transfer of shoots (Figure 2).

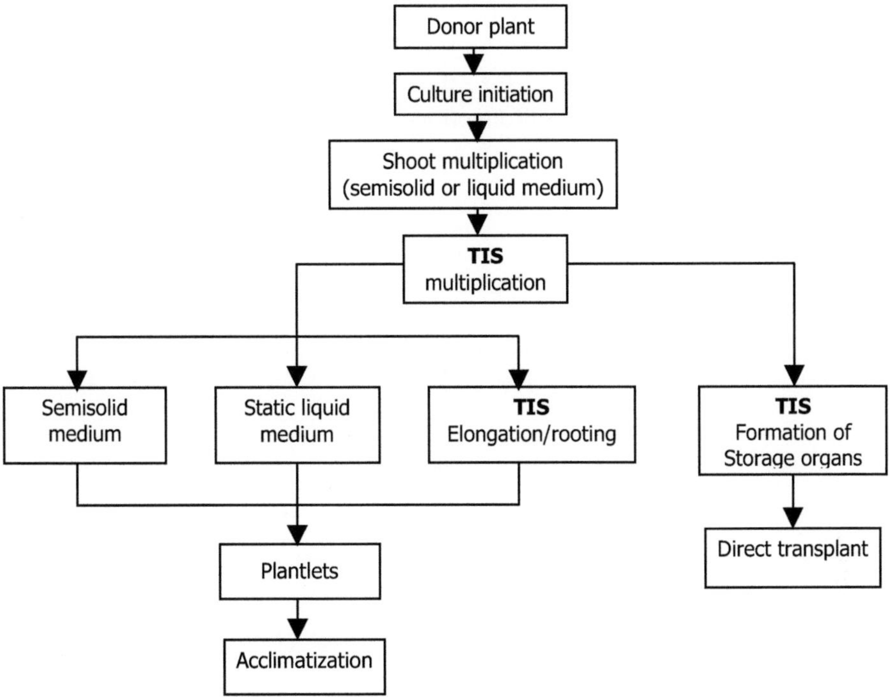

Figure 2: Scheme of TIS integration in plant propagation *via* organogenesis.

Another interesting option is the formation of storage organs like microtubers or microbulbs, since they can be afterwards directly transplanted in the greenhouse or field without acclimatization.

For the application of TIS for somatic embryo production a number of options are amenable for use (Figure 3). As in organogenic propagation, a first standard step for culture initiation and somatic embryo induction is necessary, either in semisolid medium or in liquid cell suspension culture. From that point two possibilities arise, the first is the multiplication of somatic embryos by secondary embryogenesis in TIS followed by embryo germination in semisolid medium or in TIS (Escalant et al., 1994; Etienne et al., 1997; Cabasson et al., 1997; Berthouly and Etienne, 1999; Akula et al., 2001; Vilches et al., 2002). The second possibility is the use of bioreactors for scaling up the mass production of somatic embryos and the combination with TIS for embryo germination and plant development. This option is not fully exploited up to day, but it has been already applied in banana (Gómez et al., 2002) and sugarcane (Jiménez et al., 2002).

Two plant species, *Musa* sp. and *Solanum tuberosum,* are examples to show different possibilities for the application of TIS for shoot multiplication and microtuber production.

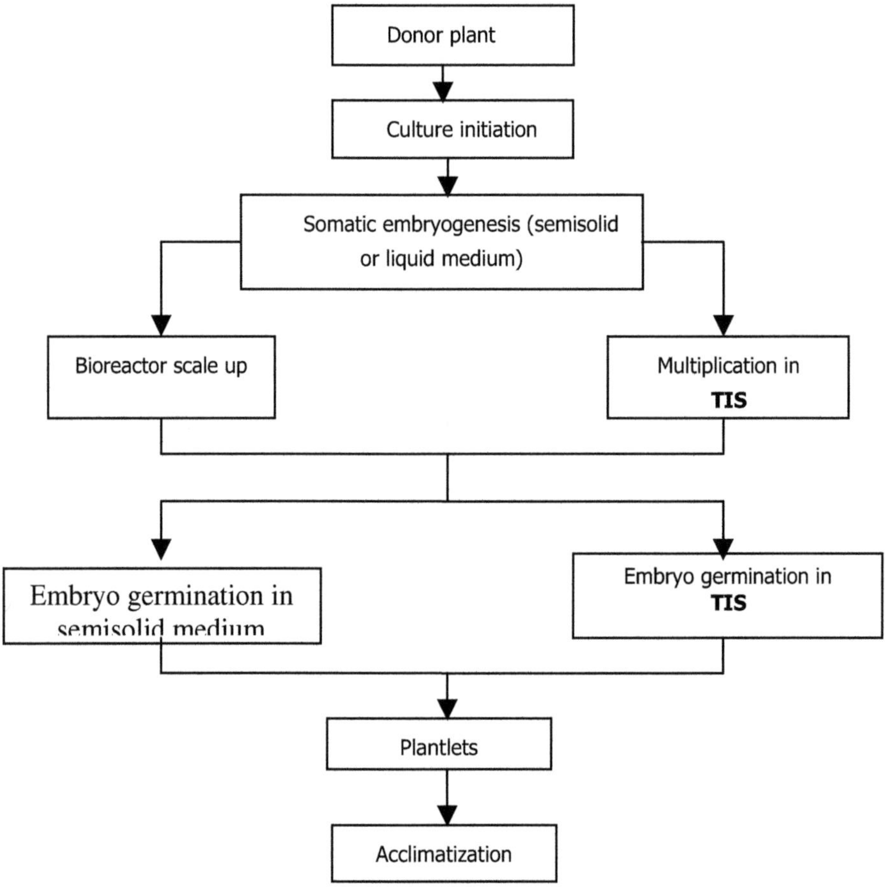

Figure 3: Scheme of TIS integration in plant propagation *via* somatic embryogenesis.

3.1 Musa sp. shoot multiplication

Banana shoot multiplication in small TIS vessels was first described by Alvard et al. (1993), who stressed the advantage of this technique for shoot development and proliferation in comparison with other liquid culture systems. However, the commercial application of TIS requires larger vessels. The use of 10-litre twin flask culture vessels resulted in an excessive enlargement of the shoots (leave surface and pseudo stem length). This limited the number of shoots produced per flask and reduced the production capacity in the growth room. Additionally, handling is more difficult when dividing and subculturing large shoots, which increases labor costs.

The solution to overcome the problems arising from large shoots is the reduction of the size of shoots by using growth retardants. Growth retardants have been successfully used to inhibit shoot length and to promote bud clusters formation in several species (Ziv, 1992; Ziv et al., 1998).

The application of PBZ and ANC in TIS (1-litre twin flasks) stimulated bud proliferation (Albany et al., 2002). Both compounds were also effective in controlling the excessive growth of the shoots and induce the formation of compact bud clusters. Shoots multiplied in TIS in the presence of PBZ (2.5 mg l^{-1}) were successfully transferred to semisolid or liquid rooting media in traditional culture vessels or in TIS. After 15 days plants were ready for acclimatization.

The protocol developed was further scaled up in 10-litre TIS using a two step procedure: first the shoots were multiplied in PBZ (2.5 mg l^{-1}) and 6-BAP (4 mg l^{-1}) containing medium for 4 weeks and, inside the same culture vessels, the multiplication medium was replaced by a hormone-free medium to promote shoot elongation and rooting (see Chapter 13, Figure 6).

3.2 Solanum tuberosum microtuber production

Potato (*Solanum tuberosum* L.) microtubers offer several advantages over *in vitro* propagated plants, since they can be stored and transplanted directly into the field without an acclimatization stage. Also handling and shipping are easier, thus facilitating commercialization and international exchange of germplasm. The main problems associated with microtuber production in conventional flasks are the low yield of tubers (1-1.5 tubers per plant) and the small tuber size that limits direct transplanting to field conditions.

At the Institute of Plant Biotechnology, Cuba, a temporary immersion system for potato microtuber production was designed using 4-litre twin flasks (Jiménez et al., 1999). In both cultivars tested, Desiree and Atlantic, an average of 3.1 and 2.8 tubers per single node cutting was achieved after 9 weeks in culture. Also the size and weight of the tubers were superior compared to cultures on solid media.

Akita and Takayama (1994) and Teisson and Alvard (1999) also stressed on the higher efficiency of TIS for potato microtuber production, not only because more tubers were produced per plant, but also size and weight of the tubers were increased.

Jiménez et al. (1999) scaled up a mass production protocol of potato microtubers in 10-litre twin flasks, with cv. Atlantic (Figure 4). Twelve TIS units were inoculated containing 150 single nodal cuttings each. An average of 2.6 tubers per inoculated cutting was obtained, with 1.3 g fresh weight per microtuber.

Field experiments were conducted to compare direct transplantation of potato microtubers into the field with plants propagated by *in vitro* cuttings (Pérez et al., 2000). Plants from microtubers produced in TIS showed an increased height and more stems per plant compared with plants from *in vitro* cuttings, which resulted in increased fresh weight and diameter of the tubers with statistical significant differences.

Figure 4: Scale up of potato microtuber production in 10-litre TIS (left) and close view of the microtuber inside the culture vessel (right).

4. Somatic embryogenesis in TIS

TIS can effectively improve the production and quality of somatic embryos. Escalant et al. (1994) described an increase in embryo number in several banana and plantain (*Musa* sp.) cultivars when comparing the yield from RITA® with that from cultures on agar medium. Somatic embryos of other tropical species have been successfully multiplied and/or germinated in TIS, such as *Camellia sinensis, Coffea arabica, Coffea canephora, Citrus deliciosa, Hevea brasiliensis, Psidium guajava* (Akula et al., 2001; Etienne et al., 1997a; Cabasson et al., 1997; Berthouly and Etienne, 1999; Vilches et al., 2002).

Standard bioreactors are also efficient for somatic embryo mass production, allowing a precise control of environmental factors inside the culture vessel (e.g. Preil, 1991). The factors limiting their practical application in plant propagation are frequently associated with the poor quality of the embryos produced and the lack of efficient germination protocols, which usually involve the use of semisolid medium, thus complicating the automation of the propagation process. However, the combination of bioreactor technology for somatic embryo production and embryo germination in TIS offers several advantages over conventional germination in semisolid medium, opening new possibilities for developing automated systems.

In our lab at the Institute of Plant Biotechnology, Cuba, experiments were conducted in order to improve the germination of sugarcane somatic embryos produced in bioreactors and to reduce manual labour costs. Embryogenic cell suspensions from the cultivar C 8751 were directly initiated in liquid medium from leaf segments of shoots derived from axillary bud proliferation medium, according to the protocol described by Freire (2001). Bioreactor culture was performed in 2-litre bioreactors with bubble free aeration systems and the oxygen concentration was controlled by gas blending. The effect of two partial oxygen pressures (40 and 80 %) on somatic embryo production was studied. The highest production of somatic embryos was obtained in the bioreactor with 80 %DO_2. However, only embryos developed under 40 %DO_2 regime germinated when transferred either to semisolid or liquid medium in TIS (1-litre twin flasks). In a second set of experiments DO_2 was controlled at 80 % from day 1 to 14 and afterwards at 40 % from day 15 to 22. A total of 31,400 somatic embryos was obtained, with a fresh weight of 44.9 g l^{-1} (Figure 5). The embryos germinated when transferred to semisolid germination medium or liquid germination medium in 1-litre TIS flasks, with an average germination of 20.6 % and 18.4 %, respectively (Figure 6).

Regenerated plants were successfully transferred for acclimatization and a total of 23,200 plants were transplanted to the field. Comparative field trials were conducted with plants regenerated from bioreactor culture, plants propagated *via* axillary bud proliferation, plants regenerated from somatic embryos from callus culture and traditional seed stalks from growers. Plants from all *in vitro* culture methods showed an increase in the number of stems per square meter and a decrease in stem diameter compared to plants from traditional seed stalks. No statistical differences were observed in sugar content (brix) as it was previously described by Jiménez (1995). Further improvements of the procedure are necessary in order to increase embryo germination rate. Extended field trials are also required to check the genetic stability of regenerated plants.

Figure 5: Sugarcane somatic embryo production in 2-litre bioreactors after 22 days in culture, embryos agitated (left), embryos settled down (right).

Figure 6: Experimental TIS unit with 1-litre twin vessels used for sugarcane somatic embryo germination (left) and plants from sugarcane somatic embryos produced in bioreactors and germinated in TIS (right).

5. Conclusions

- TIS represents an advanced technology for commercial mass propagation of tropical crops.
- Several plant species are now commercially propagated using this culture technique with different regeneration pathways and a variety of TIS designs.
- From the commercial point of view shoot multiplication systems are applied most frequently.
- Somatic embryo production and germination will be commercially applicable in near future by use of TIS or combinations of TIS and bioreactors.

Acknowledgements

I wish to thank Prof. Dr. Walter Preil for valuable discussions. Also to Victor Mederos and Jorge López from INIVIT and Dr. Maritza Escalona and Dr. Marcos Daquinta for providing data and materials used in this paper. I am also very grateful to my colleagues from the Bioreactors Laboratory at the Institute of Plant Biotechnology, which made possible the results included in this paper.

References

Akita M & Takayama S (1994) Stimulation of potato (*Solanum tuberosum* L.) tuberization by semicontinuous liquid medium surface level control. Plant Cell Rep. 13: 184-187

Akula A, Becker D & Bateson M (2001) High-yielding repetitive somatic embryogenesis and plant recovery in a selected tea clone, TRI-2025, by temporary immersion. Plant Cell Rep. 19: 1140-1145

Albany N, Vilchez J, Jiménez E, García L, de Feria M, Pérez N, Sarría Z, Pérez B & Clavelo J (2002) Use of growth retardants for banana (M*usa* AAA cv. Grand Naine) shoot multiplication in temporary immersion systems. First International Symposium on "Liquid Systems for *in vitro* Mass Propagation of Plants" (p. 108). Agricultural University of Norway. 29 May-2 June

Alvard D, Cote F & Teisson C (1993) Comparison of methods of liquid medium culture for banana micropropagation. Plant Cell Tiss. Org. Cult. 32: 55-60

Berthouly M & Etienne H (1999) Somatic embryogenesis of coffee. In: Jain SM, Gupta PK & Newton RJ (eds) Somatic Embryogenesis in Woody Plants, Vol. 5 (pp. 259-288). Kluwer Academic Publishers, London

Cabasson C, Alvard D, Dambier D, Ollitrault P & Teisson C (1997) Improvement of *Citrus* somatic embryo development by temporary immersion. Plant Cell Tiss. Org. Cult. 50: 33-37

Escalant J-V, Teisson C & Cote F (1994) Amplified somatic embryogenesis from male flowers of triploid banana and plantain cultivars (*Musa* spp.) *In Vitro* Cell. Dev. Biol. 30: 181-186

Escalona M, Lorenzo B, González B, Daquinta M, González JL, Desjardins Y & Borroto C (1999) Pineapple (*Ananas comosus* L. Merr) micropropagation in temporary immersion. Plant Cell Rep. 18: 743-748

Etienne H & Berthouly M (2002) Temporary immersion systems in plant micropropagation. Plant Cell Tiss. Org. Cult 69: 215–231

Etienne H, Lartaud M, Michaux-Ferriere N, Carron MP, Berthouly M & Teisson C (1997) Improvement of somatic embryogenesis in *Hevea brasiliensis* (Müll. Arg) using the temporary immersion technique. *In Vitro* Cell. Dev. Biol.-Plant. 33: 81-87

Freire M (2001) Nueva metodología de embriogénesis somática en caña de azúcar (*Saccharum* spp híbrido) empleando medios de cultivo líquidos. Tesis de Doctorado (p. 107). Instituto de Biotecnología de las Plantas. Universidad Central de Las Villas, Santa Clara, Cuba

Gómez R, de Feria M, Posada L, Gilliard T, Bernal F, Reyes M, Chávez M & Quiala E (2002) Somatic embryogenesis of the banana hybrid cultivar FHIA-18 (AAAB) in liquid medium and scaled-up in a bioreactor. Plant Cell Tiss. Org. Cult. 68: 21–26

Hempfling T & Preil W (2002) Application of a temporary immersion system in propagation of Phalaenopsis. First International Symposium on "Liquid Systems for *in vitro* Mass Propagation of Plants" (pp. 47-48). Agricultural University of Norway. 29 May-2 June

Hohe A, Gerth A, Jiménez E, Jordan M, Gómez R, Schmeda G & Wilken D (2002) Production of active substances applying temporary immersion systems. First International Symposium on "Liquid Systems for *in vitro* Mass Propagation of Plants" (pp. 49-50). Agricultural University of Norway. 29 May-2 June

Jiménez E (1995) Propagacion in vitro de la caña de azúcar (Saccharum spp híbrido). Tesis de Doctorado. Instituto de Biotecnología de las Plantas. Universidad Central de Las Villas, Santa Clara, Cuba. p. 95

Jiménez E, Pérez N, de Feria M, Barbón R, Capote A, Chávez M & Quiala E (1999) Improved production of potato microtubers in a temporary immersion system. Plant Cell Tiss. Org. Cult. 59: 19-23

Jiménez E, de Feria M, Freire M, Quiala E, Chávez M, Herrera I & Capote A (2002) Sugarcane somatic embryo production in bioreactors: Effect of partial oxygen pressure, germination in temporary immersion systems and field studies of regenerated plants. First International Symposium on "Liquid Systems for *in vitro* Mass Propagation of Plants" (p. 143) Agricultural University of Norway, 29 May-2 June

Lorenzo JC, González BL, Escalona M, Teisson C, Espinosa P & Borroto C (1998) Sugarcane shoot formation in an improved temporary immersion system. Plant Cell Tiss. Org. Cult. 54: 197-200

Mc Alister B, Finnie J, Watt MP & Blakeway F (2002) Use of temporary immersion system (RITA ®) for the production of commercial *Eucalyptus* clones at Mondi Forests (SA). First International Symposium on "Liquid Systems for *in vitro* Mass Propagation of Plants" (p. 79). Agricultural University of Norway. 29 May-2 June

Medero V, Rodríguez S, Borroto C, Gómez R, Lóèz J, de Feria M, García M, Ventura J & Cabrera M (2001) Sistema de inmersión temporal para producción intensiva de material de siembra de yuca. Continente Yuquero 3: 10-11

Pérez J, Jiménez E & Agramonte D (1998) Aumento de la eficiencia en la propagación masiva. In: Pérez et al. (eds) Propagación y mejora genética de plantas por biotecnología (pp. 179-192). Instituto de Biotecnología de las Plantas, Santa Clara, Cuba

Perez N, Jiménez E, de Feria M, Capote A, Chávez M & Quiala E (2000) Escalado productivo de microtuberculos de papa (*Solanum tuberosum* L var Atlantic) en sistemas de inmersión temporal y su evaluación en condiciones de campo. XIX Congreso de la Asociación Latinoamericana de la Papa, ALAP (p. 227). 28 de Febrero al 3 de Marzo. La Habana, Cuba

Preil W (1991) Application of bioreactors in plant propagation. In: Debergh PC & Zimmerman RH (eds) Micropropagation: Technology and Application (pp. 426-445). Kluwer Academic Publishers, Dordrecht

Teisson C, Alvard D, Berthouly B, Cote F, Escalant JV, Etienne H & Lartaud M (1996) Simple apparatus to perform plant tissue culture by temporary immersion. Acta Horticulturae 440: 521-526

Teisson C & Alvard D (1995) A new concept of plant *in vitro* cultivation in liquid medium: Temporary immersion. In: Terzi M et al. (eds) Current Issues in Plant Molecular and Cellular Biology (pp. 105–110). Kluwer Academic Publishers, Dordrecht

Teisson C & Alvard D (1999) *In vitro* production of potato microtubers in liquid medium using temporary immersion. Potato Research 42: 499-504

Vilches J, Albany N & Gómez R (2002) Somatic embryogenesis induction in guava (*Psidium guajava* L.). Plant Cell Tiss. Org. Cult. (in press)

Ziv M (1992) The use of growth retardants for the regulation and acclimatization of *in vitro* plants. In: Karssen CM, Van Loon LC & Vreugdenhil D (eds) Progress in Plant Growth Regulation (pp. 809-817). Kluwer Academic Publishers, Dordrecht

Ziv M, Ronen G & Raviv M (1998) Proliferation of meristematic clusters in disposable presterilized plastic bioreactors for the large-scale micropropagation of plants. *In Vitro* Cell. Dev. Biol.-Plant. 34: 152-158

Chapter 13

Use of growth retardants for banana (*Musa* AAA cv. Grand Naine) shoot multiplication in temporary immersion systems

Nilca Albany[1], Elio Jiménez González*[2], Jorge Vilchez[3], Leyanis García[2], Manuel de Feria[2], Naivy Pérez[2], Zoe Sarría[2], Blanca Pérez[2] & Justo Clavelo[2]

[1] *Dpto. Química. Facultad de Agronomía La Universidad del Zulia. AP 15205. Maracaibo (4005), Edo. Zulia.*

[2] *Institute of Plant Biotechnology, Central University of Las Villas, Carretera a Camajuaní km 5.5, Santa Clara, Cuba, CP 54830.*

[3] *Dpto. de Biología. Facultad Experimental de Ciencias. La Universidad del Zulia, Bloque A-1, 3er piso, Laboratorio de Biotecnología Vegetal (BIOVELUZ).AP 10488. Maracaibo (4002), Edo.Zulia.*

* *Phone: 53-42-281374; Fax: 53-42-281329; E-mail:* ejimenez@ibp.uclv.edu.cu

Abstract: Temporary immersion culture (TIS) offers several advantages over solid medium for banana shoot multiplication, e.g. TIS results in an increase in the multiplication rate and improves the quality of the plantlets. For a commercial application of this technique large vessels are required. When using 10-litre culture vessels an excessive growth of the shoots (leaves and pseudostem) was obtained, which limited the final number of shoots to be produced per flask and reduced the production capacity in the growth room. Labor costs also increased, since handling is more difficult when dividing and subculturing large shoots during the multiplication stage. The effect of growth retardants ancymidol (ANC), paclobutrazol (PBZ) and daminozide (DAM) in liquid shake cultures and TIS was investigated in order to reduce the size of the shoots and allow a better use of the space inside the culture vessel. In liquid shake cultures ANC and PBZ, independently of the tested concentrations, promoted bud cluster formation with reduced size and compact shape. Shoots multiplied with ANC or PBZ (2.5 mg l^{-1}) after five subcultures, recovered their normal morphology after transfer to a hormone-free medium without growth retardants. However, during the acclimatization stage, plants multiplied in ANC (2.5 mg l^{-1}) containing media showed reduced height in comparison with control plants and plants multiplied in PBZ and DAM containing medium. The application of PBZ and ANC in TIS (1-litre flasks) stimulated bud proliferation. Both compounds were also effective in controlling the excessive growth of the shoots and in inducing the formation of compact bud clusters. Shoots multiplied in TIS in presence of PBZ (2.5 mg l^{-1}) were successfully transferred to semisolid or liquid rooting media in traditional culture vessels or TIS. The developed protocol was further scaled up in 10-litre TIS vessels.

Key words: ancymidol, daminozide, paclobutrazol, propagation

A.K. Hvoslef-Eide and W. Preil (eds.), Liquid Culture Systems for in vitro Plant Propagation, 213–224.

Abbreviations: ANC – ancymidol; 6-BAP – 6-benzylaminopurine; DAM – daminozide; PBZ – paclobutrazol; TIS – temporary immersion system

1. Introduction

A broad commercial propagation of *Musaceae* by direct organogenesis (axillary shoot proliferation) has been limited due to the high costs of production, which is a result of the high number of manual operations needed. Additionally, the use of semisolid culture medium and the use of low capacity vessels limit the possibility to automate or semi-automate *in vitro* propagation processes (Ziv, 1990).

The development of more efficient protocols based on the use of liquid culture medium in some or all the stages of micropropagation can reduce the required manipulations and thus can deminish the costs of *in vitro* propagation. Temporary immersion system (TIS) is an accessible technology that allows the partial automation of some steps of *in vitro* culture with major facility for scale up; increasing the biological and productive efficiency of the propagated material without the collateral effects caused by static liquid culture medium namely hyperhydricity and hypoxia (Alvard et al., 1993; Teisson et al., 1996; Lorenzo et al., 1998; Escalona et al., 1999; Jiménez et al., 1999; Etienne and Berthouly, 2002).

Banana shoot multiplication in TIS was first described by Alvard et al. (1993), who used modified Nalgene filtration units as culture vessels and afterwards RITA® has been successfully used by many laboratories. However, the commercial application of the technique requires larger vessels (Jiménez et al., 1999).

When scaling up from RITA to 10-litre twin flask system some problems arose, e.g. an excessive growth of the shoots (leaves and pseudostem), which limited the final number of shoots per flask reducing the production capacity of the growth room. Labor costs also increase, since handling is more difficult when dividing and subculturing large shoots during the multiplication stage. The addition of growth retardants to the culture medium to reduce the size of the shoots might be a solution to overcome these problems. Growth retardants already have been successfully used to inhibit shoot length and to promote bud clusters formation in several species including bananas (Ziv, 1992; Opatrná et al., 1997; Ziv et al., 1998; Escalona et al., 1999). The aim of this study was to evaluate the effect of growth retardants ancymidol (ANC), paclobutrazol (PBZ) and daminozide (DAM) in liquid cultures and TIS in order to reduce the size of the banana shoots and allow a better use of the space inside the culture vessel.

2. Materials and methods

2.1 Plant material, culture media and culture conditions

Shoots from banana (*Musa* AAA cv. Grand Naine) were established and multiplied according to the conditions and procedures described by Orellana (1994). For shoot multiplication the MS medium (Murashige and Skoog, 1962) was used, supplemented with 1.0 mg l^{-1} of thiamine, 4.0 mg l^{-1} of 6-BAP and 3.0% (w/v) sucrose. The rooting culture medium had the same composition, but no 6-BAP. The pH was adjusted to 5.8 before sterilization at 121ºC and 1.2 kg cm^{-2}.

The cultures were kept in a growth room with natural light at a temperature of 27 ± 2ºC. Liquid cultures were shaked on a rotary shaker at 100 rpm (250 ml Erlenmeyer flasks). The immersion frequency in TIS (1- and 10-litre culture vessels) was 1 minute every 6 hours. Cultures on semisolid medium were transferred to fresh medium every 21 days. In liquid shake cultures and TIS, the medium was substituted every 15 days.

Rooted plants were transferred to the acclimatization phase in trays of poly-foam with 70 orifices (120 cm^3 each) with a substrate composed of 75% humus and 25% zeolite (v/v). The temperature ranged between 25-30ºC, under 50% shade and the irrigation was provided by micro-sprinklers with a duration and frequency of two minutes every four hours.

2.2 Shoot multiplication in liquid shake cultures

A first series of experiments were performed in order to evaluate the effect of growth retardants (ANC, DAM and PBZ) at four concentrations (0.0, 1.0, 2.5 and 5.0 mg l^{-1}). Five explants were inoculated per flask containing 20 ml of liquid multiplication medium and five flasks were used per treatment. The formation of clusters, the number of shoots per explant, the final weight (total weight of the explant at the end of the culture time), the discarded weight (weight of the leaves and tissue rejected) and the useful weight (weight of the sectioned shoots used for the following subculture) were evaluated.

A comparative study was carried out during the elongation/rooting stage and the acclimatization stage on the morphology of plants multiplied during five subcultures in ANC or PBZ (2.5 mg l^{-1}) containing media. Plants from standard semisolid medium and liquid shake medium were included as controls. After the fifth subculture, individual shoots were separated and transferred to semisolid rooting medium. Five shoots were placed per flask using a total of 20 flasks per treatment. The length and width of leaf 2, length of the petiole of leaf 2, diameter of the pseudo stem, number of leaves and roots were evaluated.

For the acclimatization stage 70 plants were evaluated 45 days after transfer to *ex vitro* conditions and the same morphological parameters described previously for the elongation/rooting stage were measured.

2.3 *Shoot multiplication in TIS*

An experimental unit with 1-litre twin flask TIS according to Jiménez et al. (1999) was used. Twentyfive bud clusters multiplied with 2.5 mg l^{-1} of ANC or 2.5 mg l^{-1} of PBZ in liquid shake medium, were inoculated per TIS unit. Each TIS unit contained 500 ml of multiplication medium supplemented with 2.5 mg l^{-1} ANC or PBZ. A control treatment was included with shoots cultivated in standard multiplication medium. Two TIS units were used per treatment and the experiment was replicated twice. The number of shoots/explant, the final weight, discarded weight and the useful weight of the shoots were evaluated. Bud clusters multiplied in TIS with 2.5 mg l^{-1} of PBZ, were transferred to semisolid and liquid rooting medium.

All the statistical analyses were done with the program "Statistix" (Copyright © 1996 Analytical software) version 1.0 for Microsoft Windows.

3. Results and discussion

3.1 *Effect of ancymidol, daminozide and paclobutrazol on shoot multiplication in liquid shake cultures*

ANC and PBZ, independently of the concentration tested, proved to be effective in reducing the size of the shoots and in inducing the formation of compact bud clusters (Figure 1). The formation of compact bud clusters was achieved in all explants treated with ANC and PBZ; whereas all the explants on DAM containing medium showed similar growth and development to the control cultures without growth retardants. Bud clusters were characterized by agglomerates of small and compact shoots with a remarkable reduction of longitudinal growth of the pseudostem, and an increase in their thickness of up to 1.0 cm in diameter. The leaves were poorly developed, folded and compact, which differed in size and development from the shoots grown on the control medium or DAM containing media. The latter, developed thinner pseudo stems with one to three completely developed and expanded leaves.

According to Ziv (1990), the absence of growth retardants allowed the leaves of *Gladiolus* to continue their growth and elongation in liquid culture medium; whereas in the presence of PBZ and ANC, a reduction in the number and length of the leaves was observed and the formation of clusters was obtained. These results confirm that the addition of growth retardants induces a shoot morphology characterized by the decrease in length of the stems and leaves.

Both compounds also stimulated bud proliferation, increasing the number of shoots produced per inoculated explant, 5.4 and 4.6 shoots for ANC and PBZ containing media respectively, with significant statistical differences with the control medium which contained only BAP (2.4 shoots per explant) (Figure 2). No statistical differences were found between all tested concentrations of both growth retardants. The significant increase in the number of shoots per explant when using ANC and PBZ could also be associated with the inhibition of gibberellin biosynthesis, which suppress the growth and dominance of the apical bud, favoring or stimulating the shooting of axillary buds (Grossmann, 1990).

DAM is described in the literature as a growth retardant in plants and though it is demonstrated that it does not exercise inhibitory effect by interruption of the synthesis of gibberellins (Smith et al., 1991), it could interfer in the action of these and favor the action of the cytokinins. In this way, an increase in the number of shoots could be obtained without morphological changes of its normal structure.

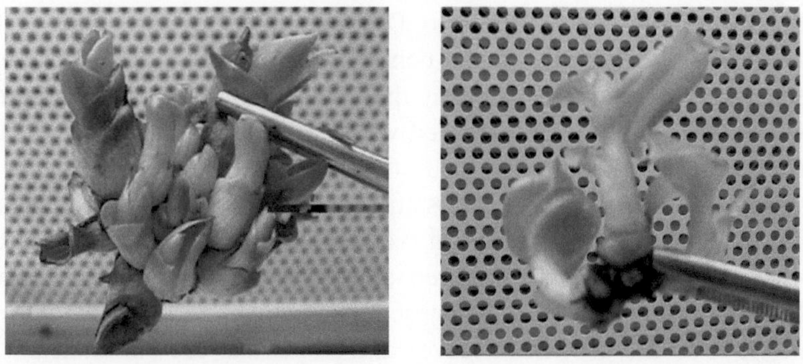

Figure 1: Bud clusters obtained from ANC- or PBZ-containing medium (left) and elongated shoots developed on DAM-containing medium or control medium free of growth retardants (right).

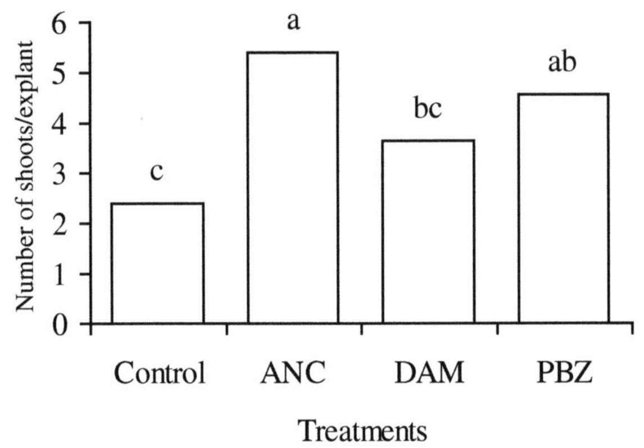

Figure 2: Effect of growth retardants (ANC, DAM and PBZ) on banana shoot multiplication in shaked liquid media.
Different letters represent significant differences according to Sheffe´s test, $p < 0.05$.

Supplementing the medium with the different growth retardants did not influence the final weight of the shoots. The final weight is composed of 'discarded weight' and the 'useful weight' (including meristematic tissue) of the shoots; these being complementary, showed an inversely proportional behavior. For the discarded weight of the shoots, the ANC and PBZ presented differences with respect to DAM and the control. The useful weight of the shoots showed a similar behavior, but inversely related (Table 1). This behavior is due to the fact that the weight of all shoots increased. However, they differed in the discarded weight and useful weight, since the shoots from the control and DAM medium increased their weight because of the growth and elongation of the stems and leaves, unlike the shoots exposed to ANC and PBZ which increased in weight due to the formation of new shoots. These differences are observed clearly in the useful weight, since both ANC and PBZ showed the greatest weight due to new shoots and not necessarily due to the growth of leaves and stems that are eliminated with the handling of the explants for the following subculture.

Ziv et al. (1998) reported a gain in total weight of 35.2 g when inoculating 10 g bud clusters of cv. Grand Naine in liquid medium in agitation with ANC, obtaining 60.7% of meristematic tissue (useful tissue) and 39.3% of expanded leaves and necrotic tissue after 25 days.

Shoots multiplied in ANC and PBZ containing media after five subcultures, recovered their normal morphology after transfer to a hormone free medium and started to form roots (Figure 3), which reaffirms there is no necessity to apply auxins to induce rooting in cv. Grand Naine; seemingly the endogenous concentration of this regulator is sufficient to restore the internal balance of growth regulators, when cytokinins are absent from the culture medium (Sandoval et al., 1999).

Additionally, it was possible to restore the normal length of the pseudostems and new leaves by elimination of the growth retardants and BAP as reported by Ziv (1989, 1992) for *Gladiolus*, ferns and bananas. This is in contrast with the results of Escalona (1999) in pineapple and Lorenzo et al. (1998) in sugarcane, who applied gibberellins to obtain plants with a suitable size for acclimatization. Such results indicate that there is no residual effects of the growth retardants ANC and PBZ, in the concentration used (2.5 mg l^{-1}), on the morphology of the plants. However, these cultures were only subjected to five consecutive cycles.

When plants were transplanted to the greenhouse, differences were observed in plant height (Table 2). Those plants coming from ANC containing media during the multiplication stage showed a reduced size compared to plants from control in semisolid or liquid medium and from plants multiplied in PBZ containing medium. This resulted in an extended acclimatization stage for the plants multiplied in ANC containing media.

The parameters evaluated indicated in field experiments that growth retardants present during five consecutive multiplication subcultures are not an apparent source or cause of somaclonal variability in plants of cv. Grand Naine. However, field evaluations are the most trustworthy to determine somaclonal variants, since the plants have expressed their phenotypic and phenologic potential (Sandoval et al., 1997). Hence the field studies will be continued with the plants coming from propagation systems with these growth retardants (ANC and PBZ).

3.2 *Effect of ancymidol and paclobutrazol on shoot multiplication in TIS*

Once the efficiency of ANC and PBZ to control excessive growth of the shoots was determined and the possibility to recover normal plants in liquid shake culture was demonstrated, a second group of experiments was conducted in 1-litre TIS vessels. The addition of ANC and PBZ also induced the formation of bud clusters in TIS (Table 3). As in liquid shake cultures ANC and PBZ promoted shoot multiplication with significant statistical differences compared to the control (Figure 4). Differences were also observed for the discarded and useful weight of the shoots in TIS, while the final weight of the shoots did not show statistical differences (data not given in Table 3).

Table 1: Influence of ANC, DAM and PBZ on the discarded and useful weight of banana shoots multiplied in shaked liquid medium

Growth retardants	Discarded weight (g)	Useful weight (g)
ANC	0.80 b	0.98 a
DAM	1.03 a	0.54 b
PBZ	0.70 b	0.91 a
Control	1.16 a	0.48 b
x ± s.e.	0.91 ± 0.05	0.73 ± 0.029

Different letters in a row represent significant differences according to Sheffe, $p < 0.05$.

Table 2: Effect of growth retardants (ANC and PBZ, 2.5 mg l^{-1}) during the multiplication stage on plant height after 10 days at the acclimatization stage

Growth retardants during the multiplication stage	Plant height (cm)
ANC (liquid)	4.92 b
PBZ (liquid)	5.60 a
Control (liquid)	5.70 a
Control (semisolid)	5.65 a
x ± s.e.	5.46±0.14

Different letters in a row represent significant differences according to Tukey's test $p<0,05$.

Table 3: Influence of ANC and PBZ on the discarded and useful weight of banana shoots multiplied in 1-litre TIS vessels

Growth retardants	Discarded weight (g)	Useful weight (g)
2.5 mg l^{-1} of ANC	1.46 b	1.39 a
2.5 mg l^{-1} of PBZ	1.40 b	1.55 a
Control	2.43 a	0.66 b
x ± s.e.	1.63 ± 0.10	1.31 ± 0.08

Different letters in a row represent significant differences according to Tukey´s test, $p < 0.05$.

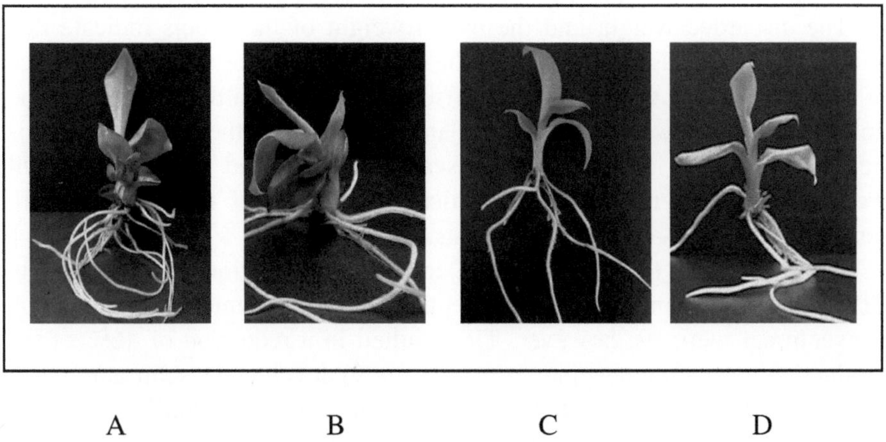

Figure 3: Plants obtained at the end of the elongation/rooting stage, after 15 days in a hormone-free medium. A: 2,5 mg l^{-1} ANC; B: 2,5 mg l^{-1} PBZ; C: Control in lĭquid medium; D: Control in semisolid medium.

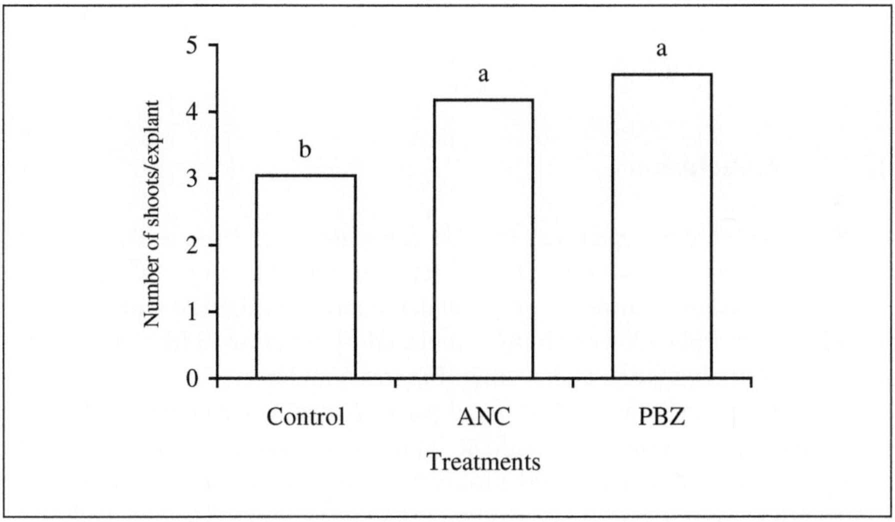

Figure 4: Shoot multiplication in 1-litre TIS on ANC and PBZ (2.5 mgl-1) containing media. Different letters represent significant differences according to Tukey´s test, $p < 0.05$.

The discarded weight and the useful weight of the shoots indicate that gain in the final weight is caused by the greater number of shoots in presence of ANC and PBZ, while the gain in weight of the control was due to the unwanted fast growth of the leaves and stem, hence the reason why this treatment resulted in the greatest discarded weight and the lowest useful weight, which determined the statistical differences of this variable with respect to the ANC and PBZ treatments.

Lorenzo et al. (1998) used PBZ in micropropagation of sugarcane in TIS, obtaining plants in the field similar to those micropropagated by conventional methods, however, TIS resulted in a reduction of 46% of the production costs. In pineapple, Escalona (1999) developed a semi-automated system more efficient than the conventional micropropagation based on the use of PBZ in TIS, which reduces the production costs by 66.7%. Besides, the plants showed better *ex vitro* growth and development in comparison to the plants obtained by conventional micropropagation.

Shoots multiplied in TIS with PBZ (2.5 mg l^{-1}) were singulated and subcultured for rooting in hormone-free medium, either semisolid or liquid medium. After 15 days plants were ready for acclimatization.

The protocol developed was suitable for scale up in 10-litre TIS vessels, using a two-stage procedure. First shoots were multiplied in PBZ and BAP containing medium during 4 weeks and secondly, inside the same culture vessel, the multiplication medium is replaced by a hormone free medium to promote shoot elongation and rooting (Figures 5 and 6).

4. Conclusion

The protocol developed was suitable for scale up in 10-litre TIS vessels, using a two-stage procedure. First shoots were multiplied in PBZ and BAP containing medium during 4 weeks and secondly, inside the same culture vessel, the multiplication medium is replaced by a hormone free medium to promote shoot elongation and rooting (Figures 5 and 6).

The addition of ancymidol (ANC) and paclobutrazol (PBZ) reduced the size of banana shoots multiplied both in liquid shake cultures and TIS. Both compounds also stimulated bud proliferation and increased the number of shoots produced per inoculated explant. Shoots multiplied in ANC and PBZ containing media, either in liquid shake cultures or 1-litre TIS, recovered their normal morphology after transfer to a hormone free medium, which indicate that there is no residual effects of the growth retardants ANC and PBZ on the morphology of the plants. Only the plants multiplied in ANC containing media showed a reduced size and an extended acclimatization period compared to plants from control in semisolid or liquid medium and plants multiplied in PBZ containing medium.

Figure 5: Scale up of banana shoot multiplication in 10-litre TIS vessels (left) and rooting (right).

Figure 6: Banana shoots multiplied in 10-litre TIS in PBZ (2.5 mg l^{-1}) containing medium (left) and subsequent elongation/rooting in a hormone-free medium (right).

References

Alvard D, Cote F & Teisson C (1993) Comparison of methods of liquid medium culture for banana micropropagation. Effects of temporary immersion of explants. Plant Cell, Tiss. Org. Cult. 32: 55-60

Escalona M, Lorenzo B, González B, Daquinta M, González JL, Desjardins Y & Borroto C (1999) Pineapple (*Ananas comosus* L. Merr) micropropagation in temporary immersion. Plant Cell Rep. 18: 743-748

Etienne H & Berthouly M (2002) Temporary immersion systems in plant micropropagation. Plant Cell Tiss. Org. Cult. 69: 215–231

Grossman K (1990) Plant Growth retardants as tool in physiological research. Physiologia Plantarum. 78: 640-648

Jiménez E, Pérez N, de Feria M, Barbón R, Capote A, Chavéz M, Quiala E & Pérez JC (1999) Improved production of potato microtubers using a temporary immersion system. Plant Cell, Tiss. Org. Cult. 59: 19-23

Lorenzo J, González B, Escalona M, Teisson C, Espinosa P & Borroto C (1998) Sugarcane shoot formation in an improved temporary immersion system. Plant Cell, Tiss. Org. Cult. 54: 197-200

Murashige T & Skoog F (1962) A revised medium for rapid growth and bioassays with tobacco tissue cultures. Physiol Plant. 15: 473-497

Opatrná J, Novák P & Opatrný Z (1997) Paclobutrazol stimulates bud regeneration in *Solanum tuberosum* L. primary explant cultures. Biologia Plantarum 39: 151-158

Orellana P (1994) Tecnología para la propagación *in vitro* de clones de Musa sp. Tesis para optar al grado científico de Doctor en Ciencias Agrícolas. Instituto de Biotecnología de las Plantas. Universidad Central de las Villas. Santa Clara, Cuba p.104

Sandoval J, Côte F & Doumas P (1999) Reconocimiento *in vitro* de variantes somaclonales de porte alto de banano (cv. Gran enano, Musa AAA). Respuesta a un tratamiento con GA3. CORBANA 24: 11-20

Sandoval J, Pérez L & Côte F (1997) Estudio morfológico y de la estabilidad genética de plantas variantes de banano (Musa AAA cv. "Gran Enano"). Etapas de cultivo *in vitro*, aclimatización y campo. CORBANA 22: 41-60

Smith E, Roberts A, Mottley J & Denness S (1991) The preparation *in vitro* of chrysanthemum for transplantation to soil. 4. The effects of eleven growth retardants on wilting. Plant Cell, Tiss. Org. Cult. 27: 309-313

Teisson C, Alvard D, Berthouly B, Cote F, Escalant J, Etienne H & Lartaud M (1996) Simple apparatus to perform plant tissue culture by temporary immersion. Acta Horticulturae 440: 521-526

Ziv M (1989) Enhanced shoot and cormlet proliferation in liquid cultured *Gladiolus* buds by growth retardants. Plant Cell, Tiss. Org. Cult. 17: 101-110

Ziv M (1990) The effect of growth retardants on shoot proliferation and morphogenesis in liquid cultured *Gladiolus* plants. Acta Horticulturae 280: 207-214

Ziv M (1992) The use of growth retardants for the regulation and acclimatization of *in vitro* plants. In: Karssen CM, Van Loon LC & Vreugdenhil D (eds) Progress in Plant Growth Regulation (pp. 809-817). Kluwer Academic Publishers, Dordrecht

Ziv M, Ronen G & Raviv M (1998) Proliferation of meristematic clusters in disposable presterilized plastic bioreactors for the large-scale micropropagation of plants. *In Vitro* Cell. Dev. Biol.-Plant. 34: 152-158

Chapter 14

Somatic embryo germination of *Psidium guajava* L. in the Rita® temporary immersion system and on semisolid medium

Rafael Gómez Kosky[1], J. Vilchez Perozo[2], N. Albany Valero[2] &
D. Agramonte Peñalver[1]
[1]*Instituto de Biotecnología de las Plantas. Universidad Central "Marta Abreu" de Las Villas (UCLV). Carretera a Camajuní Km 5 ½ Santa Clara, Villa Clara, Cuba*
Fax: 53-42-281329, E-mail: koskyrg@yahoo.es
[2]*Fundación Gran Mariscal de Ayacucho. Caracas, República Bolivariana de Venezuela*

Abstract: Germination of somatic embryos and development to plants is recognized as one of the most critical stages in the process of plant propagation via somatic embryogenesis. The objective of the study was to investigate the germination of somatic embryos of Psidium guajava cv. Cuban Red Dwarf EEA 18-40 in the RITA® system and on semisolid medium. Somatic embryos were obtained from immature zygotic embryos which were cultured on the major salts of MS medium at half strength, supplemented with 400 mg l^{-1} L-glutamine, 100 mg l^{-1} ascorbic acid, 60 g l^{-1} sucrose and 1 mg l^{-1} 2,4-dichlorophenoxyacetic acid (2,4-D). Somatic embryos at the heart and torpedo stages were transferred for germination into RITA® vessels containing liquid half strength MS medium of the major salts supplemented with 0.25 mg l^{-1} 6- benzylaminopurine (6-BAP), 10 µg l^{-1} Biobras-6 (brassinosteroid analogue) and 20 g l^{-1} sucrose or to semiliquid medium of the same composition (solidified with 2.5 g l^{-1} Gellum Gum, Spectrum®) in 250 ml glass vessels. The germination percentage, fresh weight and number of somatic embryos with complete germination were determined. After 10 weeks of culture the highest germination percentage (91%) and fresh weight (1.22 g) were achieved in the temporary immersion system, being statistically superior to those obtained from semisolid culture medium (81.79% and 1.03 g respectively).

Key words: Guava, liquid medium, somatic embryogenesis

1. Introduction

Among 150 *Psidium* species guava (*Psidium guajava* L.) presents the greatest potential from an economic point of view. Difficulties in the

A.K. Hvoslef-Eide and W. Preil (eds.), Liquid Culture Systems for in vitro Plant Propagation, 225–229.

application of conventional propagation techniques have generated considerations of other possible forms of vegetative propagation (Pontikis, 1996). Especially somatic embryogenesis, which has been investigated as a tool for genetic improvement and for the propagation of elite trees (Vilchez, 2001).

The germination of somatic embryos has proved to be one of the most critical stages in somatic embryogenesis (Merkle et al., 1995). Temporary Immersion Systems (TIS) have been successfully used in several protocols for the germination of somatic embryos of some species (Etienne-Barry et al., 1999; Gómez et al., 2000). In this research, germination of guava somatic embryos in TIS was studied and compared to cultures on semisolid medium.

2. Materials and methods

Somatic embryos of *Psidium guajava* L. cv. 'Cuban Red Dwarf EEA 18-40', were obtained from immature zygotic embryos at the torpedo and cotyledonary stages. These were cultured on the major salts of MS medium (Murashige and Skoog, 1962) at half strength, supplemented with 400 mg l^{-1} L-glutamine, 100 mg l^{-1} ascorbic acid, 60 g l^{-1} sucrose and 1 mg l^{-1} 2,4-dichlorophenoxyacetic acid (2,4-D). Subculturing was performed at three-week intervals using the same culture medium.

The operations and characteristics of the RITA®system (CIRAD, France) have been described by Etienne-Barry et al., (1999). For embryo germination each RITA® unit contained 200 ml of liquid MS culture medium with 50% of the major salts supplemented with 0.25 mg l^{-1} 6-benzylaminopurine (6-BAP), 10 µg l^{-1} Biobras-6 (brassinosteriod analogue from Nature Product Lab., Havana University) and 20 g l^{-1} sucrose. The inoculum density per RITA vessel was 800 mg fresh weight of somatic embryos at the heart or torpedo stages. The immersion frequency was one minute every 12 hours. The experiments were duplicated.

For control cultures 250 ml glass vessels were used with 30 ml of semisolid culture medium (solidified with 2.5 g l^{-1} Gellan gum, Spectrum®) consisting of the same composition as used for the RITA® units. Ten culture vessels were inoculated with the 800 mg fresh weight of somatic embryos as described for RITA® cultures.

The experiment was carried out in a growth chamber with the natural photoperiod in Cuba according to the time of the year (November-January) and a photosynthetic photon flow superior to 60 µmol m^{-2} s^{-1}. The temperature regime was 25 ± 2.0 °C. After ten weeks of culture, the percentage of germinated embryos and fresh weight were determined. The germination percentage was analyzed statistically through the proportion comparison test, complemented with Fisher's exact test and the fresh weight was compared by analysis of simple variance and the Tukey multiple range test.

3. Results and discussion

Both in the TIS and in semisolid culture medium, germination of the somatic embryos started around ten days after placement in the germination medium, when the embryo changed color from white to a shiny green. Table 1 shows the comparison between the germination of the somatic embryos in TIS and in semisolid culture medium. The statistical analysis points out significant differences for the variables germination percentage and fresh weight.

Table 1: Comparison of the germination percentages and fresh weight of the somatic embryos of *Psidium guajava* L. developed in TIS and on semisolid culture medium

	Germination percentage*	Fresh weight (g)**
TIS	91.04 a	1.22 ± 0.15 a
Semisolid	81.79 b	1.03 ± 0.09 b

*Different letters differ statistically for $p<0.05$ according to the proportion comparison test complemented with Fisher's exact test.
**Different letters differ statistically for $p<0.05$ according to Tukey multiple range test.

The highest germination percentage (91 %) and fresh weight were obtained with embryos in the TIS, being superior to the results reported in a similar system by Etienne-Barry et al. (1999) for *Coffea arabica* (60%). The somatic embryos germinated in the TIS were morphologically similar to those germinated in semisolid culture medium (Figure 1A and B). Partial germination with shoot development in some somatic embryos was found in semisolid medium only.

Hyperhydricity was not observed (Figure 1C), an aspect also pointed out by Cabasson et al. (1997) in embryos of *Citrus deliciosa* and Escalona (1999) in shoots of *Ananas comosus* multiplied in TIS.

The differences between the germination of the somatic embryos in TIS and semisolid culture medium may be explained by the limited contact between the somatic embryos and the culture medium in the TIS, which is reduced to a thin layer of culture medium that adheres to the somatic embryos and is renewed during every immersion. This layer is too thin to inhibit gaseous exchange. The aeration is also better since the atmosphere is exchanged during medium transfer. Additionally, medium exchange causes agitation of the plant material (Teisson and Alvard, 1995).

Germination of somatic embryos of several species (Citrus, Musa and Coffee) that was not possible in agitated Erlenmeyer flasks or on semisolid medium has been obtained in the TIS (Teisson and Alvard, 1995; Cabasson et al., 1997). This does not occur in the semisolid culture medium, where the contact with the culture medium is permament and the gaseous atmosphere influences the process of differentiation, due to low concentrations of oxygen which obviously reduce the number of somatic embryos per explant.

4. Conclusion

With the use of the temporary immersion systems it is possible to achieve a more efficient germination of the somatic embryos of *Psidium guajava* cv. Cuban Red Dwarf EEA 18-40. Developing plants show normal morphological characteristics without features of hyperhydricity.

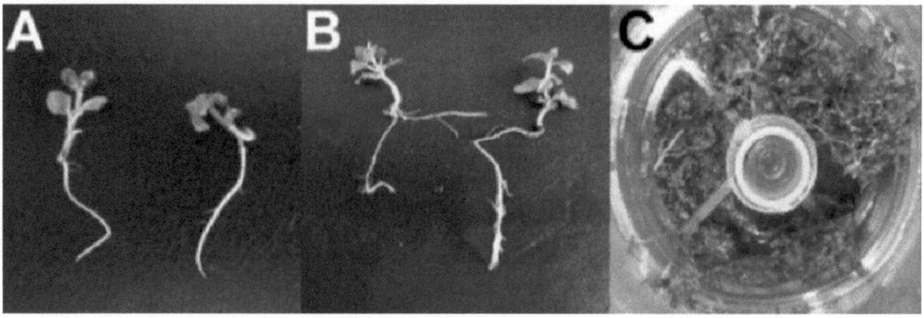

Figure 1: Morphology of germinated somatic embryos of *Psidium guajava* L. A: Embryos from RITA®. B: Embryos from semisolid culture medium. C: RITA® vessel with germinated embryos.

References

Cabasson C, Alvard A, Dambier D, Ollitraul P & Teisson C (1997) Improvement of *Citrus* somatic embryos development by temporary immersion. Plant Cell Tissue and Organ Culture 50: 33-37

Escalona M (1999) Propagación de la piña (*Ananas comosus* (L.) Merr.) en sistemas de inmersión temporal. Tesis de doctorado. Centro de Bioplantas, Ciego de Avila, Ciego de Avila, Cuba

Etienne-Barry D, Bertrand B, Vasquez N & Etienne H (1999) Direct sowing of *Coffea arabica* somatic embryos mass-produced in a bioreactor and regeneration of plants. Plant Cell Reports 19: 111-117

Gómez RK, Gilliard T, Barranco LA & Reyes M (2000) Embriogénesis somática en medios líquidos. Maduración y aumento de la germinación en el cultivar híbrido de banano FHIA-18 (AAAB). Infomusa (9) 1: 12-16

Grapin A, Ferriére M, Etienne H & Carron M (1994) Effect of hypoxia on induction and development of somatic embryos in *Hevea brasiliensis*. In: Teisson C (ed) *In Vitro* Culture of Tropical Plants (pp. 23-25). CIRAD. France

Merkle SA, Parrot WA & Flinn T (1995) Morphogenic aspects of somatic embryogenesis In: Thorpe TA (ed) *In Vitro* Embryogenesis in Plants (pp. 155-203). Kluwer Academic Publishers, Dordrecht

Murashige T & Skoog F (1962) A revised medium for rapid growth and bioassays with tobacco tissue cultures. Physiol. Plant. 15: 473-497

Pontikis CA (1996) *Psidium guajava* L. (Guava) In: Bajaj YPS (ed) Biotechnology in Agriculture and Forestry, Vol 35 Trees IV (pp. 309-319). Springer-Verlag Heidelberg, Berlin

Teisson C & Alvard D (1995) A new concept of plant in vitro cultivation liquid medium: Temporary Immersion. In: Terezi M, Cella R & Falavigna A (ed) Current Issues in Plant Molecular and Cellular Biology (pp. 105-109). Kluwer Academic Publishers, Dordrecht

Vilchez J (2001) Somatic embryogenesis in guava (*Psidium guajava* L.). Master degree Thesis. Universidad Central de Las Villas. Santa Clara, Cuba

Chapter 15

Application of a temporary immersion system in mass propagation of *Phalaenopsis*

Tino Hempfling & Walter Preil
Institute for Ornamental Plant Breeding, Bornkampsweg 31, D – 22926 Ahrensburg, Germany. E-mail: w.preil@bafz.de

Abstract: A temporary immersion system consisting of a series of five-litre twin glass vessels was successfully used for *in vitro* adventitious shoot multiplication and rooting of *Phalaenopsis*. Highest shoot multiplication rate of 25.4 was achieved after twelve weeks when eight immersions per day, of ten minutes each, were used. Fresh medium containing 0.5 mg l^{-1} TDZ was supplemented every two weeks. The different periods of TDZ exposure affect the multiplication rate, fresh weight rate and shoot size. Seven-week culture on TDZ-containing medium, followed by five weeks on TDZ-free medium, resulted in 18 % of shoots smaller than 1 cm, 56 % were of 1-3 cm and 26 % exceeded 3 cm in length, respectively. Small shoots are suitable as inoculum for the next multiplication cycle, whereas shoots larger than 1 cm can be rooted on cytokinin-free medium. Highest percentage of rooted shoots (93.8 %) and highest root number (3.7 roots per shoot) were achieved in 1.0 mg l^{-1} IAA – containing medium after exposure to six immersions per day, ten minutes each. The mean survival rate of plants rooted in TIS was 94 % under standard greenhouse conditions.

Key words: adventitious shoots, immersion frequency, *in vitro* culture, multiplication, rooting, thidiazuron

Abbreviations: BAP–6-benzylaminopurine; IAA–indole-3-acetic acid; NAA–1-naphthaleneacetic acid; PLB–protocorm-like body; TDZ–N-phenyl-N'1,2,3-thiadiazol-5-ylurea (thidiazuron); TIS–Temporary Immersion System

1. Introduction

Micropropagation of *Phalaenopsis* has increased remarkably in Europe over the last few years. In Germany 9.2 million plants were multiplied in 2000, 15.1 million in 2001 and 22.8 million in 2002 (Preil, 2004). The turnover of *Phalaenopsis*, occupying the leading position in pot plants in the

A.K. Hvoslef-Eide and W. Preil (eds.), Liquid Culture Systems for in vitro Plant Propagation, 231–242.

Netherlands, exceeded 64 million Euro in 2002. The increase was 30.9 % compared to 2001 (Anonymous, 2003).

In vitro culture procedures including 22 media formulations for *Phalaenopsis* were listed by Arditti and Ernst (1993), presenting an overview on the state of knowledge at that time. The application of standard protocols for micropropagation on solid medium on average results in a three-fold shoot multiplication rate after an eight-week subculture period. More recently, liquid systems including bioreactors or temporary immersion techniques were successfully applied giving rise to higher multiplication rates (Park et al., 1996, 2000; Samson et al., 2000).

Most of agar gelled media for *Phalaenopsis* micropropagation are supplemented with BAP as cytokinin source. A few publications stress the positive effects of TDZ on protocorm or adventitious bud regeneration (Ernst, 1994; Chen and Piluek, 1995; Chen et al., 2000). As far as we know, no experiments have been published on the effects of TDZ in liquid cultures of *Phalaenopsis*.

The present study was carried out to investigate adventitious shoot multiplication and rooting in a temporary immersion system (TIS) using five-litre twin glass vessels. The effects of medium substitution after two or four weeks were specifically determined. Further, the influence of TDZ-exposure for 7 or 12 weeks on shoot size, multiplication rate and fresh weight was investigated. During the rooting phase various immersion frequencies were applied to determine optimum conditions.

2. Material and methods

2.1 Multiplication

In vitro grown shoots of *Phalaenopsis* cv. Jaunina (Breeder: Wolfgang Bock, Bremen, Germany) were used as inoculum. Per five-litre glass vessel 70–72 shoots (1 – 3 cm length; total fresh weight: 20g) were incubated. Each vessel was connected through a silicone tube with another vessel (Figure 1 A) containing 2 litres of liquid MS-medium of ½ strength macro- and micro-elements (Murashige and Skoog, 1962), supplemented with 170 mg l^{-1} NaH_2PO_4, 100 mg l^{-1} myo-inositol, 0.5 mg l^{-1} nicotinic acid, 0.5 mg l^{-1} pyridoxine, 0.1 mg l^{-1} thiamine HCl, 2.0 mg l^{-1} glycine, 1000 mg l^{-1} peptone and 20 g l^{-1} sucrose, as published by Chen et al. (2000). Additionally 0.5 mg l^{-1} thidiazuron (TDZ) was added.

The liquid medium was drained from the 'medium-vessel' into the 'plant-vessel' and vice versa by means of compressed air being passed through a three-way solenoid valve controlled by a programmable timer as described by Escalona et al. (1999). The immersion of shoots was adjusted to eight times per day for ten minutes each. Spent medium blackened by phenolic compounds (Figure 1 A) was substituted in intervals of two or four weeks. The experiments using the twin vessels were repeated three times each. The multiplication rate and increase of fresh weight were determined twelve weeks after inoculation. An overview of the experimental design is given in figure 1 B.

Control cultures were grown on agar gelled medium (Tanaka and Sakanishi, 1985) supplemented with 10 mg l^{-1} BAP and 10 % (v/v) homogenised bananas.

2.2 *Rooting*

In the first series of experiments shoots of 4 – 7 cm size from TIS cultures were exposed to TDZ-free medium supplemented with 0.5 and 1.0 mg l^{-1} IAA or NAA. In the main experiments 1.0 mg l^{-1} IAA-containing medium was used. Each vessel was inoculated with 72 – 74 shoots of 70 g total fresh weight. The immersion frequency was adjusted to two, four or six times per day for ten minutes periods. The experiments using the twin vessels were repeated three times each. The percentage of rooted plants and the increase in fresh weight were determined after eight weeks.

Figure 1: (A) Twin glass vessels (5 l) inoculated with 20 g shoot fresh weight of *Phalaenopsis* cv. Jaunina. After two weeks the medium was blackened by phenolic compounds. (B) Experimental design of a series of twin glass vessels after 12-week culture.

3. Results

3.1 Effect of medium substitution every two or four weeks on shoot multiplication rate and fresh weight rate

After twelve weeks of growth the mean shoot multiplication rate was 25.4 when eight immersions per day were applied and the spent medium was substituted at intervals of two weeks. Four-week intervals for medium exchange resulted in a significant ($p < 0.001$) reduction of multiplication rate to 14.5 (Figure 2, first and second column from the left).

The increase in fresh weight was less affected by medium substitution intervals than the multiplication rate. Medium exchange every two weeks resulted in a 20.5-fold increase of fresh weight, whereas four-week intervals gave rise to an 18.5-fold increase (Figure 3, first and second column from the left). This difference, however, was not statistically significant ($p = 0.084$).

3.2 Shoot multiplication rate and fresh weight rate after seven or twelve -week culture on TDZ-containing medium

By testing the effects of 0.1 and 0.5 mg l^{-1} TDZ on shoot proliferation, a distinct increase in adventitious shoot production was found when the higher TDZ-concentration was used (Figure 4). For this reason 0.5 mg l^{-1} TDZ was used in the experiments described here.

A low multiplication rate of 12.0 was obtained when the shoots were grown in TDZ-containing medium for seven weeks followed by a five-week culture on TDZ-free medium. The multiplication rate was 25.4 when the shoots remained continuously for twelve weeks on the TDZ-containing medium (Figure 2, first and third column from the left).

The fresh weight increase was less affected by the duration of TDZ-application than the multiplication rate, as already described above for the effects of varying medium substitution intervals. Twelve-week culture on TDZ-medium resulted in fresh weight increase of 20.5; seven weeks on TDZ-medium gave rise to an increase of 17.4 (Figure 3, first and third column from the left). The latter difference was statistically not significant.

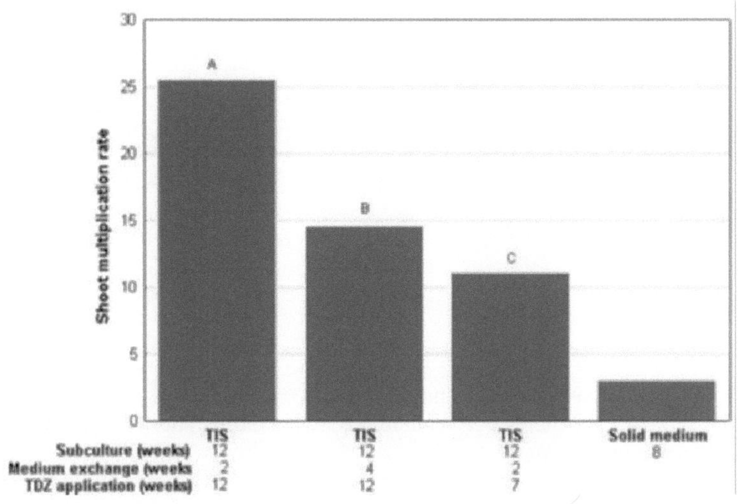

Figure 2: Mean shoot multiplication rate of *Phalaenopsis* cv. Jaunina after 12-week TIS culture. The medium was substituted every two or four weeks. TDZ exposure period was 12 or 7 weeks.
Mean rates following different letters are significantly different according to analysis of variance ($p < 0.001$). Mean rate of control culture on solid medium was excluded from statistical analysis.

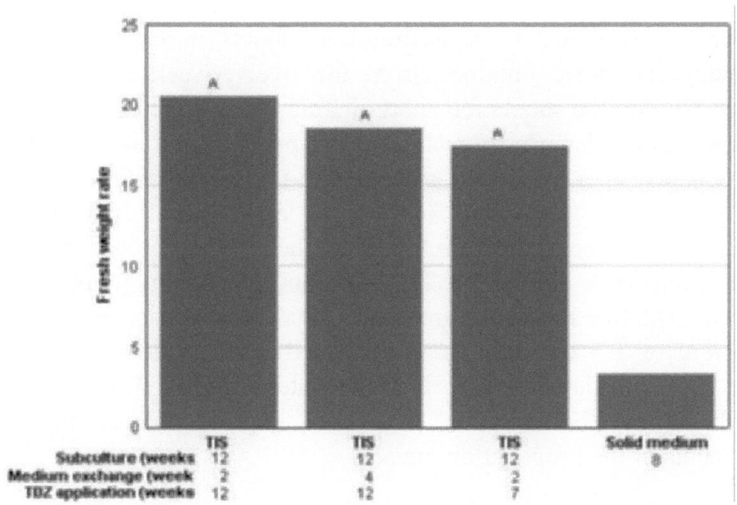

Figure 3: Mean fresh weight rate of *Phalaenopsis* cv. Jaunina after 12-week TIS culture. The medium was substituted every two or four weeks. TDZ exposure period was 12 or 7 weeks. The effects on fresh weight rate are not significantly different according to analysis of variance ($p = 0.084$). The mean rate of control culture on solid medium was excluded from statistical analysis.

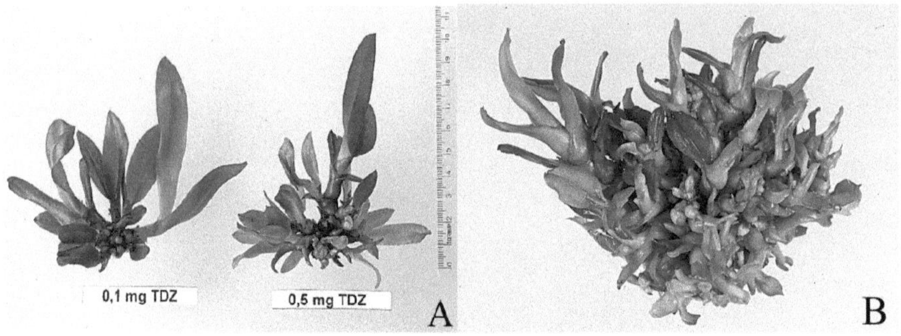

Figure 4: (A) Effect of TDZ (0.1 and 0.5 mg l^{-1}) on adventitious shoot proliferation of *Phalaenopsis* cv. Jaunina after 7-week culture. (B) Shoot cluster after 12-week culture on 0.5 mg l^{-1} TDZ-containing medium.

3.3 *Effect of TDZ-containing medium on shoot size*

Varying periods of TDZ exposure affects not only the multiplication and fresh weight rate but also the shoot size. Figure 5 shows the shoot size categories obtained from a twelve-week culture and a seven-week culture on TDZ-containing medium.

Separating the clusters of adventitious shoots into single propagules, three categories were obtained from the twelve-week culture on TDZ-medium: 50 % of the shoots fell short of 1 cm, 41 % were of 1-3 cm and 9 % exceeded 3 cm in length, respectively. For practical purposes, 50 % of the shoots smaller than 1 cm are suitable for use as inoculum for the next propagation cycle, whereas 50 % of shoots larger than 1 cm can be rooted either in TIS or on solid medium following the standard procedure.

The seven-week culture on TDZ-containing medium followed by five weeks on TDZ-free medium gave rise to 18 % of shoots smaller than 1 cm, 56 % were of 1-3 cm and 26 % exceeded 3 cm in length, respectively (Figure 5). This demonstrates that the categories of shoot size can be manipulated by the duration of TDZ- exposure depending on the requirements of the propagation process.

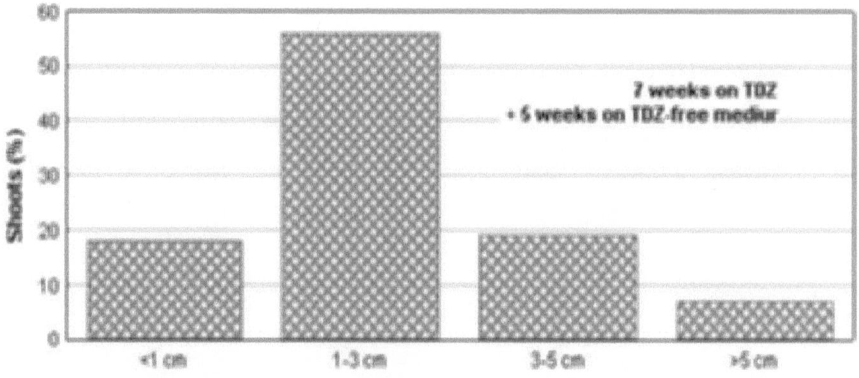

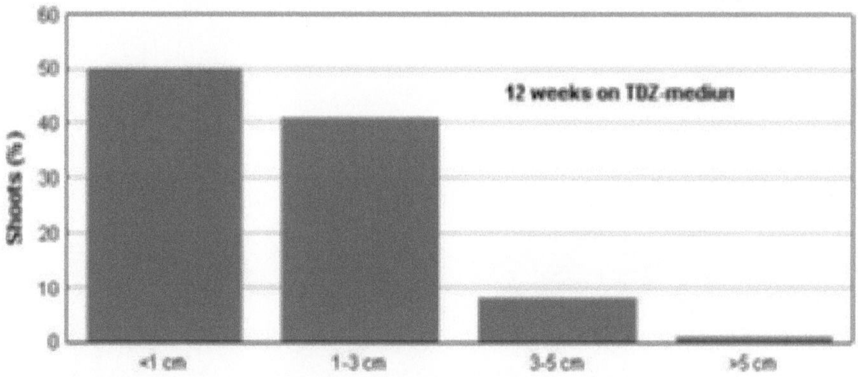

Figure 5: Shoot size categories of *Phalaenopsis* cv. Jaunina after 12-week culture on 0.5 mg l^{-1} TDZ-containing medium (below) or after 7 weeks on TDZ-medium followed by 5-week culture on TDZ-free medium.
Both size groups are significantly different according to Chi-Square Test ($p < 0.001$).

3.4 *Effect of immersion frequency on rooting*

Shoot elongation during the rooting phase and the number of roots produced per shoot can be regulated by auxin type and concentration. Application of 0.5 mg l^{-1} IAA led to elongation of the shoot axis and leaves (Figure 6 A), whereas 1.0 mg l^{-1} NAA inhibited shoot elongation significantly and induced thicker roots (Figure 6 B).

Figure 6: Effect of 0.5 mg l^{-1} IAA (A) and 1.0 mg l^{-1} NAA (B) on rooting of *Phalaenopsis* cv. Jaunina after 8-week culture. 1.0 mg l^{-1} NAA induced thick roots and inhibited shoot elongation.

The highest percentage of rooted shoots (93.8 %) and highest root number (3.7 roots per shoot) were achieved in 1.0 mg l^{-1}IAA-containing medium after exposure to six immersions per day. The increase of fresh weight was 3.7-fold.

Two immersions per day resulted in an overall average of 63.5 % of rooted shoots. Four and six immersions gave rise to 88.7 % and 87.4 % rooted shoots, respectively (Figure 7). The fresh weight increase was 2.2-fold when two immersions per day were applied and reached 3.3-fold after application of four and six immersions per day.

The survival rate of plants rooted in TIS under different experimental conditions was determined in the greenhouse after 42 days of acclimatization. On average 94 % of the plants survived. There was no significant statistical difference between plants from different TIS pre-cultures.

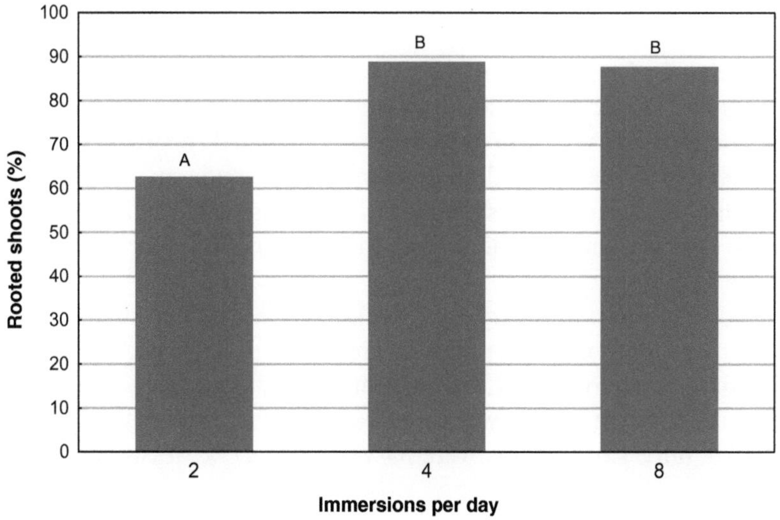

Figure 7: Percentage of rooted shoots of *Phalaenopsis* cv. Jaunina after 8-week culture differing in two, four or six immersions per day.
Different letters indicate significant differences according to analysis of variance ($p = 0.010$).

4. Discussion

Rapid micropropagation of *Phalaenopsis* on solid media has been described by several authors in the last decade. For example Tokuhara and Mii (1993) obtained more than 10,000 protocorm-like bodies (PLB) from one flower stalk within one year. Further, Duan et al. (1996) produced 2,300 plantlets per single cytokinin-treated stem after one-year-culture. Park et al. (2002) described a rapid propagation method using floral stem-derived leaves which ensures plantlet production from culture initiation to transplantation to pots after approximately six months. Somaclonal variants were rarely found. Variation in flower morphology (1.5 % of the total of 1,360 somaclones) were identified by Chen et al. (1998) in *Phalaenopsis* derived from PLBs. No variation was observed in flowering plants regenerated through somatic embryogenesis (Ishii et al., 1998).

A considerable increase in propagation of PLBs was obtained by Park et al. (2000) who harvested about 18,000 PLBs from 20 g of inoculum after eight weeks of incubation in a temporary-immersion bioreactor.

The intention of experiments presented here was to induce adventitious shoots in TIS instead of PLBs or somatic embryos, since manipulation of shoots is more practical for commercial laboratories.

4.1 *Effect of medium substitution every two or four weeks*

The release of phenolics by *Phalaenopsis* cultured *in vitro* is a common phenomenon. The accumulation depends on genotype and biomass increase in the culture vessel. The addition of activated charcoal or various antioxidants has been tested to overcome accumulation of phenolics in orchid cultures (for references see Arditti and Ernst, 1993). Park et al. (2000) reported that filtering of the liquid medium in TIS bioreactor strongly enhanced PLB growth.

Medium substitution every two weeks in TIS cultures of *Phalaenopsis* cv. Jaunina resulted in shoot multiplication rates of 25.4, i.e. 70 shoots used as inoculum gave rise to 1,780 adventitious shoots after twelve weeks. Medium substitution at four-week intervals decreased the multiplication rate to 14.5, i.e. only 1,015 shoots were harvested (= 57 % of two-week interval experiment).

Surprisingly, the fresh weight increase was less affected by accumulated phenolics than the shoot multiplication rate. Starting with 20 g inoculum, medium substitution at two-week intervals resulted in 20.5-fold increase of fresh weigth (= 410 g biomass) and at four-week intervals gave rise to 18.5 fold increase (370 g biomass; = 90 % of two-week interval experiment). Therefore, phenolic compounds more strongly inhibit initiation of adventitious shoots than biomass production.

4.2 *Effect of TDZ*

Thidiazuron (TDZ), a cytokinin-like compound, was originally registered as a cotton defoliant (Arndt et al., 1976). Later TDZ was used for *in vitro* culture of recalcitrant species. Many of them are woody plants (Huetteman and Preece, 1993). Ernst (1994) described that *Phalaenopsis* flower stem sections developed multiple shoots on TDZ-containing medium, and with higher levels also PLBs were initiated. Protocorm proliferation increased with increasing TDZ-concentration in the range of 0.23 – 1.14 μmol. Highest efficiency in adventitious bud induction was in the range of 5-10 μmol TDZ (Chen and Piluek, 1995). Protocorm-like bodies were formed in *Phalaenopsis* callus cultures when the medium was supplemented with 0.45–4.52 μmol TDZ (0.1 – 1.0 mg l^{-1} TDZ) (Chen et al., 2000).

Twelve-week TIS-culture on 0.5 mg l^{-1} (2.27 μmol) TDZ-containing medium resulted in doubling of shoot multiplication rate to 25.4 (Figure 2) compared with cultures grown for seven weeks on TDZ-medium and afterwards five weeks on TDZ-free medium, giving rise to a multiplication rate of 12.0.

The fresh weight increase was 20.5-fold after twelve weeks on TDZ-medium compared to a 17.4-fold increase after seven weeks TDZ-exposure (Figure 3). This difference was statistically not significant indicating that TDZ plays a minor role in total biomass production. In contrast, shoot size categories obtained after separating the clusters into single propagules were strongly affected by the different periods of TDZ-exposure (Figure 5). As expected more small shoots (50 % < 1 cm) developed after twelve weeks on TDZ-medium, whereas after seven-week culture on TDZ-medium in total 82 % of shoots were larger than 1 cm. This result is of practical importance for commercial propagation since the output of the required shoot size can be adjusted by the culture time on TDZ-medium.

4.3 Rooting

When two immersions per day of ten minutes each were used only 63.5 % of the shoots developed roots. An increase was achieved after four or six immersions resulting in 88.7 % and 87.4 % rooting. These frequencies are obviously optimal for root development in *Phalaenopsis* cv. Jaunina. There are no data from other cultivars. As known from TIS cultures of other plant species the recommended immersion frequencies can vary widely from seconds to hours (for references see Etienne and Berthouly, 2002). This holds true for development of shoots, somatic embryos or tubers.

5. Conclusion

TIS has proved to be an efficient tool for propagation of *Phalaenopsis*. The reason for the high multiplication rates and fresh weight increases may be not only the uptake of nutrients and hormones over the whole plant surface but also the daily multiple air exchange ventilating gaseous compounds like ethylene and CO_2. Optimum immersion frequency supports this beneficial gas exchange.

No economic data for the application of TIS *versa* conventional *in vitro* culture on solid media is yet available in *Phalaenopsis*. For sugarcane propagation, a cost reduction of 46 % was calculated for TIS compared to cultures on agar medium (Lorenzo et al., 1998). Pineapple multiplication in TIS saved 20 % of production costs per plant in comparison with conventional liquid cultures (Escalona et al., 1999). Since various kinds of automation steps are applicable to TIS, a considerable reduction of costs, especially in those for manual labour, can also be expected for *Phalaenopsis* in future.

References

Anonymous (2003) Omzettabel Kamerplanten. Vakblad voor de Bloemisterij 21 a: (p. 136)

Arditti J & Ernst R (1993) Phalaenopsis. In: Micropropagation of Orchids (pp. 467-520). John Wiley Sons, Inc., New York, Chichester, Brisbane, Toronto, Singapore

Arndt F, Rush R & Stillfried HV (1976) SN49537, a new cotton defoliant. Plant Physiol. 57: 599 (Abstr.)

Chen YC, Chang C & Chang WC (2000) A reliable protocol for plant regeneration from callus culture of *Phalaenopsis*. *In Vitro* Cell. Dev. Biol.-Plant 36: 420-423

Chen WH, Chen TM, Fu YM, Hsieh RM & Chen WS (1998) Studies on somaclonal variation in *Phalaenopsis*. Plant Cell Reports 18: 7-13

Chen Y & Piluek C (1995) Effects of thidiazuron and N^6-benzylaminopurine on shoot regeneration of *Phalaenopsis*. Plant Growth Regulation 16: 99-101

Duan J-X, Chen H & Yazawa S (1996) *In vitro* propagation of *Phalaenopsis* via culture of cytokinin-induced nodes. J Plant Growth Regul 15: 133-137

Ernst R (1994) Effects of thidiazuron on *in vitro* propagation of *Phalaenopsis* and *Doritaenopsis* (Orchidaceae). Plant Cell, Tiss. Org. Cult. 39: 273-275

Escalona M, Lorenzo JC, González B, Daquinta M, González JL, Desjardins Y & Borroto CG (1999) Pineapple (*Ananas comosus* L. Merr) micropropagation in temporary immersion systems. Plant Cell Reports 18: 743-748

Etienne H & Berthouly M (2002) Temporary immersion systems in plant propagation. Plant Cell, Tiss. Org. Cult. 69: 215-231

Huetteman CA & Preece JE (1993) Thidiazuron: a potent cytokinin for woody plant tissue culture. Plant Cell, Tiss. Org. Cult. 33: 105-119

Ishii Y, Takamura T, Goi M & Tanaka M (1998) Callus induction and somatic embryogenesis of *Phalaenopsis*. Plant Cell Reports 17: 446-450

Lorenzo JC, González BL, Escalona M, Teisson C. Espinosa P & Borroto C (1998) Sugarcane shoot formation in an improved temporary immersion system. Plant Cell, Tiss. Org. Cult. 54: 197-200

Murashige T & Skoog F (1962) A revised medium for rapid growth and bioassays with tobacco tissue cultures. Physiol. Plant. 15: 473-497

Park SY, Kakuta S, Kano A & Okabe M (1996) Efficient propagation of protocorm-like bodies of *Phalaenopsis* in liquid medium. Plant Cell, Tissue and Organ Culture 45: 79-85

Park SY, Murthy HN & Paek KY (2000) Mass multiplication of protocorm-like bodies using bioreactor system and subsequent plant regeneration in *Phalaenopsis*. Plant Cell, Tiss. Org. Cult. 63: 67-72

Park SY, Murthy HN & Paek KY (2002) Rapid propagation of *Phalaenopsis* from floral stalk-derived leaves. *In Vitro* Cell. Dev. Biol.-Plant 38: 168-172

Preil W (2004) Micropropagation in Germany (in preparation)

Samson I, Hamama L & Letouzé R (2000) The role of new synthetic cytokinins in the improvement of mass propagation of *Phalaenopsis via* protocorms regeneration. Acta Hort. 508: 265-266

Tanaka M & Sakanishi Y (1985) Regenerative capacity of *in vitro* cultured leaf segments excised from mature *Phalaenopsis* plants. Bulletin Univ. Osaka Pref. Ser. 37: 1-4

Tokuhara K & Mii M (1993) Micropropagation of *Phalaenopsis* and *Doritaenopsis* by culturing shoot tips of flower stalk buds. Plant Cell Reports 13: 7-11

Chapter 16

Propagation of *Prunus* and *Malus* by temporary immersion

C. Damiano (✉), S.R. La Starza, S. Monticelli, A. Gentile, E. Caboni & A. Frattarelli
Fruit Trees Research Institute, Ciampino Airport, 00040 Roma, Italy.
E-mail: isf.propag@mclink.it

Abstract: Temporary immersion, solid and liquid culture methods were compared to evaluate *in vitro* propagation of three *Prunus* species and a *Malus* rootstock. The growth of plants cultured at 30 and 60 minutes of immersion per day was compared to that in solid and stationary liquid conditions. After 60 days, multiplication rate, water, chlorophyll, carotenoid and fructose contents were evaluated. Stationary liquid culture negatively affected plant growth by reducing multiplication rate, chlorophyll and fructose contents and by inducing hyperhydricity and necrosis. The multiplication rate did not differ on solid medium and in temporary immersion, but hyperhydricity was present to a certain degree on solid medium but never in temporary immersion. Moreover chlorophylls, carotenoids and fructose, in the form of sucrose, increased in temporary immersion, particularly at the optimal immersion time. The accumulation of sucrose and the increase of photosynthetic pigment content could be due to a partial restoration of autotrophic activity.

Key words: carotenoids, chlorophylls, fructose, hyperhydricity, *in vitro* multiplication, liquid medium, solid medium

Abbreviations: BA - 6-benzylaminopurine; DW - dry weight; FW - fresh weight; GA_3 - gibberellic acid; IBA - indole-3-butyric acid; TIS - Temporary Immersion System; WC - water content

1. Introduction

Interest in decreasing the production costs of micropropagated plants is stimulating research on new *in vitro* culture systems useful for automation of the entire process. Many papers have reported an increase in *in vitro* shoot proliferation or improvement in explant quality with periodical immersion in the culture medium (Tisserat and Vandercook, 1985; Aitken-Christie and Davies, 1988; Etienne and Berthouly, 2002).

A.K. Hvoslef-Eide and W. Preil (eds.), Liquid Culture Systems for in vitro Plant Propagation, 243–251.

Over the last 15 years several systems for the TIS technique have been tested, and recently Teisson and Alvard (1995) described a system for plant propagation called RITA®, that could be easily automated. The RITA® system has been successfully applied to the micropropagation of several tropical plants such as banana (Alvard et al., 1993), pineapple (Escalona et al., 1999), sugarcane (Lorenzo et al., 1998) and to mass production of somatic embryos in *Musa* spp. (Escalant et al., 1994), *Hevea brasiliensis* (Etienne et al., 1997), *Citrus* (Cabasson et al., 1997), allowing a significant increase in the multiplication rate and improvement in the quality of micropropagated plants. More recently, the feasibility of the successful application of the TIS to temperate fruit trees has started to be investigated (Damiano et al., 2000).

The advantages of this technique can be related to reduced hyperhydricity of the tissues (due to the frequent renewal of the atmosphere in the culture bottle), the economics (minimal operator handling is required) and the reduction of contamination levels. In this work, the TIS technique was applied to *in vitro* propagation of apple, peach, cherry and plum, compared to stationary liquid and solid media, in order to assess its influence on multiplication response, shoot quality and physiological state of the treated plants.

2. Materials and methods

2.1 Plant material and cultural conditions

Apple (Jork 9), peach (cv. Yumyeong), cherry (cv. Bigarreau Burlat), plum (cv. Adara) shoots, *in vitro* grown for two years on solid media, were used. Plants were cultured for 60 days in: 1) solid medium (0.6% agar, B&V-Italy); 2) TIS with one immersion period of 30 min or 60 min per day; 3) stationary liquid medium, using the media reported in table 1. For all treatments the medium was renewed every 20 days. The culture conditions were: temperature 24°C ±1, photoperiod of 16 hours with light intensity of 25 μmol $m^{-2}s^{-1}$, provided by Osram fluora L58 W/77 lamps.

Figure 1: A) The "Temporary immersion" device used in the experiments. B) Temporary immersion: shoot proliferation in plum after 60 days of culture at 60 min of immersion per day. C) Temporary immersion: new shoot production in apple at 60 min of immersion. D) Temporary immersion: plantlets of peach do not show either hyperhydricity or necrosis after 60 days of culture.

2.2 Temporary immersion apparatus

The TIS was composed of two glass bottles connected with a silicone tube; one of the bottles contained the explants, the other one contained the culture medium (Figure 1A). The medium was periodically transferred from one bottle to the other by pressure created by an air pump; the air introduced into the bottle was sterilised with a 0.2 μm filter. Glass beads on the bottom of the bottle containing plants were used to isolate the plants from the thin layer of liquid medium remaining in the bottle during the period without immersion and for a better regulation of gravitropism.

2.3 Experimental plan

For each treatment 10 single shoots were used and 3 replications were carried out. After 60 days of culture, the multiplication rate (calculated as number of new shoots produced per cluster), fresh (FW) and dry weight (DW), obtained after dehydration in the oven at 110°C for 24 hrs, were recorded and the content of chlorophylls, carotenoids, and fructose (total or as sucrose) were measured. Water content (WC) percentage was calculated as $WC = [(FW - DW)/FW] \times 100$. Percentage data were analysed after normalization according to the formula arc sin$\sqrt{\%}$ and the Standard Error was calculated. Data were analysed by analysis of variance.

2.4 Chlorophyll and carotenoids extraction and determination

To analyse the a and b chlorophyll content, for each replication three samples of 100-200 mg of plant material from each treatment were prepared by soaking shoots in 12 ml of 80% acetone, according to Arnon (1949). Similarly samples for carotenoid extraction were prepared adding 50 mg of $CaCO_3$, to prevent pigment oxidation, and using pure acetone, storing it overnight at –20°C, according to Jensen (1973). The pigment was spectrophotometrically determined (Spectrophotometer Hitachi 2000), measuring light absorbance of acetone extracts of chlorophylls (λ = 663nm and 645nm, respectively for chlorophylls a and b) and carotenoids (λ = 450nm). All amounts were expressed as mg g^{-1} of dry weight.

2.5 Fructose extraction and determination

The fructose determination (total and as sucrose) was based on the estimation of a chromophoric substance formed by the reaction of resorcinol and fructose according to a modified Roe's test (Davis and Gander, 1967).

Three samples for each treatment from each replication of 5-10 mg of powdered dried tissue, were dissolved in water and after 5 minutes resorcinol (0.05 % in absolute ethanol) was added to the water solution. After addition of HCl (12 mol), the samples were heated at 77° C for 8 min and cooled immediately in ice for 5-10 min. Fructose content was determined with a spectrophotometer (Hitachi 2000) at $\lambda = 420$nm. Since the test does not distinguish between free fructose and fructose bound in sucrose, estimation of the amount bound in the sucrose was performed after initial destruction of the free fructose in the sample by hot alkaline (KOH 2N) treatment. After KOH addition the samples were boiled for 10 min and cooled immediately in ice for 5-10 min. The difference between total fructose content and the fraction bound in sucrose molecules represented the content of free fructose.

Table 1: Multiplication media

		Apple	Peach	Cherry	Plum
Macrosalts		MS	LP	MS	LP
Microsalts		MS	MS	MS	LP
Vitamins		MS*	MS	MS	MS
Calcium pantothenate	mgl^{-1}	-	-	-	1
Biotin	$mg\ l^{-1}$	-	-	-	0.1
Riboflavin	$mg\ l^{-1}$	-	-	-	0.5
Sucrose	$g\ l^{-1}$	30	30	30	30
BA	$mg\ l^{-1}$	0.5	0.4	0.5	0.25
IBA	$mg\ l^{-1}$	0.1	0.06	0.1	0.1
GA_3	$mg\ l^{-1}$	-	0.03	-	-
Adenine sulphate.	$mg\ l^{-1}$	-	3	-	-

*Thiamine ten-fold concentrated ($1mg\ l^{-1}$)
MS: Murashige and Skoog, 1962; LP: Quoirin et al., 1977.

Table 2: Multiplication rate of plants grown in different cultural conditions
(Data are the mean of three values ± s.e.). L = liquid, S = solid, 30' and 60'= TI period of 30 or 60 minutes

Cultural conditions	Plum	Cherry	Peach	Apple
L	3 ± 0.32a	2 ± 0.31a	3 ± 0.28a	3 ± 0.28a
S	30 ± 2.89b	5 ± 1.05bc	20 ± 2.06b	10 ± 1.97b
30'	n.r.	6 ± 1.17b	8 ± 1.57c	2 ± 0.41c
60'	28 ± 2.54b	4 ± 0.77c	14 ± 2.79d	8 ± 1.63b

n.r.: not recorded
Means within a column followed by different letter are significantly different ($P \leq 0.05$).

Table 3: Water (WC), chlorophyll, carotenoid, total and free fructose contents in plants grown in different cultural conditions (Data are the means of three values ± s.e.). L = liquid, S = solid, 30' or 60'= TI period of 30 or 60 minutes

		Cultural conditions	Plum	Cherry	Peach	Apple
WC *	%	L	86.36 ± 0.22a	80.82 ± 2.56a	87.86 ± 0.75a	82.21 ± 1.04a
		S	83.08 ± 0.92b	92.46 ± 0.36b	89.89 ± 0.96b	85.93 ± 0.32b
		30'	n.r.	79.79 ± 1.26a	77.76 ± 1.50c	97.56 ± 0.99c
		60'	80.72 ± 0.74c	76.66 ± 0.76c	77.27 ± 0.90c	78.53 ± 0.43d
Chloro-phylls	$mg\ g^{-1}$ DW	L	0.44 ± 0.01a	0.17 ± 0.02a	0.90 ± 0.05a	1.35 ± 0.13a
		S	1.25 ± 0.05b	3.06 ± 0.11b	2.26 ± 0.09b	3.36 ± 0.10b
		30'	n.r.	4.78 ± 0.18c	9.87 ± 0.13c	3.81 ± 0.17c
		60'	4.38 ± 0.19c	2.99 ± 0.10b	4.02 ± 0.51d	5.96 ± 0.09d
Carote-noids	$mg\ g^{-1}$ DW	L	0.39 ± 0.05a	0.13 ± 0.01a	0.43 ± 0.04a	0.54 ± 0.01a
		S	0.72 ± 0.06b	1.04 ± 0.07b	1.48 ± 0.19b	2.11 ± 0.25b
		30'	n.r.	1.82 ± 0.08c	2.69 ± 0.05c	2.05 ± 0.26b
		60'	1.72 ± 0.04c	1.67 ± 0.02d	2.19 ± 0.02d	2.59 ± 0.07c
Total fructose	$mg\ g^{-1}$ DW	L	56.30 ± 2.10a	39.51 ± 0.32a	49.02 ± 2.51a	21.54 ± 0.72a
		S	67.31 ± 1.80b	53.91 ± 2.43b	75.64 ± 1.22b	40.62 ± 1.83b
		30'	n.r.	67.03 ± 0.71c	71.43 ± 2.23c	43.23 ± 0.61c
		60'	77.11 ± 1.51c	52.54 ± 2.41b	84.21 ± 2.43d	36.71 ± 2.04d
Free fructose **	%	L	20.42 ± 0.22a	49.53 ± 4.46a	53.85 ± 4.85a	19.96 ± 2.10a
		S	24.94 ± 2.12b	12.19 ± 1.28b	33.55± 3.52b	19.83 ± 1.78a
		30'	n.r.	10.46 ± 0.89c	2.82 ± 0.27c	11.85 ± 1.01b
		60'	7.49 ± 0.71c	4.30 ± 0.47d	8.41 ± 0.88d	4.73 ± 0.43c

n.r.: not recorded
*Calculated as WC =[(FW – DW)/FW] x 100. FW: fresh weight; DW: dry weight
**Calculated as =(Free fructose/Total fructose) x 100
Means within a column followed by different letter are significantly different ($P \leq 0.05$).

3. Results

After 60 days of TI it was possible to establish the best immersion time for each species. Among the plants cultured in TIS, the multiplication rate was higher at 30 min for cherry and at 60 min for peach and apple (Figure 1C). Assuming the solid culture as control, the multiplication rate of plants at the best immersion time did not differ significantly from the solid medium, except in peach. In plum (Figure 1B) it was not possible to determine the best time of immersion because of the loss of material cultured at 30 min of immersion, due to a contamination (Table 2). On the other hand, the shoot formation was highly inhibited by liquid stationary medium.

The water content of temporarily immersed plants decreased at both immersion times, with the exception of apple at 30 min of immersion. This trend was also found in most of the plants cultured in stationary liquid medium (Table 3), in this case, probably because all the plants were necrotic, except for plum, where the plants appeared hyperhydric. After 60 days, hyperhydricity and necrosis of the explants were never detected in TIS (Figure 1D), whereas they were present in solid culture. In fact the cherry, plum and peach controls showed hyperhydricity in 7.5%, 7% and 6% of the plants respectively, while necrotic apices were present in 11%, 5% and 7% of plum, peach and apple shoot clusters.

The detrimental effects of stationary liquid culture were also evident in the content of photosynthetic pigment. On the contrary, TIS stimulated chlorophyll and carotenoid synthesis at the optimal immersion time, but, in peach, chlorophyll and carotenoid content was higher in the 30 min immersion treatment (Table 3).

The total amount of fructose was significantly higher in plants grown under the optimal immersion time, plum included, but not in apple. However, variations occurred in the fraction of free fructose depending on the genotype and the treatment. Cherry and peach plants grown in stationary liquid conditions accumulated more free fructose than the control, plum accumulated it to a lesser extent, while in apple the content was the same. On the contrary, after the cultivation in TIS the bound fructose accumulated as sucrose was increased compared to the control (Table 3).

4. Discussion

The multiplication rates of shoot cultures of the four species cultured in TIS did not differ from those cultured on solid medium as controls. However the plants cultured in TIS never showed hyperhydricity and apex necrosis. On the contrary, stationary liquid condition induced a deterioration of the plants, due to hyperhydricity and necrosis, and a critical decrease of the multiplication rate in most cases. Hyperhydricity occurred in around 10% of the controls, for all genotypes, but never in the TIS treatment. This claim is supported by the reduced water content in temporarily immersed plants *versus* the control. In contrast, the decreased water content observed in stationary liquid medium was related to necrosis and plant death after 60 days. Only in plum water content reduction was absent and the plants appeared more hyperhydric than necrotic. It is reported in the literature that the positive effects of TIS are due to a better aeration and renewal of chemical components at each immersion, otherwise limited in solid condition by the agar matrix. In this way the risk of anoxia and hyperhydricity, frequently observed in constant immersion as well as in solid culture, is significantly reduced (Etienne et al., 1997). The best immersion time varied according to the genotype.

Green tissues are not fully autotrophic in *in vitro* grown plants due to the growth conditions being unsuitable for photosynthesis. Light and CO_2 concentrations in the jars or test tubes can be limiting for the photosynthetic process (Pierik, 1987). Microshoots under *in vitro* culture conditions can have a negative net photosynthetic rate and depend on sucrose uptake from the medium for their growth (De Riek et al., 1997). Few results are available on the mechanisms of sugar uptake in shoot tissue culture systems. It is accepted that sugar uptake is an energy dependent process. Most publications dealing with sucrose uptake by cell cultures assume that sucrose molecules supplied *via* the culture medium are hydrolysed by cell wall or plasmalemma located invertases. The resulting fructose and glucose are readily taken up by the cells. Studies carried out using labelled hexoses showed the re-synthesis of sucrose inside the plants, of which 25-40 % is used for respiration. Sucrose accumulation occurs in leaves when carbohydrates derived from photosynthesis are added to the sucrose unloaded from the phloem (De Riek et al., 1997).

From the results reported here, it was shown that in plants grown in TIS the total amount of fructose significantly increased, mainly accumulated as sucrose, while in plants grown in solid and stationary liquid conditions free fructose prevailed. These results seem to support the hypothesis that the plants cultured in TIS, being in contact with the sugars in the medium only for very short period during the culture, can partially restore autotrophic ability and the capability to accumulate sugars. In fact, low or no supply of sucrose to herbaceous plants has been already shown to increase shoot growth, promoting autotrophic behaviour (Aitken-Christie and Davies, 1988). Furthermore, the content of chlorophylls and carotenoids also appeared higher in plants cultured in TIS at the specific optimal immersion time. In our experiments the TIS was shown to improve mainly the quality of the plant material. Further experiments are needed to determine the response of plants cultured in TIS to other immersion frequencies, to rooting and to the stress of acclimation.

Acknowledgements

Mr. Frattarelli has provided technical implementation of experimental plan and collection of data.

References

Aitken-Christie J & Davies HE (1988) Development of a semi-automated micropropagation system. Acta Hort. 230: 81-87

Alvard D, Cote F & Teisson C (1993) Comparison of methods of liquid medium culture for banana micropropagation. Plant Cell, Tiss. Org. Cult. 43: 55-60

Arnon DI (1949) Copper enzymes in isolated chloroplasts. Polyphenoloxidase in *Beta vulgaris*. Plant Physiol. 24 (1): 1-15

Cabasson C, Alvard D, Dambier D, Ollitrault P & Teisson C (1997) Improvement of *Citrus* somatic embryo development by temporary immersion. Plant Cell, Tiss. Org. Cult. 50: 33-37

Damiano C, Caboni E, Frattarelli A, Giorgioni M, Liberali M, Lauri P & D'Angeli S (2000) Innovative micropropagation of temperate fruit trees: the case of pear. Acta Hort. 530: 181-185

Davis JS & Gander JE (1967) A re-evaluation of the Roe procedure for determination of fructose. Anal. Biochem. 19: 72-79

De Riek J, Piqueras A & Debergh PC (1997) Sucrose uptake and metabolism in a double layer system for micropropagation of *Rosa multiflora*. Plant Cell, Tiss. Org. Cult. 47: 269-278

Escalant J-V, Teisson C & Cote F (1994) Amplified somatic embryogenesis from male flowers of triploid banana and plantain cultivars (*Musa* spp.). *In Vitro* Cell. Dev. Biol. 30P: 181-186

Escalona M, Lorenzo JC, Gonzáles B, Daquinta M, Gonzáles JL, Desjardins Y & Borroto CG (1999) Pineapple (*Ananas comosus* L. Merr) micropropagation in temporary immersion systems. Plant Cell Rep. 18: 743-748

Etienne H & Berthouly M (2002) Temporary immersion systems in plant micropropagation. Plant Cell, Tiss. Org. Cult. 69: 215-231

Etienne H, Lartaud M, Michaux-Ferrière N, Carron MP, Berthouly M & Teisson C (1997) Improvement of somatic embryogenesis in *Hevea brasiliensis* (Müll. Arg.) using the temporary immersion technique. *In Vitro* Cell. Dev. Biol. – Plant, 33: 81-87

Jensen A (1973) Chlorophylls and catotenoids. In: Hellebust JA & Craigie JS (eds) Handbook of Phycological Methods. Physiological and Biochemical Methods (pp. 59-70).Cambridge University Press

Lorenzo JC, Gonzáles BL, Escalona M, Teisson C, Espinosa P & Borroto C (1998) Sugarcane shoot formation in an improved temporary immersion system. Plant Cell, Tiss. Org. Cult. 54: 197-200

Murashige T & Skoog F (1962) A revised medium for rapid growth and bioassay with tobacco tissue cultures. Physiol. Plant. 15: 473-497

Pierik RLM (1987) *In Vitro* Culture of Higher Plants. Martinus Nijhoff Publishers, Dordrecht

Quoirin M, Lepoivre P & Boxus Ph (1977) Un premier bilan de 10 annèes de recherches sur les cultures de meristèmes et la multiplication *in vitro* de fruitiers ligneux. In: Compte Rendu des Recherches 1976-1977 (pp. 93-117). Station des cultures fruitières et maraichères, Gembloux

Teisson C & Alvard D (1995) A new concept of plant *in vitro* cultivation liquid medium: temporary immersion. In: Terzi M, Cella R & Falavigna A (eds) Current Issues in Plant Molecular and Cellular Biology (pp. 105-109). Kluwer Academic Publishers, Dordrecht

Tisserat B & Vandercook CE (1985) Development of an automated plant culture system. Plant Cell, Tiss. Org. Cult. 5: 107-117

Chapter 17

Optimisation of growing conditions for the apple rootstock M26 grown in RITA containers using temporary immersion principle

Li-Hua Zhu*, Xue-Yuan Li & Margareta Welander
Swedish University of Agricultural Sciences, Department of Crop Science, P.O. Box 44, SE-230 53 Alnarp, Sweden
**Corresponding author: tel: +46 40 415373, fax: +46 40 460442;*
e-mail: Li-Hua.Zhu@vv.slu.se

Abstract: The use of bioreactors may provide an efficient and economic tool for mass clonal propagation of plants if technical problems can be solved. In this paper, we report the results of experiments aimed at optimising conditions for apple rootstock M26 growth in RITA containers using the temporary immersion principle. We tested different types and sizes of explants, different concentrations of plant growth regulators (BAP, kinetin and IBA) in the multiplication and elongation phases, and medium exchange during the shoot elongation period. The results show that the higher concentrations of cytokinins were required during the shoot multiplication phase, while the lower concentrations were better during the shoot elongation phase. Hyperhydricity was increased with increasing concentrations of cytokinins during both shoot multiplication and shoot elongation phases. The best shoot production in terms of shoot number and shoot quality was obtained using 4.4 μmol BAP and 0.5 μmol IBA during the shoot multiplication phase and 1.1 μmol BAP and 0.25 μmol IBA during the shoot elongation phase. Medium exchange twice during the shoot elongation phase resulted in higher shoot production compared with no exchange of the medium. However, it also resulted in increased hyperhydricity. Immersion frequency of 16 times per day gave a higher multiplication rate and longer shoots than 8 times per day. The explant size of 0.5 cm or 1 cm resulted in a significantly higher shoot production rate compared with that of 1.5 cm, but shoot length and hyperhydricity were not affected by the explant size. Shoot cultures from the liquid media rooted normally in the RITA containers with more than 90 % rooting and the rooted plantlets acclimatised well in the greenhouse.

Key words: apple, bioreactor, micropropagation, RITA, temporary immersion

Abbreviations: BAP – benzylaminopurine; IBA – indole butyric acid; MS – Murashige and Skoog medium (1962)

A.K. Hvoslef-Eide and W. Preil (eds.), Liquid Culture Systems for in vitro Plant Propagation, 253–261.

1. Introduction

Clonal propagation by conventional tissue culture techniques is limited in commercial production due to high input of manual labour and a low degree of automation (Chu, 1995; Maene and Debergh, 1985; Simonton et al., 1991, Sluis and Walker, 1985). In contrast to conventional tissue culture techniques, the use of bioreactors may provide a promising tool for mass clonal propagation. Due to the use of liquid media, bioreactors have the following advantages compared with agar media: 1) a large number of plantlets can be more easily produced due to more uniform culture conditions and the ease with which, the explants can take up the nutrients; 2) time- and labour-saving in the handling of cultures because of the semi-automatic operation; 3) better growth and biomass production because of good aeration by forced oxygen supply; 4) the decrease of apical dominance and the stimulation of lateral bud growth, which is probably due to the loss of culture orientation. However, bioreactors developed in the past were mainly for bacterial cultures and not suitable for plant micropropagation because of the mechanical damage and hyperhydricity (Teisson et al., 1999). Recent years, researchers have developed different semi-automated systems using temporary immersion principle and without the impeller (Aitken-Christie and Jones, 1987; Simonton et al., 1991; Tisserat and Vandercook, 1985) to eliminate hyperhydricity and mechanical damage, respectively. The RITA temporary immersion system is one of those developed lately (Teisson and Alvard, 1995). The system provides a temporary contact between explants and liquid media to avoid culture hyperhydricity. Studies on micropropagation using this system have been reported in *Citrus* (Cabasson et al., 1997), pineapple (Escalona et al., 1999), potato (Jiménez et al., 1999), and sugarcane (Lorenzo et al., 2001). The aim of this study was to optimise micropropagating conditions for the apple (*Malus domestica*) rootstock M26 grown in RITA containers using the temporary immersion principle.

2. Materials and methods

The experiments were conducted in RITA (VITROPIC, Saint-Mathieu-de-Tréviers, France) containers using temporary immersion principle. *In vitro* propagated shoot cultures of the apple rootstock M26 were used. The cultures were originally grown on the solid MS (Murashige and Skoog, 1962) medium supplemented with 4.4 μmol BAP (benzylaminopurine), 0.5 μmol IBA (indole butyric acid), 30 g l^{-1} sucrose and 0.7 % agar at pH 5.5.

In the RITA containers, 150 ml of liquid MS medium supplemented with 30 g l^{-1} sucrose at pH 5.5 was used for all treatments. In total, five experiments were included in this study. The first experiment aimed to improve shoot production by using different combinations of growth regulators and different types of explants (for details see Table 1). Based on experiment 1, the second and third experiments were conducted aimed at optimising concentrations of growth regulators for shoot elongation. In the forth experiment, the influence of medium exchange on shoot production was studied, and the influence of explant size in the fifth experiment. The detailed treatments and conditions used in experiments 2-5 are presented in table 3. In all experiments, two phases were included, namely, the shoot multiplication phase with higher cytokinin concentrations (15 days for experiment 1 and 10 days for experiments 2-5) and the shoot elongation phase with lower cytokinin concentrations for 3-4 weeks. The multiplication media contained 4.4 μmol BAP and 0.5 μmol IBA except for experiment 1. The frequency of medium immersion was 8 times per day for experiments 1-3 and 16 times per day for experiments 4-5. The duration of medium immersion was 5 minutes per time for experiments 1-2 and 2 minutes per time for experiments 3-5. The media were not exchanged during the shoot elongation phase except for experiment 4.

Shoots of 1-1.5 cm in length from four treatments in experiment 1 were rooted in RITA containers using the rooting medium consisting of Lepoivre (Quoirin et al., 1977) macro and microelements, Walkey (1972) vitamins, 1.2 μmol IBA and 30 g l^{-1} sucrose at pH 5.5. The shoots were kept in the dark for 4 days, and then transferred to the identical rooting medium without IBA in light for 3 weeks.

Shoot number (≥ 0.5 cm) and shoot quality, measured as shoot length and hyperhydricity were recorded at the end of the experiments for experiments 1-5, and the root number and root length for experiment 1. All data were subjected to ANOVA analysis with Duncan's multiple range test using the Statgraphics program.

Rooted plantlets from the RITA containers were planted in pots in a mixture of peat and perlite, and acclimatised in the greenhouse conditions.

3. Results and discussion

3.1 Experiment 1

The results from experiment 1 are presented in table 1 and figure 1. Using 1.1 μmol BAP and 1.2 μmol IBA in the shoot multiplication medium resulted in a significantly higher shoot number compared with 4.4 μmol BAP

A B C D E

Figure 1: Shoot cultures of the apple rootstock M26 grown in the RITA containers after 5 weeks from experiment 1. A and B: Shoot tip and stem segment explants, respectively, 4.4 µmol BAP and 0.5 µmol IBA during the shoot multiplication phase and 0.5 µmol kinetin and 0.05 µmol IBA during the shoot elongation phase for both. C and D: Shoot tip and stem segment explants, respectively, 8.8 µmol BAP and 1.0 µmol IBA for the shoot multiplication phase and no growth regulators during the shoot elongation phase for both. E: Shoot tip explants, 1.1 µmol BAP and 1.2 µmol IBA during the shoot multiplication phase and no growth regulators during the shoot elongation phase.

and 0.5 µmol IBA, and 8.8 µmol BAP and 1.0 µmol IBA when shoot tips were used as explants. Stem segments gave a higher shoot number than shoot tips at low BAP and IBA concentrations. The shoot length decreased at the concentrations of 8.8 µmol BAP and 1.0 µmol IBA for both explant types. The percentage of hyperhydricity increased with increasing the concentration of the growth regulators. Stem segments resulted in higher hyperhydricity than shoot tip explants at the same concentration of the growth regulators.

The above results show that higher concentrations of plant growth regulators resulted in a higher multiplication rate, but this was often accompanied by increased hyperhydricity. This means that the apple cultures in liquid media are more sensitive to plant growth regulators compared with those grown on solid medium where 4.4 µmol BAP and 0.5 µmol IBA are routinely used during the whole shoot culture period without hyperhydricity. The reason for the higher sensitivity of the cultures to the growth regulators is possibly due to an easy access of the cultures to growth regulators in liquid media. The experiment has also shown that stem segments showed more hyperhydricity than shoot tips at the same concentration of the growth regulators. This might be because lateral buds are already differentiated in stem segments when they are cultured in the medium and there is no need for a high concentration of growth regulators. On the other hand, higher concentrations of growth regulators are required for meristems from shoot tips to induce new buds. It is well known that higher concentrations of cytokinins are necessary for the differentiation of new shoots, but that shoot elongation is often inhibited by a high concentration of cytokinins.

Table 2 shows that the rooting percentage was generally high, ranging from 91 to 100 %. Among the four different treatments, the lowest rooting percentage was obtained from the medium containing the highest BAP and IBA concentrations in the shoot multiplication medium. The root number and root length were also lower when the previous medium contained higher concentrations of BAP and IBA. Addition of kinetin in combination with IBA in the shoot elongation phase had no negative effects on rooting percentage and root number. These results suggest that high concentrations of cytokinins during the shoot multiplication phase inhibit rooting, which is also common in micropropagation of plants. The survival of the rooted plantlets in the greenhouse was 100% (data not shown).

3.2 Experiment 2 and 3

Based on experiment 1, experiments 2 and 3 were carried out to further optimise growth conditions and the results are presented in table 3. The highest kinetin concentration resulted in the highest shoot number and shoot length among the three kinetin concentrations, but it also caused the highest percentage of hyperhydricity. The increase of BAP concentration from 1.1 to 2.2 µmol did not give a better multiplication rate, but resulted in a higher percentage of hyperhydricity. These results further confirm that the high percentage of hyperhydricity is closely related to high BAP or kinetin concentrations in culture medium. In order to obtain more shoots with better quality from the RITA system, the concentration of cytokinin needs to be below a threshold level. The results also revealed that, at a similar concentration, BAP and kinetin gave a similar result for shoot production and shoot quality. Based on this, only BAP was used in the later experiments.

Table 1: Results of shoot multiplication and elongation of the apple rootstock M26 grown in RITA containers from experiment 1. Explant size was 0.5 cm for tips and 0.6-0.8 cm for stem segments, and no medium exchange during the shoot elongation period for all treatments. The shoot cultures were in the multiplication media for 15 days, and then transferred to the shoot elongation media

Exp.	Explant type	Treatment		Shoot No.	Shoot length (mm)	H (%)
		Multiplication	Elongation			
		BAP/IBA (µmol)	Kinetin/IBA (µmol)			
1	Shoot tip	4.4/0.5	0.5/0.05	2.5 a	13.4 b	0
	Stem seg.	4.4/0.5	0.5/0.05	3.5 b	13.3 b	33
	Shoot tip	8.8/1.0	0/0	3.2 ab	9.7 a	25
	Stem seg.	8.8/1.0	0/0	3.8 bc	9.6 a	50
	Shoot tip	1.1/1.2	0/0	4.3 c	13.3 b	50

Figures followed by different letters in each column differ significantly at P=0.05 (n=40). Exp.=experiment. seg.=segment. H.=Hyperhydricity.

Table 2: Rooting results of the apple rootstock M26 grown in the RITA containers where the shoots were derived from experiment 1. The shoots were rooted in the rooting medium consisting of Lepoivre macro- and micro nutrients, Walkey vitamins and 1.2 µmol IBA for 4 days in the dark, and then in the identical IBA-free rooting medium in light for 3 weeks

Explant type	Treatment prior to rooting		Root No. per shoot	Root length (cm)	Rooting %
	Multiplication	Elongation			
	BAP/IBA (µmol)	Kinetin/IBA (µmol)			
Shoot tip	4.4/0.5	0.5/0.05	6,48 b	2,5 b	96
Stem seg.	4.4/0.5	0.5/0.05	6,58 b	2,0 b	100
Shoot tip	8.8/1.0	0/0	4,95 a	1,2 a	100
Shoot tip	1.1/1.2	0/0	4,65 a	1,5 a	91

Figures followed by different letters in each column differ significantly at P=0.05 (n=20). seg.=segment.

Table 3: Results of shoot production and shoot quality of the apple rootstock M26 grown in the RITA containers from experiments 2-5. The cultures were grown in the multiplication medium (basal MS plus 4.4 µmol BAP and 0.5 µmol IBA) for 10 days, and then in the shoot elongation media for 4 weeks as shown in the table. The frequency of medium immersion was 8 times per day for experiments 2-3 and 16 times per day for experiments 4-5. ANOVA analysis with Duncan's multiple range test was carried out within one experiment

Exp.	Treatment					Shoot No. per explant	Shoot length (mm)	H. (%)
	Kinetin/ BAP (µmol)		IBA (µmol)	Times of exchanging medium [a]	Explant size (cm)			
	K	0.5	0.05	0	0.5	2.2 a	11.3 a	5
2	K	2.3	0.25	0	0.5	3.3 b	18.8 b	43
	BAP	2.2	0.25	0	0.5	2.8 ab	16.5 b	50
	K	0.5	0.05	0	0.5	1.6 a	16.5 a	5
3	K	1.2	0.25	0	0.5	1.8 a	20.1 a	6
	BAP	1.1	0.25	0	0.5	2.1 a	17.1 a	0
	BAP	1.1	0.25	0	0.5	6.4 a	23.4 a	20
4	BAP	1.1	0.25	1	0.5	8.6 b	25.0 ab	25
	BAP	1.1	0.25	2	0.5	9.0 b	28.4 b	44
	BAP	1.1	0.25	0	0.5	7.2 b	25.9 a	0
5	BAP	1.1	0.25	0	1.0	8.1 b	25.8 a	0
	BAP	1.1	0.25	0	1.5	5.2 a	23.8 a	0

Figures followed by different letters in each column differ significantly at P=0.05 (n=40). Exp.=experiment. K=kinetin. H.=Hyperhydricity. [a] 0=no medium exchange; 1=one time of medium exchange with the interval of 15 days; 2=two times of medium exchange with the interval of ten days during one month of subculture.

3.3 *Experiment 4*

Experiment 4 was conducted to study the influence of medium exchange on shoot production and quality during the shoot elongation phase. The results are presented in table 3. Exchanging the medium once or twice during the elongation phase increased the shoot number significantly, and the shoot length non-significantly. However, this was accompanied by an increase in hyperhydricity when the medium was exchanged twice. This is clearly associated with the persistent presence of cytokinin in the medium due to the more frequent exchange of the fresh media. Therefore, we can conclude that exchanging the medium during the shoot elongation period has no obvious advantages for higher quality shoot production and from the economic point of view. As shown in table 3, shoot number and shoot length were higher in this experiment than those in experiments 2 and 3. This is possibly due to the increased frequency of tissue immersion to the culture medium since the tissue was immersed 16 times per day in experiment 4, while only 8 times in experiments 2-3. The multiplication rate in this experiment was much higher than that on agar medium where 1-3 shoots can usually be produced from normal shoot cultures in our case.

3.4 Experiment 5

Experiment 5 was carried out to study the influence of explant size on shoot production and quality. Table 3 shows that the shoot number was not significantly affected when the explant size was within 1 cm. However, when the explant size was 1.5 cm, shoot number was significantly reduced. Hyperhydricity was not affected by the explant size. The decreased shoot number with larger explant size was likely due to poor transport of cytokinin from the cut surface to buds and tips.

4. Conclusion

In conclusion, the micropropagation of the apple rootstock M26 in the RITA system included two steps, i.e. shoot multiplication and shoot elongation. The optimised conditions include explant size of 0.5-1.0 cm, 4.4 μmol BAP and 0.5 μmol IBA in the shoot multiplication phase, 1.1 μmol BAP and 0.25 μmol IBA in the shoot elongation phase, no exchange of medium during the shoot elongation period and tissue immersion 16 times per day. No obvious differences in shoot production and shoot quality between BAP and kinetin when they were used at similar concentrations during shoot elongation phase. The multiplication rate in the RITA system was higher than than on agar media. The shoots produced from the RITA system could root normally and the rooted plantlets could easily acclimatise under greenhouse conditions.

Acknowledgements

We wish to thank Swedish Foundation for Innovation Centre (Stiftelsen för Innovationscentrum) for the financial support to this research.

References

Aitken-Christie J & Jones C (1987) Towards automation: radiata pine shoot hedges *in vitro*. Plant Cell Tiss. Org. Cult. 8: 185-196

Cabasson C, Alvard D, Dambier D, Ollitrault P & Teisson C (1997) Improvement of Citrus somatic embryo development by temporary immersion. Plant Cell, Tiss. Org. Cult. 50: 33-37

Chu I (1995) Economic analysis of automated micropropagation. In: Aitken-Christie J, Kozai T & Smith MAL (eds) Automation and Environmental Control in Plant Tissue Culture (pp.19-26). Kluwer Academic Publishers, Dordrecht

Escalona M, Lorenzo JC, González B, Daquinta M, González JL, Desjardins Y & Borroto CG (1999) Pineapple (*Ananas comosus* L. Merr) micropropagation in temporary immersion systems. Plant Cell Reports 18: 743-748

Jiménez E, Pérez N, Feria M de, Barbón R, Capote A, Chávez M, Quiala E & Pérez JC (1999) Improved production of potato microtubers using a temporary immersion system. Plant Cell, Tiss. Org. Cult. 59: 19-23

Lorenzo JC de los Angeles Blanco M, Peláez O, González A, Cid M, Iglesias A, González B, Escalona M, Espinosa P & Borroto C (2001) Sugarcane micropropagation and penolic excretion. Plant Cell, Tiss. Org. Cult. 65: 1-8

Maene L & Debergh P (1985) Liquid medium additions to established tissue cultures to improve elongation and rooting *in vitro*. Plant Cell, Tiss. Org. Cult. 5: 23-33

Murashige F & Skoog F (1962) A revised medium for rapid growth and bioassys with tobacco tissue cultures. Physiol. Plant. 15: 473-492

Quoirin M, Lepoivre P & Boxus P (1977) Un premier bilan de dix années de recherche sur les cultures de méristèmes et la multiplication *in vitro* de fruitiers ligneux (in French). Compte rendu des recherches, Station des Cultures Fruitières et Marrîchères de Gembloux 1976-1977, 93-117

Simonton W, Robacker C & Krueger S (1991) A programmable micropropagation apparatus using cycled medium. Plant Cell, Tiss. Org. Cult. 27: 211-218

Sluis CJ & Walker KA (1985) Commercialization of plant tissue culture propagation. International Association for Plant Tissue Culture. Newsletters 47: 2-12

Teisson C & Alvard D (1995) A new concept of plant *in vitro* cultivation liquid medium: temporary immersion. In: Terzi M et al. (eds) Current Issues in Plant Molecular and Cellular Biology (pp. 105-110). Kluwer Academic Publishers, Dordrecht

Teisson C, Alvard D, Lartaud M, Etienne H, Berthouly M, Escalona M & Lorenzo JC (1999) Temporary immersion for plant tissue culture. In: Plant Biotechnology and *In Vitro* Biology in the 21st Century, Proceedings of the IXth International Congress of Plant Tissue and Cell Culture, Section H: Novel micropropagation methods (pp. 629-632). Jerusalem

Tisserat B & Vandercook CE (1985) Development of an automated plant culture system. Plant Cell, Tiss. Org. Cult. 5: 107-117

Walkey DG (1972) Production of apple plantlets from axillary bud meristems. Canadian Journal of Plant Science 52: 1085-1087

Chapter 18

Experimental use of a novel temporary immersion system for liquid culture of olive microshoots

Katerina Grigoriadou[1], Miltiadis Vasilakakis[2], Theofilos Tzoulis[3] & Eleftherios P. Eleftheriou[4]
[1]VITRO HELLAS S.A., Nisseli Imathias, 593 00, Greece, tel:+302333024888, fax: +302333026417, e-mail: vitro2@otenet.gr (author for correspondence).
[2]Pomology Department, School of Agriculture, Aristotle Univ. of Thessaloniki, 541 24, Greece.
[3]B. Hirs 11, 546 41, Thessaloniki, Greece. [4]Department of Botany, School of Biology, Aristotle University of Thessaloniki, 541 24, Greece

Abstract: A novel temporary immersion system (TIS), designed at the laboratory of VITRO HELLAS S.A., was used for the liquid culture of olive microshoots during their proliferation phase. The results were compared with those obtained in agar solidified medium, which was used as control, and with other bioreactor systems including the LifeReactor©, liquid culture in Erlenmeyer flasks under agitation and liquid culture on filter paper bridges. After 30 days of culture, results derived from the novel TIS (1.93 new microshoots per explant, 0.95 cm shoot length) were statistically the same as the control (1.75 microshoots per explant, 1.22 cm shoot length respectively). Equally high results were obtained with the agitated Erlenmeyer flasks, while LifeReactor© gave the poorest ones (0.35 new microshoots per explant, 0.40 cm shoot length). Cultures on filter paper bridges developed large masses of callus at the explant bases. However, all the liquid cultures suffered from hyperhydricity, the most serious was observed in the Erlenmeyer flasks and the least in the novel TIS. In order to reduce hyperhydricity, the TIS was combined with cultures on agar solidified medium, which eliminated the problem but shoot proliferation was also reduced. It seems that the novel TIS provides a promising device, although some technical problems need to be overcome for large scale production.

Key words: bioreactor, *in vitro* culture, liquid culture, *Olea europaea* L.

Abbreviations: GA - gibberellic acid; NAA - naphthaleneacetic acid

1. Introduction

The use of liquid media for *in vitro* culture is considered an ideal technique for mass propagation. However, the main disadvantage of liquid

A.K. Hvoslef-Eide and W. Preil (eds.), Liquid Culture Systems for in vitro Plant Propagation, 263–274.

culture is hyperhydricity, a physiological abnormality of cultivated tissues causing anatomical deformities and failure of shoot development or rooting (Ziv and Ariel, 1992; Miguens et al., 1993). Different procedures and devices have been developed including temporary immersion systems (TIS) designed especially to overcome these problems (Teisson et al., 1996; Escalona et al., 1999). Many vessels, mainly consisting of more or less complex containers, have been used for periodical immersion of the explants. Some of them are commercially available. This technique has been successfully used for micropropagation of tropical crops such as banana, pineapple, sugarcane, coffee and rubber tree, while its efficiency for other woody species including pear and apple rootstocks has been reported to be promising, providing the advantages of increased proliferation rate and the possibility of automation (Teisson et al., 1996; Escalona et al., 1999; Damiano et al., 2000; Welander et al., 2001). Furthermore, the formation of shoot clusters in disposable presterilized plastic bioreactors for large-scale micropropagation of plants has recently become a reality as commercially available devices, easy to set up and operate in any laboratory, have been already developed by manufacturers involved in micropropagation industry (Ziv et al., 1998; Peak et al, 2001).

The *in vitro* culture of olive tree has not been very successful, mainly due to the problems of poor microshoot formation, which is cultivar dependent (Dimassi, 1999; Rugini et al., 1999; Grigoriadou et al., 2002). Moreover, hyperhydricity problems are usual in olive tissues, which accumulate a mucus material in the intercellular spaces inhibiting the normal development of the cultures (Grigoriadou, 2003). The use of a TIS might improve microshoot development during the proliferation phase and overcome hyperhydricity problems.

The aim of this work was to study the effect of a new TIS, designed in VITRO HELLAS S.A. laboratory, on olive microshoot formation, in comparison with other liquid culture techniques and devices successfully applied to different woody species.

2. Materials and methods

2.1 Plant material and culture conditions

Stock *in vitro* cultures of the Greek olive tree (*Olea europaea* L.) cv. Chondrolia Chalkidikis were maintained in 500 ml glass jars containing 125 ml of Woody Plant Medium (Lloyd and McCown, 1981), supplemented with 20 μmol zeatin, 10 μmol GA_3, 0.3 μmol NAA, 2% (w/v) sucrose and 0.5% (w/v) agar (B&V S.r.L., type S 1000). The pH was adjusted to 5.2 before agar addition and autoclaving. Cultures were kept at 22±2°C under a 16-h photoperiod of cool white fluorescent light (40 μmol m^{-2} s^{-1}) and subcultured every 8 weeks (Grigoriadou et al., 2002).

2.2 *Effect of temperature*

A preliminary experiment on agar-solidified medium to estimate a favourable temperature for microshoot development was carried out. For this purpose the effect of high (28±2°C) and low (22±2°C) temperatures was studied in test tubes (100 x 20 mm) containing 10 ml of the medium mentioned above. All other conditions remained the same. The culture continued till all plant material had senesced (105 days). Every 15 days the number of new microshoots and their length were recorded. At the end of the culture period the number of nodes per microshoot and internode length were also determined.

2.3 *Liquid medium culture*

The effect of liquid medium on olive microshoot formation using several systems designed for this purpose was examined. The devices compared were:

- LifeReactor©, the bioreactor system produced by Osmotek Ltd, Israel. Disposable plastic vessel of 1.5-litre working volume filled with 1-litre of medium was used (Ziv et al., 1998) (Figure 3a).
- Erlenmeyer flasks of 100 ml, filled with 30 ml of liquid medium, shaken on a rotary shaker under 100 rpm (Figure 4a).
- Test tubes equipped with filter paper bridges, containing 10 ml of liquid medium (Figure 4c).
- A *novel* TIS (Figures 5a and 5b). Immersion period was 15 min every 8 hours (Escalona et al., 1999; Damiano et al., 2000; Welander et al., 2001). Three periodical sequences of culture were examined: a) 10 days in TIS followed by 20 days in test tubes containing 10 ml of agar solidified medium, b) 20 days in TIS followed by 10 days in agar solidified medium and c) 30 days continuous culture in TIS.
- Cultures of explants in test tubes in agar solidified medium (0.5% w/v) were used as control.

Cultures were kept at 28±2°C, which proved to be an adequate temperature for *in vitro* olive microshoot growth. The experiments lasted for 30 days. At the end of the culture period the number of new microshoots per explant and the shoot length were measured, while the percentage of buds that developed into microshoots was determined.

2.4 Description of the new TIS

The system consisted of a 500 ml polycarbonate vessel in the top of which a pneumatic cylinder was positioned. The piston of the cylinder was connected to a plastic basket with holes through which the liquid medium passes when immersed (Figures 5a and 5b). Plant material was placed in the basket, which periodically was immersed in the medium. The pneumatic cylinder was connected with an air pump and its movement was controlled by compressed air that did not enter into the vessels, intended to reduce the possibility of contamination. The frequency and duration of immersions were controlled by a timer connected to an electro-pneumatic 5/2 valve. The compressed air passed through two 0.2 μm micropore filters providing additional protection. All components were autoclavable.

2.5 Experiments

Single node segments, bearing two opposite buds taken from the above-mentioned stock cultures, were used as explants for all treatments.The experiments followed a complete randomised design, with at least 40 replications. The Student's *t*-test was used to compare data from the temperature treatments. For liquid medium treatments, analysis of variance was performed with the General Linear Model procedure (SPSS 8.0 statistical package) and mean separation with Duncan's Multiple Range Test. Significance was recorded at $P \leq 0.05$.

3. Results

3.1 Effect of temperature

Cultures grown at 28°C formed 1.75 new microshoots per explant, while those at 22°C produced 1.40 shoots. Shoot length was 3.41 and 3.08 cm respectively. Neither parameters was significantly different (Table 1). However, the number of nodes was significantly higher at 28°C (11.22) than that at 22°C (8.95). The internodal length was reduced from 0.34 cm at 22°C to 0.30 cm at 28°C (Table 1). Microshoot length increased faster at 28°C than that at 22°C, but shoot tip necrosis was noticed in microshoots developed at 28°C after 60 days of culture, while those at 22°C continued to grow till the 105th day (Figures 1, 2a and 2b). Hyperhydricity did not appear either at 22°C or at 28°C.

Table 1: Olive cv. Chondrolia Chalkidikis microshoot development at 28° C and 22° C in agar solidified medium, after 105 days of culture

	No. of new micro-shoots per explant	Shoot length (cm)	No. of nodes	Internodal length (cm)
T 28°C	1.75±0.66	3.41±1.75	11.22±3.43*	0.30±0.10*
T 22°C	1.40±0.60	3.08±2.13	8.95±2.99*	0.34±0.26*

* Statistically significant difference by Student's *t*-test ($P \leq 0.05$).

Table 2: Effect of agar solidified and liquid medium on olive cv. Chondrolia Chalkidikis microshoot development, after 30 days of culture at 28° C

	No. of new shoots/explant	Percentage of buds developed into shoots	Average shoot length (cm)
Control (agar solidified)	1.75 ab	89	1.22 ab
LifeReactor©	0.35 c	18	0.40 cd
Filter bridges	0.43 c	22	0.32 d
Erlenmeyer flasks	1.93 a	97	0.40 cd
TIS 10 days + 20 days agar solidified medium	0.64 c	32	1.40 a
TIS 20 days + 10 days agar solidified medium	1.41b	71	0.70 c
TIS 30 days	1.93 a	97	0.95 bc

Different letters within columns indicate significant differences ($P \leq 0.05$).

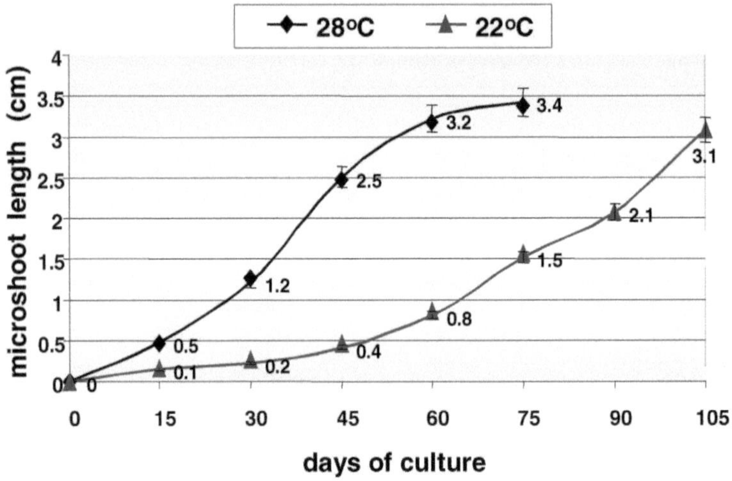

Figure 1: Increase of olive cv. Chondrolia Chalkidikis microshoot length, developed at 28ºC and 22ºC in agar solidified medium.

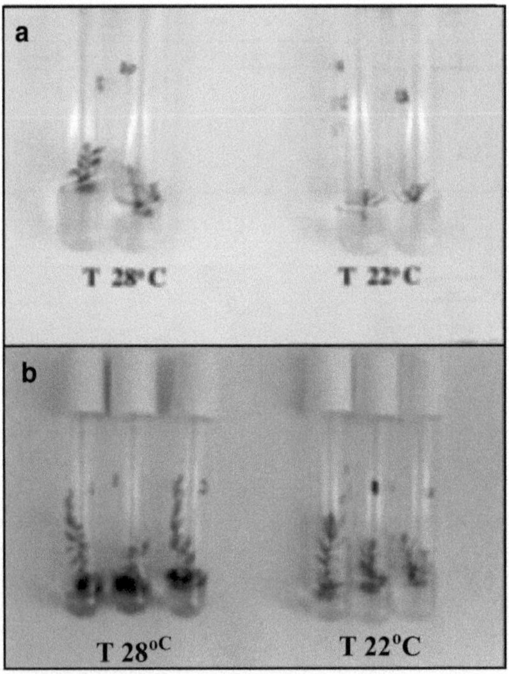

Figure 2: Differences between olive microshoots developed at 28ºC and 22ºC in agar solidified medium, a) 30th day of culture, b) 60th day of culture.

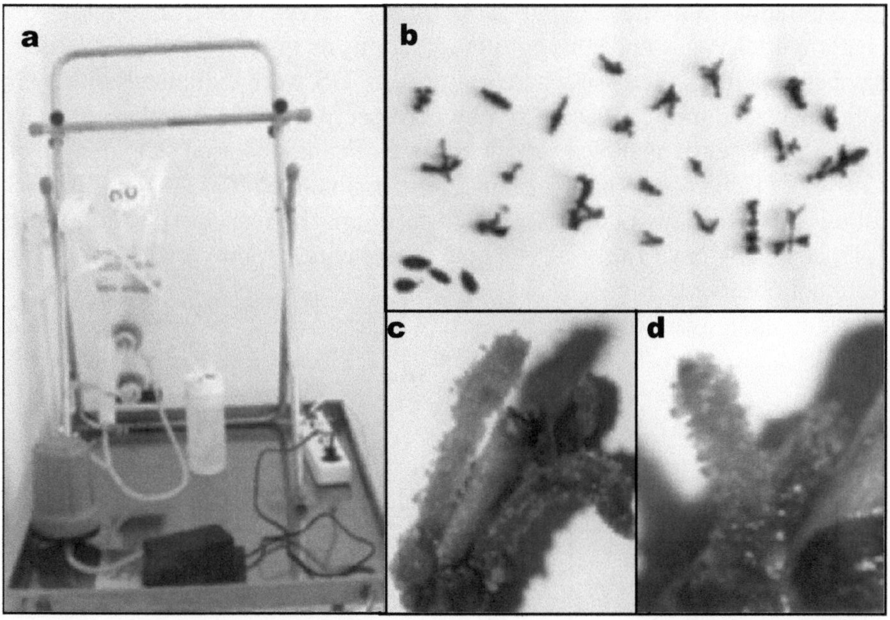

Figure 3: Culture of olive microshoots in LifeReactor©: a) the system, b) sample of plant material produced, c and d) callus formation on the surface of the shoots and leaves.

3.2 *Liquid medium culture*

Culture in LifeReactor© resulted in 0.35 new microshoots per explant, with an average length of 0.40 cm (Table 2). These small microshoots were intensively hyperhydrized (macroscopically determined) and small callus masses were present on the surface of the shoots and leaves (Figures 3b, 3c and 3d).

Cultures on filter paper bridges formed 0.43 new microshoots per explant with an average shoot length of 0.32 cm (Table 2). Extensive callus formation was noticed at the base of the explants (Figure 4d).

Liquid culture of olive explants in Erlenmeyer flasks under agitation resulted in the highest number of new microshoots per explant (1.93) with an average shoot length of 0.40 cm (Table 2, Figure 4b). These microshoots suffered from the most serious hyperhydricity compared to the other systems tested.

The novel TIS proved to be quite effective as far as the development of new microshoots in liquid media is concerned. The number of new microshoots per explant was 1.93 and shoot length was 0.95 cm in the case of continuous culture in the TIS for 30 days (Table 2, Figure 6). Hyperhydricity was apparent but not as serious as in the other liquid system devices tested. In the case where cultures in TIS were combined with agar solidified medium (10 and 20 days, respectively) the number of new microshoots per explant was 0.64 with a significant increased shoot length of 1.40 cm (Table 2). When explants were maintained for 20 days in TIS followed by 10 days in agar solidified medium, the respective results were 1.41 and 0.70 cm (Table 2). In both treatments hyperhydricity problems were not observed.

Cultures in agar-solidified medium, which were used as control, resulted in 1.75 new microshoots per explant of 1.22 cm average shoot length (Table 2).

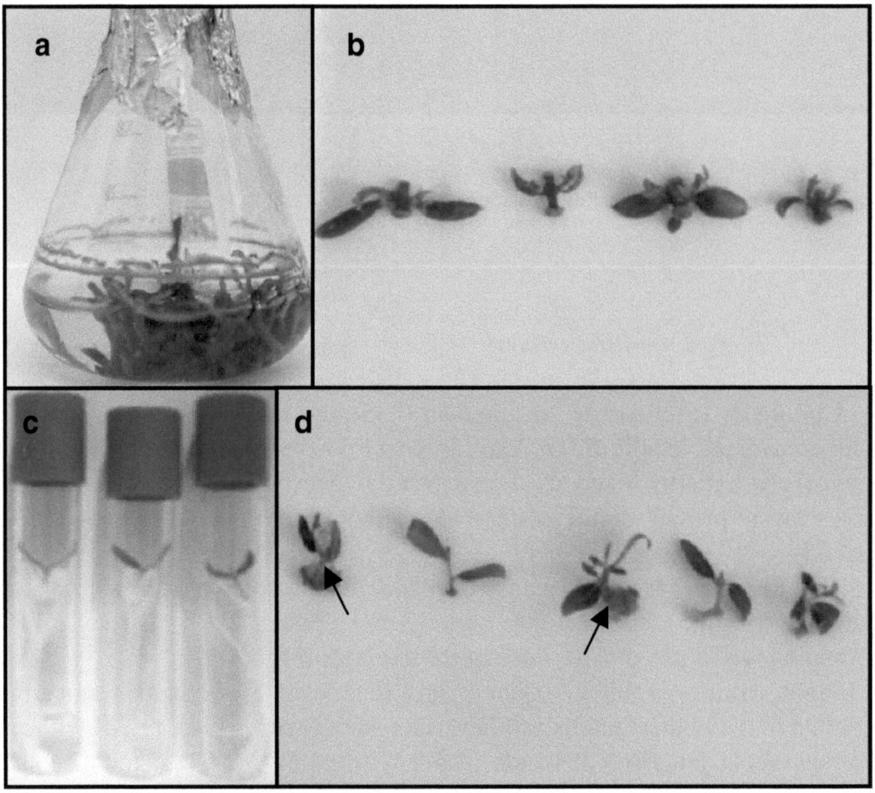

Figure 4: Culture of olive microshoots in Erlenmeyer flasks under agitation (a and b) and on filter paper bridges in test tubes (c and d). In d, arrows point to callus masses.

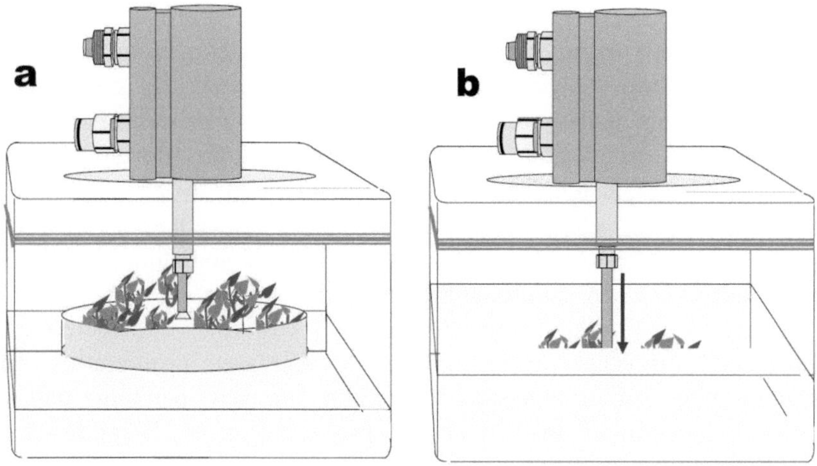

Figure 5: How the novel TIS works: a) position of piston when plant material is out of liquid medium, b) position of piston when plant material is immersed in the medium.

Figure 6: Olive microshoots developed after 30 days of culture in liquid medium using the novel TIS.

4. Discussion

Olive trees are mainly grown in the Mediterranean basin, where the average temperature is high during the vegetative period. The development of new shoots starts usually in May and continues all over summer when the temperature is occasionally extremely high. Faster development of microshoots at high temperature (28°C) compared to 22° C may be explained due to the adaptation of the olive plants to such an environment. Nevertheless, quicker degeneration of plant material sometimes could be a disadvantage for *in vitro* culture at high temperature.

The use of a liquid medium for *in vitro* culture is considered quite advantageous for mass production due to increased proliferation rates combined with low production cost based on easy handling and the simplicity of medium replacement (Teisson et al., 1996; Escalona et al., 1999). Different methods have been developed to increase the efficiency of liquid culture technique, such as the use of a raft to support plants over liquid, addition of liquid medium to cultures already established in agar and mist cultures. Among them, the TIS has been reported to be successfully used for many species, the majority of which are from tropical regions (Teisson et al., 1996; Escalona et al., 1999). Some of the TIS are now available on the market, such as the RITA system (Teisson et al., 1996), while others have been used only for research in laboratories (Damiano et al., 2000; Welander et al., 2001).

The novel TIS used in the present study is based on the same principle as that used in other similar systems but it was designed in such a way that the compressed air, which moves the piston of the pneumatic cylinder and controls immersion movement, is prevented from entering the vessel. Thus, theoretically the possibility of contamination was reduced. In practice, contamination problems were initially noticed and for that reason micropore filters were added. The mechanism that controls the movement was adjusted only in closed vessels but gas exchange could also be easily achieved *via* openings at the top of the vessel protected by sterile filters.

This system worked without problems during the period of the experiments, presenting the advantages of reliability, easy change of the medium and easy assembling and disassembling for autoclaving, without these procedures affecting its functioning.

In olive tissue culture, where low propagation rates and slow microshoot growth seem to be the main problem (Rugini et al., 1999), the effect of liquid medium using different culture methods was tested aiming at easier and faster absorption of nutrients by the cultivated tissues. Even though the percentage of buds developed into microshoots remained high, the main disadvantage of liquid media culture, namely the hyperhydricity, proved to be a problem also in the case of olive as in other species (Damiano et al., 2000; Peak et al., 2001; Welander et al., 2001). The use of the novel TIS produced better results when compared with the other liquid culture systems tested, but they were not significantly different than those obtained on agar solidified medium. The addition of growth retardants to the liquid media may have positive effects on reduction of hyperhydricity (Ziv et al., 1998; Peak et al., 2001) and deserves further experimentation.

Even though the novel TIS described has only been tested for olive culture, it seems promising, after some technical problems have been overcome and with the introduction of improvements, it may have application for other species with the potential for effective application in large scale micropropagation.

Acknowledgements

Special thanks are due to our colleague of VITRO HELLAS S.A., Mr. Nikos Leventakis for the critical reading and the designs of this manuscript.

References

Escalona M, Lorenzo JC, Gonzalez B, Daqinta M, Gonzalez JL, Dejardins Y & Borroto CG (1999) Pineapple (*Ananas comosus* L. Merr) micropropagation in temporary immersion systems. Plant Cell Reports 18: 743-748

Grigoriadou K (2003) A study on the *in vitro* culture of greek olive cultivars (*Olea europaea* L.) PhD Thesis, School of Biology, Aristotle University of Thessaloniki, Greece (in Greek with an English summary): 1-239

Grigoriadou K, Vasilakakis M & Eleftheriou EP (2002) *In vitro* propagation of the greek olive cultivar "Chondrolia Chalkidikis". Plant Cell, Tiss. Org. Cult. 71: 47-54

Damiano C, Caboni E, Frattarelli A, Giorgioni M, Liberali M, Lauri P & D'Angeli S (2000) Innovative micropropagation of temperate fruit trees: the case of pear. Acta Hortic. 530: 181-185

Dimassi K (1999), Micropropagation studies of the cv. Kalamon olives (*Olea europaea* L.). Acta Hortic. 474: 83-86

Lloyd G & McCrown B (1981) Commercially-feasible micropropagation of mountain laurel, *Kalmia latifolia*, by use of shoot-tip culture. Proc. Intern. Plant Propag. Soc. 30: 421-427

Miguens FC, Louro RP & Machado RD (1993) A scanning electron microscope study of normal and vitrified leaves from *Datura insignis* plantlets cultured *in vitro*. Plant Cell, Tiss. Org. Cult. 32: 109-113

Peak KY, Han EJ & Son SH (2001) Application of bioreactors for large-scale micropropagation systems of plants. *In Vitro* Cell. Dev. Biol.-Plant 37: 149-157

Rugini E, Gutierrez-Pesce P & Sampinato PL (1999) New perspective for biotechnologies in olive breeding: morphogenesis, *in vitro* selection and gene transformation. Acta Hortic. 474: 107-110

Teisson C, Alvard D, Berthouly B, Cote F, Escalant JV, Ethienne H & Lartaud M (1996) Simple apparatus to perform plant tissue culture by temporary immersion. Acta Hortic. 440: 521-526

Welander M, Li XY & Zhu LH (2001) Improved micropropagation of apple rootstocks by temporary immersion system. Cost 843, WG2: Quality enhacement of plant production through tissue culture. 2nd Meeting Thessaloniki, 22-25 September, Extended Abstracts: (pp. 7-8)

Ziv M & Ariel T (1992) On the relation between vitrification and stomatal cell wall deformity in carnation. Acta Hortic. 314: 121-129

Ziv M, Ronen G & Raviv M (1998) Proliferation of meristematic clusters in disposable presterilized plastic bioreactors for the large-scale micropropagation of plants. *In Vitro* Cell. Dev. Biol.- Plant 34: 152-158

Chapter 19

Shoot regeneration from nodules of *Charybdis* sp.: A comparison of semisolid, liquid and temporary immersion culture systems

Ch. Wawrosch*, A. Kongbangkerd, A. Köpf & B. Kopp
*Institute of Pharmacognosy, University of Vienna, Althanstr. 14, A-1090 Vienna, Austria. *requests for offprints; Tel: +43-1-4277-55286; Fax: +43-1-4277-9552; E-mail: christoph.wawrosch@univie.ac.at*

Abstract: Nodules of *Charybdis numidica* maintained in liquid Murashige and Skoog (MS) medium with 20 µmol BA in the dark were subjected to different treatments under continuous light for shoot regeneration. A high regeneration rate without hyperhydration of the shoots was observed on semisolid basal MS medium with 1 % sucrose. The use of liquid MS medium (1 % sucrose, no growth regulators) resulted in a significantly lower amount of shoots per gramme of nodules under both submerged and temporary immersion (TI) conditions. Shoot hyperhydration was lowest in a TI system with one 5 min immersion every 24 hours. When compared on a per container base, large amounts of shoots could be produced in the TI system with less labour input than in the system with semisolid medium.

Key words: bufadienolides, *Hyacynthaceae*, hyperhydration, meristematic nodules, micropropagation, squill

Abbreviations: AM – semisolid medium; BA – 6-benzyladenine; LM – liquid medium; MS – Murashige and Skoog medium (1962); TI – temporary immersion

1. Introduction

The genus *Charybdis* (formerly *Urginea*, family *Hyacynthaceae*; common name: squill) comprises a number of bulbous, perennial species, which contain a number of steroidal glycosides. Besides the centuries-old use against heart diseases some of these compounds have other interesting biological activities. Scilliroside, found in *Charybdis numidica*, has a pronounced rodenticidal activity (Verbiscar et al., 1986). Recently it has been demonstrated that the glycoside proscillaridin A (*Ch. maritima*)

A.K. Hvoslef-Eide and W. Preil (eds.), Liquid Culture Systems for in vitro Plant Propagation, 275–280.

possesses strong immunoregulatory activities (Terness et al., 2001). For various reasons large scale collection of wild-growing plants can not be recommended and field cultivation would therefore be the preferred way for the production of crude drug for further processing. However, the vegetative multiplication by bulb offsets is very low (Van Horn and Domingo, 1950). *In vitro* propagation of *Charybdis* sp. has been performed e.g. through bulblet induction (El Grari and Backhaus, 1987) or by adventitious shoot formation (Stojakowska, 1993).

Nodules, cell aggregates with a high regenerative potential, have been first described for poplar by McCown et al. (1988) and have since then been reported for a few other plants, too. Micropropagation through nodule culture is also an interesting alternative for *Charybdis* sp., although the use of liquid medium, as one of the advantages of nodule culture, is restricted due to hyperhydration (Wawrosch et al., 2001). In the present communication the results obtained on semisolid and in submerged liquid culture are compared to those achieved in temporary immersion systems.

2. Materials and methods

Shoot cultures were established from adult bulbs and the formation of meristematic nodules was induced on leaf cuttings (Wawrosch et al., 2001). Nodules of the clone UN26 (*Charybdis numidica*) were used for the experiments (Figure 1A). Nodules were multiplied in liquid MS basal medium (Murashige and Skoog, 1962) with 3 % sucrose and 20 μmol BA (50 ml of medium in 250 ml Erlenmeyer flasks) in the dark. Shoot regeneration was studied in the following three systems:

As semisolid medium (AM) we used MS basal medium with 0.8 % agar (Merck) and 1 % sucrose. 40 ml of medium were dispensed in baby food jars (9.5 cm height and 6 cm diameter) closed with Magenta B-caps®. The nodule inoculum weight was approximatily 1 g. Submerged liquid culture (LM) was carried out in 250 ml Erlenmeyer flasks containing 50 ml of MS basal medium with 1 % sucrose. The bottles were closed with cellulose plugs, plug and bottleneck were wrapped with aluminium foil. Flasks were inoculated with ca. 2 g of nodules and kept on a gyratory shaker at 80 rpm. The temporary immersion (TI) system used in our studies was based on the simple device described by Damiano et al. (2001). Basically, two 1000 ml Schott Duran® bottles were connected through glass and silicone tubings. One bottle contained 500 ml of the nutrient medium (MS with 1 % sucrose), the other one approximately 10 g of nodules on top of a layer of glass beads. The medium was moved periodically by applying compressed air, which was sterilized through a 0.2 μm filter. Two of these units were operated serially

(Figure 1). In the first setup (TI 1) the time of immersion was 5 min every 24 hours while in the second setup (TI 2) the frequency of immersion was doubled, i.e. 5 min every 12 hours. During the 5 min immersion a constant air flow was kept in order to remove gaseous metabolites and to supply fresh air (Damiano et al., 2001). The cultures were kept at 25±1 °C under continuous light of 60 μmol m^{-2} s^{-1} (cool white fluorescent tubes) for 4 (LM, TI 1, TI 2) and 8 weeks (AM), respectively. Subsequently, the number of regenerated shoots was counted on every single nodule, differing between shoots (length > 2 mm) and buds. Hyperhydration was rated with a score of 0 (no hyperhydration) to 3 (all shoots hyperhydrated). Statistical analysis (ANOVA followed by Duncan´s multiple range test) was performed using the software Statistica® v5.0 by StatSoft, Inc.

Figure 1: Shoot regeneration from nodules of *Charybdis numidica* clone UN26. (A) Typical nodules prior to inoculation in regeneration systems (bar=10 mm). (B) Shoot regeneration on MS medium after 8 weeks of culture under continuous light (bar=10 mm). (C) Hyperhydrated shoots regenerated in submerged culture in MS medium after 4 weeks under continuous light (bar=10 mm). (D) Two serially coupled temporary immersion systems after Damiano et al. (2001) (bar=5 cm). (E) Mostly normal shoots regenerated after 4 weeks with 5 min immersion with MS medium every 24 hours (bar=10 mm). (F) Hyperhydrated shoots regenerated after 4 weeks with 5 min immersion with MS medium every 12 hours (bar=10 mm).

3. Results and discussion

As previously reported the main factor, which triggers the formation of adventitious shoots on nodules of *Charybdis* sp. is light (Wawrosch et al., 2001). The highest regeneration frequency (208.7±23.5 per gram of initial nodule explants) was observed in the AM system with semisolid medium. However, about 7 % of the regenerants were small buds, which were not suitable for further processing. For this reason the number of regenerated shoots is presented separately (Table 1), which was 194.3±23.0 for the AM system. Shoots regenerated on semisolid medium were typically of healthy appearance and did not display any symptoms of hyperhydration (Figure 1B). In contrast, the results obtained in submerged liquid culture were significantly different. In the LM system 59.7±3.6 shoots were formed which is about 29 % of the regeneration frequency in the AM system. Not only were the values per container lower than those achieved with the AM system, but practically all shoots were severely hyperhydrated, too (Figure 1C).

It is well known that one of the advantages of temporary immersion culture systems lies in the higher multiplication rates when compared to semisolid medium. However, this could not be verified for our *Charybdis* nodule micropropagation system. In the TI setup (Figure 1D) an immersion of 5 min every 24 hours (TI 1) resulted in 25.4±3.2 shoots per g which was only about half of the regeneration frequency under LM conditions and was significantly lower than on semisolid medium. However, only a minor portion of the shoots showed symptoms of hyperhydration (Figure 1E). A comparison of the total number of shoots obtained from one container or culture unit (Table 1) reveals that the use of the TI 1 system resulted in significantly higher amounts of regenerants. Although the inoculum was approximately 5-fold that of the AM system it should be taken into consideration that the TI system is more labour saving. Indeed the nodules are just loaded into the bottle (using e.g. a sterile funnel) whereas for the AM systems containers have to be opened, nodules have to be placed firmly onto the medium, and the containers have to be closed again. In order to increase the regeneration rate the immersion frequency was doubled in the TI 2 system. Although a significant increase in the number of shoots per container was observed it was noted that just a doubling of the time of immersion resulted in completely hyperhydrated shoots (Figure 1F). It seems evident that shoots of *Charybdis numidica* are very sensitive to contact with liquid medium, on the other hand the more frequent immersion did not significantly increase the number of regenerated shoots per gram of nodules (Table 1).

Table 1: Regeneration rates (average±SE) from nodules of *Charybdis numidica* clone UN26 and hyperhydration on semisolid (AM), submerged in liquid medium (LM), and in temporary immersion (TI) systems

Culture method	Total regenerants per g inoculum*	Shoots per g inoculum	Shoots per container**	Hyperhydration (score 0-3)
AM	208.7±23.5 [a]	194.2±23.0 [a]	194.2±23.0 [ab]	0
LM	59.7±3.6 [b]	55.0±3.8 [b]	130.4±7.5 [a]	3
TI1 (5 min every 24h)	38.3±1.6 [b]	25.4±3.2 [b]	270.5±34.5 [b]	1
TI2 (5 min every 12h)	48.0±2.4 [b]	36.6±1.2 [b]	437.0±1.0 [c]	3

Mean separation within columns by Duncan´s multiple range test at p=0.05
* shoots + buds
** AM: baby food jar/c. 1 g inoculum; LM: 250ml Erlenmeyer flask/c. 2 g inoculum; TI1 and TI2: 1-litre flask/c. 10 g inoculum/500 ml medium

Micropropagation of *Charybdis* sp. through nodule culture offers a convenient alternative to other *in vitro* methods. Growth of the nodules (without regeneration) can be achieved in liquid MS medium supplemented with 3 % sucrose and 20 μmol BA in the dark, with an up to 8-fold increase in biomass depending on the clone (Wawrosch et al., 2001). Two different procedures can be proposed for the regeneration of shoots. Recently a multiplication protocol described for a *Phalaenopsis* hybrid suggested the multiplication of protocorm-like bodies in liquid culture systems followed by conversion into plantlets on semisolid medium (Park et al., 2000). Similarly, nodules of *Charybdis* sp. can be multiplied in liquid medium and subsequently transferred to semisolid medium for shoot regeneration. In order to achieve higher shoot numbers per unit, larger containers and/or higher inoculum density could be tested. As an alternative, using a temporary immersion system with low immersion frequency and duration, large amounts of shoots per container can be produced with less labour input. Because the shoots can be rooted *ex vitro* during acclimatization (unpublished results) the procedure is labour and time saving. Studies are currently in progress on the suitability of a TI system for nodule growth. Using TI culture for both nodule growth and shoot regeneration would allow for a production system with least expenditure of labour: at the end of the growth period only nutrient medium and light conditions would have to be changed for shoot regeneration.

References

Damiano C, Gentile A, La Starza SR, Fratarelli A & Monticelli S (2001) Automation in plant propagation through temporary immersion systems. 1st International Symposium 'Acclimatization and Establishment of Micropropagated Plants', Sani-Halkidiki, Greece, Abstract Book (p. 90)

El Grari R & Backhaus RA (1987) *In vitro* propagation of red squill, *Urginea maritima* Baker. Plant Cell, Tiss. Org. Cult. 10: 65-71

McCown BH, Zeldin EL, Pinkalla HA & Dedolph R (1988) Nodule culture: a developmental pathway with high potential for regeneration, automated micropropagation, and plant metabolite production from woody plants. In: Hanover JW & Keathley DE (eds) Genetic Manipulation of Woody Plants (pp. 149-166). Plenum Press, New York

Murashige T & Skoog F (1962) A revised medium for rapid growth and bioassays with tobacco tissue cultures. Physiol. Plant. 15: 473-497

Park SY, Murthy HN & Paek KY (2000) Mass multiplication of protocorm-like bodies using bioreactor system and subsequent plant regeneration in *Phalaenopsis*. Plant Cell, Tiss. Org. Cult. 63: 67-72.

Stojakowska A (1993) Micropropagation of *Urginea maritima* (L.) Baker s. str. Acta Soc. Bot. Pol. 62: 11-15

Terness P, Navolan D, Dufter Ch, Kopp B & Opelz G (2001) The T-cell suppressive effect of bufadienolides: structural requirements for their immunoregulatory activity. Int. Immunopharmacol. 1: 119-134

Van Horn DL & Domingo WE (1950) Comparison of seed and vegetative propagation methods for red squill. Econ. Bot. 4: 350-353

Verbiscar AJ, Atel J, Banigan TF & Opelz G (1986) Scilliroside and other Scilla compounds in red squill. J. Agric. Food Chem. 34: 973-979

Wawrosch Ch, Kongbangkerd A & Kopp B (2001) Micropropagation of *Charybdis* sp. (Hyacynthaceae) through nodule culture. COST843 Working Group 2 2nd Meeting, Thessaloniki, Greece, Abstract Book (pp. 41-42)

III. Somatic Embryogenesis and Shoot Initiation

Chapter 20

Propagation of Norway spruce *via* somatic embryogenesis

Sara von Arnold[1], Peter Bozhkov, David Clapham, Julia Dyachok, Lada Filonova, Karl-Anders Högberg[2], Mathieu Ingouff & Malgorzata Wiweger
[1] *Department of Forest Genetics, SLU, PO Box 7027, S-750 07 Uppsala, Sweden. Phone: +46 18 673230, Fax: +46 18 6732, E-mail: Sara.von.Arnold@sgen.slu.se*
[2] *The Forestry Research Institute of Sweden, Ekebo, S-268 90 Svalöv, Sweden.*

Abstract: Somatic embryogenesis combined with cryopreservation is an attractive method to propagate Norway spruce (*Picea abies*) vegetatively both as a tool in the breeding programme and for large-scale clonal propagation of elite material. Somatic embryos are also a valuable tool for studying regulation of embryo development. Embryogenic cell lines of Norway spruce are established from zygotic embryos. The cell lines proliferate as proembryogenic masses (PEMs). Somatic embryos develop from PEMs. PEM-to-somatic embryo transition is a key developmental switch that determines the yield and quality of mature somatic embryos. Withdrawal of plant growth regulators stimulates PEM-to-somatic embryo transition accompanied by programmed cell death (PCD) in PEMs. This PCD is mediated by a marked decrease in extracellular pH. If the acidification is abolished by buffering the culture medium, PEM-to-somatic embryo transition together with PCD is inhibited. Cell death, induced by withdrawal of PGRs, can be suppressed by extra supply of lipo-chitooligosaccharides (LCOs). Extracellular chitinases are probably involved in production and degradation of LCOs. During early embryogeny, the embryos form an embryonal mass surrounded by a surface layer. The formation of a surface layer is accompanied by a switch in the expression pattern of an *Ltp*-like gene (*Pa18*) and a homeobox gene (*PaHB1*), from ubiquitous expression in PEMs to surface layer-specific in somatic embryos. Ectopic expression of *Pa18* and *PaHB1* leads to an early developmental block. Transgenic embryos and plants of Norway spruce are routinely produced by using a biolistic approach. The transgenic material is used for studying the importance of specific genes for regulating plant development, but transgenic plants can also be used for identification of candidate genes for use in the breeding programme.

Key words: conditioning factors, development of somatic embryos, embryogenic cell suspension, gene transformation, genetic regulation, Norway spruce, programmed cell death

Abbreviations: ABA – abscisic acid; AGP – arabinogalactan protein; LCO – lipo-chitooligosaccharide; PCD – programmed cell death; PEM – proembryogenic mass; PGR – plant growth regulator

A.K. Hvoslef-Eide and W. Preil (eds.), Liquid Culture Systems for in vitro Plant Propagation, 283–293.

1. Introduction

Forest trees play a vital role in the lives of humans and functioning ecosystems. They provide renewable sources of wood, fibres and chemicals for human societies. They provide habitats for numerous organisms and essential ecological functions such as water purification and carbon storage. Forests are managed in a diversity of ways ranging from intensively managed short rotation tree-farms to old-growths reserves. Whatever the goals for forest management and conservation, the methods of clonal propagation and especially somatic embryogenesis provide powerful options for breeding and management. The pressure to increase productivity of the forest will increase dramatically in the future. At the same time, pressure will be brought to bear to increase forest conservation and sustainability. The possibility to meet the increased need for wood products by intensive forestry in highly productive plantations (fibre farming) and thereby making it possible to release large areas as natural forests, are an attractive alternative. Before this alternative can be accepted, it is important to develop methods and to establish field trials showing that this alternative is safe and sustainable.

The possibility to propagate trees vegetatively creates significant advantages both for the deployment of selected genotypes through mass-propagation and for capturing and enhancing the genetic gain in the breeding programme. To ensure that maximum genetic gain is achieved, the influence of environmental factors on field performance of the genotypes has to be determined. Today the common way to propagate plants vegetatively is *via* cuttings. However, large scale cutting propagation may be limited in some species owing to problems with rooting, aging of mother trees and survival of cuttings, as well as high costs. Some of the problems can be overcome by using tissue culture techniques and especially somatic embryogenesis combined with cryopreservation. An embryogenic cell line established from one seed can generate a high number of somatic embryos, therefore, it is possible to produce a large number of genetically identical plants within a short period.

It has been shown that somatic embryogenesis combined with cryopreservation is an attractive method to propagate Norway spruce vegetatively both as a tool in the breeding programme and for large-scale clonal propagation of elite material (Högberg et al., 2001). Somatic embryos of Norway spruce are also valuable as a tool for studying regulation of embryo development. In addition, the somatic embryos can be used for producing transgenic plants of Norway spruce (Clapham et al., 2000; Brukhin et al., 2000).

2. Somatic embryos as a tool for breeding

In conventional breeding programmes, with several forest tree species, by the time the superior genotypes have been identified in field trials, they are too old to be propagated vegetatively. Consequently, the identified genotypes are lost and they can only be used as parents for the next generation. If genotypes that go into field tests in the breeding programme are cryopreserved, a tested superior genotype can immediately be re-cultured and mass propagated. The close connection to the breeding programme ensures that the best genotypes are available for mass propagation.

3. How somatic embryo development proceeds and how it is regulated

In technological terms, plant regeneration through somatic embryogenesis in Norway spruce is comprised of a sequence of steps including initiation, proliferation, early embryo formation, embryo maturation, desiccation, germination and plant development (Figure 1). To execute this pathway efficiently, a number of critical physical and chemical treatments should be applied with proper timing.

Recently we described the developmental pathway of somatic embryo formation and development in Norway spruce by employing time-lapse tracking technique that involved continuous observation of individual preselected single cells and few-celled aggregates isolated from embryogenic cell suspensions and embedded in thin agarose layers under assigned trophic and hormonal conditions (Filonova et al., 2000a). The pathway involves two broad phases, which in turn are divided into more specific developmental stages. The first phase is represented by proembryogenic masses (PEMs) – proliferating cell aggregates which can pass through a series of three characteristic stages distinguished by cellular organisation and cell number (stages I, II and III) but can never develop directly into an embryo. The second phase encompasses development of somatic embryos, which arise *de novo* from PEM III, and then proceed through the same stereotyped sequence of stages as described for zygotic embryogeny of *Pinaceae* (Singh, 1978). Plant growth regulators (PGRs), auxins and cytokinins, are necessary to maintain PEM proliferation, whereas embryo formation from PEM III is triggered by the withdrawal of PGRs. Once early somatic embryos have formed, their further development to mature forms requires addition of abscisic acid (ABA).

We have previously noticed an important difference between developmental pathways in liquid medium and under immobilisation (Filonova et al., 2000a). Immobilisation of PEMs in the presence of auxin and cytokinin permits both multiplication by new PEM I formation and successive growth to more advanced levels (PEMI to PEMII and PEMII to PEMIII). In contrast, the predominant response of PEMs plated at low density in well aerated liquid medium with high levels of auxin and cytokinin is unequal division of embryogenic cells with dense cytoplasm leading to the restart of the process from the PEMI level. When auxin and cytokinin are being gradually depleted, the average level of the whole system is biased forward to PEMII or PEMIII levels. This occurs towards the end of the liquid culture passage.

Comprising a link between the unorganised proliferation phase (i.e. PEMs) and highly organised embryonic development phase, and at the same time holding them apart, PEM-to-embryo transition plays a pivotal role in Norway spruce somatic embryogenesis (Filonova et al., 2000a) just as it does during somatic embryogenesis of angiosperms. It seems likely that the inability of many embryogenic cell lines to form well-developed cotyledonary embryos is in large part associated with disturbed or arrested PEM-to-embryo transition that might be a consequence of inappropriate culture conditions.

Developmental dynamics experiments performed with whole suspension cultures have strengthened the significance of PEM-to-embryo transition (Bozhkov et al., 2002). Three major points to consider for efficient regulation of somatic embryogenesis were identified. First, PEM-to-embryo transition occurs within a short time after withdrawal of PGRs. The time for this process can probably vary for different cell lines but usually the switch from proliferation to development occurs after about 24 hours. Second, ABA is incapable of inducing PEM-to-embryo transition. This may be responsible for the failure to stop proliferation once embryogenic cultures are directly transferred from medium containing auxin and/or cytokinin to ABA-containing medium. Third, since newly formed somatic embryos could develop in PGR-free medium for at least 7 days and thereafter retain the ability to respond to ABA with continuing development until the cotyledonary stage, it is evident that early somatic embryogeny does not require exogenous ABA which, however, is important for promoting late embryogeny. This provides the possibilities to avoid prolonged contact with ABA during somatic embryo maturation, which otherwise inhibits the growth of somatic embryo plants (Bozhkov and von Arnold, 1998; Högberg et al., 2001). Taken together, these three points have significantly improved the biotechnology of somatic embryogenesis in terms of yield and quality of somatic embryos.

Ex vitro growth of somatic embryo plants is under a cumulative influence of a number of previously applied treatments. The time of contact with ABA during somatic embryo maturation and the duration of continuous growth during the first growth period strongly affect the height growth during two successive growth periods. In both cases longer treatments exerted negative effects (Högberg et al., 2001). The maturation step can be shortened (not exceeding 5 weeks) and synchronised by giving the cultures a pre-maturation treatment in growth regulator-free medium. The period of continuous growth during the first growth period can be shortened by a two-phase germination treatment, first on solidified medium and then in liquid medium. Another advantage of the two-phase germination treatment is a better developed root system possessing lateral roots. Somatic embryo plants produced according to this method can be transferred directly to the greenhouse (Högberg et al., 2001).

4. Programmed cell death during somatic embryogenesis

Two successive waves of programmed cell death (PCD) occur during formation and development of somatic embryos of Norway spruce (Filonova et al., 2000b). The first wave of PCD is responsible for the degradation of PEMs when they give rise to somatic embryos. The second wave of PCD eliminates terminally differentiated embryo-suspensor cells at the end of early embryogeny. During dismantling phase of PCD, PEMs and embryo-suspensor cells exhibit progressive autophagy, resulting in the formation of a large central vacuole. Autolytic degradation of the cytoplasm is accompanied by lobing and budding-like segmentation of the nucleus. Nuclear DNA undergoes fragmentation into both large fragments of about 50 kb and multiples of approximately 180 bp. The tonoplast rupture is delayed until lysis of the cytoplasm and organelles, including the nucleus, is almost complete. The protoplasm then disappears, leaving a cellular corpse represented by only the cell wall.

Programmed cell death (PCD) is an important component of PEM-to-embryo transition in Norway spruce (Filonova et al., 2000b). Triggered by withdrawal of PGRs, somatic embryo formation is accompanied by massive cell death in PEMs. Furthermore, strong positive correlation has been shown between the frequency of somatic embryo formation and the percentage of PEM cells fragmenting DNA suggesting that PCD in PEMs and somatic embryo formation are closely interlinked processes both stimulated upon withdrawal or partial depletion of PGRs (Filonova et al., 2000b). This type of PCD is also accompanied by a marked decrease in extracellular pH (Bozhkov et al., 2002). If extracellular acidification is artificially abolished by buffering PGR-free medium, PEM-to-embryo transition together with concomitant PCD is inhibited. Our results point to a rigid pH-control in developmental PCD associated with plant embryogenesis.

Figure 1: Propagation of Norway spruce through somatic embryogenesis. The system includes (from the top, clockwise) selection of donor trees, harvesting female cones, isolating and culturing zygotic embryos, establishment of proliferating cell suspension culture followed by somatic embryo formation, maturation, desiccation (not shown) and germination, and finally growing somatic embryo plants as seedlings in the nursery and then in the field (from Bozhkov et al., 2002).

5. Conditioning factors regulating somatic embryogenesis

It has long been known that conditioned medium from embryogenic cultures can promote embryogenesis. The ability of conditioned medium to sustain or stimulate somatic embryogenesis implies that secreted soluble signal molecules play an important role. Several components in conditioned medium have been found to promote somatic embryogenesis. In Norway spruce we have shown that extracellular chitinases (Egertsdotter et al., 1993; Dyachok et al., 2000), arabinogalactan proteins (AGPs) (Egertsdotter and von Arnold 1995) and lipo-chitooligosaccharides (LCOs) (Dyachok et al., 2000, 2002) affect somatic embryogenesis.

Chitinases from sugar beet and *Streptomyces griseus* stimulate early development of somatic embryos in Norway spruce (Egertsdotter and von Arnold 1998; Dyachok et al., 2002). However, we have also shown that chitinases from *S. griseus* degrade LCOs (Dyachok et al., 2002). Taken together, our results suggest that chitinases can regulate embryogenesis in different ways, both by degrading LCOs and by formation of LCOs. It has previously been shown that enzymes that form and degrade oligosaccharides are largely responsible for when and where oligosaccharides are active in the plant tissue (Albertsheim et al., 1994). Chitinases might therefore be a part of such a regulatory mechanism involving production and degradation of LCOs.

AGPs are a heterogeneous group of structurally complex macromolecules composed of a polypeptide and a large branched glycan chain (Majewska-Sawka and Notnagel 2000). Some AGPs also have a lipid chain. AGPs isolated from seeds of Norway spruce promote formation of more developed somatic embryos in Norway spruce (Egertsdotter and von Arnold, 1995).

LCOs are a class of signalling molecules that promote division of plant cells. Nod factors, LCOs, produced by different *Rhizobium* species uniformly consist of an oligosaccharide backbone of 1,4-linked N-acetyl-D-glucosamine residues varying in length between 3 and 5 sugar units, and always carry an N-acyl chain at the non-reducing terminus. This basic structure is essential for the infection leading to formation of nitrogen-fixing nodules. At the same time, several lines of evidence suggest the involvement of LCOs in regulating somatic embryo development. Extracts of media conditioned by embryogenic cultures stimulate development of PEMs in auxin-deficient media in Norway spruce. Partial characterisation of the conditioning factor has shown that it is a lipophilic, low molecular weight molecule, which is sensitive to chitinase and contains GlcNAc residues (Dyachok et al., 2002). Our conclusion is that the conditioning factor is a LCO. The amount of LCO correlates to the developmental stages of PEMs and somatic embryos, with the highest level in media conditioned by developmentally blocked cultures. LCOs are not present in non-embryogenic cultures. Cell death, induced by withdrawal of auxin, is suppressed by extra supply of LCO. Taken together, our data suggest that endogenous LCOs act as signal molecules in embryogenic cultures of Norway spruce.

6. Genetic regulation of somatic embryogenesis

Singh (1978) divided the gymnosperm development process into three phases: proembryogeny (stages before elongation of the suspensor), early embryogeny (stages after elongation of the suspensor and before the establishment of the root meristem) and late embryogeny (establishment of the root and shoot meristems and further development of the embryo until maturity). During early embryogeny, the embryo forms a distinct embryonal mass (analogous to the embryo proper in angiosperms). Later, the embryonal mass is surrounded by a surface layer. Late embryogeny in gymnosperms corresponds to the "post-globular" embryo development in angiosperms. Early during this period, the root and the shoot meristems are delineated and the plant axis is established. A root organising centre is first formed which gives rise to the root meristem. The cotyledon primordia arise in a ring around the distal end of the embryo. Following the differentiation of the inner primary tissues, the embryo shoot apex is formed at the top of the embryo (Romberger et al., 1993).

In order to determine if tissue specification occurs in the embryonal mass of somatic embryos of Norway spruce comparable to the differentiation of protoderm in angiosperms we isolated a *Ltp*-like gene (*Pal8*) which is expressed in somatic embryos of Norway spruce (Sabala et al., 2000). During development of somatic embryos there is a switch from ubiquitous to restricted localisation of mRNA to the surface layer. We also showed that a correct expression pattern of *Pal8* is required for normal embryo development and for plant survival (Hjortswang et al., 2002). Our data demonstrate that differentiation of a surface layer occurs early during embryo development.

Thereafter, we addressed the question of whether a protoderm layer, typical of angiosperm embryos, is defined during somatic embryogenesis in Norway spruce. We isolated the *PaHB1* (for *Picea abies* Homeobox 1), which is expressed in somatic embryos of Norway spruce (Ingouff et al., 2001). *PaHB1* exon/intron organisation and its corresponding protein are highly similar to those of HD-GL2 angiosperm counterparts. A phylogenetic analysis indicated that the *PaHB1* is strongly associated with one subclass consisting of protoderm/epiderm-specific angiosperm genes. *PaHB1* expression switches from a ubiquitous expression in PEMs to an outer cell layer-specific expression later during embryo development. Ectopic expression of *PaHB1* in somatic embryos leads to an early developmental block. The transformed embryos lack a smooth surface. These findings show that the *PaHB1* expression pattern is highly analogous to angiosperm HD-GL2, suggesting similarities in the definition of the outer cell layer in seed plants.

7. Gene transformation of somatic embryos

We have developed a biolistic method to produce transgenic plants of Norway spruce (Clapham et al., 2000). Somatic embryos are bombarded with gold particles coated with a reporter gene (*gusA*) and a selectable marker gene (*bar*), responsible for Basta resistance. Embryogenic cell lines resistant to Basta appear about two months after bombardment. In a standard procedure, 40 filter papers with embryogenic cells are bombarded per experiment and from these, between five and fifty independent putative stable transformants, i.e. Basta-resistant sublines, are obtained. At least 65% of these putative transformants express the reporter gene. More than 300 independent stably transformed sublines have been produced. Of eleven transformants analysed, four contained transgenes in low copy number (1-3), the others contained transgenes with up to 15–20 copies.

The *bar* gene giving resistance to the herbicide Basta was further used for screening for stable expression of transgenes (Brukhin et al., 2000). A simple biotest for screening for Basta tolerance based on the colour change of detached needles induced by Basta was developed. In total, eighty three 9-month-old transgenic plants from six transformed sublines, were analysed for continued tolerance to Basta. The tolerance for Basta varied among the plants from the different sublines. Needles from four of the sublines were resistant to 100 mg l^{-1} phosphinothricin, a concentration inducing yellowing in control needles, while plants from the two other sublines were, on average, two to four times as resistant as untransformed control plants. The same plants were analysed for Basta tolerance after two years. The biotest enables rapid semi-quantitative monitoring for continued transgenic expression in long-lived tree species.

We have now reached the stage when we routinely can produce transgenic plants of Norway spruce. The method is used for studying the importance of specific genes for regulating embryo development, but we have also shown that transgenic plants of Norway spruce can be used for identification of candidate genes for use in molecular breeding (Elfstrand et al., 2001 a, b and 2002).

References

Albertsheim P, An JH, Freshour G, Fuller MS, Guillen R, Ham KS, Hahn MG, Huang J, O'Neill M & Whitcombe A (1994) Structure and function studies of plant cell wall polysaccharides. Biochem. Soc. Trans. 22: 374–378

Bozhkov PV & von Arnold S (1998) Polyethylene glycol promotes maturation but inhibits further development of *Picea abies* somatic embryos. Physiol. Plant. 104: 211–224

Bozhkov PV, Filonova LH & von Arnold S (2002) A key developmental switch during Norway spruce somatic embryogenesis is induced by withdrawal of growth regulators and is associated with cell death and extracellular acidification. Biotechnology and Bioengineering Vol 77 (6): 658–667

Brukhin VB, Clapham D, Elfstrand M & von Arnold S (2000) Basta tolerance as a selectable and screening marker for transgenic plants of Norway spruce. Plant Cell Reports 19: 899-903

Clapham D, Demel P, Elfstrand M, Koop H-U, Sabala I & von Arnold S (2000) Effective biolistic gene transfer to embryogenic cultures of *Picea abies* and the production of transgenic plantlets. Scand. J. For. Res. 15: 151–160

Dyachok J, Tobin A, Price N & von Arnold S (2000) Rhizobial Nod factors stimulate somatic embryo development in *Picea abies.* Plant Cell Reports 19: 290–297

Dyachok J, Wiweger M, Kenne L & von Arnold S (2002) Endogenous Nod-factor-like signal molecules suppress cell death and promote early somatic embryo development in Norway spruce. Plant Physiol. 128: 1–11

Egertsdotter U, Mo H & von Arnold S (1993) Extracellular proteins secreted to the culture medium of embryogenic suspension cultures of Norway spruce (*Picea abies*). Physiol. Plant. 88: 315–321

Egertsdotter U & von Arnold S (1995) Importance of arabinogalactan proteins for the development of somatic embryos of *Picea abies*. Physiol. Plant. 93: 334–345

Egertsdotter U & von Arnold S (1998) Development of somatic embryos in spruce. Exp. J. Bot. 49: 155–162

Elfstrand M, Fossdal, C-G, Swedjemark G, Clapham D, Olsson O, Sitbon F, Sharma P, Lönneborg A & von Arnold S (2001a) Identification of candidate genes for use in molecular breeding – A case study with the Norway spruce defensin-like gene, *spi1*. Silvae Genetica 50(2): 75–81

Elfstrand M, Fossdal C-G, Sitbon F, Olsson O, Lönneborg A & von Arnold S (2001b) Overexpression of the endogenous peroxidase-like gene *spi2* in transgenic Norway spruce plants results in increased total peroxidase activity and reduced growth. Plant Cell Rep. 20: 596–603

Elfstrand M, Sitbon F, Lapierre C, Bottin A & von Arnold S (2002) Altered lignin structure and resistance to pathogens in *spi 2*-expressing tobacco plants. Planta 214: 708-716

Filonova LH, Bozhkov PV & von Arnold S (2000a) Developmental pathway of somatic embryogenesis in *Picea abies* as revealed by time-laps tracking. J. Exp. Bot. 51 /343: 249-264

Filonova LH, Bozhkov PV, Brukhin VB, Daniel G, Zhivotovsky B & von Arnold S (2000b) Developmental programmed cell death in plant embryogenesis: exploring a model system of Norway spruce somatic embryogenesis. Journal of Cell Science 113(24): 4399–4411

Hjortswang H, Filonova LH, Vahala T & von Arnold S (2002) Modified expression of the *Pa18* gene interferes with somatic embryo development. Plant Growth Reg. 38: 75-82

Högberg K-A, Ekberg I, Norell L & von Arnold S (1998) Integration of somatic embryogenesis in a tree-breeding programme – a case study with *Picea abies*. Can. J. For. Res. 28 (10): 1536–1545

Högberg K-A, Bozhkov PV, Grönroos R & von Arnold S (2001) Dissection of critical factors affecting *ex vitro* performance of somatic embryo plants of Norway spruce. Scand. J. For. Res. 16 (4): 295–304

Ingouff M, Farbos I, Lagercrantz U & von Arnold S (2001) PaHB1 is an evolutionary conserved HD-GL2 homeobox gene defining the protoderm during Norway spruce embryo development. Genesis 30: 220–230

Majewska-Sawka A & Nothnagel EA (2000) The multiple roles of arabinogalactan proteins in plant development. Plant Physiol. 122: 3–9

Romberger JA, Hejnowicz Z & Hill JF (1993) The embryo. In: Plant structure, function and development. Springer-Verlag, Berlin

Sabala I, Clapham D, Elfstrand M & von Arnold S (2000) Tissue-specific expression of *Pa18*, a putative lipid transfer protein gene, during embryo development in Norway spruce (*Picea abies*). Plant Mol. Biol. 42: 461–478

Singh H (1978) Embryology of Gymnosperms. In: Zimmerman W, Carlquist Z, Ozenda P & Wulff HD (eds) Handbuch der Pflanzenanatomie (pp. 187–241). Berlin, Stuttgart

Chapter 21

Norway spruce somatic embryogenesis: membrane rafts as a compromise between liquid and solidified media

M. Vágner, Z. Vondráková, L. Fischerová & J. Opatrná
Institute of Experimental Botany AS CR, Rozvojová 135, CZ-165 02 Prague 6, Czech Republic. E-mail: vagner@ueb.cas.cz

Abstract: Embryogenic cultures of Norway spruce (*Picea abies*) were cultivated either on solidified media, in liquid media, or on polypropylene membrane rafts (LifeRaft). The cultivation on rafts was found to be the most successful way: the number of developed somatic embryos increased, synchronization of the development was enhanced, and the time necessary for embryo development and maturation was shortened. It was shown that the process could be further improved by insertion of a pre-maturation phase on PGR-free medium between proliferation and maturation steps. Germination frequency remained unchanged. PEG 4000 added to the maturation medium increased the number of developed somatic embryos. PEG in lower concentration (1.87 % (w/v)) still had the significant beneficial effects on embryo numbers and development, but a decrease of germination frequency or increased aberrations of developing root and shoot were not found. On the other hand, PEG in the concentration 5 and 7.5 % (w/v) had a negative effect on the germination of somatic embryos and, in sensitive cell line, 7.5 % (w/v) PEG decreased somatic embryo yield.

Key words: conifer, membrane raft, Norway spruce, PEG, *Picea abies*, somatic embryogenesis

Abbreviations: ABA-abscisic acid; 2,4-D - 2,4-dichlorophenoxyacetic acid; BAP - 6-benzylaminopurine, N6-benzyl-adenine; IAA - indole-3-acetic acid; PEG - polyethylene glycol; PGR - plant growth regulator

1. Introduction

The majority of cultivation protocols for somatic embryogenesis of different coniferous species use solidified media, at least for part of the process. The quality of somatic embryos grown on solidified medium is still higher than that of their counterparts cultivated in liquid medium; germination frequencies thus differ considerably (Tautorus and Dunstan,

A.K. Hvoslef-Eide and W. Preil (eds.), Liquid Culture Systems for in vitro Plant Propagation, 295–302.

1995). In liquid proliferation medium, cultures usually grow even more rapidly than on agar-solidified medium. Maturation of somatic embryos is considered as the more problematic step. The ability to produce mature somatic embryos in liquid medium differs widely across species and even across cell lines. On the other hand, the use of somatic embryogenesis as a large-scale micropropagation tool in commercial applications depends mainly on reduction of labour costs through the automation of the process. Automated techniques clearly demand protocols based on liquid media (Paques et al., 1992).

There are a few examples of successful use of liquid cultures for somatic embryo production of Norway spruce (Paques et al., 1992, Gorbatenko and Hakman, 2001). The use of liquid medium is usually limited to the proliferation phase, maturation cannot be easily achieved in the liquid medium (Paques et al., 1995). Many trials thus have been conducted to combine proliferation in liquid medium with maturation on a solidified one. Although the production of mature embryos and emblings was successful in a few cases, these protocols are rather complicated to be automated.

Successful use of the polypropylene membrane rafts (LifeRafts, Osmotec, Israel) was published for a number of tissue cultures (Luckett et al., 1991, Watad et al., 1995, Paek et al., 2001). Cultivation on membrane rafts floating on liquid medium could have a great potential to improve the development of coniferous somatic embryos. Beside the numerous problems with the cultivation of Norway spruce embryogenic cultures in liquid media, long vacuolated suspensor cells of embryogenic cultures are sensitive to the damage made by forceps during transfer to fresh medium. Therefore, as entire rafts with cultures are transferred to fresh liquid medium, embryogenic cultures are not touched from the proliferation stage untill the stage of mature embryos. Moreover, membrane rafts could combine the advantages both of solidified and liquid media: the cultures are sufficiently aerated, the exchange of compounds on the tissue-medium interface is enhanced, and medium can be rapidly replenished or replaced with minimal disturbance of embryogenic tissue.

Firstly, this work focused on the possibility of the use of polypropylene membrane rafts for the cultivation of Norway spruce embryogenic cultures and the production of mature somatic embryos. Different cultivation techniques were compared. The cultures grown on membrane rafts were further used for the evaluation of benefits and drawbacks of the insertion of a pre-maturation phase on PGR-free medium. Finally, we studied the effects of PEG in maturation medium on the embryo yield, quality and germination frequency.

2. Material and methods

Embryogenic cultures of Norway spruce (*Picea abies* L. [Karst.]) were obtained as a gift from laboratories in France (AFOCEL, Nangis, Dr. Paques) and Austria (ÖF Seibersdorf, Dr. Wilhelm), or induced from immature or mature zygotic embryo in our lab. The embryogenic cultures proliferated at 24 °C in total darkness on agar-solidified GD medium (Gupta and Durzan, 1986), containing 5 μmol 2,4-D, 2 μmol kinetin, and 2 μmol BAP. Twelve cell lines were selected for this study that differed in their embryogenic capacity (the ability to develop mature somatic embryos able to germinate). Both the age of embryogenic cultures (0.5 to 12 years since induction), and the geographical origin of primary explant (lowlands to alpine region) varied.

Cultures induced and maintained on agar-solidified media were transferred onto membrane rafts (Sigma), agar-solidified and liquid media (Vágner et al. 1998, 2000, 2001). Development and maturation of somatic embryos were started either with ABA alone, or with ABA and 1.87 – 5 % (w/v) polyethylene glycol (PEG 4000). This phase followed directly after the proliferation phase (during which the medium contained 2,4-D and cytokinins), or after the pre-maturation phase (1 week, no plant growth regulators), which was inserted between proliferation and maturation stages. Development and maturation of somatic embryos depended on the maturation treatment and cell line used, and took 4 – 7 weeks. Mature somatic embryos then were desiccated for 3 weeks in high relative humidity (HRH > 97 %, 18 °C, 12h/12h dark/light). Desiccated somatic embryos germinated on ¼-strength agar-solidified GD medium with 0.5 % (w/v) sucrose and 0.5 % (w/v) activated charcoal.

Development of somatic embryos (number, speed of development, and quality expressed as a germination ratio), and germinating plantlets was evaluated with the use of computer image analysis (Lucia, var. 4.61, Laboratory Imaging, Czech Republic).

3. Results and discussion

3.1 Membrane rafts compared to other cultivation techniques

The use of membrane rafts compared to solidified media brought a number of benefits. A number of mature embryos significantly increased in 7 of 12 tested cell lines (Figure 1). Synchronization of the process was enhanced, and the time necessary to reach the stage of mature embryo was shortened (the difference was 6 days on average). Just the shortening of the maturation period, during which the cultures are exposed to high exogenous ABA concentration, is supposed to be crucial for further 'normal' seedling development. No difference in germination frequency was observed in cell lines with higher embryogenic capacity (Figure 2). Germination frequency of cell lines with low embryogenic capacity increased. Probably, this phenomenon could be ascribed to the better quality of somatic embryos grown on membrane rafts.

Compared to solidified media, routine passage of the cultures grown on membrane rafts is less time consuming, as the cultures are not transferred to fresh media individually, but with the whole raft. Moreover, there is probably a better exchange at the medium – embryogenic tissue interface in this system, because the cultures could be transferred at slightly longer subcultivation intervals. The relatively high price of membrane rafts together with their low durability remained the only drawback of this system.

The ability of different cell lines to proliferate in liquid medium varied markedly; a few of them did not change morphological appearance even after a long cultivation in liquid medium. On the other hand, cultivation of all embryogenic cell lines in liquid maturation medium resulted in severe decrease of the number of developed somatic embryos, which, after desiccation, were able to germinate.

3.2 Pre-maturation (PGR-free) phase

Insertion of a pre-maturation phase (no PGR in the medium) (Bozhkov et al., 2002) between proliferation (+ 2,4-D, BAP, and kinetin) and maturation phase (+ ABA) resulted in further improvement of synchronization, the enhancement of speed of somatic embryo development, and an increase in the number of developed embryos (Figure 3). This improvement was more pronounced in cultures grown on membrane rafts compared to solidified media. We suppose that after pre-maturation phase the endogenous levels of auxins (2,4-D, IAA) could be lower in cultures grown on rafts due to facilitated flow from the culture to the liquid medium. In a similar way after pre-maturation stage, the endogenous levels of cytokinins in embryogenic tissue grown on membrane rafts were found to be lower compared to cultures grown on solidified medium (data not shown). The ABA is thus supplied to the embryogenic cultures with lower levels of endogenous cytokinins and IAA in the moment of ABA application. These endogenous hormonal levels (high level of ABA, low levels of auxins and cytokinins) seem to be beneficial for embryo development.

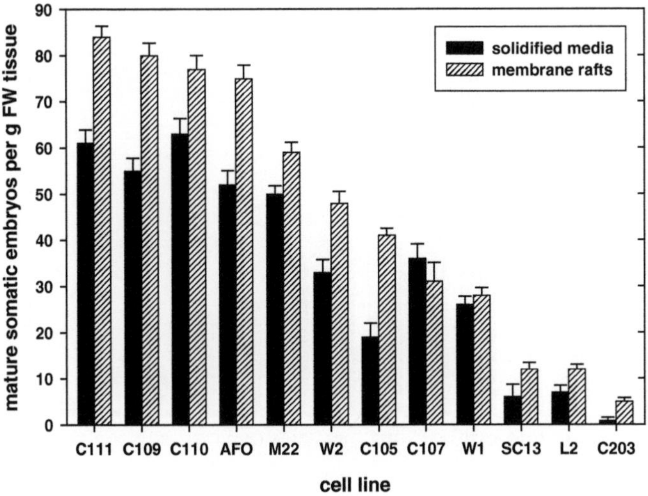

Figure 1: Comparison of yield of mature Norway spruce somatic embryos grown on solidified medium and on membrane rafts. 12 cell lines were used for the experiment. Bars indicate S_E.

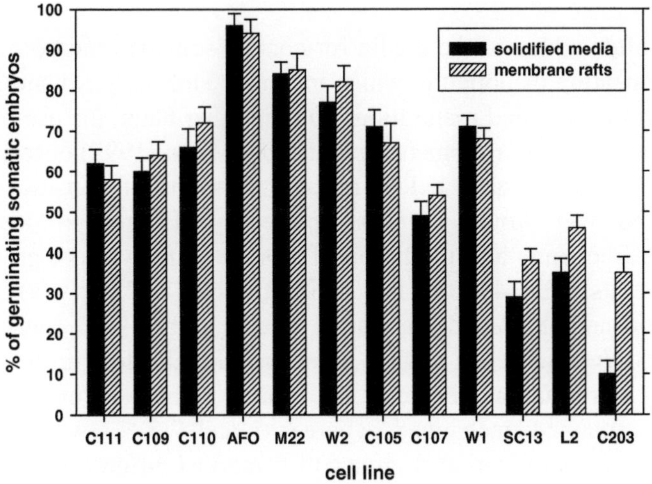

Figure 2: Comparison of germination frequencies of desiccated Norway spruce somatic embryos previously grown on solidified medium and on membrane rafts. The same cell lines were used as in figure 1. Bars indicate S_E.

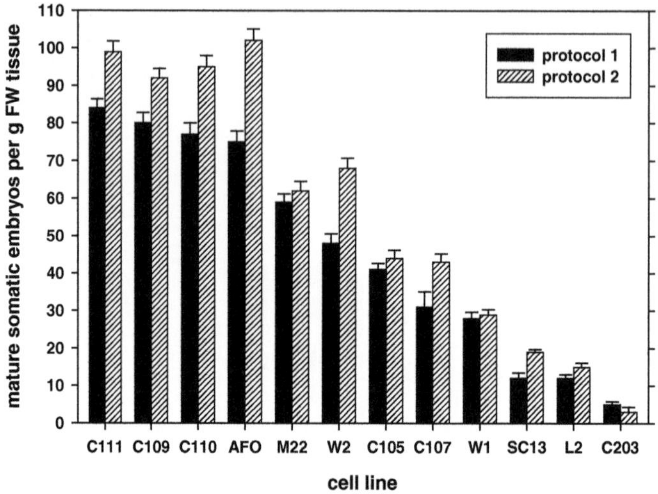

Figure 3: The yield of mature Norway spruce somatic embryos grown on membrane rafts. Protocol 1: Cultures transferred from proliferation medium (containing 5 μmol 2,4-D, 2 μmol kinetin and 2 μmol BAP) directly to maturation medium (20 μmol ABA).
Protocol 2: Cultures were transferred from proliferation medium to pre-maturation medium (no PGR, 1 week), and then to maturation medium. Bars indicate S_E.

3.3 Use of PEG in maturation medium

PEG is often added to the media for conifer somatic embryo maturation as a nonpermeating osmoticum, which increases embryo yield and enhances embryo development and maturation. On the other hand, the use of PEG in the maturation medium is sometimes criticized. Find (1997) observed PEG-related intercellular spaces below apical meristems of somatic embryos. Bozhkov and von Arnold (1998) reported a PEG-related decrease of germination frequency and inhibition of post-germinative root growth. In our experiments, PEG 4000 at low concentration (1.87 % (w/v) increased embryo yield and shortened maturation phase: higher PEG concentrations (up to 5 % (w/v)) had even slightly more beneficial effect in these aspects (Figure 4). Germination frequency of somatic embryos grown on 3.75 % (w/v) PEG remained unchanged, whereas 5 and 7.5 % (w/v) PEG 4000 slightly inhibited germination (Figure 5). A marked difference in sensitivity to PEG treatment was observed between different cell lines. The study of microscopic sections revealed no morphological aberrations of developing somatic embryos, which were grown on medium containing 3.75 % (w/v) PEG, and the resultant germinating plantlets. However, we did not study development of plantlets after 8 weeks of germination. Thus we cannot exclude the possibility that higher concentration of PEG could interfere with further plantlet development. On the other hand, our results suggest that low concentration of PEG can improve Norway spruce somatic embryogenesis without negative, or with minimal impact, on embling performance.

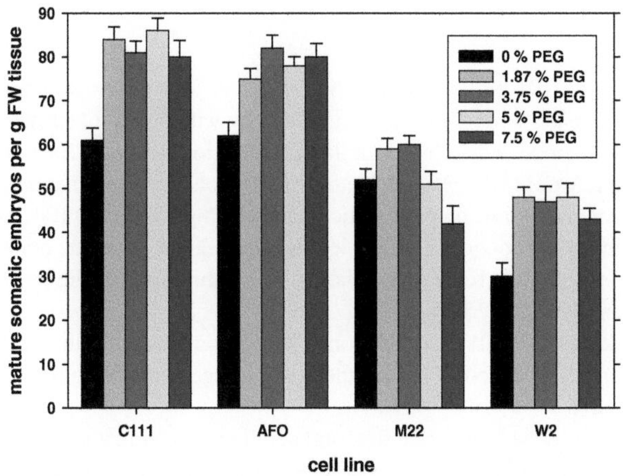

Figure 4: The effect of PEG on the yield of mature Norway spruce somatic embryos grown on membrane rafts. Maturation medium was supplemented with 0 – 7.5 % (w/v) PEG 4000. Bars indicate S_E.

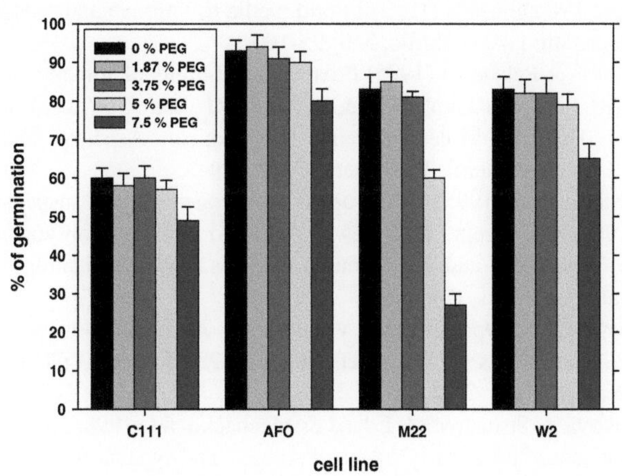

Figure 5: The effect of PEG on the germination of desiccated Norway spruce somatic embryos grown on membrane rafts. Maturation medium was supplemented with 0 – 7.5 % (w/v) PEG 4000. Bars indicate S_E.

Acknowledgement

This work was supported by grant No. OC 843.50 (COST) and Z5 038910.

References

Bozhkov PV, Filonova LH & von Arnold S (2002), A key developmental switch during Norway spruce somatic embryogenesis is induced by withdrawal of growth regulators and is associated with cell death and extracellular acidification. Biotech. Bioeng. 77: 658-667

Bozhkov PV & von Arnold S (1998) Polyethylene glycol promotes maturation but inhibits further development of *Picea abies* somatic embryos. Physiol. Plant. 104: 211-224

Find JI (1997) Changes in endogenous ABA levels in developing somatic embryos of Norway spruce (*Picea abies* (L.) Karst) in relation to maturation medium, desiccation and germination Plant Sci. 128: 75-83

Gorbatenko & Hakman I (2001) Desiccation-tolerant somatic embryos of Norway spruce (*Picea abies*) can be produced in liquid cultures and regenerated into plantlets. Int. J. Plant Sci. 162: 1211-1218

Gupta PK & Durzan DJ (1986) Plantlet regeneration via somatic embryogenesis from subcultured callus of mature embryos of *Picea abies* (Norway spruce). *In Vitro* Cell. Dev. Biol. - Plant 22: 685-688

Luckett DJ, Venkatanagappa S, Darvey NL & Smithard RA (1991) Anther culture of Australian wheat germplasm using modified C-17 medium and membrane rafts. Aust. J. Plant Physiol. 18: 357-367

Paek KY, Hahn EJ & Son SH (2001) Application of bioreactors for large-scale micropropagation systems of plants. *In Vitro* Cell. Dev. Biol.- Plant 37: 149-157

Paques M, Bercetche J & Dumas E (1992) Liquid media to improve and reduce the cost of *in vitro* conifer propagation. Acta Hortic. 319: 95-100

Paques M, Bercetche J & Palada M (1995) Prospects and limits of somatic embryogenesis of *Picea abies*. In: Jain SM, Gupta PK & Newton RJ (eds) Somatic Embryogenesis in Woody Plants: History, Molecular and Biochemical Aspects, and Applications (pp. 399-414), Kluwer Academic Publishers, Dordrecht

Tautorus TE & Dunstan DI (1995), Scale-up of embryogenic plant suspension cultures in bioreactors. In: Jain SM, Gupta PK & Newton RJ (eds) Somatic Embryogenesis in Woody Plants: History, Molecular and Biochemical Aspects, and Applications (pp. 265-292), Kluwer Academic Publishers, Dordrecht

Vágner M, Kumstýřová L, Opatrná J & Vondráková Z (2001) Norway spruce somatic embryogenesis: Bottlenecks of the method. (pp. 28-29) Proc. COST 843, WG 2, Thessaloniki, September 22.-25.2001

Vágner M, Vondráková Z, Strnadová Z, Eder J & Macháčková I (1998) Endogenous levels of plant growth hormones during early stages of somatic embryogenesis of *Picea abies*. Adv. Hort. Sci. 12: 11-18

Vágner M, Vondráková Z, Kumstýřová L, Opatrná J, Cvikrová M, Eder J, Vaněk T & Macháčková I (2000) Norway spruce somatic embryogenesis. (pp. 15-16) Proc. COST 843, WG 2 Meeting, July 2000, Tampere, Finland

Watad AA, Kochba M, Nissim A & Gaba V (1995) Improvement of *Aconitum napelus* micropropagation by liquid culture on floating membrane rafts. Plant Cell Rep. 14: 345-348

Chapter 22

Picea abies somatic embryo development from suspension cultures and agar-based cultures: a comparison

Christopher Hunter & Lionel Levy
University of the West of England, Frenchay, Bristol BS16 1QY, England. E-mail: Christopher.Hunter@uwe.ac.uk

Abstract: Embryogenic suspensor mass derived from a zygotic embryo cotyledon of *Picea abies* (L.) Karst. was propagated either in suspension culture or on agar-based 'proliferation medium' (Vagner, 1998), then transferred to agar-based 'maturation medium' for the development of somatic embryos. The 'maturation' medium was supplemented with silver nitrate at either 0, 1, 10 or 100μmol. Cultures were transferred to fresh medium weekly, for 10 weeks, during which numerical and photographic records of somatic embryo development were kept for each culture (n = 14 for each treatment). For cultures originating from both suspension and agar-based systems, there was approximately a two to three-fold increase in the number of Stage IV somatic embryos in media that had been supplemented with silver nitrate.

Key words: embryogenic suspensor mass (ESM), mass-propagation, silver nitrate

Abbreviations: COST - EU Action in Co-operation in Science and Technology; DW – dry weight; ESM - embryogenic suspensor mass; FW – fresh weight; SE - somatic embryo

1. Introduction

Picea abies (L.) Karst. remains the backbone of much European forestry both for timber and paper pulp. Whilst selected tree provenances provide seed-tree populations which give rise to seedling-derived forests with enhanced agronomic properties, the mass-propagation of elite genotypes of *P. abies* through somatic embryogenesis remains a target for large-scale plantings of mixed clones. Systems that enhance the numbers and quality of propagules contribute to that objective and much work has been undertaken investigating various media and growing techniques *in vitro* to enhance propagule yields.

A.K. Hvoslef-Eide and W. Preil (eds.), Liquid Culture Systems for in vitro Plant Propagation, 303–312.

Suspension-based cultures offer potential for the greatest biomass increase, but it is often the somatic embryos that develop entirely, or subsequently, on gelled media that yield the highest quality emblings suitable for field growth. Different workers publish conflicting experiences with regard to the full or partial development and conversion to emblings, and subsequently into field-grown maturing trees. The full and seedling-like development of a sizeable population of somatic embryos to produce vigorous soil-grown emblings continues to be a challenge for many.

Whilst somatic embryo recalcitrance and embling losses continue to be a problem, researchers continue to modify, empirically, standard tissue-culture propagation protocols by the addition of media supplements, with the aim of increasing embryo yields. For example, supplementation of media with silver nitrate has shown both growth promotive (e.g. Kong and Yeung, 1995) and non-promotive effects (Selby et al., 1996) on various *Picea* species. In this paper, the addition of silver nitrate is evaluated as a stimulant to somatic embryogensis and development in *P. abies.* in relation to cultures grown in suspension or gelled media.

2. Materials and methods

Embryogenic suspensor mass (ESM) of *P. abies* cell line AF0541.1, donated by Afocel of Nangis, France, originated from a zygotic embryo and had been used for the COST 822 Working Group 2 programme. In each COST-participant's lab it was maintained routinely by subculture at 7 to 10-day intervals on gelled 'proliferation medium'. Somatic embryos (SEs) developed when the ESM was transferred and subcultured weekly to gelled 'maturation medium'. Both media are based on that of Gupta and Durzan (1986) with plant growth regulators and culture protocol as described by Vagner (1998).

In this work proliferation was both from suspension culture (referred to as SA cultures) and from agar-based medium (referred to as AA cultures): aliquots of the resultant cell mass or callus was inoculated to 15 ml agar-based proliferation medium in a 9 cm diameter Petri dish supplemented with $AgNO_3$ at 0, 1, 10 or 100µmol (14 replicate cultures per treatment) and subcultured weekly to fresh medium for 10 weeks: these $AgNO_3$ treatments are referred to as T0, T1, T10 and T100. For SA cultures, 750 mg blotted ESM cells were spread over a square template 2 x 2 cm; for AA cultures *c.* 340 mg was spread over a circular template *c.* 15mm diameter. Petri dishes were sealed with Parafilm M. Sample inocula for SA and AA cultures were oven-dried to determine the fresh to dry weight ratios (FW:DW).

Plate 1: Somatic embryo development in *Picea abies* callus after 69 days in medium supplemented with $AgNO_3$. Scale in mm.

Plate 2: Somatic embryo development in *Picea abies* callus after 69 days in medium without $AgNO_3$. Scale in mm.

A digital camera and Clickometer™ software were used to record, weekly, SE numbers and development in each replicate culture. At each observation, each SE was ascribed to a 'development stage' (Hakman and von Arnold, 1988) ranging from Stage I and II (immature, < 2 mm long), to III (intermediate, length > 2mm), to IV in which cotyledons were well developed and *c.* 3mm long. Stages are referred to as S1, SII, SIII and SIV.

3. Results

FW:DW ratio (n=20) for SA was 4.93% and 4.95% for AA. As these means were not statistically different, results of SE development in this paper are able to be expressed, with some confidence, on the basis of FW (g) of inoculum.

At the time of inoculation only proembryos and slightly later stage SEs were visible microscopically in both SA and AA cultures. Following transfer to maturation medium, SI and SII SEs differentiated: by week 6, SA cultures (Figure 1a) had 75 SEs g^{-1} (T0) to 172 SEs g^{-1} (T10) which decreased to 14 SEs g^{-1} (T0) to 53 SEs g^{-1} (T10) by week 10. SI and SII AA cultures (Figure 1b) were maximal at week 7 (between 161 and 174 SEs g^{-1}), decreasing to between 71 and 104 SEs g^{-1} at week 10: throughout the period there was no substantial treatment difference. Plates 1 and 2 show the effect of the culture response to the presence of silver nitrate at week 10.

In SA cultures, numbers of SIII SEs (Figure 2a) increased from 6 to 14 (T0), 5 to 20 (T100), 25 to 45 (T1 and T10). In AA cultures (Figure 2b) there was little change overall in yield during the 5-week period (23 increasing to 26 SEs g^{-1} for T0, 41 increasing to 47 for T1, 52 decreasing to 50 for T10 and 62 decreasing to 41 for T100): the yield for T0 was significantly less than the other concentrations.

SIV SE numbers for SA cultures (Figure 3a) were zero at week 6, but thereafter increased linearly to week 10. SIV SE numbers for AA cultures (Figure 3b) ranged from 7 to 23 at week 6, to between 18 and 59 at week 10. The yields of T0 were always significantly lower that the other three silver nitrate treatments.

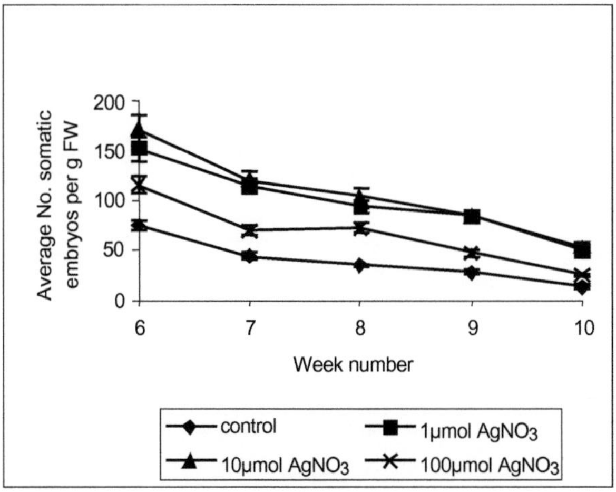

Figure 1a: Average number of stages I plus II somatic embryos during a 5-week period from ESM previously grown in suspension. Bars represent ± standard error (n = 14).

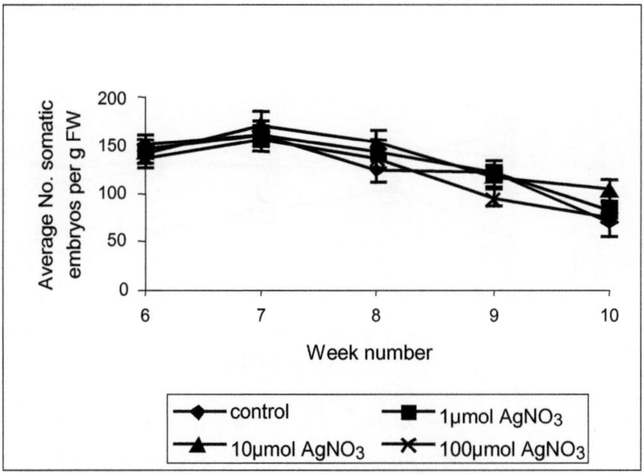

Figure 1 b: Average number of stages I plus II somatic embryos during a 5-week period from ESM previously grown on agar. Bars represent ± standard error (n = 14).

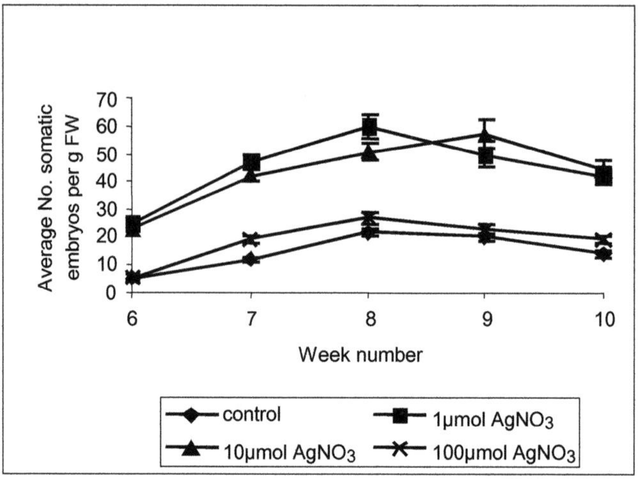

Figure 2 a: Average number of stage III somatic embryos during a 5-week period from ESM previously grown in suspension. Bars represent ± standard error (n = 14).

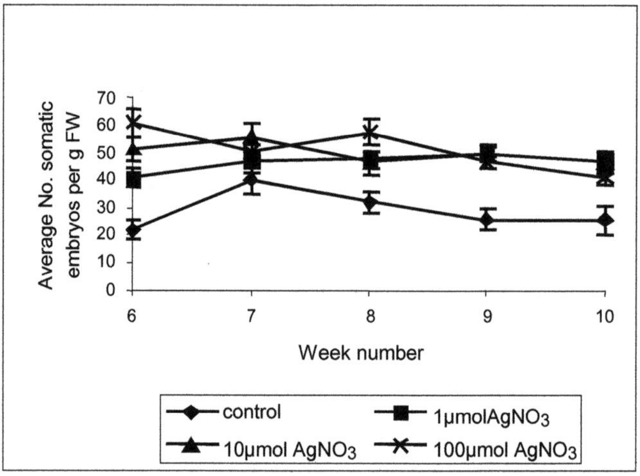

Figure 2 b: Average number of stage III somatic embryos during a 5-week period from ESM previously grown on agar. Bars represent ± standard error (n = 14).

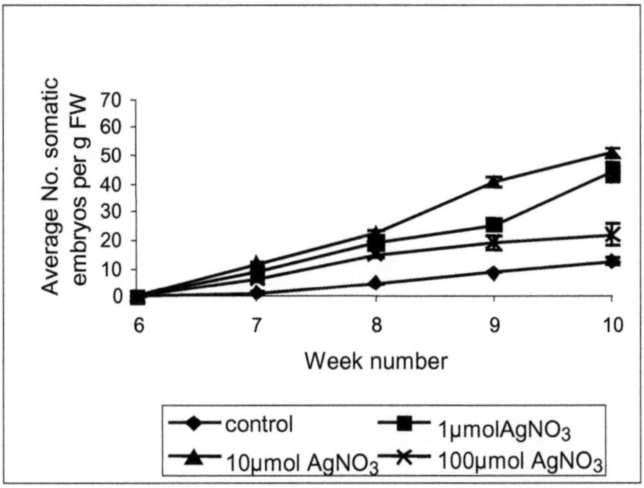

Figure 3 a: Average number of stage IV somatic embryos during a 5-week period from ESM previously grown in suspension. Bars represent ± standard error (n = 14).

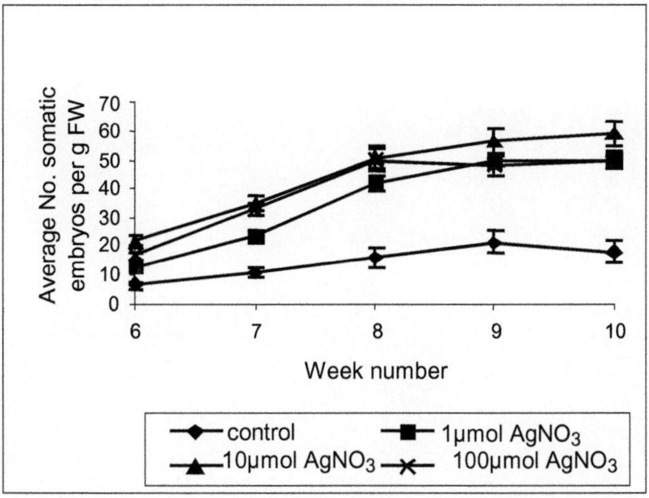

Figure 3 b: Average number of stage IV somatic embryos during a 5-week period from ESM previously grown on agar. Bars represent ± standard error (n = 14).

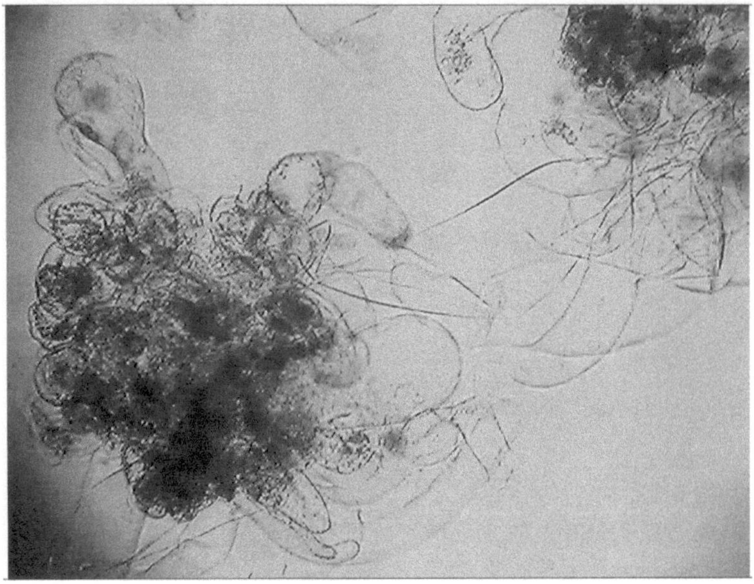

Plate 3: Embryogenic suspensor mass of *Picea abies* grown in suspension for 17 days in 'proliferation medium'.

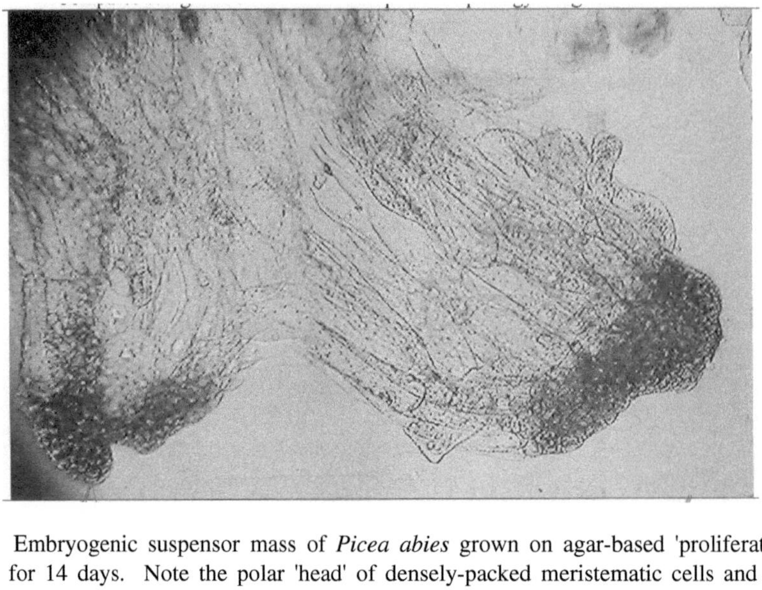

Plate 4: Embryogenic suspensor mass of *Picea abies* grown on agar-based 'proliferation medium' for 14 days. Note the polar 'head' of densely-packed meristematic cells and the highly vacuolated embryo suspensor cells.

4. Discussion

The mechanism for the stimulation of SIV SEs by all concentrations of $AgNO_3$ remains unclear. The microbial suppressant properties of the silver salt may have had an effect, but the 'master cultures' had been tested and no microbial colonies had been recovered. Silver nitrate is generally regarded as an inhibitor of ethylene synthesis (e.g. Kawano et al., 2002), but recent work by Lim et al. (2002) reported that S-adenosylmethionine synthetase, which catalyses the conversion of methionine to S-adenosylmethionine, a precursor for ethylene synthesis, was upregulated by $AgNO_3$. Nevertheless, Mandal et al. (2001) reported increased somatic embryo production in $AgNO_3$–treated safflower cultures. Similarly, Achar (2002) reported increased somatic embryo yields from anther cultures of cabbage. Kvaalen (1994) reported that ethylene, supplied *via* ethephon to *P. abies* embryogenic cultures did not inhibit growth. In relation to exogenous application of $AgNO_3$, it could be that the additional nitrate content is stimulatory or that the silver ion is itself a stressor which induces abscisic acid synthesis in the cultures and thereby promotes SE maturation.

The differences relating to the origin of the cultures (suspensions *vs* agar) may reflect the high numbers of 'early-stage' pro-embryos in the suspension culture compared with the later stage pro-embryos which exhibit clear signs of polar development, thus whilst the SA *vs* AA cultures were notionally of the same age, because of the developmental differences (Plates 3 and 4), the AA cultures were at a more advanced early stage embryo and these differences were reflected throughout the subsequent period of observation.

References

Achar PN (2002) A study of factors affecting embryo yields from anther culture of cabbage Plant Cell, Tissue and Organ Culture 69: 183-188

Gupta PK & Durzan DJ (1986) Plantlet regeneration *via* somatic embryogenesis from subcultured callus and mature embryos of *Picea abies*. *In Vitro* Cell. Dev. Biol 22: 685-688

Hakman I & von Arnold S (1988) Regulation of somatic embryo development in *Picea abies* by ABA. J. Plant Physiol. 132: 164-169

Kong L & Yeung E (1995) Effect of silver nitrate and polyethylene glycol on white spruce (*Picea glauca*) somatic embryo development: enhancing cotyledon embryo formation and endogenous ABA content. Physiol. Plant. 93: 298-304

Kvaalen H (1994) Ethylene synthesis and growth in embryogenic tissue of Norway Spruce - effects of oxygen, 1-aminocyclopropane-1-carboxyllic acid, benzyladenine and 2,4-dichlorophenoxyacetic acid. Physiol. Plant. 92: 109-117

Kawano N, Ella E, Ito O, Yamauchi Y & Tanaka K (2002) Metabolic changes in rice seedlings with different submergence tolerance after desubmergence. Environ. Exp. Bot 47: 195-203

Lim CC, Liu JZ & Pua EC (2002) Characterization of S-adenosylmethionine genes and its expression associated with ethylene synthesis in mustard (*Brassica juncea*). Physiol. Plant. 16: 522-530

Mandal AKA Gupta SD & Chatterji AK (2001) Factors affecting somatic embryogenesis from cotyledonary explants of safflower. Biol. Plantarum (Prague) 44: 503-507

Selby C, McRoberts WC, Hamilton JTG & Harvey BMR (1996) The influence of culture vessel head-space volatiles on somatic embryo maturation in Sitka Spruce (*Picea sitchensis*). Plant Growth Regulation 20: 37-42

Vagner M (1998) Protocol for cultivation of embryogenic cultures of *Picea abies*. Pers. comm.

Chapter 23

Somatic embryogenesis by liquid culture of epidermal layers in sunflower: from genetic control to cell development

Michel Petitprez, A. Sarrafi, E. Flores-Berrios, X. XuHan, C. Briere & L. Gentzbittel

Biotechnology and Plant Breeding (BAP), INP-ENSAT, Avenue de l'Agrobiopole, F-31326 Castanet-Tolosan. France. E.mail : petitprez@ensat.fr

Abstract: Embryos were obtained using liquid medium culture of sunflower hypocotyl epidermis layers according to the Pélissier et al. (1990) method. In the present work we identified genetic factors controlling somatic embryogenesis and we evidenced the role of ionic channels in embryogenic tissues. Two traits, the number of embryogenic explants (EE) and the number of embryos (EM) were scored in 74 recombinant inbred lines (RILs) from a cross between lines PAC-2 and RHA-266. Analysis of variance indicated the existence of highly significant differences among the parental genotypes and their RILs. Heritability for the somatic embryogenesis traits studied were high (0.64 for EE and 0.77 for EM). Four quantitative trait loci (QTLs) for EE and seven for EM were detected using composite interval mapping. The QTLs for EE explained 48% of the phenotypic variation while the QTLs for EM explained about 89% of the variation, thus revealing several genomic regions related to somatic embryogenesis control in sunflower. In order to study the distribution of ion channels in somatic embryos as compared to zygotic ones, we used a fluorescent-labelled phenylalkylamine, DM-Bodipy PAA, as a probe. Fluorescence labelling was determined by confocal microscopy. The probe intensively labelled the protoderm and epidermis cells in both zygotic and somatic embryos. Callus exhibited labelling on sites where somatic embryos developed. Considering that the location of phenylalkylamine (PAA) binding sites is related to the distribution of ion channels, the high intensity in the protoderm and epidermis of embryos, point to similar properties and functions and their key role in embryo development.

Key words: calcium, embryo, ion channels, protoderm, quantitative trait loci (QTLs), recombinant inbred lines

Abbreviations: EE-embryogenic explants; EM-number of embryos; PAA-phenylalkylamine; QTLs-quantitative trait loci; RILs-recombinant inbred lines

A.K. Hvoslef-Eide and W. Preil (eds.), Liquid Culture Systems for in vitro Plant Propagation, 313–322.

1. Introduction

The ability to regenerate large numbers of plants from tissue culture is important for the successful application of most biotechnological techniques, such as genetic engineering. During the last few years, regeneration methods have been developed for sunflowers (*Helianthus annuus*). There are two main types of regeneration methods: Organogenesis and somatic embryogenesis (Pelissier et al., 1990; Jeannin et al., 1995). Embryogenesis capacity is influenced by cultural conditions, genotype and their interaction. In sunflower, embryogenic events increase with increasing sucrose concentration (Jeannin et al., 1995) and darkness (Carola et al., 1997). Direct somatic embryogenesis could be obtained either from immature embryos or from epidermal layers; both responses are highly variable depending upon the genotype (Pelissier et al., 1990; Bolandi et al., 2000). At present, the number of reports about the genetic control of regeneration in sunflower remains limited. Additive and dominant effects of genes controlling embryogenesis traits have been reported by Bolandi et al. (2000) in this species.

The construction of genetic maps has provided a tool for identification of the number, significance and location of quantitative trait loci (QTLs) associated with a variety of phenotypic characteristics. In sunflower, maps have been developed and linkage of molecular markers with resistance genes have been identified (Gentzbittel et al., 1998, 1999). The utilization of molecular markers linked to different traits would help to identify the genes involved. Moreover, estimates of genetic variation and determination of chromosomal regions that control somatic embryogenesis can be used to determine the value of genotypes in a breeding program.

Beside genotype, culture conditions of explant are known to assume particular importance in somatic embryo development. Tissues, which are at the interface between the organ and the external medium are essential sites for communication between plant and environment. This is particularly crucial for the protoderm of the young embryo. Recent investigations have revealed that specific genes are expressed during the development of embryonic protoderm and epidermis (Vroemen et al., 1996) and embryonic protoderm shows a calcium-binding pattern, which differs from that in the inner embryonic cells (Timmers et al., 1996). These findings call attention upon the functions of the plant epidermis system, especially in signal transduction pathways. Although it is well known that ion channels play an important role in signalling and control of morphogenesis in plants, their distribution in higher plant tissues has not been described yet. Phenylalkylamines (PAAs) are pharmacological drugs able to block specifically the L-type Ca^{2+} -channel activity in animal cells (Norris and

Bradford, 1985) and has been shown to block the entry of calcium into plant cells (Thuleau et al., 1990). A fluorescently labelled PAA, DM-Bodipy PAA, has been used as a probe for labelling Ca^{2+}-channels in animal cell membranes (Knaus et al., 1992) and in sunflower protoplasts (Vallée et al., 1997). It appears to be a good tool to study the distribution of Ca^{2+} channels antagonist binding sites in higher-plant tissues. In the present work we have studied the distribution of PAA binding sites in epidermal layer systems giving rise to somatic embryos as compared to non-embryonic explants.

The objective of the investigation presented here was to carry out a QTL mapping analysis to characterize the genomic regions involved in somatic embryogenesis in order to localise more precisely the genes involved and to clone them. The coupled approaches of genetic mapping and cell biology could allow us to decipher both genetic control and cellular mechanisms involved in somatic embryo determination.

2. Material and methods

2.1 Scoring of somatic embryogenesis

A population of 74 recombinant inbred lines (RILs) developed by the SSD method from the cross between lines PAC-2 and RHA-266 were used in this experiment. Surface-sterile seeds were germinated on agar-gelled MS basal medium (Murashige and Skoog, 1962) pH 5.7. Cultures were maintained at 24°C under a light flux of $50\mu Em^{-2}s^{-1}$ (16-h light, 8-h dark). Epidermal strips from 7-day-old hypocotyls were peeled, cut in 2cm sections and transferred to MS basal medium for 5 days, then to B5-90 medium for 8 days, according to Pelissier et al. (1990). The strips, including the epidermis and about 4 sub-epidermal layers, were cultured at 24°C in the dark with shaking at 120rpm. After this period, explants were transferred to MS-120 embryo-developing medium for 15-20 days at 26°C in the dark. The embryos were separated from the layers and transferred to B-60 medium in order to develop secondary embryos for 10 days. The experiment was designed as a randomized complete block with 76 genotypes (74 RILs and 2 parents) and three replicates. Each replicate consisted of three Erlenmeyer flasks each with 40 epidermal strips. The following traits were determined for each genotype per replicate: the number of embryogenic layers per 40 plated strips, and the number of embryos per 40 strips. Variance analysis was performed and the means separated using a Newman-Keuls-test (P=0.05). Additive, environmental variances and heritability were calculated according to Kearsey and Pooni (1996), using least-square estimates of the genetic parameters.

2.2 *QTL mapping*

This set of 74 RILs and 2 parents was used for DNA extraction, AFLP products screened with 333 markers and a linkage map was constructed based on 254 linked loci, as previously described (Flores-Berrios et al., 2000). The chromosomal location of QTLs for embryogenic traits were resolved by composite interval mapping (CIM) using QTL cartographer v1.13 model 6 software (Basten et al., 1999). Inclusion of the background makes the analysis more sensitive to the presence of a QTL in the target interval. A window size of 10cM and 15 markers were chosen to account for the background. At each marker locus, the significance of the association was tested by the likelihood ratio statistic (LRS) (Haley and Knott, 1992).

2.3 *Probe loading*

Fluorescently labelled PAA (Molecular Probes, Eugene, Oregon, USA) was prepared as a 2.5mmol stock solution in dimethyl sulfoxide. Before use, the DM-Bodipy PAA stock solution was diluted to 1mmol in MS medium supplemented with 120g sucrose per litre (MS-120). Embryos and epidermal layers were hand-cut in 0.5-1mm thick sections, incubated in 1mmol DM-Bodipy PAA solution at 25°C for 30 min, rinsed in MS-120 medium and mounted in an observation chamber. A confocal laser-scanning microscope equipped with an argon-ion laser (488nm/510nm) was used to analyse the labelling. For image acquisition a x25/0.8 Plan objective was used with a 32% attenuation filter. Each image corresponds to the average of four frames.

3. Results and discussion

We have previously shown that the fluorescent probe DM-Bodipy PAA labels sunflower cell plasma membranes (Vallée et al., 1997), and is selectively removed by Ca^{2+}-antagonist treatment. Thus we may speculate that the location of labelling is highly related to the distribution of Ca^{2+}-channels, and to a lesser extent, K^{+}-channels (Xu XuHan et al., 1999).

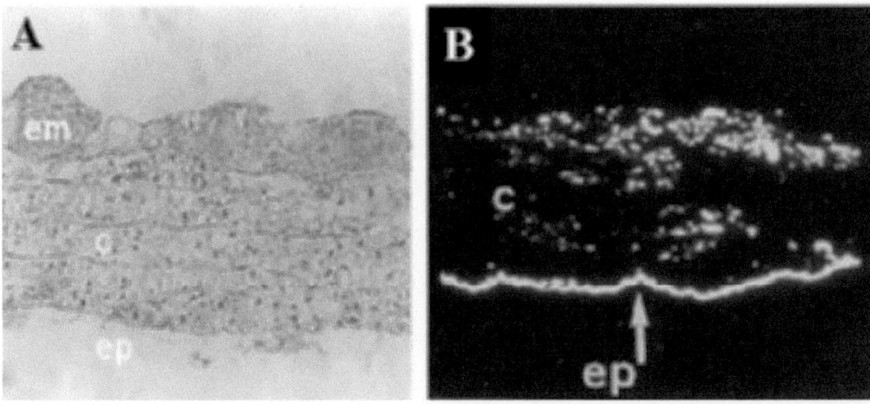

Figure 1: (A) Thin section of an epidermal strip showing the initiation sites of somatic embryos (em) in the outermost part of the strip in relation to the epidermis (ep). (B) Confocal optical section of an epidermal strip in culture, labelled with DM-Bodipy PAA. The original epidermis (ep) is strongly fluorescent, whereas the inner cells of the callus (c) are weakly labelled. The outermost cell layers of the callus are intensively labelled, corresponding to the embryo-forming cells.

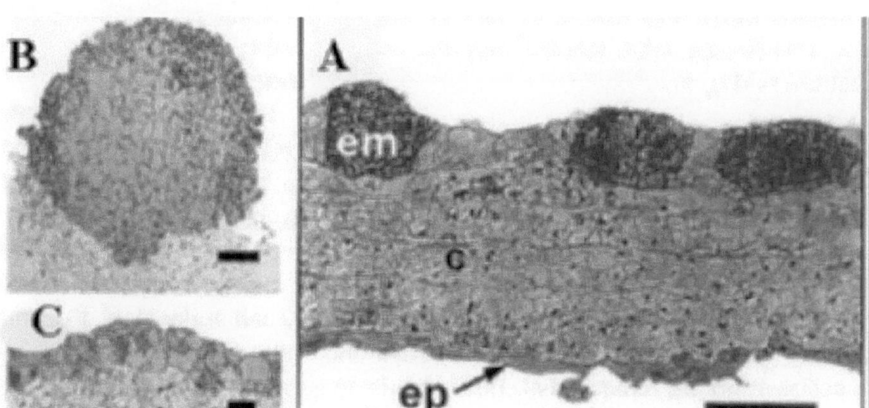

Figure 2: (A) On this section of an epidermal strip presented in false colour, the original hypocotyls epidermis (red) was still attached to the subepidermal cell layers that became callogenic (green). In the outermost cell layers of the callus there appeared initiation sites of somatic embryos (blue) (Bar= 100µm). (B) Thin section of a somatic embryo at the globular stage (blue) surrounded by callus cells at its base (Bar= 100µm). (C) Part on a somatic embryo at the heart-shaped stage, showing aligned protodermal cells (red) covering the embryonic ground tissue (green) (Bar= 20µm).

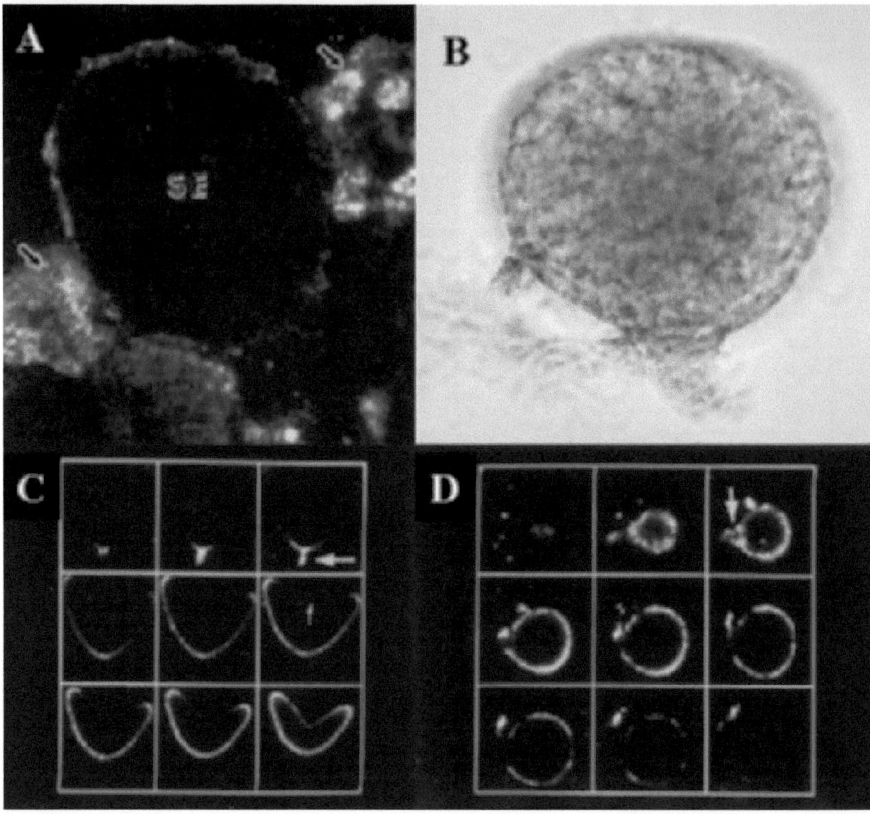

Figure 3: (A, B) Somatic embryo as observed under confocal scanning microscope, with (A) or without (B) labelling with DM-Bodipy PAA; note the intensive labelling of embryonic protoderm. (C, D) Series of optical sections of a zygotic embryo of sunflower at the globular (D) or heart-shaped (C) stage. Note in both cases the strong labelling of the outer cell layers corresponding to protoderm.

In epidermal strips cultured in liquid medium, callus developed on the outermost cells – subepidermal cells. On this callus, mainly at the side opposite from the epidermis, embryogenic masses appeared (Figures 1A, 2A) developing next in a rounded "globular" structure (Figure 2B). Such globular somatic embryos develop further a protoderm layer (Figure 2B) and take a classical heart shape. In cultured epidermal strips, the original epidermis, which could still be attached to the callogenic subepidermal tissue, showed a strong labelling by DM-Bodipy PAA (Figure 1B). Most callus cells exhibited no, or a very weak, fluorescence, except in the areas where somatic embryos were to be formed (Figure 1B). Such somatic embryos developed in the upper cell layers of the explant (Figure 2A). Calli, which did not develop protoderm and epidermis did not show DM-Bodipy

PAA labelling. When somatic embryos developed further, they showed a fluorescence pattern, which was similar to that of zygotic embryos: DM-Bodipy PAA labelled only the outermost cell layers (Figure 3). Such a similar labelling strongly argues in favour of common developmental pattern in somatic and zygotic embryos. Moreover the major role of protoderm is brought out: These cells are characterized by a high density of K^+ and Ca^{2+}-channels. Such a specificity has been reported concerning the ML1 gene, an homeobox gene, which expressed very early in the 8-celled embryo and, further, only in protoderm cells up to heart-shape embryo (Lu et al., 1996).

Analysis of variance for the number of embryogenic explants and the number of embryos per explant are presented in table 1. The parental genotype "PAC-2" showed higher values as compared with RHA-266 for the two tested traits. Bolandi et al. (2000) have also demonstrated that embryogenic parameters are highly genotype-dependent. The difference between all recombinant inbred lines and their parents was not significant, indicating that the RILs obtained are representative of the total possible recombinant lines from the cross PAC-2 X RHA-266 (Table 1). The best parent (PAC-2) compared with the best RIL or with the mean of the best 10% of the RILs, presented significant differences for the 2 traits studied (Table 1). This genetic gain might be due to the accumulation of favourable alleles for embryogenic ability. Narrow-sense heritability was 0.64 and 0.77 for EE and EM, respectively indicating that selection for these embryogenic traits will be possible in progeny of this cross.

Significant peak values of LOD score, the position of the peaks, the percentage of phenotypic variation explained and the estimate of QTL effects based on a composite interval mapping analysis for embryogenic traits studied are shown in table 2. Four QTLs were detected for embryogenic explants and seven for embryo development. The effect of each QTL is moderate (ranging from 7 to 20%). The transgressive phenotypes observed could be explained by the presence of QTLs of opposite sign in each parent. For the two components of somatic embryogenesis capacity, the detected QTLs explain together 48 and 89% of the phenotypic variation respectively. Wan et al. (1992) performed RFLP analysis on regenerable calli formed from embryo-like structures in maize. They hypothesized that some regions found might be related to the induction of embryos and the ability to produce embryogenic calli. Histological studies showed that division occurred within the different layers and that embryos were produced directly at the surface of the epidermal layers (Nonohay et al., 1999).

Table 1: Genetic gains and heritabilities for somatic embryogenesis traits in recombinant inbred lines (RILs) of sunflower

	EE	EM
PAC2 (P1)	46.67	352.5
RHA266 (P2)	11.67	20
P1-P2	35 *	332.5 *
Xp=(P1-P2)/2	29.17	186.25
XRILs	14.15	70.95
XRILs-Xp	-15.02 ns	-115.3 ns
BRIL (Best RIL)	66.67	450.82
GG= BRIL-BP	20 *	98.32 *
10%SRIL	63.12	427.05
GG=10%SRIL-BP	16.45 *	74.55 *
Heritability	0.64	0.77

*$P<0.05$; ns, not significant at $P<0.05$
EE, number of embryogenic explants per 100 explants plated
EM, number of embryos per 100 explants plated
BP, best parent (PAC2)
XRILs, mean of all recombinant inbred lines
GG, genetic gain when either the best RIL (BRIL) or the selected 10% (10%SRIL) are compared with the best parent (BP)

Table 2: Mapping position and effects of the QTLs detected in recombinant inbred lines for somatic embryogenesis traits in sunflower

Trait	QTL	Linkage group	Position (a)	Log-likelihood	Variance explained (b)	Additive effect (b)
EE	ee1.1	I	29	5.6	0.15	-3.56
	ee3.1	III	33	3.3	0.07	2.24
	ee13.1	XIII	103	6.2	0.15	3.74
	ee15.1	XV	93	4.3	0.11	2.89
EM	em1.1	I	58	6.2	0.16	30.37
	em4.1	IV	1	8.3	0.20	-27.23
	em6.1	VI	27	4.9	0.09	-23.23
	em11.1	XI	62	4.1	0.07	-16.50
	em16.1	XVI	8	7.0	0.15	21.97
	em17.1	XVII	1	4.0	0.07	-16.33
	em17.2	XVII	95	7.1	0.15	23.84

a) expressed in Kosambi cM, from north of the linkage group (Flores-Berrios et al., 2000)
b) values determined by QTL Cartographer, version 1.13 (Basten et al., 1999)

A model where somatic embryogenesis is divided into two different steps can be considered. The first step, an induction phase of the explant, can be approached by the number of embryogenic explants, the second step, expression of embryogenic potential, can be evaluated by the number of embryos formed. The large number of detected QTLs together with the fact that only one region (on linkage group I) is associated with both induction and expression parameters, suggest that the genetic control of this trait is probably complex. It also suggests that the two components of the model would be supported by different genetic controls. Although the interesting region on linkage group needs to be more precisely mapped, the available information should help the transfer of embryogenic ability to genotypes that respond poorly.

The first step corresponds to cell re-programming, giving the subepidermal layers the ability to differentiate, divide and initiate an embryo. During this induction phase, some cells strongly expressed Ca^{2+}-channel proteins on their plasma membrane. Such an ionic channel activity seems to be determinant in their cell fate. The second step consists of the expression of the embryogenic potential, leading to the development of embryos on the induced explant. In this expression phase, setting up of the protoderm is a critical event. In the protodermal cells, ionic channel activity appears to characterize their specificity early in controlling the relationships between the embryo proper and the surrounding medium. This focuses on the determinant role of calcium, not only as a compound of the culture medium, but mainly as a second messenger in signal transduction pathways.

These results should provide a starting point for the deciphering of molecular mechanisms leading to somatic embryogenesis. The coupled approaches of genetic mapping and cell biology could allow us to unravel both genetic control and cellular events involved in somatic embryo determination.

References

Basten CJ, Weir BS & Zeng ZB (1999) QTL Cartographer Version 1.13: programme in statistical genetics. Department of statistics North Carolina State University

Bolandi AR, Branchard M, Alibert G, Gentzbittel L, Berville A & Sarrafi A (2000) Combining ability analysis of somatic embryogenesis from epidermis layers in sunflower. Theor. Appl. Genet. 100: 621-624

Carola M, Trabace T & Sunseri F (1997) High frequency of plant regeneration in sunflower from cotyledon *via* somatic embryogenesis. Plant Cell Rep. 16: 295-298

Flores-Berrios E, Gentzbittel L, Alibert G & Sarrafi A (2000) Genotypic variation and chromosomal location of QTLs for somatic embryogenesis revealed by epidermal layers culture of recombinant inbred lines in sunflower. Theor. Appl. Genet. 101: 1307-1312

Gentzbittel L, Mouzeyar S, Badaoui S, Mestries E, Vear F, Tourvielle de Labrouhe D & Nicolas P (1998) Cloning of molecular markers for disease resistance in sunflower Theor. Appl. Genet. 96: 519-525

Gentzbittel L, Mestries E, Mouzeyar S, Mazeyrat F, Badaoui S, Vear F, Tourvielle de Labrouhe D & Nicolas P (1999) A composite map of over expressed sequences and phenotypic traits of the sunflower genome. Theor. Appl. Genet. 99: 218-234

Jeannin G, Bronner R & Hahne G (1995) Somatic embryogenesis and organogenesis induced on the immature zygotic embryo of sunflower cultivated *in vitro*: role of sugar. Plant Cell Rep. 15: 200-204

Haley CS, Knott SA (1992) A simple regression method for mapping quantitative trait loci in line crosses using flanking markers. Heredity 69: 315-324

Kearsey MJ, Pooni HS (1996) The genetical analysis of quantitative traits. Chapman & Hall., London, UK

Knaus HG, Moshammer T, Kang HC, Haugland RP & Glossmann H (1992) A unique fluorescent phenylalkylamine probe for L-type Ca^{2+}-channels. J. Biol. Chem. 267: 2179-2189

Lu P, Porat R, Nadeau JA & O'Neill SD (1996) Identification of a meristem L1 layer specific gene in Arabidopsis that is expressed during embryonic pattern formation and defines a new class of homeobox genes. The Plant Cell 8: 2155-2168

Murashige T, Skoog F (1962) A revised medium for rapid growth and bioassays with tobacco tissue cultures. Physiol. Plant. 15: 473-497

Nonohay JS, Mariath JEA & Winge H (1999) Histological analysis of somatic embryogenesis in Brazilian cultivars of barley. Plant Cell Rep. 18: 929-934

Norris DK & Bradford HF (1985) On the specificity of verapamil as a calcium channel blocker. Biochem. Pharmacol. 34: 1953-1956

Pelissier P, Bouchefra O, Pépin R & Freyssinet G (1990) Production of isolated somatic embryos from sunflower thin cell layers. Plant Cell Rep. 9: 47-50

Thuleau P, Graziana A, Canut H & Ranjeva R (1990) A 75kDa polypeptide, located primarily at the plasma membrane of carrot cell-suspension cultures, is photoaffinity labelled by the calcium channel blocker LU 49888. Proc. Natl. Acad. Sci. USA 87: 10000-10004

Timmers ACJ, Reiss HD, Boshung J, Traxel K & Schel JHN (1996) Localization of calcium during somatic embryogenesis of carrot. Protoplasma 190: 107-118

Vallée N, Brière C, Petitprez M, Souvré A & Alibert G (1997) Studies on ion channel antagonist binding sites in sunflower protoplasts. FEBS Lett. 411: 115-118

Vroemen CW, Langeveld S, Mayer U, Ripper G, Jürgens G, Van Kammen A & de Vries SC (1996) Formation in the *Arabidopsis* embryo revealed by position-specific lipid transfer protein gene expression. The Plant Cell 8: 783-791

Wan Y, Rocheford TR & Widholm JM (1992) RFLP analysis to identify putative chromosomal regions involved in the anther culture response and callus formation in maize. Theor. Appl. Genet. 85: 360-365

Xu XuHan, Brière C, Vallée N, Borin C, van Lammeren A, Alibert G & Souvré A (1999) In vivo labelling of sunflower embryonic tissues by fluorescently labelled phenylalkylamine. Protoplasma 210: 52-58

Chapter 24

Development of photoautotrophy in *Coffea* somatic embryos enables mass production of clonal transplants

F. Afreen, S.M.A. Zobayed[1] & T. Kozai
Department of Bioproduction Science, Chiba University, Matsudo Chiba 271-8510, Japan.
[1]Present address: Department of Plant Agriculture, University of Guelph, Guelph Ontario N1G 2W1, Canada.
Corresponding author: T. Kozai, Department of Bioproduction Science, Chiba University, Matsudo, Chiba 271-8510, Japan. E-mail: kozai@faculty.h.chiba-u.jp

Abstract: Somatic embryogenesis offers the promise of a cost-effective, large-scale propagation method and is considered as a unique alternative technique to overcome some of the limitations of conventional clonal propagation methods. Production of somatic embryos from cell, tissue and organ cultures may occur directly which involves the formation of an asexual embryo from a single cell or a group of cells on a part of the explant tissue without an intervening callus phase. In this study, the photosynthetic ability of different stage coffee (*Coffea arabusta*) somatic embryos and the development of photoautotrophy are reported. Results revealed that cotyledonary and converted somatic embryos have the ability to photosynthesise and can be grown under photoautotrophic conditions (with no supply of sugar from the culture medium). The development of photosynthetic ability can be accelerated by placing the somatic embryos in a photosynthetic photon flux of 100 µmol $m^{-2}s^{-1}$ for at least 14 days. Cotyledonary stage somatic embryos were cultured under photoautotrophic conditions in three different growing systems to develop an optimized protocol for a large-scale embryo-to-plantlet conversion and propagation system. Our results demonstrated that the use of a newly developed temporary root zone immersion bioreactor is effective for the embryo-to-plantlet conversion and enhanced growth under photoautotrophic conditions.

Key words: bioreactor, CO_2 enrichment, embryo-to-plantlet conversion, photoautotrophy, photosynthesis, somatic embryo, stomata

Abbreviations: F_m: maximal fluorescence; F_o: minimal fluorescence; F_s: fluorescence value at steady state condition; PPF – photosynthetic photon flux; PS II – photosystem II

A.K. Hvoslef-Eide and W. Preil (eds.), Liquid Culture Systems for in vitro Plant Propagation, 323–335.

1. Introduction

Somatic embryogenesis is defined as a process in which a bipolar structure, resembling a zygotic embryo, develops from a non-zygotic cell without vascular connection with the original tissue (von Arnold et al., 2002). Since the first observation of somatic embryo formation in *Daucus carota* cell suspensions by Steward et al. (1958) and Reinert (1958), the potential for somatic embryogenesis has been shown in a wide range of plant species. Production of somatic embryos from cell, tissue and organ cultures may occur directly which involves the formation of an asexual embryo from a single cell or a group of cells on a part of the explant tissue without an intervening callus phase. Such occurrences are notable in many species -for instance in coffee (Figure1) where globular, heart-shape, torpedo shape, precotyledonary, cotyledonary and converted somatic embryos developed from leaf discs after 14 weeks of culture (Afreen et al., 2002a). The most promising application of somatic embryo is in the field of genetic engineering where, using somatic embryos, specific and directed changes are introduced into elite individuals. As an embryo originates from a single cell or a group of cells, plants derived from somatic embryos tend to be genetically alike (Yasuda et al., 1986).

Somatic embryogenesis, which offers the promise of a cost effective, large-scale mass propagation method, is therefore considered as a unique alternative technique to overcome some of the limitations of conventional clonal propagation methods. In general, the production cost of plant propagation *via* somatic embryogenesis is potentially lower than that of microcuttings especially when bioreactors and automation procedures are introduced into the production process. One of the challenges preventing the wider commercial application of somatic embryogenesis to mass production of clonal transplants is the low percentages of embryos that convert and develop into plantlets. The quality of a somatic embryo, when used for the commercial mass production of clonal transplants, is determined by its maturation and conversion ability. In the conventional system using sugar-containing medium in the airtight vessel for embryo-to-plantlet conversion and growth, a number of steps are involved (Gupta et al., 1993): embryo maturation, embryo selection and transfer on the conversion medium, converted and rooted plantlet selection and transfer to soil and acclimatization *ex vitro*. Generally, in each of the above steps, cotyledonary, late cotyledonary or converted somatic embryos are selected individually by hand often under a stereomicroscope. These procedures are still time consuming and involve high labour costs, and more importantly the conversion percentage is low. One method to increase the percent conversion and the vigour of seedlings from somatic embryos is to provide a synthetic endosperm as a coating to the somatic embryos (Redenbaugh and Walker, 1990; Redenbaugh et al., 1988). This approach has achieved little success, perhaps because of the poor uptake of the added nutrients by the embryo axis, leaching of the nutrients during conversion, or toxicity of the coating.

Coffee plays a major role in the economy of many African, American and Asian countries. *Coffea arabusta* has been clonally propagated to obtain genetically uniform transplants using microcuttings. However, the growth of microcuttings *in vitro* is slow (Dublin, 1980) therefore somatic embryogenesis is considered to be an effective method for the mass clonal multiplication of *C. arabusta* (Dublin et al., 1991).

This review describes the recent research on photoautotrophic culture of coffee somatic embryos by Afreen et al. (2002a, b) to overcome the problems in conventional propagation system of somatic embryos. We also discussed the photosynthetic ability of different stage coffee somatic embryos, the development of photoautotrophy in somatic embryos, optimization of the photoautotrophic growth and advantages of photoautotrophic culture of somatic embryos for mass production of clonal transplants.

2. Photosynthetic ability in coffee somatic embryos

Naturally, somatic embryos developed in sucrose-containing medium are heterotrophic or mixotrophic and use sugar in the medium as carbon and chemical energy source for their dry mass accumulation. Afreen et al. (2002a) investigated some of the physiological variables in relation to the photosynthetic ability of coffee somatic embryos.

1. The CO_2 concentration in the culture headspace during the photoperiod, was normally less than ambient (370 μmol mol^{-1}) in converted embryos and little above ambient in cotyledonary stage embryos. This indicates that these embryos have photosynthetic ability and that the CO_2 uptake rate, or carbon assimilation rate, of converted embryos was positive (Figure 2b). In the case of torpedo and precotyledonary stage embryos, the occurrence of comparatively higher CO_2 concentrations than the ambient air indicated that plant respiratory activity masked any photosynthetic assimilation. This was probably because of there being few chlorophyll-containing tissues and a significant amount of chlorophyll-free tissues without stomatal development.
2. The development of stomata is essential for the physiological processes of plants because stomata act as portals for entry of CO_2 into the leaf for photosynthesis (Willmer, 1983). Anatomical studies revealed that, generally, stomata did not form in torpedo (Figure 1a, b) and

precotyledonary-stage embryos (Figure 1c), whereas well-developed stomata were seen in the cotyledonary (Figure 1d) and converted embryos (Figure 1e). Stomatal density was highest in the converted embryos (Figure 2c). The absence of stomata in these stages was reflected in the limited ability of these embryos to photosynthesize.

3. Chlorophyll which is essential for plant photosynthesis, uses light energy to synthesise chemical energy. Low chlorophyll concentrations were recorded in torpedo and precotyledonary-stage embryos. In the cotyledonary stage embryos, chlorophyll concentrations were higher compared with those in the torpedo or precotyledonary stage embryos; chlorophyll a and b were highest in the converted embryos (Figure 2d, e)
4. The potential activity of PSII (ΦpMAX) as estimated in the dark was low in the torpedo stage but increased in precotyledonary stage embryos (ΦpMAX = 0.69) followed by cotyledonary stage embryos (ΦpMAX = 0.84) and was highest in converted embryos (ΦpMAX = 0.88) (Figure 2f). As expected from their low PSII activity, torpedo stage embryos exhibited low electron-transport activity (Φp = 0.1). An increase in the quantum yield for electron-transport was observed in the cotyledonary and converted embryos (Figure 2g).

The results revealed that the cotyledonary and converted embryos showed photosynthetic ability, stomata did not fully develop in the precotyledonary stage embryos and were absent in the torpedo stage. Low chlorophyll concentrations were noted in the torpedo and precotyledonary stage embryos, but the concentration increased in the cotyledonary and converted embryos. Therefore, we suggest that the cotyledonary stage is the earliest stage, which can be cultured photoautotrophically to develop into plantlets.

3. Technique to enhance photosynthetic ability of coffee somatic embryos

To develop an optimised protocol for the regeneration of somatic embryos from coffee leaf discs, cultures have to be incubated in darkness or at a low light intensity (PPF: 20-30 μmol m^{-2} s^{-1}) (Afreen et al., 2002a). This is because endogenous plant growth regulator concentrations and/or the sensitivity of cells to growth regulators are known to differ in dark or light conditions and thus affect embryogenesis. Hutchinson et al. (2000) reported that a continuous light treatment significantly reduced the amount of the endogenous plant growth substances in geranium (*Pelargonium x hortorum*) tissues with concomitant inhibition of somatic embryogenesis. Croke and Cassells (1997) also observed promotive effect of dark incubation resulting

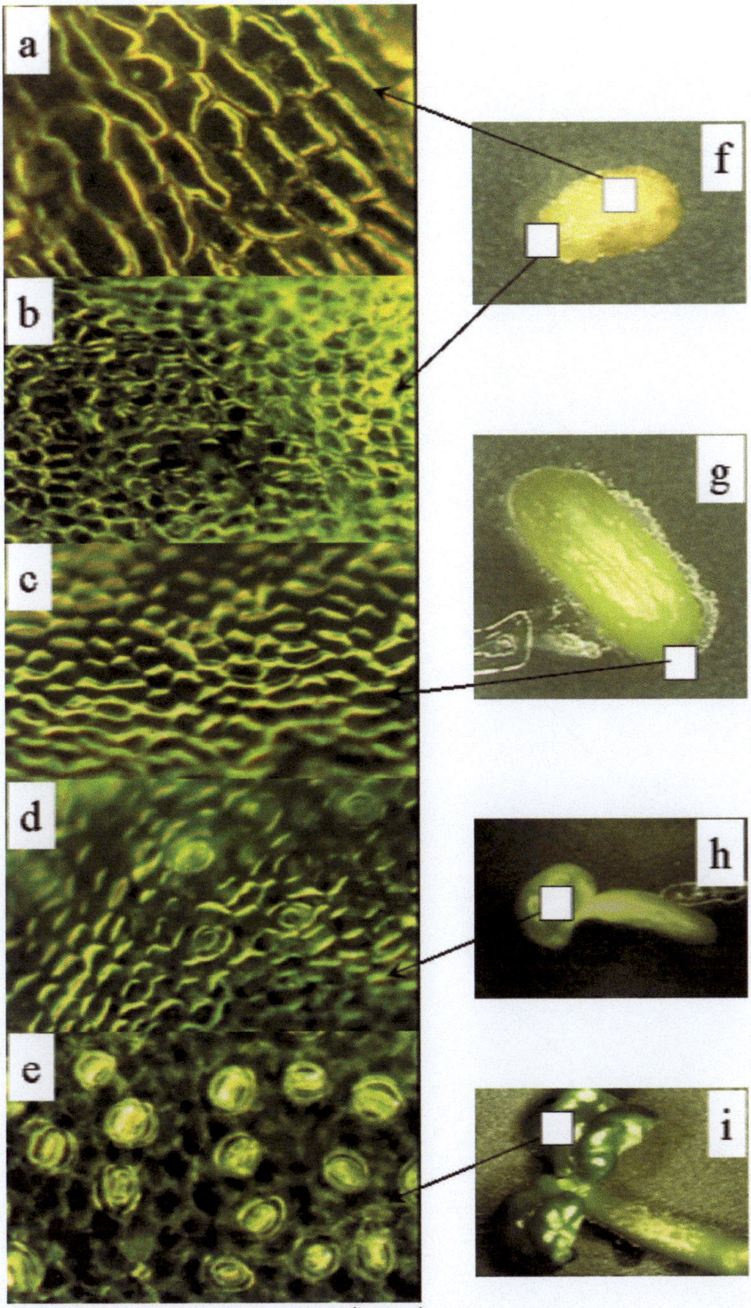

Figure 1: Coffea arabusta. a-e), Microscopic study showing the development of stomata in a and b) torpedo, c) precotyledonary, d) cotyledonary and e) converted embryos.
Different stage coffee somatic embryos: f) torpedo (x16), g) precotyledonary (x10), h) cotyledonary (x8) and i) converted embryos (x11). (Afreen et al., 2002a).

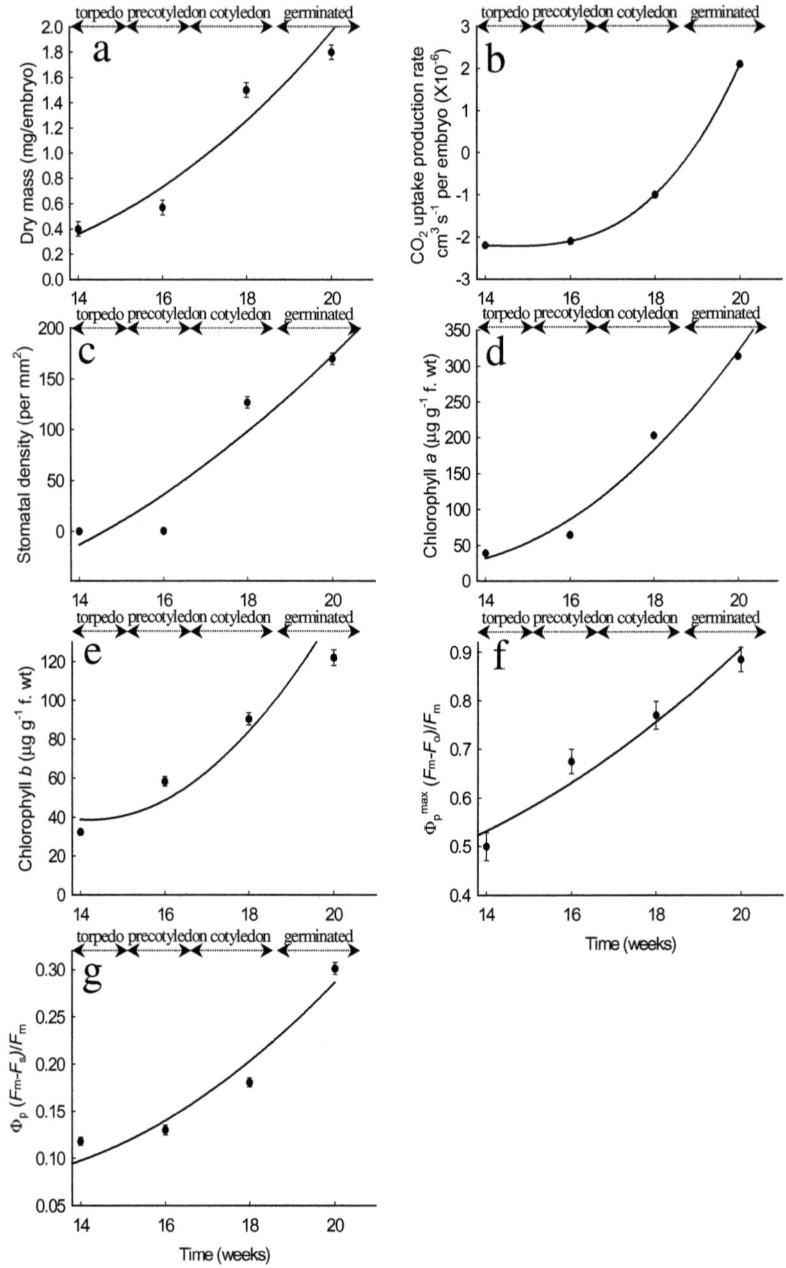

Figure 2: Growth (dry mass) and physiology of different stage coffee somatic embryos (data from Afreen et al., 2002a).
ΦpMAX is the maximal quantum yield (in dark-adapted somatic embryos) and (Φp) is the actual quantum yield of PSII photochemistry.

in doubling of the embryo frequency in zonal geraniums. The positive influence of a dark treatment to induce bud break, and thus increased *in vitro* multiplication, was reported for evergreen azaleas (Hsia and Korban, 1998).

However, growing somatic embryos in dark or at low light intensity significantly inhibits or delays the development of photosynthetic pigments. In coffee, we observed that placing the somatic embryos under a relatively high light intensity (PPF: 100 μmol m^{-2} s^{-1}) for 14 days resulted in the development of photosynthetic pigments, functional stomata and subsequent enhancement of photosynthetic ability. Placing the embryos under high light intensity increased the chlorophyll concentrations significantly for all stages of embryos (Afreen et al., 2002a). In general, light-pre-treated embryos exhibited higher stomatal density than those without light pre-treatment. Well-developed stomata were observed in the converted embryos irrespective of light pre-treatment. Both the pre-treated and non-pre-treated converted embryos photosynthesized, but in the cotyledonary stage embryos, CO_2 assimilation was recorded only in the pre-treated ones (Afreen et al., 2002a). From these findings we concluded that light pre-treatment of the somatic embryos at high PPF (100 μmol m^{-2} s^{-1}) increased the photosynthetic ability in almost all embryonic stages, but the cotyledonary stage embryo is the earliest stage to grow photoautotrophically.

4. Photoautotrophic culture of different stage coffee somatic embryos

Pre-treated somatic embryos (torpedo, precotyledonary, cotyledonary and converted) were cultured in plastic petridishes containing MS medium excluding sugar and under CO_2 enrichment. Overall greater performance in terms of growth occurred in the embryos grown under photomixotrophic conditions compared with photoautotrophic (Afreen et al., 2002a). After 60 days of culture, under photoautotrophic conditions, torpedo and precotyledonary stage embryos lost at least 25 and 20% respectively, of their initial dry mass. The most likely reason for this loss could be that the low photosynthetic ability of the plant materials coupled with continuous respiration, probably led the plantlets to depend completely on their own reserve food material. On the contrary, under photomixotrophic conditions, the dry mass of each of the torpedo and precotyledonary stage embryos increased by upto 190 and 200% respectively, of their initial dry mass. Under photoautotrophic conditions in the later stages i.e. cotyledonary and converted embryos, the dry mass of each of the embryos was increased by upto 10% and 50%, respectively, of their initial dry mass (Afreen et al., 2002a). The probable reason for the dry mass increment in the later stages under photoautotrophic treatment could be that at the later stages, embryos are capable of photosynthesizing more than those at the early stages. These results confirm the above findings and are in agreement with our previous observation where the cotyledonary and converted embryos showed stomatal development and scavenging of CO_2.

5. Optimization of the photoautotrophic production of cotyledonary stage coffee somatic embryos

5.1 Supporting medium and environmental conditions

For optimizing the photoautotrophic culture of cotyledonary stage somatic embryos, three different types of supporting media were investigated: a) agar (8 g l^{-1}) b) vermiculite and c) Florialite (a mixture of vermiculite and cellulose fibre; as described by Afreen et al. (2000). Results revealed that the use of Florialite and/or vermiculite can improve root and shoot growth during the development of plantlet from cotyledonary stage coffee somatic embryos under photoautotrophic conditions.

For optimization of the environmental condition the following treatments were investigated (Afreen et al., 2002a):

a) PPF was 50 µmol m^{-2} s^{-1} and ambient CO_2 concentration was 400 µmol mol^{-1}
b) PPF was 100 µmol m^{-2} s^{-1} and CO_2 concentration was 400 µmol mol^{-1}
c) PPF was 150 µmol m^{-2} s^{-1} and CO_2 concentration was 400 µmol mol^{-1}
d) PPF was 50 µmol m^{-2} s^{-1} and CO_2 concentration was 1100 µmol mol^{-1}
e) PPF was 100 µmol m^{-2} s^{-1} and CO_2 concentration was 1100 µmol mol^{-1}
f) PPF was 150 µmol m^{-2} s^{-1} and CO_2 concentration was 1100 µmol mol^{-1}.

The results suggested that to develop plantlets from the cotyledonary stage somatic embryos under photoautotrophic conditions, high PPF (100-150 µmol m^{-2} s^{-1}) and increased CO_2 concentration (1100 µmol mol^{-1}) were necessary (Afreen et al., 2002a).

5.2 Different growing systems

As noted above, in the multi-stage somatic embryogenesis of coffee, the cotyledonary stage is the earliest stage embryo, capable of photosynthesizing. However, the extent of heterotrophy, photomixotrophy or photoautotrophy is dependent not only on photosynthetic ability but also on medium composition, volume of culture vessels, their cross-sectional areas, and mode and amount of aeration of the vessel (Solarova et al., 1995). Therefore, we cultured cotyledonary stage coffee somatic embryos under photoautotrophic conditions in different growing systems with the aim of developing an optimized protocol for large-scale embryo-to-plantlet conversion and growth system.

Pre-treated cotyledonary stage embryos were cultured under photoautotrophic conditions (in sugar-free medium with CO_2 enrichment in the culture headspace and high PPF) in three different types of culture systems to optimize the plantlet conversion and growth: i) Magenta vessel ii) RITA-bioreactor (a temporary immersion system) modified for photoautotrophic micropropagation, and iii) a newly developed bioreactor with a temporary root zone immersion system (TRI-bioreactor). The design of the TRI-bioreactor has been described by Afreen et al. (2002b). The planting density for all the treatments was 2.4×10^3 plantlets per m^2 vessel surface area.

After 60 days of culture, results revealed that, in the TRI-bioreactor, almost 84% of the embryos produced plantlets, whereas in Magenta vessel and in modified RITA-bioreactor the conversion percentages were 53% and 20% respectively (Afreen et al., 2002b). Embryos cultured in the TRI-bioreactor produced more vigorous shoots and normal roots than those grown in modified RITA-bioreactor and in Magenta vessel (Figure 3). The TRI-bioreactor grown plantlets exhibited a greater number of leaves and larger leaf area per plantlet than those noted in the modified RITA-bioreactor and Magenta vessel. Maximum leaf, stem and root dry mass were recorded in the plantlets grown in TRI-bioreactor (Afreen et al., 2002b). In general, most of the growth variables of the plantlets grown in Magenta vessel were marginally different from those grown in the modified RITA-bioreactor. The most remarkable difference observed among the treatments was in the percentage of rooting. In the TRI-bioreactor, 90% of the plantlets developed roots; some roots produced lateral roots (Afreen et al., 2002b). In the modified RITA-bioreactor, the roots, which developed in few plantlets, remained very small.

In general, the chlorophyll concentration based on the fresh mass of the leaves was highest in the TRI-bioreactor grown plantlets (Afreen et al., 2002b). In case of the Magenta vessel, both the chlorophyll *a* and *b* concentrations of the leaves exhibited an intermediate value between those of the TRI- and modified RITA-bioreactors. Among the treatments, the highest net photosynthetic rate was observed in plantlets grown in TRI-bioreactor (Afreen et al., 2002b). The results clearly pointed out that, in this treatment, the forced ventilation system provided the best condition throughout the experiment for the assimilation of CO_2 and also the high rate of air exchange was promotive of the development of functional stomata.

Figure 3: a) Cotyledonary stage embryos used as an experimental material; b-d) different growing systems used to optimise the growth and conversion of cotyledonary stage embryos under photoautotrophic conditions: b) Magenta vessel (x0.8), c) modified RITA-bioreactor (x0.22) and d) TRI-bioreactor (x0.3) (after Afreen et al. 2002b).

The results showed that for plantlet conversion from cotyledonary stage embryos under photoautotrophic conditions, Magenta vessel and modified RITA-bioreactor resulted in the lowest growth regime. Our results also highlighted that for the embryo-to-plantlet conversion under photoautotrophic conditions the use of the modified RITA-bioreactor was less effective at promoting shoot and root growth compared with the newly developed TRI-bioreactor system. This is most likely to be because in the modified RITA-bioreactor after every immersion of the plant material with nutrient solution, the entire plant became wet and, because the relative humidity inside the vessel is normally high (95-99%), the plant material either is never completely dried out or takes a long period to dry out. Thus, this thin layer of water surrounding the plant material acts as a liquid boundary layer, which impedes the exchange of gases between the plant and the surrounding environment and possibly prevents the CO_2 fixation in the chlorophyll-containing zones - clearly a key factor for the photoautotrophic growth of embryos. In case of conventional photomixotrophic systems, the media contain sugar and therefore the lack of air exchanges may not be as serious a consequence as it is for the plantlets which completely depend on CO_2 in the atmosphere for their photoautotrophic growth.

Again, it is emphasized that the RITA-bioreactor system has not been developed for culturing plantlets under photoautotrophic conditions. Also, in this study, the RITA-bioreactor was modified by attaching three gas permeable filter membranes on the lid, as was done for Magenta vessels. Thus, a significantly different result might be expected if the original RITA-bioreactor with sugar-containing nutrient solution were to be used.

6. Advantages of photoautotrophic culture of somatic embryos

Photoautotrophic micropropagation is the propagation of plantlets using relatively small chlorophyllous explants in a sugar-free nutrient medium under pathogen-free conditions where the plantlets can photosynthesise and produce their own carbohydrate for growth (Kozai et al., 1988). The major benefits of photoautotrophic micropropagation of somatic embryos are that:

1. Photoautotrophic growth can improve the quality of somatic embryos (it can reduce the hyperhydricity and the development of abnormal embryos). It can possibly shorten and simplify the conversion and plantlet development procedures and improve the percentage of emblings that are able to grow after *ex vitro* acclimatization (Afreen et al., 2002b).
2. Relatively uniform growth in size and shape and uniform development are expected.
3. Application of growth regulators and other organic substances such as amino acids and vitamins to the culture medium can be eliminated or minimised. Use of some growth regulators is sometimes crucial in the conventional culture system to convert somatic embryos.
4. There is minimization of the risk of loss of cultures due to microbial contamination, and reduced production costs (Kozai et al., 1999; Zobayed et al., 2000). Problems related to synthetic seeds containing sugar and other organic nutrients can be overcome by using photosynthetically active somatic embryos.
5. Whilst considerable effort has been devoted to full automation of somatic embryo development and micropropagation (Cervelli and Senaratna, 1995), the ability of somatic embryos to grow photoautotrophically will be beneficial for automation, robotization and computerization.
6. Asepsis in the culture vessel for embryo-to-plantlet conversion may not be required if pathogen-free status in the culture vessel is certified.
7. Photoautotrophic culture of somatic embryos will contribute to reduce production costs specifically by reducing the labour input and improving

the plant quality. In commercial micropropagation, labour usually accounts for about 70 % of the total *in vitro* and *ex vitro* costs (Aitken-Christie et al., 1991).

7. Conclusion

For the first time we have shown the photosynthetic ability of somatic embryos of *Coffea arabusta* and successfully grew the somatic embryos photoautotrophically. The important consequences of the study are that the cotyledonary and converted embryos have photosynthetic ability and that the light pre-treatment (PPF: 100 μmol m^{-2} s^{-1}) speeds up the process. Therefore, it is concluded that the cotyledonary stage is the earliest stage embryo, which can be cultured photoautotrophically (in the absence of sugar in the culture medium) to develop coffee plantlets. We hope that the photoautotrophic culture system described here might also provide the basis of a useful model for the *in vitro* propagation by somatic embryogenesis of other important plant species.

References

Afreen F, Zobayed SMA, Kubota C, Kozai T & Hasegawa O (2000) A combination of vermiculite and paper pulp supporting material for the photoautotrophic micropropagation of sweet potato. Plant Sci. 157: 225-231

Afreen F, Zobayed SMA & Kozai T (2002a) Photoautotrophic culture of *Coffea arabusta* somatic embryos: Photosynthetic ability and growth of different stage embryos. Ann. of Bot. 90: 11-19

Afreen F, Zobayed SMA & Kozai T (2002b) Photoautotrophic culture of *Coffea arabusta* somatic embryos: Development of a bioreactor for the large-scale plantlet conversion from cotyledonary embryos Ann. of Bot. 90: 21-29

Aitken-Christie J, Kozai T & Takayama S (1991) Automation in plant tissue culture – general introduction and overview In*:* Aitken-Christie J, Kozai T & Smith L, (eds) Automation and Environmental Control in Plant Tissue Culture. (pp. 1-18). Kluwer Academic Publishers, Dordrecht

Cervelli R & Senaratna T (1995) Economic aspects of somatic embryogenesis. In: Aitken-Christie J, Kozai T, Smith ML (eds), Automation and Environmental Control in Plant Tissue Culture (pp. 29-64). Kluwer Academic Publishers, Dordrecht

Croke JT & Cassells AC (1997) Dark induction and genetic stability of somatic embryos of zonal geraniums (*Pelargonium x hortorum* Baily). J. Appl. Bot. 71: 119-124

Dublin P (1980) Multiplication vegetative *in vitro* de l ' *Arabusta.* Café – Cacao – The. Vol. WWIV 4: 281-290

Dublin P, Enjalric F, Lardet L, Carron MP, Trolinder N & Pannetier C (1991) Estate crops. In: Debergh PC, Zimmerman RH (eds) Micropropagation - Technology and Application (pp: 337-361). Kluwer Academic Publishers, Dordrecht

Gupta PK, Timmis R & Carlson WC (1993) Somatic embryogenesis: A possible tool for large-scale propagation of forestry species. In: Soh WY, Liu JR & Komamine A (eds) Advances in Development Biology and Biotechnology of Higher Plants (pp. 18-37).The Korean Society of P. T. C., Korea

Hsia C & Korban SS (1998) Effect of growth regulators, dark treatment and light intensity on shoot organogenesis from leaf tissues of evergreen azalea. J. Hort. Sci. Biotech. 73: 53-60.

Hutchinson MJ, Sanaratna T, Sahi SV & Saxena PK (2000) Light mediates endogenous plant growth substances in thidiazuron-induced somatic embryogenesis in *Geranium* hypocotyl cultures. J. Plant Biochem. Biotec. 9: 1-6

Kozai T, Koyama Y & Watanabe I (1988) Multiplication of potato plantlets *in vivo* with sugar free medium under high photosynthetic photon flux. Acta Hort. 230: 121-127

Kozai T, Kubota C, Zobayed SMA, Nguyen QT, Afreen F & Heo J (1999) Developing a mass-propagation system of woody plants. In: Watanabe K, Komamine A. (eds) Challenge of Plant and Agriculture Sciences to the Crisis of Biosphere on the Earth in the 21st Century. Chapter 28 (pp. 293-306). Landes Bioscience, Texas, USA

Reinert J (1958) Untersuchungen über die Morphogenese an Gewebenkulturen. Ber. Dtsch. Bot. Ges. 71:15

Redenbaugh K & Walker K (1990) Role of artificial seeds in alfalfa breeding. In: Bhojwani SS (ed) *Plant Tissue Culture.* Application and Limitations (pp. 102-135). Elsevier Publishing, New York

Redenbaugh K, Fuji JA & Slade D (1988) Encapsulated plant embryos. In: Biotechnology in Agriculture (pp: 226-229). Great Britain Agriculture Food Research Council

Solarova J, Souckova D, Ullmann J & Pospisilova J (1995) *In vitro* culture: environmental conditions and plantlets growth as affected by vessel type and stopper material. Zahradnictvi 21: 61-72

Steward FC, Mapes MO & Hears K (1958) Growth and organized suspended cells. Am. J. Bot. 45: 705-708

Von Arnold S, Sabala I, Bozhkov P, Dyachok J & Filonova L (2002) Developmental pathways of somatic embryogenesis. Plant Cell, Tiss. Org. Cult. 69: 233-249

Willmer CM (1983) Stomatal responses to environmental factors. In: Willmar, CM (ed) *Stomata.* (pp. 64-90). Longman Inc., New York

Yasuda T, Fujii Y & Yamaguchi T (1986) Embryogenic callus induction from *Coffea arabica* leaf explants by benzyladenine. Plant Cell Physiol. 26: 595-597

Zobayed SMA, Afreen F, Kubota C & Kozai T (2000) Mass propagation of *Eucalyptus in* a scaled-up vessel under *in vitro* photoautotrophic condition. Ann. Bot. 85: 587-592

Chapter 25

Potential of flow cytometry for monitoring genetic stability of banana embryogenic cell suspension cultures

Nicolas Roux [1], Hannelore Strosse[2], Arsenio Toloza[1], Bart Panis[2] & Jaroslav Doležel[3]

[1]*Plant Breeding Unit, FAO/IAEA Agriculture and Biotechnology Laboratory, International Atomic Energy Agency Laboratories, A-2444 Seibersdorf, Austria. Now working for the International Network for the Improvement of banana and Plantain (INIBAP), Parc Scientifique Agropolis 2, 34397 Montpellier Cedex 5, France. E-mail: N.Roux@cgiar.org.*

[2]*Laboratory of Tropical Crop Improvement, Katholieke Universiteit Leuven, Kasteelpark Arenberg 13, B-3001 Heverlee, Belgium.*

[3]*Laboratory of Molecular Cytogenetics and Cytometry, Institute of Experimental Botany, Sokolovská 6, CZ-77200 Olomouc, Czech Republic.*

Abstract: Cell suspensions are the material of choice for rapid multiplication and for genetic engineering strategies such as *in vitro* mutagenesis and genetic transformation. Effective use of cell suspension cultures relies on the knowledge of several key parameters, which include genetic stability, kinetics of the cell cycle and a mode of plant regeneration. Here we report on the use of DNA flow cytometry for quality monitoring of banana cell suspension cultures. The method facilitates detection of ploidy changes and the occurrence of aneuploidy, which result in somaclonal variation of cell-suspension-derived plants. Flow cytometry could also be used to analyse the cell cycle kinetics by calculating the ratio of cells in the G2 and G1 phase of the cell cycle. This is important to determine the most appropriate moment for mutagenic treatment or for genetic transformation but also as an indicator on the proportion of cycling cells. In addition, the unicellular origin of somatic embryos was verified by treating embryogenic cell suspensions with colchicine and by determining the ploidy of regenerated plants by flow cytometric analysis. None of the plants regenerated from colchicine-treated embryogenic cell suspensions were mixoploid (chimeric). The application of flow cytometry will be discussed in relation to (a) the monitoring of genetic instability in DNA content of cell suspensions (b) the analysis of cell cycle and (c) the origin of somatic embryos of bananas and plantains.

Key words: cell cycle, chimerism, DNA content, *Musa acuminata*, somatic embryogenesis

Abbreviations: ECS - embryogenic cell suspensions; ITC - INIBAP International Transit Center-International Network for the Improvement of Banana and Plantain

A.K. Hvoslef-Eide and W. Preil (eds.), Liquid Culture Systems for in vitro Plant Propagation, 337–344.

1. Introduction

The use of shoot-tip culture in *Musa* has allowed great progress in mass propagation (micropropagation), conservation (medium-term conservation and cryopreservation), elimination of virus diseases and exchange of germplasm (Van den Houwe et al., 1995; Panis et al., 1996). Unfortunately, the use of banana shoot tips as target tissues for genetic engineering strategies such as *in vitro* mutagenesis and genetic transformation leads to chimeric plants (Roux et al., 2001). Somatic embryogenesis is usually a preferred mode of regeneration over organogenesis because of the presumed single cell origin of the regenerants (Thorpe, 1988; Peschke and Phillips, 1992). The occurrence of off-types among regenerants has delayed the wide-spread industry acceptance of micropropagated bananas (Smith and Hamill, 1993). Before adopting somatic embryogenesis as a new method to support genetic improvement and mass propagation of bananas and plantains, further studies need to be undertaken to better understand the phenomenon of somaclonal variation. The aim of this study was to verify the ploidy level stability of embryogenic cell suspensions prior to the formation of embryos or plants.

2. Material and methods

Embryogenic cell suspensions (ECS) of four triploid (2n=3x=33) *Musa* cultivars were kindly provided by the Laboratory of Tropical Crop Improvement at the Katholieke University of Leuven, Belgium. The suspensions were maintained in 100 ml Erlenmeyer flasks containing 15 ml of liquid maintenance media (Dhed'a et al., 1991). Erlenmeyer flasks were placed on an orbital shaker at 70 rpm. The medium was renewed every two weeks.

For monitoring the genetic stability of ECS, flow cytometric analysis was performed according to Dolezel et al. (1997). The flow cytometric assay involved the use of nuclei isolated from chicken red blood cells (CRBC), which served as internal reference standard. The gain of the instrument was adjusted so that the G_0/G_1 peak of CRBC nuclei was positioned approximately at channel 100. The relative DNA content of ECS was expressed as a ratio of DNA content of CRBC and *Musa* (DNA index). By flow cytometric analysis, it was also possible to measure the proportion of cells in the G_1 and G_2 phase of the cell cycle since the DNA content during G_2 is doubled compared to G_1 phase. Through a combination of mixoploidy induction using colchicine and flow cytometry detection, Roux et al. (2001) could monitor the effectiveness of three micropropagation techniques to dissociate chimeras. This method was also applied on ECS to verify the unicellular origin of somatic embryos. After colchicine treatment, the ploidy of two cell lines were determined using flow cytometry before transfer to the regeneration medium and after complete regeneration into plantlets.

3. Results

Abnormalities in DNA content could be detected at an early stage during *in vitro* culture. Two non-regenerable banana suspensions, characterized by their extremely fine structure and white colour were growing twice as fast as compared to regenerable embryogenic cell suspensions. Flow cytometric analysis revealed that these unusual suspensions were in fact polyploid. One was hexaploid (6x) and the other decaploid (10x) compared to the normal triploid (3x) cultivars from which the ECS were originating (Figure 1). Other aneuploid banana ECS with 2n=28 could be detected at an early stage (Figure 2). Cells from four *Musa* accessions were analyzed every two or three days during 30 days of culture. Peaks on graphs displaying the ratio of frequency of cells in G_2 and G_1 indicated waves of mitotic activity, i.e., a certain degree of cell cycle synchrony. It seems that when the cells were transferred to fresh medium, the cell cycle is stimulated in a relatively synchronous manner. The highest proportion of cells in G_2 phase was observed 8 days after subculture in three of the four *Musa* ECS analysed (Figure 3).

Nuclear DNA content distribution in ECS cultures of two cell lines (WIL-124C and WIL-124T) was determined using flow cytometry before transfer to the regeneration medium R1 (Figure 4) and after complete regeneration into plantlets (Table 1). Even though the two cell lines (were initiated from the same accession "Williams", the effect of colchicine on polyploidy induction was different in both cell lines. The response to colchicine was clearer for cell line WIL-124C than for WIL-124T especially when high concentrations are applied (0.1% and 0.2%). The proportion of polyploidized cells increased to about 50% for WIL-124C, while this was only 15% for WIL-124T. Among the regenerated plants, none were mixoploid. This suggest that embryos are of single cell origin (Table 1).

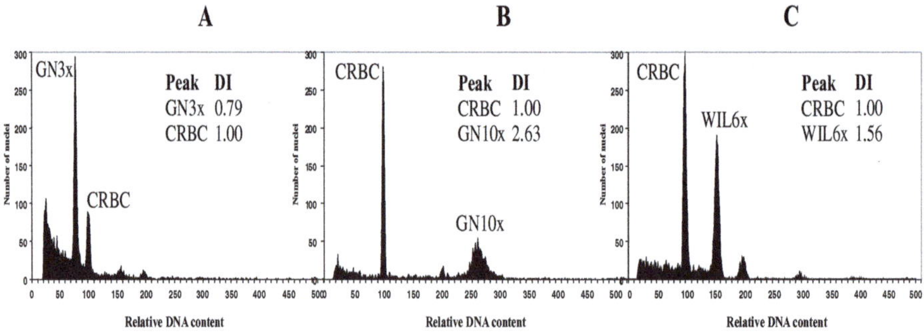

Figure 1: Histograms of relative nuclear DNA content obtained after simultaneous analysis of nuclei isolated from *Musa* embryogenic cell suspensions and chicken red blood cell nuclei (CRBC, used as internal standard). Flow cytometer was adjusted so that CRBC nuclei peak appeared at channel 100. (A) 'Grande Naine', 2n=3x (GN3x); (B) Polyploid 'Grande Naine', 2n=10x (GN10x); (C) Polyploid 'Williams', 2n=6x (WIL6x). Relative DNA content of cell lines was expressed as a DNA index (DI) calculated according to:

$$DI = \frac{\text{Mean of the relative DNA content of the } G_0/G_1 \text{ nuclei of the sample}}{\text{Mean of the relative DNA content of the } G_0/G_1 \text{ of CRBC nuclei}}$$

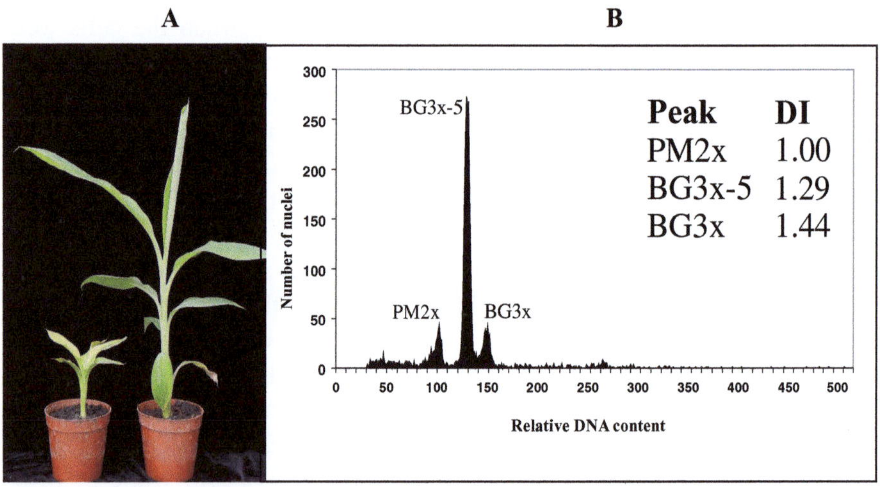

Figure 2: **(A)** An off-type plant (chlorotic and dwarf) regenerated from a 9 year-old cell suspension culture (left) and a true to type plant regenerated from shoot tip culture (right) of "Bluggoe"; **(B)** Histogram of relative nuclear DNA content obtained after simultaneous analysis of nuclei isolated from *Musa* embryogenic cell suspensions (BG3x-5), leaves of a diploid banana plant "Pisang Mas" (PM2x) and a triploid true to type 'Bluggoe' (BG3x), which served as internal standards. Flow cytometer was adjusted so that the peak representing PM2x nuclei was localized at channel 100. DI represents the DNA index.

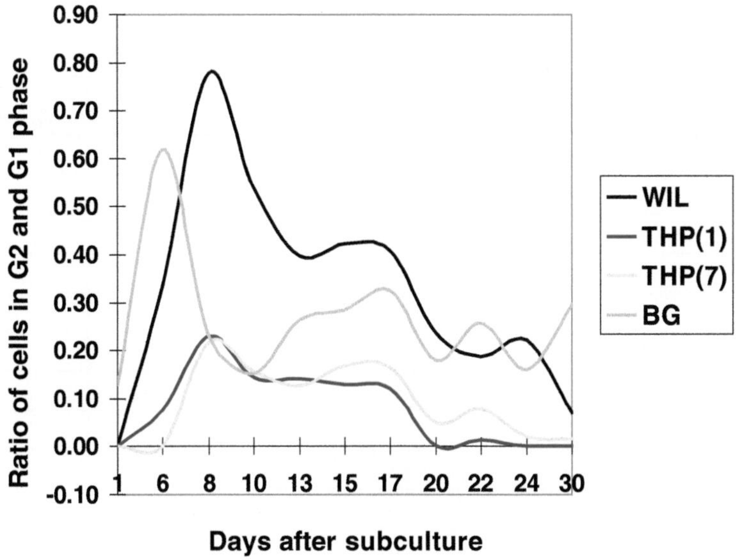

Figure 3: Changes in cell cycle distribution during culture of embryogenic cell suspensions of 4 cell lines of cultivars Williams (WIL), Three Hand Planty (THP1 and THP7) and Bluggoe (BG). The curves were obtained by measuring the ratio of cells in G2 and G1 phase of the cell cycle during 30 days without medium renewal. The G2/G1 ratio was measured for five 100ml Erlenmeyer flasks per cell line. The lines represent the mean.

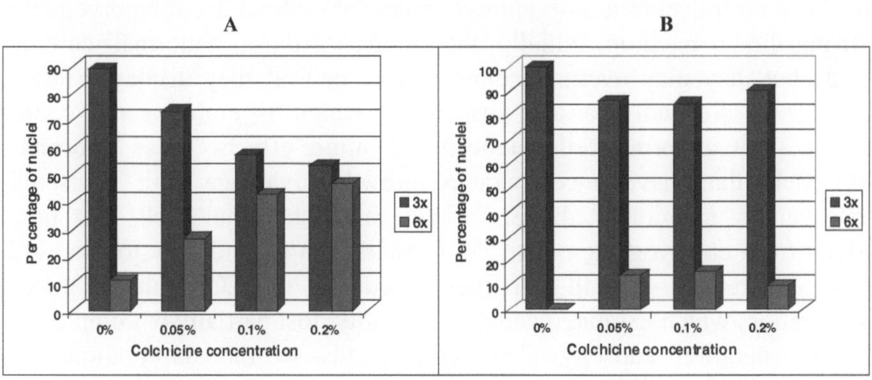

Figure 4: Frequency distribution of "Williams" suspension cells with triploid (3x) and hexaploid (6x) nuclear DNA content observed 15 days after colchicine treatment: (**A**) WIL-124C and (**B**) WIL-124T.

Table 1: Ploidy level distribution of regenerated plants from colchicine-treated embryogenic cell suspensions of two triploid cell lines of"Williams"

Cell Line	Colchicine concentration (%)	Regenerated plants			
		total	3x	6x	3x+6x (Mixoploidy)
WIL-124C	0	5	5	0	0
	0.05	12	12	0	0
	0.1	5	3	2	0
	0.2	0	0	0	0
Sub-total		22	20	2	0
WIL-124T	0	108	108		0
	0.05	63	58	5	0
	0.1	37	36	1	0
	0.2	88	84	4	0
Sub-total		296	286	10	0
TOTAL		318	306	12	0

4. Discussion

In this report we show that flow cytometry is a useful method to monitor genetic stability of ECS cultures. We assume that somaclonal variation from cell suspensions can have two causes. Firstly, the transfer of some non-embryogenic cells (with possible abnormal chromosome numbers) could have been co transferred with embryogenic cell cultures from embryogenic calli into liquid medium. Initially, the amount of these 'abnormal' cells is limited but they may overgrow embryogenic cells if they display a faster growth rate. Secondly, a cell suspension which is embryogenic could produce a few abnormal cells due to tissue culture effects. These 'abormal' cells could than have a comparative growth advantage with the same consequences mentioned above. In this case 'abnormal' cells are thus derived from embryogenic cells. The method is rapid and easy to use, and gives good results with all cell lines tested. All the ECS cultures under investigation, which exhibited genetic instability, lost partially or completely their regeneration capacity. This corresponds to the observations of Kubaláková et al. (1996) who showed that polyploidization of cucumber cultures was accompanied by a gradual loss of regeneration ability. In case of a 9-year-old cell suspension of cultivar "Bluggoe", the regeneration decreased considerably over the years. Since we found that 100% of the cells are aneuploid, we can conclude that during the extended period of subculture, there was a selection towards abnormal cells. Our results showed that even though regeneration decreases, aberrant plants might still be regenerated. Screening for off-types is therefore necessary at an early stage of the somatic embryogenesis process.

Flow cytometry could also be used to analyse the cell cycle kinetics by calculating the ratio of cells in the G_2 and G_1 phase of the cell cycle. The majority of cells are in the G_1 phase during the first 3 days after subculture. This is important to determine the most appropriate moment for mutagenic treatment or for genetic transformation but also as an indicator on the proportion of cycling cells. In addition, the unicellular origin of somatic embryos was verified by treating embryogenic cell suspensions with colchicine and by determining the ploidy of regenerated plants by flow cytometric analysis.

To conclude, our results prove that ploidy levels of *Musa* embryogenic cell suspension cultures can easily be determined using flow cytometry. The results of this study also indicate that *in vitro* cultures may become mixoploid in a relatively short period and that abnormal ploidy levels coincide with poor regeneration abilities. To develop protocols, which avoid variation at least in DNA content, we have now a simple and high-throughput assay at hand that can be applied immediately after tissue culture initiation and that gives representative data on ploidy level of cultured cells.

Acknowledgements

This work was supported by a Joint FAO/IAEA/DGDC (Belgian General Direction for International Cooperation) Coordinated Research Project. The study was undertaken as part of the Global Programme for *Musa* Improvement (PROMUSA).

References

Dhed'a D, Dumortier F, Panis B, Vuylsteke D & De Langhe E (1991) Plant regeneration in cell suspension cultures of the cooking banana cv. 'Bluggoe' (*Musa* spp. ABB group). Fruits 46: 125-135

Doležel J (1997) Cytological evidence of spontaneous and induced mutations and cytogenetics techniques. 15th IAEA/FAO Interregional training course on advances in technologies for induced mutations in crops. IAEA, Vienna

Kubalakova M, Dolezel J & Lebeda A (1996) Ploidy instability of embryogenic cucumber (*Cucumis sativus* L.) callus culture. Biol. Plant. 38: 475-480

Panis B, Totte N, Van Nimmen K, Withers LA & Swennen R (1996) Cryopreservation of banana (*Musa* spp.) meristems cultures after preculture on sucrose. Plant Sciences 21: 95-106

Peschke VM & Phillips RL (1992) Genetic implication of somaclonal variation in plants. Advance Genetics 30: 41-75

Roux N, Doležel J, Swennen R, & Zapata-Arias FJ (2001) Effectiveness of three micropropagation techniques to dissociate cytochimeras in *Musa* spp. Plant Cell, Tiss. Org. Cult. 66: 189-197

Smith M & Hamill S (1993) Early detection of dwarf off-types from micropropagated Cavendish bananas. Australian Journal of Experimental Agriculture 33: 639-44

Thorpe TA (1988) *In vitro* somatic embryogenesis. In: Atlas of Science: Animal and Plant Science, ISI, Philadelphia: 81-88

Van den Houwe I, De Smet K, Tezenas de Montcel H & Swennen R (1995) Variability in storage potential of banana shoot cultures under medium term storage conditions. Plant Cell, Tiss. Org. Cult. 42: 269-274

Chapter 26

Somatic embryogenesis of *Gentiana* genus IV.: Characterisation of *Gentiana cruciata* and *Gentiana tibetica* embryogenic cell suspensions

Anna Mikuła[1]*, Jan J. Rybczyński[1], Jerzy Skierski[2], Monika J. Latkowska[3] & Agnieszka Fiuk[1.]

[1]*Botanical Garden - Center for Biological Diversity Conservation, Polish Academy of Sciences, Prawdziwka 2, 02-973 Warsaw, Poland.*

[2]*Drug Institute, Flow Cytometry Laboratory, Chełmska 30/34, 00-725 Warsaw, Poland.*

[3]*Department of Ornamental Plants, Warsaw Agricultural University, Nowoursynowska 166, 02-787 Warsaw, Poland.*

* *Corresponding author: Phone: + 48 22 648 38 56;* E-mail: *obpan@ikp.atm.com.pl*

Abstract: Experiments to characterise long-term embryogenic suspension cultures of *Gentiana cruciata* (L.) and *G. tibetica* (King) are reported. Cell suspensions of both species differed in the percentage of five selected fractions of cell aggregates, as well as in fresh and dry mass during three years of culture. In *G. tibetica* the ratio of cells in phase G_2 to G_1 was higher than in *G. cruciata*. The response of suspension cultures to GA_3 (at 0, 1.49 or 2.89 μmol), kinetin (at 0, 2.32, 4.64 or 9.28 μmol) and adenine sulphate (at 0 or 434 μmol) was studied. The increase of kinetin concentration stimulated embryo production in suspensions of *G. tibetica*. Somatic embryo production in *G. tibetica* was significantly higher than in *G. cruciata*. In *G. tibetica*, the aggregate fraction >450 μm was at least four times more productive than the same fraction in *G. cruciata* suspensions.

Key words: cell aggregate, flow cytometry, gentian, long-term suspension culture, plant growth regulators, somatic embryo

Abbreviations: AS - adenine sulphate; BAP – benzylaminopurine; Dic – dicamba; DW – dry weight; ECM - embryo conversion medium; FW – fresh weight; GA_3 - gibberellic acid; G_1 and G_2 phases of cell cycle; IM – initiation medium; KIN - kinetin; MM – maintaining medium MS – Murashige and Skoog medium; NAA – naphthaleneacetic acid; PEM – proembryogenic mass; PGR - plant growth regulator; TDZ - thidiazuron

A.K. Hvoslef-Eide and W. Preil (eds.), Liquid Culture Systems for in vitro Plant Propagation, 345–358.

1. Introduction

Most gentians originate from alpine climatic zones: about 190 species were discovered in the Himalayas and Alps, and 27 in the Pyrenees. More than 100 species are protected by law in their native habitats. Many gentians are used in gardens as ornamental plants, some species have also pharmaceutical value. Because of the various uses of *Gentiana* species, highly efficient methods for their propagation are required.

Micropropagation of gentians is usually *via* organogenesis (Mikuła and Rybczyński, 1999): shoot or axillary meristem cultures are initiated from *in vivo* or *in vitro* cultured plants. Somatic embryogenesis appears to be an effective system of plant propagation in tissue cultures, already described in three *Gentiana* species: *G. cruciata*, *G. pannonica* and *G. tibetica* (Mikuła and Rybczyński, 2001). The initiation (Mikuła and Rybczyński, 2001;Mikuła et al., 2002b; Mikuła et al., 2003), proliferation and maintenance of embryogenic cell suspension cultures of gentians (Mikuła et al., 2001; 2002a) have been described in our previous papers. Embryogenic cell suspensions appear to be more productive than callus cultures of gentians grown on gelled medium, and provide a long-term source of somatic embryos for plant regeneration (Mikuła et al., 2002).

Many trials have been undertaken to obtain genetically modified gentians – new ornamental varieties and plant material for secondary metabolite production (Momčilović et al., 1997). Long-term embryogenic suspension cultures can provide protoplasts for somatic hybridisation and transformation. The long-term preservation of suspension cultures in liquid nitrogen (-196°C) creates opportunities for experiments to be carried out on the same plant material during several years.

The aim of this experiment was to characterise the well-established highly embryogenic cell suspensions of *Gentiana cruciata* and *G. tibetica*. Selected growth parameters of the suspension cultures were evaluated. Additionally, the influence of different PGRs applied in the gelled medium on PEM development and embryo production was studied.

2. Material and methods

Callus cultures of *G. cruciata* (L) and *G. tibetica* (King) were induced on agar-gelled MS (Murashige and Skoog, 1962) supplemented with 4.52 μmol 2,4-D and 0.53 μmol KIN (IM) (Mikuła et al., 1996b; Mikuła and Rybczyński, 2001). After six months, callus pieces were transferred to liquid MM composed of MS mineral salts and supplemented with 4.52 μmol Dic, 0.45 μmol NAA, 5.77 μmol BAP and 434 μmol AS. Cultures were maintained under continuous diffused light (3.5 μE m^{-2} s^{-1}) at 23°C on an INFORS gyratory shaker at 130 rpm. Each 250 ml Erlenmeyer flask contained 80 ml of the cell suspension.

Studies on the cell cycle, the percentage of cell aggregate fractions, the fresh and dry mass of cultures as well as flow cytometric analysis and light and scanning electron microscopic examinations, were carried out to characterise the established cell suspensions. The methods described in the previous paper were used (Mikuła et al., 2002).

To characterise the cell suspensions, their morphogenic potential in different culture conditions was additionally taken into consideration. To study the effect of selected PGRs on gentian somatic embryogenesis, three-year-old suspensions were implanted onto the agar-gelled MS supplemented with GA_3 (at 0, 1.49 or 2.89μmol), KIN (at 0, 2.32, 4.64 or 9.28 μmol) and AS (at 0 or 434 μmol). Before the implantation, cell aggregates were divided into two fractions depending on their size (240-450 μm and >450 μm).

The effect of cell aggregate fraction on somatic embryo production was analysed by means of 1-factor variance analysis, whereas the effect of GA_3 and KIN - by 2-factor analysis (ANOVA).

3. Results

In the MM supplemented with Dic, NAA, BAP, and AS, proembryogenic masses of *G. cruciata* underwent cyclic morphogenic changes. Single embryogenic cells and small cell aggregates (phase I) (Figure 1a) formed larger aggregates with proembryos and somatic embryos at the globular stage (phase II) (Figure 1b, c). Under the same culture conditions the embryos developed to phase III, initiated by the degradation of the proepidermal cells (Figure 1d). Finally the structure disintegrated into aggregates and single cells. The particular phases followed each other during the two-week subculture period. The phenomenon described above has not been observed in *G. tibetica* cell suspensions. Further somatic embryo development required implantation of the cultures onto the agar-gelled medium with the same PGRs.

Growth phase strongly affected the regeneration competence of PEM in *G. cruciata*. Cell aggregates of 240-450 μm (phase I and II) produced *c.*200 somatic embryos from 100 mg FW of tissue when only embryos at the cotyledonary stage were counted. Fraction >450 μm gave *c.*105 and *c.*100 embryos for phase I and II, respectively. Cultures, which originated from phase III did not form any somatic embryos (Figure 2). PEM in phase II formed numerous somatic embryos after six weeks of culture, i.e. two weeks earlier than PEM in phase I (Figure 3). Only PEM in phase II was selected for further experiments.

Cell suspensions of *G. cruciata* and *G. tibetica* differed in the percentage of five cell aggregate fractions studied (Table 1). In both species the most numerous were the aggregates from fraction 70-120 µm (54% in *G. cruciata,* and 36% in *G. tibetica).*

Growth parameters of the established cell suspensions were analysed during three consecutive years of culture at 12-month intervals. Both FW and DW decreased with extension of the culture age, but the FW/DW ratio increased (the highest value was obtained in the third year; the lowest, in the first year) (Table 2).

For both species flow cytometric DNA content in control plants, PEM and regenerants was studied (Table 3). PEM mitotic activity was evaluated by the comparison of the number of cells in the cell-cycle phases G_2 and G_1. The G_2/G_1 ratio was 6% and 18% for *G. cruciata* and *G. tibetica,* respectively.

Embryogenic capacities differed for both aggregate fractions of the species studied. In the presence of KIN, GA_3 and AS in the medium, cell suspension of *G. cruciata* was less productive than *G. tibetica*, and its fraction > 450 µm appeared to be superior. Fraction > 500 µm of *G. tibetica* gave at least four times more embryos than the same fraction of *G. cruciata* suspension.

The enrichment of the medium with AS resulted in the increased somatic embryo production in both species studied (Figures 4, 5). The statistical analysis revealed a significant effect of PGR concentration on embryo yield. Tables 4 and 5 show that the increased embryo production was correlated with the increase in KIN concentration up to 4.64 µmol. This concentration was found to be optimal for both species and fractions used. Its double increase did not significantly affect embryo production in *G. tibetica*, but it strongly reduced embryogenic competence in *G. cruciata* suspension.

Gibberellic acid played an important role in development and maturation of gentian somatic embryos. Although the most efficient embryo production was observed in *G. cruciata* in the absence of GA_3, many of embryos showed developmental disorders. Thus in *G. cruciata* GA_3 seemed to control the development of somatic embryos (Table 4). In both aggregate fractions of *G. tibetica,* GA_3 concentration did not affect the yield of somatic embryos (Table 5), however no embryos were obtained in the medium without GA_3 (Figure 6).

Table 1: Percentage of aggregate fractions in 3-year-old cell suspension of *G. cruciata* and *G. tibetica* (in 1.0 ml of suspension, after 5 days of subculture)

Aggregate fraction (μm)	*G. cruciata*	*G. tibetica*
70 –120	54 ± 0.59	36± 0.70
120-240	29 ± 0.54	30 ± 0.67
240-450	11 ± 0.37	14 ± 0.51
450-500	6 ± 0.27	17 ± 0.55
>500	0.6 ± 0.09	3 ± 0.24

Table 2: Ratio (R) of coefficient of fresh (FM) and dry (DM) mass of cell suspensions cultured in MS supplemented with Dic + NAA + BAP + AS

Age of culture	*Gentiana cruciata*			*Gentiana tibetica*		
	FM*	DM*	R	FM*	DM*	R
1-year-old	4.0	3.1	0.78	3.8	3.2	0.84
2-year-old	4.1	3.4	0.83	3.4	3.8	1.12
3-year-old	2.2	2.0	0.91	1.4	2.6	1.86

* ratio of final to initial cell mass

Table 3: 2C value describe by cytometic DNA (pg) content in control plant, 3-year-old PEM and regenerants of *G. cruciata* and *G. tibetica.* Mitotic activity of PEM - the ratio of cells in phase G_2 to phase G_1 (in bold)

Species	Control plants	PEM	Regenerants
G. cruciata	2.45 ± 0.13	1.5 ± 0.03 **(6%)**	2.77 ± 0.65
G. tibetica	2.51 ± 0.036	1.15 ± 0.35 **(18 %)**	2.74 ± 0.68

Table 4: Effect of GA_3 and kinetin (KIN) on the number of mature somatic embryos formed by 100 mg of *G. cruciata* embryogenic cells from suspension culture plated on agar medium supplemented with 434 µmol AS*. Data were collected between 6th - 10th week of the culture

KIN (µmol)	GA_3 (µmol)							
	0.0	1.49	2.89	Average SE_1=3.4	0.0	1.49	2.89	Average SE_1=3.26
	Fraction 240-450 µm a			SE_3=5.965	Fraction >450 µm b			SE_3=5.646
0.0	257 a	189 c	146 de	197 a	103 cde	85 f	87 ef	92 b
2.32	139 ef	161 d	126 f	142 b	128 ab	82 f	81 f	97 b
4.64	231 b	202 c	150 de	194 a	135 a	105 cd	83 f	107 a
9.28	138 ef	129 f	146 de	137 b	117 bc	89 def	76 f	94 b
Average SE_2=2.98	191 a	170 b	142 c		120 a SE_2=2.82	90 b	82 c	

Table 5: Effect of GA_3 and kinetin (KIN) on the number of mature somatic embryos formed by 100 mg of *G. tibetica* embryogenic cells from suspension culture plated on agar medium supplemented with 434 µmol AS*. Data were collected between 6th - 10th week of the culture

KIN (µmol)	GA_3 (µmol)							
	0.0	1.49	2.89	Average SE_1=3.27	0.0	1.49	2.89	Average SE_1=4.059
	Fraction 240-450 µm *b*			SE_3=5.659	Fraction >450 µm *a*			SE_3=7.030
0.0	0 *e*	201 *d*	213 *cd*	138 *c*	0 *d*	510 *c*	524 *c*	345 *c*
2.32	0 *e*	223 *bc*	230 *b*	151 *b*	0 *d*	581 *b*	592 *b*	391 *b*
4.64	0 *e*	291 *a*	288 *a*	193 *a*	0 *d*	634 *a*	629 *a*	421 *a*
9.28	0 *e*	288 *a*	290 *a*	193 *a*	0 *d*	620 *a*	630 *a*	417 *a*
Average SE_2=2.8	0 *b*	251 *a*	255 *a*		0 *b* SE_2=3.5	586 *a*	594 *a*	

* Data represent the average of 6 replicates (two independent experiments with 3 Petri plates). Effect of the aggregate fraction on embryo production was analysed by means of 1-factor variance analysis, effect of GA_3 and KIN and correlation between them was analysed by 2-factor analysis. SE_1 – standard error for kinetin; SE_2 – for GA_3; SE_3 – for aggregate fractions.

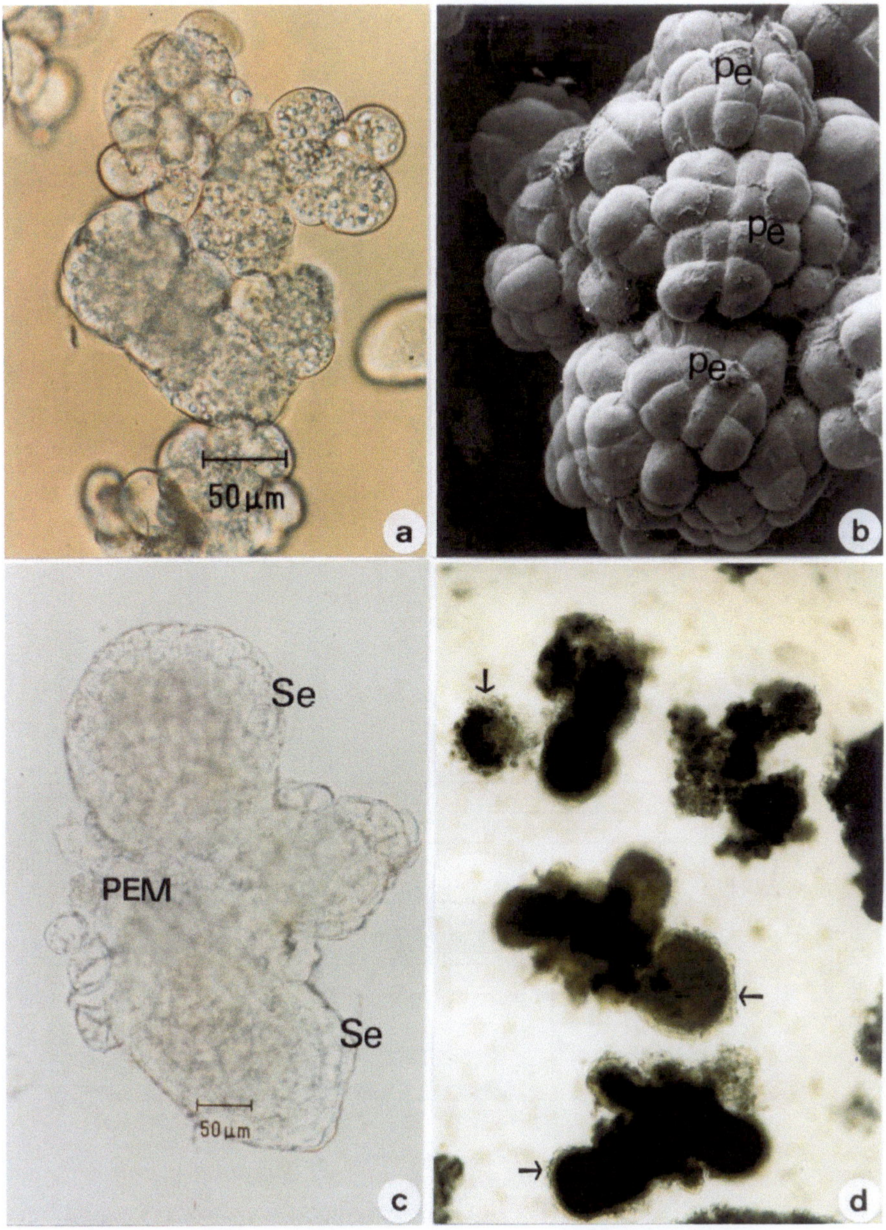

Figure 1: Phases of PEM development in *G. cruciata* suspension observed during 6 weeks of culture (MS+4.52 µmol Dic, 0,45 µmol NAA, 5.77 µmol BAP and 434 µmol AS) (intravital light microscope and SEM analysis):

a) cytoplasmatically rich embryogenic cells of PEM (Phase I)

b) scanning electonogram of PEM showing proembryos (SEM 720x)

c) PEM and globular embryos (Phase II) (SEM 600x)

d) Cell proliferation of epidermis of globular embryos (←) as the effect of extended auxin treatment of culture (Phase III).

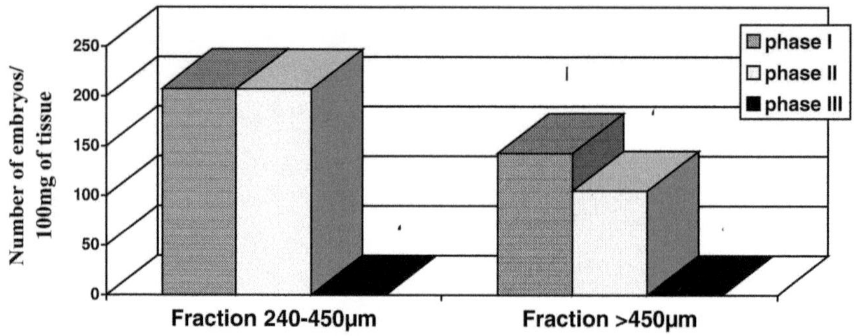

Figure 2: The effect of the cell growth phase on the total number of regenerated somatic embryos in *G. cruciata* (MS supplemented with 1.49 µmol GA_3 + 4.64 µmol KIN + 434 µmol AS) (SE=3.47).

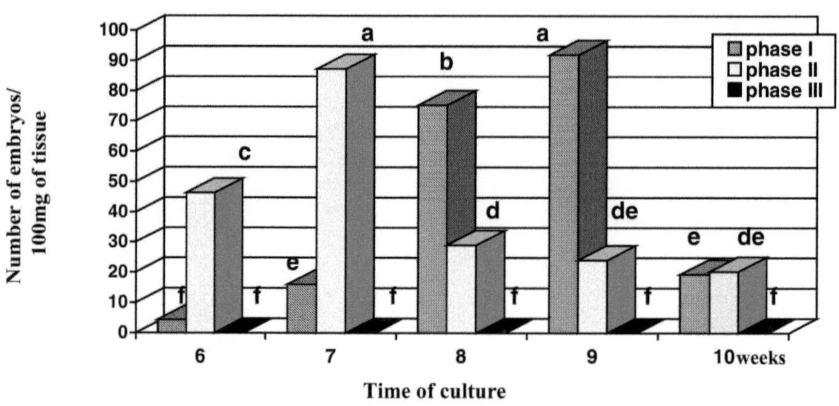

Figure 3: The effect of the cell growth phase on somatic embryo production in *G. cruciata* culture on MS supplemented with 1.49 µmol GA_3 + 4.64 µmol KIN + 434 µmol AS (SE = 3.134) in ten-week period (PEM fraction 240-450µmol).

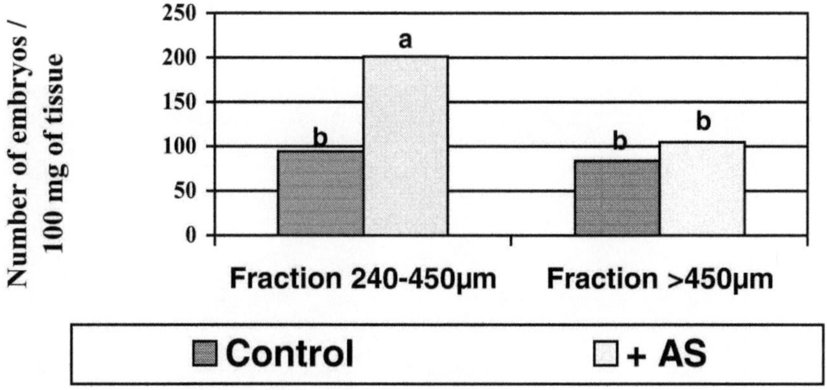

Figure 4: The effect of adenine sulphate (AS) on the total number of somatic embryos of *G. cruciata* on MS supplemented with 1.49 µmol GA_3 + 4.64 µmol KIN or without GA_3 and KIN (control), with (+AS) or without 434 µmol AS (SE=10.38).

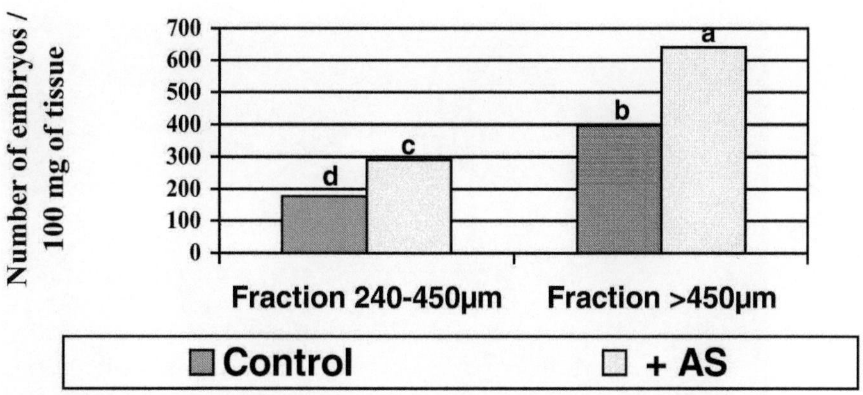

Figure 5: The effect of adenine sulphate (AS) on the total number of somatic embryos of *G. tibetica* on MS supplemented with 1.49 µmol GA_3 + 4.64 µmol KIN or without GA_3 and KIN (control), with (+AS) or without 434 µmol AS (SE=22.18).

Figure 6: The development of PEM in *G. tibetica* on the solid medium supplemented with AS (434 µmol), GA_3 and KIN (different concentrations; µmol).

4. Discussion

Several differences have been found between the cell suspensions of *G. cruciata* and *G. tibetica*. Cell suspensions were described by the following parameters: percentage of five fractions of cell aggregate, ratio of fresh to dry mass, cytometric DNA content and cell cycle phase. Differences in the PEM development as well as in mitotic activity of cells have been observed. Flow cytometric analysis of nuclei DNA content revealed the differences between two suspensions studied. In both gentian cultures the flow cytometric analysis showed that the DNA content decreased almost by half in comparison to the leaf tissue of field-grown plants and regenerants (Table 3). The mitotic activity of suspensions could be expressed as the ratio of cells being in phases G_2 and G_1 of the cell cycle (Galbraith et al., 1983). In *G. tibetica* the G_2/G_1 ratio was threefold more than for *G. cruciata* cell suspension.

The most advanced development stage obtained in liquid cultures of *G. cruciata* and *G. tibetica* was the globular embryo. This stage was recognised in both aggregate fractions: 240-450 μm and >450 μm. These two fractions appeared to play an important role also in embryogenic cell suspension cultures of *Lisianthus russellianus* (*Gentianaceae*) (Ruffoni and Massabo 1996). In this species the highest yield of somatic embryos was obtained from the bigger fraction (>500μm) in the light. In darkness, however, the cell aggregates smaller than 200 μm also retained the embryogenic capacities. In cultures of *Exacum affine* (*Gentianaceae*) the cell fraction >100 μm was superior in comparison to smaller fractions and produced a large number of well-developed embryos (Ørnstrup et al., 1993). Cultures of *G. tibetica* appeared to be superior to *G. cruciata*. Fraction >450 μm in *G. tibetica* was almost three times more productive than the best fraction of 240-450 μm in *G. cruciata*.

The development of somatic embryos in gentian cultures requires implantation of cell suspensions onto an agar-gelled medium. It has been proved already that gibberellic acid plays an important role in morphogenesis of *Gentianaceae*. At a concentration of 1.99 μmol together with 13.86 μmol zeatin, 9.28 μmol BAP and 5.37 μmol NAA, it stimulated the growth of protocolonies of embryogenic calli of *Gentiana crassicaulis* (Meng et al., 1996). Additionally, GA_3 was required to stimulate shoot formation and their multiplication in *G. scabra* and *G. corymbifera* (Morgan et al., 1997; Yamada et al., 1991). In *G. punctata* GA_3 at a concentration of 0.289 - 2.89 μmol affected strongly the elongation of already existing and newly formed nodes (Vinterhalter and Vinterhalter, 1998). Also, GA_3 played an important role in rooting of the shoots of some gentians (Morgan et al., 1997).

Results presented here confirmed the important role of gibberellic acid in somatic embryogenesis in *Gentiana* species. Statistically significant differences were found in somatic embryo production in *G. cruciata* subjected to different concentrations of GA_3. In these cultures gibberellic acid controlled embryo development and conversion. In contrast to cultures of *G. cruciata*, PEM of *G. tibetica* did not formed any embryos on the medium lacking GA_3. The increase in the embryo production was correlated with the increase in GA_3 and KIN concentration. Similarly, in cultures of *G. pannonica,* PGRs used in the medium influenced embryo development (Mikuła et al., 2002a).

Cytokinins play a crucial role in plant morphogenesis. To induce shoot regeneration in such explants as leaf, shoot and root of *G. triflora* and *G. scabra* 90.8 µmol TDZ, was used (Hosokawa et al., 1996, Nakano et al., 1995). High concentration of BAP stimulated shoot differentiation in *G. scabra* (Takahata et al., 1995). Adenine sulphate supports the effect of other cytokinins used in the medium, but is not often used in tissue cultures of both mono - and dicotyledonous plants, as concentration usually does not exceed 27.1 µmol (Pradhudesai et al., 1972). In cultures of *Gentiana* species for callus initiation and bud formation, AS at concentrations of 434 and 217 µmol was previously used (Wesołowska et al., 1985). In our present study, cytokinins (KIN and AS) appeared to be indispensable for the long-term maintenance of embryogenic cell suspensions and for somatic embryo production. In both species of gentian studied, embryo production decreased when AS was not included in the medium. In cultures of *G. pannonica, AS at* 868 µmol in the medium showed the same inhibitory effect to when it was absent (Mikuła et al., 2002a).

5. Conclusions

Cultures in liquid and on agar-gelled medium were used in the experiment to characterise embryogenic cell cultures of two species of gentian. They differed in the response to culture conditions, however, they retained embryogenic capacities for a long time. The size of PEM aggregates had an influence on the regeneration abilities: for *G. cruciata,* the highest results were obtained for the size fraction 240-450 µm and for *G. tibetica,* fraction >450 µm was best. Kinetin and GA_3 promoted regular embryo development.

References

Galbraith DW, Harkins KR, Maddox JM, Ayres NM, Sharma DP & Firoozbady F (1983) Rapid flow cytometric analysis of cell cycle in intact plant tissue. Science 220: 1049-1051

Hosokawa K, Nakano M, Oikawa Y & Yamamura S (1996) Adventitious shoot regeneration from leaf, stem and root explants of commercial cultivars of *Gentiana*. Plant Cell Rep. 15: 578 –581

Meng Y, Gao Y & Jia J (1996) Plant regeneration from protoplasts isolated from callus of *Gentiana crassicaulis*. Plant Cell Rep. 16: 88–91

Mikuła A & Rybczyński J (1999) Present status of *Gentiana* taxa biotechnology – A review. Biotechnologia 4: 1-16. (in polish)

Mikuła A & Rybczyński JJ (2001) Somatic embryogenesis of *Gentiana* genus I. The effect of the preculture treatment and primary explant origin on somatic embryogenesis of *Gentiana cruciata* (L.), *G. pannonica* (Scop.) and *G. tibetica* (King). Acta Physiol. Plant. 23: 15-25

Mikuła A, Rybczyński JJ, Tykarska T & Kuraś M (2001) Somatic embryogenesis of *Gentiana* genus II. Scanning and ultrastructural analysis of early stages of somatic embryogenesis in liquid medium. Biol. Bull. of Poznań 38: 47-53

Mikuła A, Skierski J & Rybczyński JJ (2002a) Somatic embryogenesis of *Gentiana* genus III. Characterization of three-year-old embryogenic suspensions of *G. pannonica* originated from various seedling explants. Acta Physiol. Plantarum 24: 311-322

Mikuła A, Tykarska T, Kuraś M, Iwanowska A & Rybczyński J (1996a) Studies on the morphogenic potential of Gentianas in cell suspension. Proceedings International Symposium. "Breeding research on medicinal and aromatic plants" (pp. 290-295). June 30-July 4, 1996, Quedlinburg, Germany

Mikuła A, Tykarska T, Rybczyński JJ, & Kuraś M (2002b) Ultrastructural analysis of initial stages of differentiation of root explants of *Gentiana cruciata* seedlings. Acta Soc. Bot. Pol. 71: 287-297

Mikuła A, Tykarska T, Zielińska M, Kuraś M & Rybczyński JJ (2003). Ultrastructural changes in zygotic embryos of *Gentiana punctata* (L.) during initiation of callus tissue and somatic embryogenesis. Acta Biol. Cracoviensia (accepted)

Mikuła A, Wesołowska M, Kapusta J, Skrzypczak L & Rybczyński JJ (1996b) Cytomorphological studies on somatic embryogenesis of *Gentiana tibetica* (King) and *G. cruciata* (L.). Acta Soc. Bot. Pol. 65: 47-51

Momčilović L, Grubišić D, Kojić M & Nešković M (1997) *Agrobacterium rhizogenes*-mediated transformation and plant regeneration of four *Gentiana* species. Plant Cell, Tiss. Org. Cult. 50: 1-6

Morgan ER, Butler R M & Bicknell R A (1997) *In vitro* propagation of *Gentiana cerina* and *Gentiana corymbifera*. New Zealand of Crop and Horticultural Science 25: 1-8.

Murashige T & Skoog F (1962) A revised medium for rapid growth and bioassay with tobacco tissue cultures. Physiol. Plant. 15: 473 – 497

Nakano M, Hosokawa K, Oomiya T & Yamamura S (1995) Plant regeneration from protoplasts of *Gentiana* by embedding protoplasts in gellan gum. Plant Cell, Tiss. Org. Cult. 41: 221-227

Ørnstrup H, Molgaard JP & Farestveit B (1993) Somatic embryogenesis and plant regeneration from cell suspension of *Exacum affine*. Plant Cell, Tiss. Org. Cult. 35: 37–41

Pradhudesai RR & Narayanaswamy S (1972) Differentiation of cytokinin-induced shoot buds and embryoids on excised petioles of *Nicotiana tabacum*. Phytomorphology 23: 133–135

Ruffoni B & Massabo F (1996) Plant production by somatic embryogenesis in cell suspension cultures of *Lisianthus russellianus* Hook. Plant. Tiss. Cult. Biotech. 2: 194–198

Takahata Y, Jomori H, Miyano S, Kunitake H & Mii M (1995) Regeneration of plants from protoplasts of *Gentiana* species (Gentian). In: Bajaj Y P S (ed) Biotechnology in Agriculture and Forestry, Vol. 34. Plant Protoplasts and Genetic Engineering VI (pp. 55-62). Springer-Verlag Berlin Heidelberg

Vinterhalter B & Vinterhalter D (1998) *In vitro* propagation of spotted gentian *Gentiana punctata*. Arch. Biol. Sci. Belgrade 50: 177 – 182

Wesołowska M, Skrzypczak L & Dudzińska R (1985) Rodzaj *Gentiana* w kulturze *in vitro*. Acta Polonica Pharmaceutica XLII: 79-83 (only English summary)

Chapter 27

Induction and growth of tulip 'Apeldoorn' bulblets from embryo cultures in liquid media

Anna Bach & Agata Ptak
Department of Ornamentals, Agricultural University, al. 29 Listopada 54, Kraków, Poland.
E-mail: robach@cyf-kr.edu.pl

Abstract: Bulbing has been induced in somatic embryos formed in ovary tissue cultures of *Tulipa gesneriana* 'Apeldoorn'. A high frequency of adventitious bulb formation took place on the stolons developing from embryos converted into plantlets in the liquid medium lacking growth regulators and containing 6% of sucrose. Number of bulblets per explant and individual fresh weight significantly increased in liquid medium (160 mg per bulblet) as compared to control (49 mg per bulblet). Plantlets must be chilled at 5°C in dark during 12 weeks period before formation and growth of adventitious bulbs. Plantlets, which were exposed to white light and cultured at temperature of 25°C (before chilling period), resulted in a higher crop of bigger bulblets (149 mg per bulblet, an average fresh weight) as opposed to embryos cultivated in darkness or irradiated by red light (59 mg per bulblet, an average fresh weight). After chilling period bulbs grew well in darkness and a temperature of 25°C. The temperature of 20°C of the liquid medium was less suitable for bulblet formation and growth, resulting in a lower percentage of large 200 mg bulblet (2.2%) in comparison to a medium temperature of 25°C (25.4%). Small bulbs, weighing less then 200 mg per bulblet, did not survive during the acclimatisation. Analysis of the level of ploidy of regenerated tulip plants did not reveal any changes.

Key words: bulb formation, *in vitro*, liquid media, somatic embryo, tulip

Abbreviations: BA-6-benzylaminopurine; MS-Murashige & Skoog's (1962) medium; NAA-naphthalene acetic acid

1. Introduction

In vitro propagation of tulip includes a 2-steps- process: the induction of adventitious shoots or somatic embryos and the induction of bulbing. As initial explants, flower stems (Wright at al., 1982; Gude and Dijkema, 1997), bulb scales (Nishiuchi, 1986; Baker at al., 1990; Koster, 1993), and ovary

A.K. Hvoslef-Eide and W. Preil (eds.), Liquid Culture Systems for in vitro Plant Propagation, 359–364.

tissues (Bach and Ptak, 2001) can be used. Most of the studies concern the induction of bulbing of the shoots formed *in vitro*. The presence of a meristem appears to be a very important factor for further propagation and bulblet formation (Kuijpers and Langens-Gerrits, 1997).

This is the first report about adventitious bulb regeneration from *in vitro* somatic embryos of tulip. Here, a series of experiments was carried out to investigate the influence of medium type, light quality and temperature on bulb induction.

2. Materials and methods

Tulipa gesneriana L. 'Apeldoorn' somatic embryos were generated in ovary tissues cultures under the influence of Picloram and BA on solid Murashige and Skoog (1962) medium according to the method described earlier (Bach and Ptak, 2001). Embryos at the globular stage developed into torpedo stage under influence of 5 μmol BA and 0.5 μmol NAA. The regenerated plantlets (shoots with meristems) were exposed to 20°C and 25°C, and they were cultured under a 16h photoperiod under light of different spectra: white (390-760 nm, Tungsram lamp 40 w F33) and red (647-770 nm, Philips TLD 36 W) during 10 weeks. Afterwards they were transferred to liquid and solid MS medium lacking growth regulators and containing 6% of sucrose. For bulblets induction the shoots were cultured for 12 weeks at 5°C in darkness while for bulblets growth the cultures were transferred to 25°C to darkness and to white and red light. Well-formed bulblets were planted in soil at 9°C. After 10 weeks they were transferred to the greenhouse.

Flow cytometric analysis of the regenerants was carried out using a Partec CA II instrument (Munster, Germany). All biometrical data were evaluated statistically by analysis of variance using Duncan's test to determine significance for differences at $\alpha < 0,05$.

3. Results

A high frequency of adventitious bulblets induction took place from the stolons developing in the liquid medium (Figure 1 and 2). As a result, number of bulblets per explant and individual fresh weight significantly increased in liquid medium (an average of 160 mg per bulblet) as compared to control (an average of 49 mg per bulblet) (Table 1). The morphological features of the bulblets formed in liquid medium were similar to those which

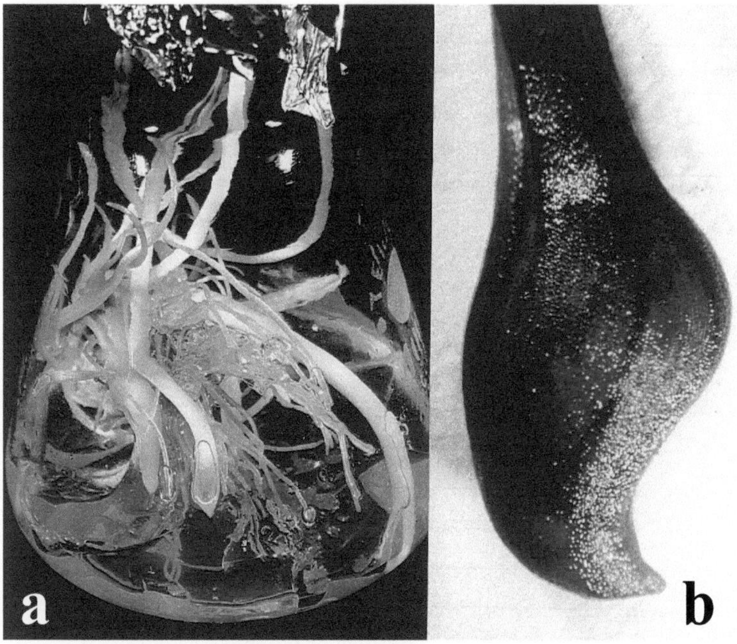

Figure 1: Development of tulip stolons: a) liquid cultures of tulip; b) detail of tulip stolon.

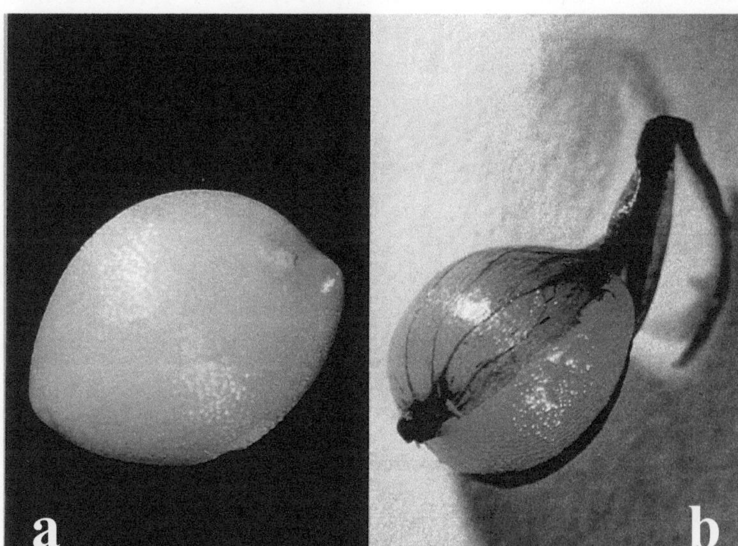

Figure 2: Tulip adventitious bulblets formed in: a) liquid medium; b) solid medium.

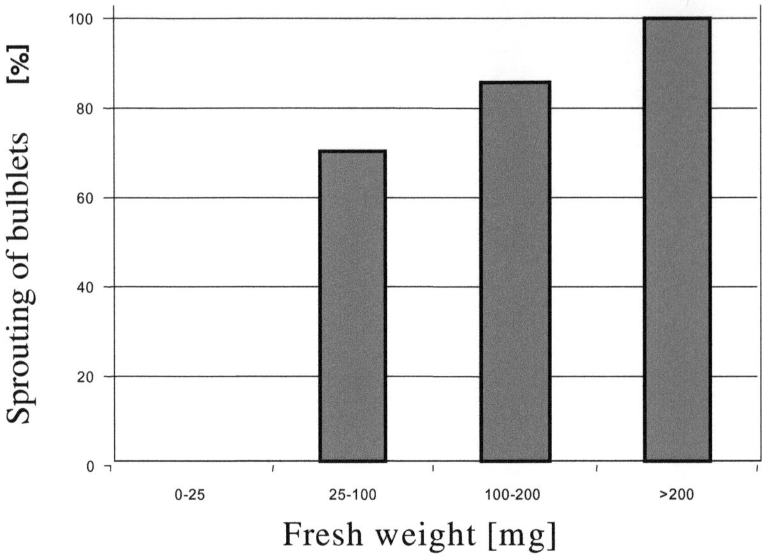

Figure 3: Effect of bulblets weight on the sprouting process

developed in agar medium but a higher percentage of sprouting in 'tunic-less' bulblets was observed in the case of liquid system (Figure 3).
The induction and formation of bulblets was affected by the temperature of liquid medium and the quality of light conditions. It was observed that plantlets must be chilled at 5°C in dark during 12-weeks-period, before formation of adventitious bulbs. Plantlets which were exposed to white light and cultured at a temperature of 25°C before the chilling period resulted in a higher crop of bigger bulblets (11.6% of bulblets with an average weight of 149 mg) as opposed to tulips cultivated in red light (1.3% of bulblets with an average weight of 59 mg) (Table 2). The temperature of 20°C was less suitable for bulblet formation and growth, resulting in a lower percentage of large (>200 mg) bulblet in comparison to a medium temperature of 25°C (Table 3). Two weeks after the chilling period, adventitious bulbs began to grow and developed well in darkness at 25°C (Table 4). Small bulbs, weighing less then 200 mg per bulblet, did not survive during the acclimatisation. Flow cytometric analysis did not show any differences in ploidy levels between the bublets and the parent plant.

Table 1: Effect of medium type on tulip 'Apeldoorn' bulblet formation

Medium	% of embryos formed bulbs	Fresh weight of bulblets [mg]
solid	30.4 a*	49 a*
liquid	78.5 b	160 b

*-means followed by the different letters are significantly different at α =0,05

Table 2: Effect of light quality and temperature before chilling period on tulip bulb formation

Light, temperature	% of embryos formed bulbs	Fresh weight of bulblets (mg)
white, 20°C	1.3 a*	59 a*
white, 25°C	11.6 b	149 b
red, 25°C	1.3 a	49 a

*-means followed by the different letters are significantly different at α =0,05

Table 3: Effect of medium temperature on percent of tulip embryos formed bulblets

Temperature	Fresh weight of bulblets (mg)			
	<25	25-100	100-200	>200
20°C	4.3 a*	87.4 b*	6.1 a*	2.2 a*
25°C	15.2 a	35.4 b	24.0 b	25.4 b

*-means followed by the different letters are significantly different at α =0,05

Table 4: Effect of light quality after chilling period on tulip bulb formation

Light quality	% of embryos formed bulbs	Fresh weight of bulblets (mg)
darkness	4.7 a*	160 b*
white	2.3 a	20 a
red	1.5 a	114 b

*-means followed by the different letters are significantly different at α =0,05

4. Discussion

In vitro cultures of ornamental bulbous plants in liquid media are largely employed for bulb vegetative propagation of some species e.g. *Lilium* (Kim et al., 2001). The present studies focused on the application of liquid culture in regeneration of tulip bulblet derived from somatic embryos. It was observed that liquid medium increased the number and fresh weight of 'Apeldoorn' adventitious bulblets. Koster (1993) also demonstrated the quick development of tulip stolons and bulblets in liquid medium. Stolon formation in tulip was observed in seedlings and small bulbs (Le Nard and de Hertogh, 1993) as well as in large bulbs, which grow in wet soil. This can explain the formation of stolons in liquid medium. It was also noticed that medium temperature of 25°C and white light applied to embryos prior induction of bulb, strongly improved quality of bulblets. Such results, on the effect of light quality in the development of bulblets have not been reported until now. To induce bulbs, embryos must be chilled (5°C) during 12-weeks-period in dark while the growth of bulbs was stimulated by liquid media, at 25°C, in darkness. Our observation on the sequence of temperature, which was necessary for the development of tulip bulbs in embryo cultures confirmed previous data of bulb formation in seedling grown in the field (Le Nard and De Hertogh, 1993). The results obtained indicate that it is possible to control the process of tulip bulblet formation in liquid medium and the quality of adventitious bulblets by the optimal temperature and light conditions in embryo cultures.

References

Bach A & Ptak A (2001) Somatic embryogenesis and plants regeneration from ovaries of *Tulipa gesneriana* L. in *in vitro* cultures. Acta Hortic. 560: 390-394

Baker CM, Wilkins HF & Ascher P D (1990) Comparison of precultural treatments and cultural condition on *in vitro* response of tulip. Acta Hortic. 266: 83-90

Gude H & Dijkema M (1997) Somatic embryogenesis of tulip. Acta Hortic. 430: 275-300

Koster J (1993) *In vitro* propagation of tulip. Thesis, Univ. of Leiden, The Netherlands

Kuijpers AM & Langens-Gerrits M (1997) Propagation of tulip *in vitro*. Acta Hortic. 430: 321-324

Le Nard M & De Hertogh A (1993) Tulipa. In: The Physiology of Flower Bulbs. (pp. 617-683). Elsevier.

Murashige T & Skoog F (1962) A revised medium for rapid growth and bioassays with tobacco tissue cultures. Physiol. Plant. 15: 473-479

Nishiuchi Y (1979) Studies on vegetative propagation of tulip. Formation and development of adventitious buds in the excised bulb scale cultured *in vitro*. J. Japan. Soc. Hortic. Sci. 48: 99-105

Kim YS, Hahn EJ & Paek KY (2001) A large scale production of lilium bulblets through bioreactor culture. Acta Hortic. 560: 383-386

Wright N & A Alderson PG (1980) The growth of tulip tissue *in vitro*. Acta Hortic. 109: 263-270

Chapter 28

Micropropagation of *Ixia* hybrids in liquid medium

Barbara Ruffoni*, Marco Savona, Silvia Doveri, Manuela Pamato, Silvana Carli°

*Experimental Institute for Floriculture, Propagation Dept., Corso Inglesi, 508, I-18038 Sanremo-Italy; Tel.:+39 0184 667251; Fax:+39 0184 695072, E-mail: propagazione@istflori.it; istflori@sistel.it. (*Corresponding author)*
°University of Genoa, DIP. TE. RIS., Corso Dogali 1C, I-16136 Genoa, Italy. Tel.: +39 010 2099375.

Abstract: *Ixia* is a monocotyledonous bulbous plant native of the South African Cape Region grown commercially as an ornamental plant for the production of flower stems. Commercial propagation is by the vegetative production of corms, which may suffer from a number of serious phytosanitary problems. In order to increase production uniformity and quality and to produce virus-free plants, a study of an alternative vegetative propagation system was investigated. Corms of two cultivars: "Panorama" and "Rose Emperor" were surface-sterilised and the meristems of the sprouting buds were excised and multiplied in semisolid medium in the presence of BA at 2.22 μmol. For liquid culture initiation, the explants were transferred to glass jars containing the liquid media supplemented with BA (at 0, 2.22 or 4.44 μmol) in the presence or in absence of ANC (3.90 μmol). At the end of the second subculture, in the presence of BA at either 2.22 or 4.44 μmol, the biomass was at least doubled and significantly increased when compared to the control treatment. The addition of ANC appeared to inhibit biomass proliferation although in the medium containing 2.22 μmol BA + ANC, a high number of bulblets was counted per biomass (callus) unit. At the end of the second subculture (60 days) all the material was transferred to hormone-free medium and, after 30 additional days, the culture parameters were evaluated. In the cultures of both genotypes, several shoot primordia and complete small plantlets developed from the callus and the cell aggregates. The histological analysis of the calli and the behaviour of the neoformed propagules indicated the activation of the embryogenetic pathway in the presence of ANC. The plantlets were successfully acclimatised in the greenhouse.

Key words: ancymidol, benzylaminopurine, histological analysis, morphogenesis, somatic embryogenesis

Abbreviations: ANC - ancymidol; BA - benzylaminopurine; PAN - cv. Panorama; PPFD - photosynthetic photon flux density; RE - cv. Rose Emperor; RH - relative humidity; rpm- rotations per minute

A.K. Hvoslef-Eide and W. Preil (eds.), Liquid Culture Systems for in vitro Plant Propagation, 365–372.

1. Introduction

Ixia is a monocotyledonous plant native of the South African Cape Region and belongs to the *Iridaceae* family. *Ixia* is considered to be a bulb plant of minor importance, compared to the well-known *Lilium*, *Tulipa*, *Freesia* and others, but enters the ornamental plant market for the production of particularly attractive flower stems. Seed propagation in this species is traditionally relegated to breeding programmes while the vegetative production of cormels is the preferred propagation methods for all other applications (Le Nard and De Hertogh, 1993). Vegetative propagated plants often suffer from serious phytosanitary problems, thus it was considered useful to investigate the possibility of developing an alternative propagation system that would provide virus-free and *Fusarium*-free plants and increase production uniformity, as well as the quality of the mother plants. *In vitro* multiplication results were published by Dinkelman and Van Staden (1988) only for *Ixia maculata*. Corm surface-sterilisation and *in vitro* multiplication attempts with three commercial hybrids were reported by Ruffoni et al. (1998). No data were found on growth in suspension cultures, nor on their multiplication in liquid culture systems. In the present work, the application of liquid media to morphogenetic callus culture is reported.

Other bulbous monocotyledonous plant species which development in liquid medium are strongly affected by the use of growth retardant agents (Ziv et al., 1994; Chen and Ziv, 2001): in this work, the influence of the growth retardant ancymidol was investigated on two cultivars of Ixia grown under diverse culture conditions.

2. Material and methods

Corms of two cultivars "Panorama" and "Rose Emperor" were surface-sterilised by a fungicide pre-treatment followed by dip in 100°C water for 30 s, 70% ethanol for 30 s and NaOCl (3% available chlorine) for 50 min according to Ruffoni et al. (1998). The meristems of the buds sprouting from the aseptic corms were excised and cultured in semisolid basal medium comprising Murashige and Skoog (1962) macro and microelements, vitamins, 30 g l^{-1} sucrose, 8 g l^{-1} agar, in the presence of BA (2.22 μmol). The developed shoots were multiplied in the same medium.

These "initial explants" were transferred to glass jars containing the liquid media with the above composition, supplemented with BA (0, 2.22, or 4.44 μmol) in the presence or in absence of ANC (3.90 μmol). Three replications of ten initial explants were used for each treatment. The average fresh weight per glass jar of these initial explants was 2.5 g for RE and 3.5 g for PAN. The cultures were grown with continuous agitation (80 rpm) at 23±1 °C, in a photoperiod of 16 h of light at a P-P-F-D of 30 μE m^{-2} s^{-1} supplied by Osram Lumilux growth lamps. Data were analysed with ANOVA; mean comparison ($p<0.05$) was performed with the Student-Neuman-Keuls test. Media were sterilised by autoclaving at 120°C for 20 min.

Every 30 days the media were renewed and some data were recorded (total biomass fresh weight, shoot number and height, bulblets formation, presence and quality of the callus). At the end of the second subculture (60 days) all the plant material was transferred to hormone-free medium and, after 30 additional days, the same parameters were measured and observations were recorded and evaluated.

The complete plantlets that developed were transferred to the greenhouse for acclimatisation for two weeks under mist at 25 °C, with 80 % of RH and using a sterile peat-perlite substrate (1:1, v/v). Finally, the plants were placed in lower RH conditions (50-60 %) to complete their hardening.

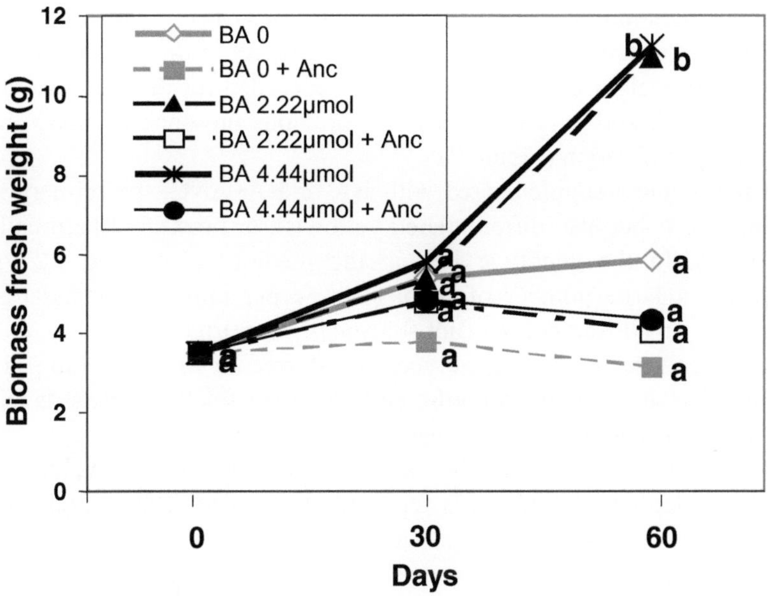

Figure 1: *Ixia*, cv. PAN, biomass fresh weight increase after 30 and 60 days related to the hormonal concentration of the media (different letters indicate significant differences at p=0.05, SNK test).

3. Results

After 30 days of culture 60 % of the excised meristems showed asepsis. Vegetative development was achieved from 23 % of these aseptic explants. Continuous shoot proliferation was achieved after 2 months in culture in semisolid medium; in some cases white-reddish callus was recorded at the base of the shoots.

From the explants transferred in liquid medium, biomass proliferation was obtained with both genotypes tested and some differences were recorded. In the cv. PAN, after the first subculture (30 days) the biomass weight values were similar in all treatments. At the end of the second subculture (60 days), in the presence of either 2.22 and 4.44 μmol BA, the fresh weight was at least doubled and significantly increased with respect to the control treatment. The addition of ANC seemed to inhibit biomass proliferation (Figure 1). After transfer to the hormone-free medium, the biomass production of the cv. PAN remained low for the cultures coming from ANC-supplemented medium, while the fresh weight of the cultures coming from 4.44 μmol BA increased significantly (Figure 2).

In the cv. RE, the callus showed an homogeneous behaviour among the treatments both during culture in the presence of growth regulators and after the transfer to hormone-free medium. With respect to the control treatment, greater fresh weight biomass increase was recorded for the cultures coming both from 2.22 and 4.44 μmol BA; the previous presence of ANC did not affect this proliferation (Figure 2).

In the cultures supplemented with BA or BA+ANC, the primary shoot explants grew, but also differentiated yellowish or brown/red hard callus at their bases. In the genotype RE, in the medium containing 2.22 μmol BA+ANC, a large number of bulblets grew per unit of biomass (callus) (Figure 3). This difference was not statistically confirmed.

In the cultures of both genotypes transferred to media lacking growth regulators, several shoot primordia and complete small plantlets developed from the callus and the cell aggregates.

The mean comparison of the data related to the shoot development in growth regulator-free medium is reported in figure 4. The response of the two genotypes was similar: the previous presence of ANC strongly inhibited shoot development and the greatest number of shoots was observed in cultures previously grown with 2.22 BA (7.25 shoots per callus unit).

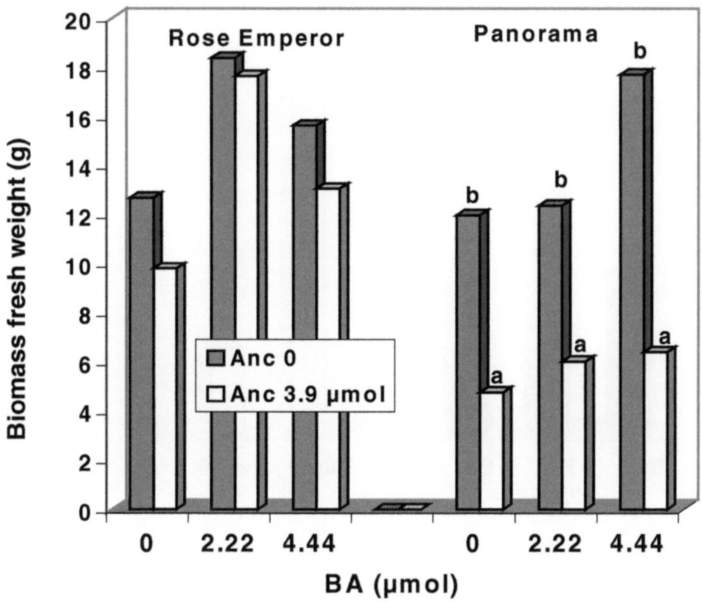

Figure 2: *Ixia*, cv. RE and PAN, biomass proliferation (fresh weight) in hormone-free medium related to the previous hormonal concentration of the media (different letters indicate significant differences at p=0.05, SNK test).

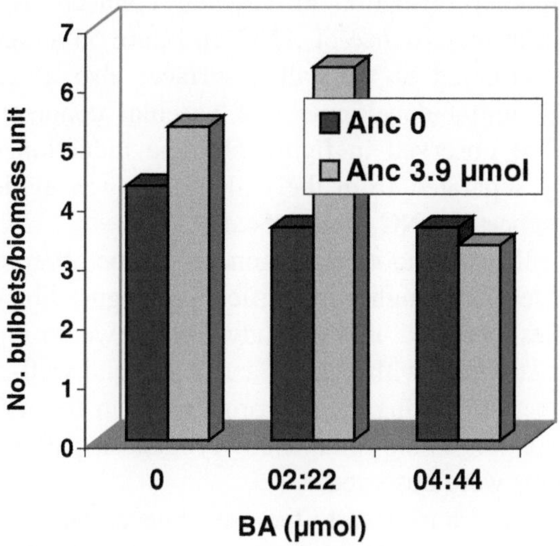

Figure 3: *Ixia*, cv. RE, cormels (bulblets) production at the end of the second subculture related to the hormonal concentration of the media.

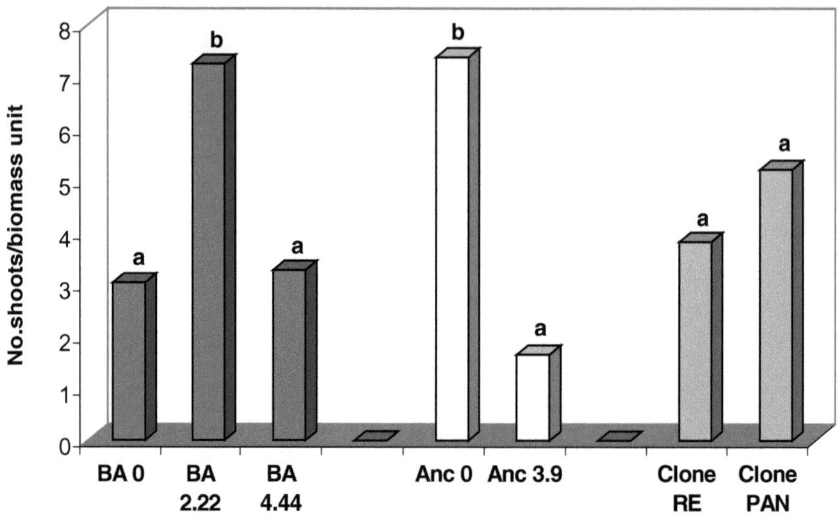

Figure 4: *Ixia*, cv. RE and PAN, efficiency of shoot differentiation related to the BA concentration and ANC presence in the media (different letters indicate significant differences at p=0.05, SNK test).

The histological analysis confirmed the development of adventitious shoots with a regular formation of vascular connections from the calli previously grown in the absence of ANC. In figure 5a a clear meristematic activity can be observed at the callus surface: also a leaf-shape single structure with an initial development of vascular connections within the callus cells can be observed in figure 5b. The induction of embryo-like structures clearly separated from the callus cells was evident in the calli grown in the presence of ANC (Figure 5c,d).

In figure 6a calli with shoot organogenesis can be observed, and a single structure can be seen at a higher magnification (Figure 6b, 120x). The best quality propagules obtained in the study are shown in figure 6c. These present a bipolar structure with a green aerial part and well-connected small roots. This production of singularised propagules in the medium, shows a characteristic belonging to the somatic embryogenesis pathway. In few cases abnormal structures were observed.

When plants were transferred to the greenhouse, they grew for about 30 days: then the leaves senesced and the corm diameter increased up to 1 cm. The corms then became dormant.

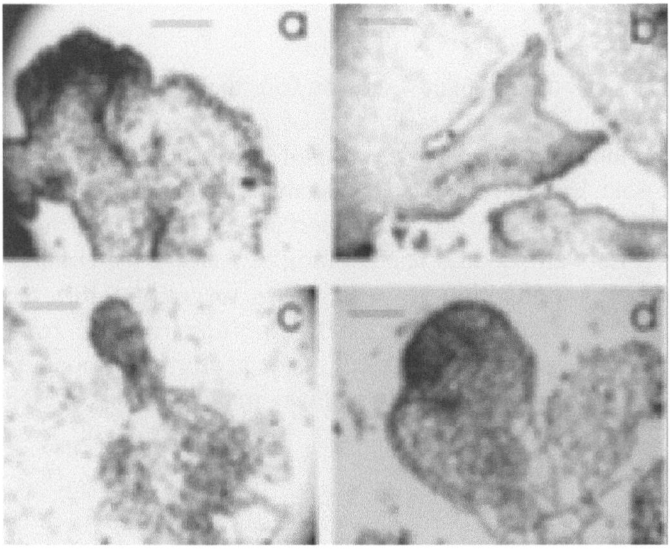

Figure 5: Histological analysis of *Ixia calli*; a) meristematic tissue on the callus surface (250x); b) leaf-shoot primordia with vascular connections (250x); c) embryo-like structure, globular phase (250x); d) embryo-like structure (400x).

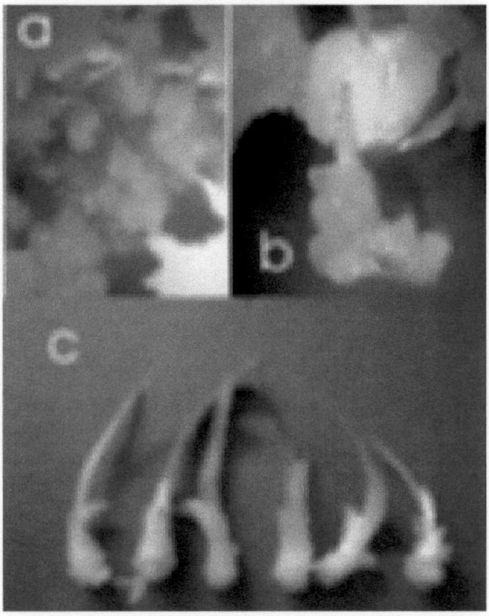

Figure 6: Ixia liquid culture; a) organogenetic calli; b) neoformed shoot (120x); c) propagules developed after ANC culture.

4. Discussion

Both in other bulbous plants and for *Ixia,* the liquid culture seems to be particularly promising for the scale-up of mass-production of healthy and virus-free plants. In fact, the presence of *Fusarium oxisporum* and Bean Yellow Mosaic Virus in the propagated material is a serius bottle neck for the exploitation of the commercial importance of this ornamental plant (De Hertogh and Le Nard, 1993). The addition of ANC in the cultures seems to inhibit biomass proliferation but also seems to promote differentiation as reported also by Chen and Ziv (2001) in *Narcissus*. In *Ixia*, the calli histological analysis and the behaviour of the neoformed propagules indicate the activation of the embryogenetic pathway in the presence of ANC.

References

Chen J & Ziv M (2001) Ancymidol effects on oxidative stress and the regeneration potential of *Narcissus* leaves in liquid culture. Acta. Hort. 560: 299-301

De Hertogh A & Le Nard M (1993) General chapter on spring flowering bulbs. In: The Physiology of Flower Bulbs (pp. 723-726). Elsevier

Dinkelman K & VanStaden J (1988) Micropropagation of *Ixia maculata*. South African Journal of Science 84: 589

Murashige T & Skoog F (1962) A revised medium for rapid growth and bioassay with tobacco cultures. Physiologia Plantarum 15: 473-497

Ruffoni B, Semeria L & Dalla Guda C (1998) Moltiplicazione *in vitro* di *Ixia*. Atti Giornate scientifiche SOI: 33-34

Ziv M, Kahany S & Lilien-Kipnis H (1994) Scaled-up proliferation and regeneration of Nerine in liquid cultures. Part I. The induction and maintenance of proliferating meristematic clusters by paclobutrazol in bioreactors. Plant Cell, Tiss. Org. Cult. 39: 111-117

Chapter 29

Micropropagation of *Rosa damascena* and *R. bourboniana* in liquid cultures

*Pratap Kumar Pati, **Madhu Sharma, **Anil Sood & **Paramvir Singh Ahuja

**Department of Botanical Sciences, Guru Nanak Dev University, Amritsar-143 005, India, E-mail: pkpati@yahoo.com.*

*** Institute of Himalayan Bioresource Technology, Palampur-176 061 (H.P.), India.*

Abstract: A liquid culture system using nodal segments was used for shoot proliferation and root induction in *Rosa damascena* and *R. bourboniana*, two commercially-important species of scented rose. For efficient and large scale induction of roots in microshoots, a rooting vessel was designed and developed to facilitate the micropropagation protocol. The present work highlights the significance of osmotic potential in relation to enhanced growth and development in liquid cultures, vis-à-vis agar-gelled cultures, especially in relation to root induction during micropropagation. An additional significant feature of the protocol developed, was the high success rate of hardening the micropropagated plants in low-cost hardening chambers, up to ca. 96.7% for *R. damascena* and 100% for *R. bourboniana*.

Key words: liquid medium, micropropagation, root induction, rooting vessel, scented rose, shoot multiplication

Abbreviations: BAP- 6-benzylaminopurine; IBA- indole-3-butyric-acid; MS- Murashige & Skoog (1962) medium; NAA-α-naphthalene acetic acid; PGR- plant growth regulation

1. Introduction

High economic value and widespread cultivation of roses make them one of the commercially-important ornamental crops. Among the existing species of rose, only a few exhibit delightful fragrance and are categorized as "scented roses". *Rosa damascena* Mill. and *R.. bourboniana* Desp. are two such commercially-important species. The former is generally preferred for production of rose oil and a whole range of other valuable products (Pati et al., 2001).

A.K. Hvoslef-Eide and W. Preil (eds.), Liquid Culture Systems for in vitro Plant Propagation, 373–385.

Roses are generally propagated by vegetative methods such as cuttings, layering, budding and grafting. Seeds are also used for propagation of species, new cultivars and for production of rootstocks (Horn, 1992). However, *in vitro* propagation methods have attracted attention of many rose growers for having enormous potential for mass multiplication of elite clones and production of healthy, and disease-free planting material, especially in ornamental roses, namely *Rosa hybrida* (Short and Roberts, 1991; Horn, 1992; Taslim and Patel, 1995; Rout et al., 1999). Among oil-bearing roses there is no report on *in vitro* propagation in *R. bourboniana* and only fragmentary information is available for *R. damascena* (Khosh-Khui and Sink, 1982; Ishioka and Tanimoto, 1990; Koronova and Michailova, 1994; Kumar et al., 2001).

In rose micropropagation, agar-gelled medium is most frequently used. However, the use of liquid media during different stages of *in vitro* propagation and improvement within the existing tissue culture protocols has been successfully demonstrated for multiplication of *R. damascena* and *R. bourboniana*. The different stages, namely, initiation of aseptic culture, shoot proliferation, rooting of microshoots, hardening and transfer to soil have been standardized for production of tissue culture raised plants in a cost effective manner. The present work also highlights the significance of osmotic potential for enhanced performance of rooting or microshoots in liquid cultures compared to agar-gelled cultures.

2. Materials and methods

2.1 Initiation of aseptic cultures and shoot proliferation on agar-gelled and liquid media

Nodal explants (2.5-3.0 cm), taken from field-grown plants of *R. damascena* and *R. bourboniana* were thoroughly washed with Tween-80 for 20 minutes, followed by surface sterilization with an aqueous solution of 0.04% (w/v) mercuric chloride for 7-8 minutes and subsequent rinsing (3-4 times) with sterilized distilled water. These were inoculated and maintained for 3-4 weeks on Murashige and Skoog (1962) medium (MS) containing 3.0% (w/v) sucrose and 0.8% (w/v) agar for initial screening. The pH of the medium was adjusted to 5.8 before autoclaving. Cultures were incubated at a photosynthetic photon flux density (PPFD) of 20 μmol $m^{-2}s^{-1}$ from cool white fluorescent lamps at 25±2°C. Daylength was 14h in a 24h light/dark cycle.

The BAP concentration (5.0 µmol), optimized in our laboratory for shoot proliferation in *R. damascena* (Pati et al., 2001), was used for determining the rates of shoot proliferation on both agar-gelled (100 ml) and static liquid (10, 20 and 40 ml) basal MS media. Other parameters such as shoot length, stem thickness were recorded with respect to the initial explant after 6 weeks of culture in both agar-gelled and static liquid media.

In order to optimize the liquid medium requirement for growth and shoot proliferation, different volumes of basal MS medium (10, 20 and 40 ml) supplemented with BAP (5.0 µmol) and sucrose (3.0% (w/v)) were tested in glass jars (Kasablanka, Mumbai; 350 ml capacity). The shoots were maintained in liquid medium by subculture into fresh medium every 4-6 weeks.

2.2 *Rooting of microshoots in agar-gelled and liquid medium*

The existing protocol for efficient rooting of microshoots of *R. damascena* (Pati et al., 2001) from our laboratory involving culture of microshoots for one week in half-strength MS medium supplemented with IBA (10.0 µmol) and sucrose (3% (w/v)), followed by transfer to MS medium without PGRs with unchanged sucrose concentration was used to study the effect of agar-gelled and liquid media on rooting of microshoots. Initial incubation of microshoots for rooting was conducted in the dark. Means for root initiation, rooting percentage, root length and root number were calculated from three replications (15 plants/replicate) and standard error of these means were determined. In order to study the effect of osmoticum on rooting, different concentrations (0.25mol-1.25mol) of mannitol were added to agar-gelled medium. The osmotic potential of the medium was measured using a dew point microvoltmeter, model HR-33T (Wescor, USA).

2.3 *Culture vessel for efficient rooting of microshoots and transfer to soil*

For efficient rooting of microshoots, a culture vessel was designed (Figure 1). It was comprised of an outer transparent autoclavable container with a tight lid in which a perforated platform with holes of suitable sizes (5 mm diameter) was placed. The excised shoots were inserted in the holes, ensuring that their cut ends constantly touched the rooting medium in the container. Different steps involved in the process were i) preparation of a perforated platform (A) in order to provide a support system for the microshoots, ii) placement of the platform in a container (B), iii) pouring of root-inducing medium into the box and medium sterilization, iv) inserting microshoots into the holes of the perforated platform, v) sealing the lid (C) of the container with Parafilm, and vi) incubating the cultures in appropriate conditions of temperature and light.

The rooted microshoots before transfer to soil were categorized as follows on the basis of root number and root length (Figure 5):

A) Root numbers >10 and root length less than 0.2-1.0 cm
B) Root numbers >10 and root length more than 1.0-5.0cm
C) Root numbers <10 and root length more than 1.0-5.0 cm
D) Root numbers <10 and root length less than 0.2-1.0 cm

In another set of experiment, the effect of the duration of incubation of microshoots in the rooting vessels and their subsequent transfer and survival in greenhouse conditions, was studied. Plants were transferred every week up to 6 weeks and means were calculated from three replications of 15 plants each and standard errors of these means were determined.

After 6 weeks, the rooted microshoots were transplanted in Hikko trays/pots filled with soil mix (sand : garden soil, 1 : 1). These Hikko trays/pots were placed in specially designed hardening chambers (Sharma et al., 1999) enriched with CO_2 (20/11x10^{-5} mol l^{-1} to 80/13x10^{-7} mol l^{-1}), which was measured using Infra Red Gas Analyser (IRGA; Model Li-6400, LiCor, Lincoln, USA). Relative humidity (RH) was maintained at 70-80% and provided with a light intensity of 35 μmol m^{-2} s^{-1} from the top.

2.4 *Comparative study of micropropagation of R. damascena and R. bourboniana*

In a separate experiment, a parallel study on different stages of micropropagation, such as the initiation of aseptic culture, shoot proliferation, rooting and plantlet establishment for *R. damascena* and *R. bourboniana* was made and their comparative responses recorded. Agar-gelled medium was used for initiation of aseptic cultures whereas, liquid medium was used for shoot proliferation and rooting.

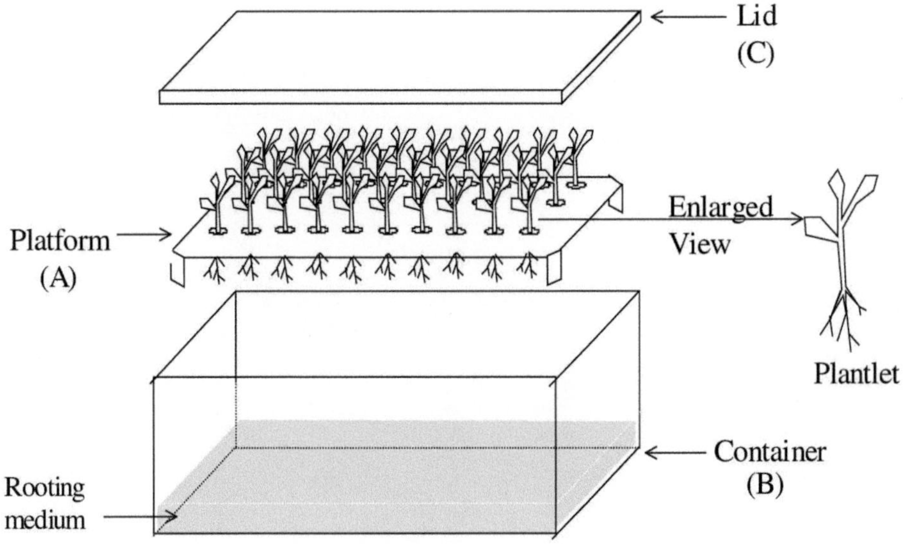

Figure 1: Rooting vessel.

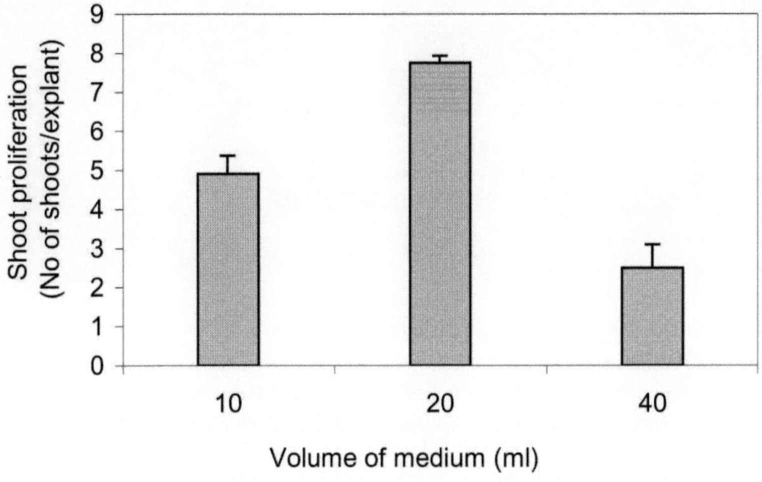

Figure 2: Shoot proliferation after 4 weeks in different volumes of liquid MS medium supplemented with BAP (5μmol).

Figure 3 a-f: Micropropagation of *R. damascena*.
a: Axillary bud proliferation
b: Shoot proliferation in liquid and agar gelled medium, necrosis of shoots in gelled medium
c: Polypropylene box with bunches of shoots inserted in perforated platform with their cut ends touching the medium
d: Perforated platform turned upside down to show profuse and healthy rooting of microshoots
e: Rooted microshoots after 42 days of incubation in rooting vessel
f: Hardened plants ready for transfer to field.

3. Results

3.1 *Shoot proliferation*

The axillary buds sprouted (Figure 3a) and were transferred to both agar-gelled and liquid media. On agar-gelled medium shoots showed some necrosis (Figure 3b) and only a four-fold increase of shoots was recorded in 6 weeks, whereas in liquid medium, an eight-fold increase in the number of shoots was recorded during the same time period (Table 1). Shoots thus formed in liquid media were sturdy with thicker stems compared to those produced on agar-gelled media. It was observed that the different volumes of liquid medium in the culture vessel resulted in varied responses. After 4 weeks of culture, optimum shoot proliferation was observed in 20 ml medium (Figure 2). The period of a subculture cycle could be extended up to 6 weeks without affecting the optimal growth. However, vitrification and desiccation of shoots was evident when medium volumes of 40 and 10 ml, respectively.

3.2 *Rooting, hardening and transfer to soil*

Agar-gelled and liquid media induced different rooting responses (Table 2). On agar-gelled and liquid media containing 3% (w/v) sucrose, the rooting response was 5% and 85.8%, respectively and it took 12 days for root induction in the former, compared to 8 days in the latter medium. In half strength MS liquid medium with IBA (10 µmol), rooting percentage was 85.8, whereas in agar gelled medium with same medium composition, it decreased drastically to 5%. However, in agar gelled medium there was a marked increase in the percent rooting with the addition of mannitol (0.25mol to 1.25mol) and at 0.5mol concentration, the rooting response increased from 5% to 81.8% (Figure 4). Beyond 0.5mol of mannitol, there was a gradual decline in rooting.

Based on root number and root length, four categories of plantlets were obtained. These four categories differed in their survival under greenhouse conditions (Figure 5). The data recorded after 4 weeks of incubation in greenhouse conditions indicated that the survival percent was significantly higher (73.2 ±1.8) in category B, with more, longer roots compared to category C with fewer roots but of greater length. The lowest survival percent was in category D with lowest root number and shortest length.

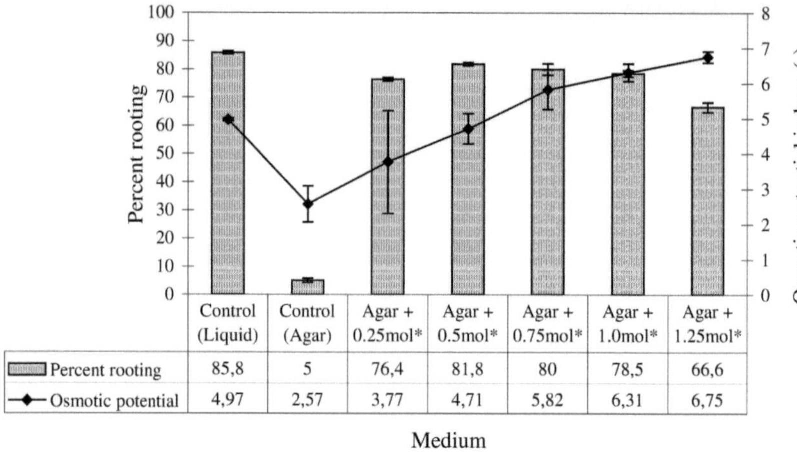

Figure 4: Effect of osmotic potential on rooting of microshoots in *R. damascena* [Control liquid: ½ MS+IBA (10μmol)+sucrose (3%); Control agar: ½ MS+IBA (10μmol)+sucrose (3%)+agar (0.8%)]. Vertical bars are Mean± SE (n=45).
*addition of mannitol (0.25 - 1.25 mol) to medium

Figure 5: Effect of plantlet category [A. Root number more (>10) and root length less (0.2-1.0 cm); B. Root number more (>10) and root length more (1.0-5.0cm); C. Root number less (<10) and root length more (1.0-5.0 cm); D. Root number less (<10) and root length less (0.2-1.0 cm)] on percent survival of *R. damascena* plantlets after 4 weeks of incubation under greenhouse conditions.

The rooted microshoots (Figure 3c-d) were incubated in rooting vessel for different times before their transfer to greenhouse conditions. It was observed that highest survival percentage (96.7% in *R. damascena* and 100% in *R. bourboniana*) was recorded after 6 weeks of incubation (Table 3 and 4). However, at this stage, most of the shoots were touching the top of the box (Figure 3e) and therefore, required immediate transplantation. These plantlets were kept in a greenhouse for 6-8 weeks before their transfer to larger pots (20cm) (Figure 3f) and eventually, to the fields.

3.3 *Comparative response of micropropagation of R. damascena and R. bourboniana*

A comparative analysis of different stages of micropropagation of *Rosa damascena* and *R. bourboniana* indicates that these two species did not have much difference in respect to different test parameters under the *in vitro* culture systems tested (Table 4).

Table 1: Development of cultured shoots of *R. damascena* in agar-gelled and liquid MS+BAP (5μmol sucrose (3% (w/v))

Medium	No. of shoots		Shoot length (cm)		Stem thickness (mm)	
	Initial	Final*	Initial	Final*	Initial	Final*
Gelled (Agar: 0.8 % w/v))	7.2 ± 0.7	30.4 ± 4.1	1.08 ± 0.03	1.62 ± 0.03	0.90 ± 0.04	1.54 ± 0.1
Liquid	7.2 ± 0.37	58.0 ± 2.60	1.04 ± 0.02	2.96 ± 0.2	0.92 ± 0.1	1.96 ± 0.05

* Data recorded after 6 weeks of culture

Table 2: Comparison of rooting response* in liquid and agar-gelled media**

	Liquid medium	Agar-gelled medium
Time to root induction	8 days	10 days
Rooting (%)	85.8	5%
Root length (mm)	3.5-4.0	1.5-2.5
Root number per shoot	8-10	3-6

*Data recorded after 3weeks

**1/2 MS+ IBA (10.0 μmol) + sucrose (3% (w/v)) for 1week, then transfer to 1/2 strength MS without PGRs

Table 3: Effect of *R. damascena* microshoot incubation period in rooting vessels, on the percent survival after transfer to greenhouse

Week of transfer	Survival (%)
1	3.3±3.3
2	23.3±3.3
3	60.0±5.7
4	66.7±8.8
5	73.3±8.8
6	96.7±3.3

Table 4: Comparative response of *R. damascena* and *R. bourboniana to* micropropagation protocols

Stage of micropro-pagation	Parameters	Response		Remarks
		R. damascena	*R. bourboniana*	
* Initiation of aseptic culture	Contamination rate	29.6 ± 1.9	18.6 ± 3.8	Surface of nodal explants of *R. bourboniana* was smooth and had fewer / smaller thorns compared to *R. damascena*
** Shoot proliferation	Proliferation rate (No. of shoots per explant)	8.1 ± 0.2	7.5 ± 0.2	*R. bourboniana* shoots were sturdy, leaves dark green with greater surface area compared to *R. damascena*
*** Rooting	Root induction (days)	7-8	7-8	Rooting is easier in *R. bourboniana.* Basal MS medium produces rooting in *R. bourboniana* unlike *R. damascena*
	Rooting (%)	85.8 ± 0.6	94.9 ± 2.1	
	Root length (mm)	4.6 ± 0.4	5.2 ± 0.4	
	Root number per shoot	9.0 ± 0.5	11.4 ± 0.4	
**** Hardening and plantlet establishment	Survival (%)	96.7	100	Hardening is easier in *R. bourboniana* compared to *R. damascena*

*Data recorded after 4 weeks of culture in MS medium without PGRs

**Data recorded after 6 weeks of culture in liquid MS medium + BAP (5 μmol) + sucrose (3% (w/v))

***Data recorded after 3 weeks of culture; 1 week in liquid ½ MS + IBA (10 μmol) + sucrose (3% (w/v)) and 2 weeks in MS + sucrose (3% (w/v))

****Data recorded after 3 weeks of transplantation in potting mix (sand : garden soil, 1: 1)

4. Discussion

The establishment of cultures in liquid media has several advantages (Smith and Spomer, 1994; Chu et al., 1993) and is an important step towards automation (Aitken-Christie et al., 1995). Realizing this fact, in the present micropropagation protocol, elimination of agar in the multiplication medium was tested whereby, a substantial cost reduction in raising multiple shoots was achieved (data not presented). Further, a comparison of shoot proliferation rates in agar-gelled and liquid media indicates an 8-fold increase in the number of shoots compared to a 4- fold increase in agar-gelled media. The higher proliferation rate, as well as the development of sturdy shoots with thicker stem in liquid media could be ascribed to: (i) increased availability of cytokinins and other nutrients in liquid medium (Debergh, 1983), (ii) dilution of any exudates from the explants (Ziv and Halevy, 1983), and (iii) greater aeration of the cultures (Ibrahim, 1994).

A close scrutiny of literature indicates that rooting response with different auxins is cultivar–dependent, also it is difficult to induce rooting in oil-bearing rose cultivars (Kirichenko et al., 1991). Further, Chu et al. (1993) reported that only shoots of *R. chinensis* cultured in liquid medium without BAP developed roots. However, in the present study, the experiments on root induction were done with great consistency and ease and, moreover, used microshoots cultured in liquid medium supplemented with BAP.

In the present study, it was established that agar greatly contributes to the osmotic potential of the medium and in turn affects rooting of microshoots. A similar effect of agar on osmotic potential of the medium was also reported earlier by Ghashghaie et al. (1991) while studying agar concentration on water status and growth of rose plants cultured *in vitro*. The other finding, which relates to osmotic potential of the medium on root induction is also supported by the earlier report by Lakes and Zimmerman (1990) for apple.

Rooting of microshoots is an important step for hardening and subsequent establishment in soil. For consistency in the rooting of microshoots, a rooting vessel was designed and developed in our laboratory (Figure1). The major advantages of such a vessel were: i) mass scale rooting of shoots, ii) elimination of agar resulting in substantial savings in cost, iii) ease of operation, iv) minimisation of damage to roots as rooted plantlets can be easily pulled out of the containers.

The micropropagated roses are known to be a difficult system for hardening and acclimatization as they undergo rapid desiccation. Such plants are also susceptible to diseases due to high relative humidity (Messeguer and Melle, 1986). Moreover, it was reported by Tanimoto and Ono (1994) that multiple shoots of rose, established in liquid medium failed to acclimatize in soil. In the present study, these limitations were overcome by keeping the transplanted microshoots in specially designed low-cost hardening chambers enriched with CO_2 and with high relative humidity, permitting up to 96.7% survival in *R. damascena* and 100% in *R. bourboniana*.

In conclusion, a workable cost-effective, simple and efficient protocol has been evolved for micropropagation for both *R. damascena* and *R. bourboniana* using liquid culture media. Importantly, a simple culture vessel for efficient and large-scale rooting of microshoots was designed, developed from locally available autoclavable boxes and tested for its performance.

References

Aitken-Christie J, Kozai T & Takayama S (1995) Automation in plant tissue culture. General introduction and overview. In: Aitken- Christie J, Kozai T & Smith MAL (eds) Automation and Environmental Control in Plant Tissue Culture (pp. 1-15). Kluwer Academic Publishers, Dordrecht

Chu CY, Knight SL & Smith MAL (1993) Effect of liquid culture on the growth and development of miniature rose (*Rosa chinensis* Jacq. 'Minima'). Plant Cell, Tiss. Org. Cult. 32: 329-334

Debergh PC (1983) Effects of agar brand and concentration on the tissue culture medium. Physiol. Plant. 59: 270-276

Ghashghaie J, Brenckmann F & Saagier B (1991) Effects of agar concentration on water status and growth of rose plants cultured *in vitro*. Physiol. Plant. 82: 73-78

Horn WAH (1992) Micropropagation of rose (*Rosa* L.). In: Bajaj YPS (ed) Biotechnology in Agriculture and Forestry, Vol. 20, High- Tech and Micropropagation IV (pp. 320-342). Springer, Germany

Ibrahim AI (1994) Effect of gelling agent and activated charcoal on the growth and development of *Cordyline terminalis* cultured *in vitro*. In: Proceedings of the First Conference of Ornamental Horticulture Vol 1 (pp. 55-67)

Ishioka N & Tanimoto S (1990) Plant regeneration from Bulgarian rose callus. Plant Cell, Tiss. Org. Cult. 22: 197-199

Khosh-Khui M & Sink KC (1982) Micropropagation of new and old world rose species. J. Hort. Sci. 57: 315-319

Kirichenko EB, Kuz'-mina TA & Kataeva NV (1991) Factors in optimizing the multiplication of ornamental and essential oil roses *in vitro*. Byulleten'-Glavnogo-Botanicheskogo-sada. 159: 61-67

Kornova KM & Michailova J (1994) Study of the *in vitro* rooting of Kazanlak oil-bearing rose (*Rosa damascena* Mill.). J. Essent. Oil Res. 6 (5): 485-492

Kumar A, Sood A, Palni UT, Gupta AK & Palni LMS (2001) Micropropagation of *Rosa damascena* Mill. from mature bushes using thidiazuron. J. Hort. Sci. Biotech. 76: 30-34

Lakes C & Zimmerman RH (1990) Impact of osmotic potential on *in vitro* culture of apple. Acta Hort. 280: 417-424

Messeguer J & Mele E (1986) Acclimatization of *in vitro* micropropagated roses. In: Somers DA, Gegenbrach BG, Biesboer DD, Hackett WP & Grenn CE (eds) Abstracts of the VI International Congress of Plant Tissue and Cell Culture (p 236). Univ. Minnesota, MN

Murashige T & Skoog F (1962) A revised medium for rapid growth and bioassays with tobacco tissue cultures. Physiol. Plant. 15: 473-479

Pati PK, Sharma M & Ahuja PS (2001) Micropropagation, protoplast culture and its implications in the improvement of scented rose. Acta Hort. 547: 147-158

Rout GR, Samantaray S, Mottley J & Das P (1999) Biotechnology of the rose: a review of recent progress. Sci. Hort. 81: 201-228

Sharma M, Sood A, Nagar PK, Prakash O & Ahuja PS (1999) Direct rooting and hardening of tea microshoots in the field. Plant Cell, Tiss. Org. Cult. 58: 111-118

Short KC & Roberts AV (1991) *Rosa* spp. (Roses): *In vitro* culture, micropropagation and production of secondary products. In: Bajaj YPS (ed) Biotechnology in Agriculture and Forestry Vol 15 Medicinal and Aromatics Plants III (pp. 376-397). Springer, Berlin

Skidmore DI, Simons AJ & Bedi S (1988) *In vitro* culture of shoots of *Pinus caribaea* on a liquid medium. Plant Cell, Tiss. Org. Cult. 14: 129-136

Smith MAL & Spomer LA (1994) Vessel, gels, liquid media and support systems. In: Aitken-Christie J, Kozai T & Smith, MAL (eds) Automation and Environmental Control in Plant Tissue Culture (pp. 371-404). Kluwer Academic Publishers, Dordrecht

Tanimoto S & Ono R (1994) Development of plant propagation method in rose. Bull. Fac. Agri. Saga Univ. 77: 17-26

Taslim A & Patel BA (1995) Micropropagation of rose – a review. Agric. Rev. 16: 211-218

Ziv M & Halevy AH (1983) Control of oxidative browning and *in vitro* propagation of *Strelitzia reginae*. Hort. Sci. 18 (4): 434-436

IV. Commercial Process Development andCulture Environment

Chapter 30

Mass propagation of conifer trees in liquid cultures - progress towards commercialization

Pramod K. Gupta & Roger Timmis
Weyerhaeuser Co., Federal Way, WA 98063 USA

Abstract: At Weyerhaeuser Company, somatic embryo production in liquid medium and manufactured seed delivery has been developed to reduce labor costs and increase efficiency of mass clonal propagation. We have scaled-up embryonal suspensor masses (ESM) of Douglas-fir in 1-litre liquid medium flasks. Over 10,000 somatic embryos have been produced from a single flask. We have cryostored ESM of over 700 genotypes in liquid nitrogen. Somatic embryos of Douglas-fir have also been produced from liquid medium in a bioreactor. Different types of bioreactors are required for embryo multiplication and for cotyledonary embryo development. Over 250,000 Douglas-fir somatic seedlings from a large number of genotypes have been produced for clonal field tests.

Key words: bioreactor, Douglas-fir, manufactured seed, somatic embryos

Abbreviation: ABA – abscisic acid, ESM – embryonal suspensor mass, GA – gibberellic acid, PEG – polyethylene glycol

1. Introduction

Mass propagation *via* tissue culture is more expensive in many plant species than conventional vegetative propagation such as by cuttings and this often limits its commercial application. Labor costs generally account for 60% or more of the total production costs of *in vitro* propagation. These can be lowered by robotic automation of a few steps of the micropropagation process by robots or by use of somatic embryogenesis.

Somatic embryogenesis is the process for development of embryos from somatic cells. It is accomplished by taking cultures through a series of developmental stages that are similar to zygotic embryogenesis. Somatic embryo development in liquid medium with automation offers tremendous

A.K. Hvoslef-Eide and W. Preil (eds.), Liquid Culture Systems for in vitro Plant Propagation, 389–402.

potential as a method for mass propagation of superior conifer genotypes. It potentially provides many production advantages:

1. A large number of plantlets can be produced inexpensively.
2. Production of both root and shoot meristem occur in the same process step.
3. Easy and quick scale-up can be achieved *via* liquid culture.
4. Long term storage *via* cryopreservation can be utilized.
5. There are opportunities for use of manufactured seeds for direct delivery to a nursery and easy handling.

In forestry, the production of manufactured seeds throughout the year provides a complementary technology, which will reduce risks relative to seed orchards where seed production is limited and variable.

Considerable progress has been made over the last decade in the development of somatic embryogenesis systems for large-scale clonal propagation of conifers. Since the first report in 1985, a large number of papers have been published on development of protocols for somatic embryo development, maturation, germination and cryopreservation of several conifer species (Gupta et al., 1993, Gupta and Grob, 1995, Timmis 1998). Several patents have been granted to forest industries and universities on conifer somatic embryogenesis technology. Weyerhaeuser Company also has several patents on this technology. Commercialization of this technology consists of the following steps:

1. Initiation of ESM (embryonal suspensor mass) from a large number of genotypes of selected families.
2. Multiplication of ESM cultures in liquid medium.
3. Cryopreservation of ESM cultures.
4. Production of somatic seedlings for clonal field-testing.
5. Development of an efficient low cost delivery system to nurseries.
6. Collection of data from clonal field-tests.
7. Retrieving the ESM of field-tested clones from cryostorage.
8. Scale-up of ESM cultures for large-scale production of somatic seedlings and clonal reforestation.

This paper describes all of the above steps and scale-up in liquid media for mass clonal propagation using somatic embryogenesis technology.

2. Culture initiation and maintenance

Induction of embryo suspensor masses (ESMs) has been reported from immature embryos, mature embryos, hypocotyls, cotyledons, and explants of somatic and zygotic seedlings of Norway spruce (*Picea abies*). Recently, ESM induction has also been reported from explants of 10-20-year-old trees of *Pinus radiata*, *Pinus pinaster* and *Picea abies* (Timmis, 1998). At Weyerhaeuser, we have initiated ESM from immature embryos (pre-dome and dome stages before development of cotyledons) of Douglas-fir (*Pseudotsuga menziesii*) and loblolly pine (*Pinus taeda*). ESM cultures were initiated onto solid medium with auxin and cytokinin, details of which has been published earlier. There is no callus stage in ESM induction from immature embryos. ESMs form directly from heads of early-stage embryos through cleavage polyembryony. Cleavage polyembryony is natural in several conifer species, and multiple embryos develop inside the megagametophyte through this process. ESM cultures are multiplied true-to-conifer-type cleavage polyembryony in the presence of auxin and cytokinin (Durzan and Gupta, 1987). Induction of ESM has not been achieved in liquid medium. Semi-solid medium is necessary for ESM initiation, and a relatively low gelrite concentration (0.1 – 0.2 %) has been found best for the induction process (Becwar, 1994).

After initiation, ESM cultures are transferred to maintenance medium for continuation of true-to-conifer-type cleavage polyembryony under lower concentrations of auxin and cytokinin. At this time osmolality is increased, from 90-100 to 190-200 mmol kg^{-1}, with 5 g l^{-1} myo-inositol and 30 g l^{-1} maltose (compared with 0.1 g l^{-1} myo-inositol and 15 g l^{-1} sucrose in initiation medium). We have found that the type of sugar is very important for subsequent embryo development on plates. Early-stage embryos were able to fully mature only when grown in maltose maintenance medium (Gupta, 1996). Without increased osmolality treatment, which leads to larger embryonal heads of early-stage embryos, many genotypes of Douglas-fir and loblolly pine did not develop good quality cotyledonary embryos (Gupta and Pullman, 1990).

3. Culture establishment in liquid medium

Liquid cultures offer a number of technical advantages over solid cultures. Cultures grown in liquid medium have shown a faster rate of growth. Cultures are bathed in nutrients, which allows rapid uptake of nutrients by cells and speedy nutrient replacement at the cell surface by diffusion and movement from outlying liquid. In a gel, this is a slow process, generating gradients of concentration for each nutrient in the zone of gel next to the cells, and slowing growth. For these reasons, lower concentrations of nutrients are usually optimal, compared to those used in gel media formulation. Due to the faster diffusion rate in liquid systems, exuded growth inhibitors such as phenolics are rapidly diluted to innocuous levels. Negative effects on growth are thereby minimized. Liquid media can be filter sterilized, facilitating aseptic transfer into large closed vessels such as bioreactors. Handling of plant tissue for harvest and /or transfer is more amenable to mechanization, which will save labor and time. Scale-up of liquid cultures also requires less space than for their solid counterparts.

For establishment in liquid culture (suspension culture), 2-3 g ESM are transferred to 250 ml Erlenmeyer flasks with 20-25 ml of liquid medium. Flasks are placed on a rotary shaker (90-110 rpm) in darkness at 23°C. After 7-8 days, old medium is replaced with fresh. Twenty to twenty-five days after establishment of suspension culture, ESM liquid cultures are poured into sterile 100 ml measuring cylinders and allowed to settle for 30 minutes. The supernatant is discarded, and settled ESM cultures measured for volume. Settled ESM is subcultured in fresh medium at a density of 1:9 (v/v) by transferring 5 ml settled ESM to a 250 ml Erlenmeyer flask containing 45 ml of fresh liquid medium. ESM suspension cultures are maintained by regular weekly subculture.

4. Embryo development, maturation and germination

Embryo development is achieved with the combination of activated charcoal (1.25 mg l^{-1}) and ABA (30 mg l^{-1}) (Gupta and Pullman, 1990). Higher concentration of ABA is needed due to adsorption by activated charcoal. Abscisic acid alone did not inhibit the precocious germination of cotyledonary embryos. An increased osmolality of the medium is found to be necessary for late-stage embryo development and maturation. The osmolality of the late embryo development and maturation medium is increased to 300-600 mmol kg^{-1} by adding polyethylene glycol (PEG) MW 4,000-8,000, which was found to be the best osmoticum for good quality cotyledonary embryo development (Gupta and Pullman, 1991). Several workers have reported improved conifer embryo development with PEG (Attree et al, 1994). The quality of cotyledonary embryos was further improved by a combination of an increased osmolality with ABA (5-50 mg l^{-1}), $GA_{4/7}$ (5-50 mg l^{-1}) and activated charcoal (1.25 g l^{-1}) (Pullman and Gupta, 1994). Embryo development and maturation has been achieved using liquid medium soaked into pads. High osmolality of the medium on the pad has been maintained by adding a second pad (soaked with higher osmolality medium) beneath the first. Twenty to fifty cotyledonary embryos were produced from 1 ml settled ESM suspension culture on development medium after 5-6 weeks incubation in the dark (Figure 1). The variation in both quality and number of cotyledonary embryos produced from 1 ml settled ESM cultures was due to genotypic differences.

For germination, good quality cotyledonary embryos were selected individually (as looking similar to zygotic embryos) by hand from development medium under a stereo microscope. After cold treatment, selected embryos were transferred onto semi-solid medium for germination. Culture plates were incubated for the first 5-7 days in the dark followed by transfer to light. Percentage germination varied considerably among genotypes. Germinated embryos (Figure 2) bearing an epicotyl were then selected by hand and transplanted into 10 cubic inch Supercell pots containing a mixture of peat, vermiculite and perlite, and incubated in the greenhouse with frequent misting for acclimatization and growth. After establishment in soil, somatic seedlings were transplanted to Weyerhaeuser forest regeneration sites (Figure 3).

Figure 1: Somatic embryos of Douglas-fir from liquid medium.

Figure 2: Germinating somatic embryos on semi-solid medium.

5. Cryopreservation

Cryopreservation is at present considered essential for the large-scale commercial use of somatic embryogenesis technology for clonal forestry because it permits genetic material to be saved at low cost until its genetic value can be measured in field tests (see implementation section). We and others have found that almost 100% of embryogenic culture lines, frozen under controlled conditions and stored in liquid nitrogen, can be recovered after 5 years. Trees produced from them are visually indistinguishable from the products of unstored cultures, and there is no evidence for genetic change. Space requirements for cryogenic storage are minimal, since even a 45-litre storage unit typically can accommodate 8-10 thousand vials. A completely reliable system for maintenance of liquid nitrogen service, complete with alarm systems and backups, is essential. We have cryostored ESM of over 750 clones of Douglas-fir from superior full-sib families.

Figure 3: Douglas–fir trees from somatic embryos growing at Weyerhaeuser forest regeneration site, Washington.

6. Field performance

Somatic seedlings have now been growing for 9 years on a typical forest regeneration site in Washington State for clonal demonstration purposes. Strikingly uniform growth is apparent for somatic seedlings within a clone (Figure 3) compared with the less uniform zygotic seedlings. No morphological differences have been observed between somatic and zygotic seedlings. Recently, we have produced over 100,000 somatic seedlings of Douglas-fir from a large number of genotypes and established them in soil at Weyerhaeuser forest regeneration sites for clonal field tests.

7. Culture scale-up for production

Shake flask cultivation is a very effective method of growing embryogenic cultures for laboratory scale production of clones for field-testing. We have established ESM cultures of Douglas-fir and loblolly pine in liquid suspension in 1-litre Erlenmeyer flasks. Cultures have been bulked–up by weekly subculture. Each flask contains over 10,000 early-stage embryos which may double or triple in number weekly. However, the occurrence of culture variation from flask to flask, long-term stability of cultures and lack of control over the culture environment restrict the usefulness of shake flasks.

For large-scale production of somatic embryos, a bioreactor is one of the most promising ways for scaling-up the system. Bioreactors offer various advantages over shake flasks due to the possibilities for automation, continuous monitoring and control of growth conditions (agitation, pH, oxygen, and carbon dioxide), larger volume, and maintenance of homogeneous culture. Many configurations and sizes of vessels have been used to grow plant cells. Bioreactors for plant cell culture can be classified according to the type of agitation system used: mechanically agitated bioreactors, including aeration agitation bioreactors, rotating drum and spin-filter bioreactors, pneumatically agitated bioreactors, and simple aeration bioreactors. Non-agitated bioreactors, include gaseous phase (mist) bioreactors, oxygen permeable membrane bioreactors, overlay aeration bioreactors and perfusion bioreactors. Aeration-agitation bioreactors are often referred to as stirred tank and use impellers such as turbines, screws, paddles and helical ribbons (Ibaraki and Kurata, 2001).

There are several reports of ESM cultures of conifers in bioreactors (Tautorus et al., 1994, Gupta and Timmis, 1999, Ibaraki and Kurata, 2001). For different conifer species, different types of bioreactors have been used. (Table 1). At Weyerhaeuser, several ESM lines of Douglas-fir were grown in a 1.5-litre stirred tank bioreactor equipped with low speed, winged, magnetically driven stirrers without any cell damage (Timmis et al., 1998). After plating onto the pads imbibed with liquid development medium as described for the shake flask cultures, the embryo yields showed no significant difference from ESM grown in shaker flasks. However, embryos from ESM grown in bioreactors tended to be of larger size and germinated better, although the smaller embryos accounted for the improved germination (Timmis et al., 1998).

Moorhouse et al. (1996) used a Biostat BF2 (magnetic stirrer base) bioreactor for development of Sitka spruce (*Picea sitchensis*) somatic embryos. The maturation medium was perfused in the bioreactor to replace the initial proliferation medium in order to induce the development of somatic embryos in a submerged cell culture. However, embryos failed to develop beyond the globular stage. Cotyledonary embryo development has not been achieved in a submerged cell culture bioreactor.

Table 1: Embryonal suspensor mass (ESM) growth in bioreactor

Bioreactor type	Species	Author
1. Airlift-bioreactor	Black spruce (*P. mariana*) Interior spruce (*P. glauca-engelmannii*)	Tautorus et al. (1994)
2. Stirred-bioreactor	Black spruce (*Picea mariana*) Interior spruce (*P. glauca-engelmannii*)	
3. Stirred-bioreactor	Monterey pine (*Pinus radiata*)	Smith et al. (1994)
4. Stirred-bioreactor (Biostat BF2)	Sitka spruce (*Picea sitchensis*)	Moorhouse et al. (1996)
5. Stirred-bioreactor	Douglas-fir (*Pseudotsuga menziesii*)	Timmis et al. (1998)
6. Stirred-bioreactor, Bubble-bioreactor	Sitka spruce (*Picea sitchensis*)	Ingram & Mavituna (2000)

Table 2: Cotyledonary embryo development in a bioreactor or shake flask

Vessel / Bioreactor	Species	Authors
1. Perfusion bioreactor	White spruce	Attree et al. (1994)
2. Semi-continuous perfusion bioreactor	Douglas-fir	Gupta & Timmis (1999)
3. Regular immersion bioreactor	Norway spruce	Paques et al. (1995)
4. Shake flask	Norway spruce	Gorbatenko & Hakman (2001)

There are few reports on cotyledonary embryo development in a bioreactor using some support for ESM (Table 2). Cotyledonary embryo development and maturation of White spruce (*Picea glauca*) has been achieved in a bioreactor on flat absorbent pads above the surface of a liquid medium. Medium was continuously supplied to the one end of the pad, while the spent medium exited by gravity from the opposite end (Attree et al., 1994). Development medium was pumped from the reservoir into the bioreactor using a peristaltic pump at the rate of 60 ml per day.

Paques et al. (1995) produced cotyledonary embryo development of Norway spruce (*Picea abies*) in a bioreactor using polyurethane layers in liquid medium. ESM was immobilized in polyurethane layers and placed vertically in liquid maturation medium. The liquid medium was retained in polyurethane layers by capillary action, and replaced frequently with fresh medium. The early-stage embryos in direct contact with liquid medium (submerged) turned brown while those not in contact (above the surface of liquid medium) produced cotyledonary embryos.

At Weyerhaeuser, we have also used a bioreactor for cotyledonary embryo development of Douglas-fir. In the bioreactor, the medium was supplied semi–continuously from the lower surface of the pads to the developing embryos on the top. Development medium was pumped from the reservoir into the bioreactor until it made contact with the lower surface of pads. Medium was absorbed in the pads by capillary action, and after few hours, medium was pumped out to the reservoir. This was repeated at regular intervals until mature cotyledonary embryos developed. Higher yields of good quality embryos were produced in a bioreactor.

Now it is possible to produce somatic embryos of conifers from liquid medium in bioreactors. However, different types of bioreactor are currently required for early-stage embryo growth and cotyledonary embryo development and maturation. Recently Gorbatenko and Hakman (2001) produced desiccation-tolerant embryos of Norway spruce in liquid medium in a shake flask after 8 weeks by subculturing weekly at high density.

8. Embryo selection

The selection of good embryos from among the poorly developed ones and residual ESM can be a costly bottleneck in production. It is a necessary step, however, in order to ensure that (1) embryos in the initial mass on a plate or bioreactor are physically separated for one-by-one processing, and (2) only those with the highest probability of vigorous germination ("quality embryos") move forward in the production process. These two steps are typically carried out as a single hand motion in which a technician picks out each good-looking embryo and transfers it with sterile forceps to another vessel for germination. The costs of doing this on a commercial scale for reforestation (*versus* clonal testing), or of not selecting at all, are unacceptably high.

We have found that the physical separation process, "singulation", can be accomplished *en masse* by various methods of washing and sieving (Gupta et al., 1993) to eliminate both the residual ESM and under- or over-sized embryos. Such processes are quite amenable to large-scale mechanization, and do not reduce germination if done within defined operating limits.

To select quality embryos, we have developed automated selection algorithms based on embryo digital images. The process extends the earlier work of other groups (Iberaki and Kurata, 2001) by utilizing shape, color and texture information from three camera viewpoints perpendicular to one another. Classification models are based on germination truth data from training sets that have been imaged and tracked through the germination process. The models are developed by the computer itself, starting with all pixel data in all images of the training populations. The best 3- or 4-variable models are validated on independent test sets. Once a (genotype-specific) model has been chosen, its speed at classifying new embryos is much greater than manual classification, and the results more accurate (Timmis et al., 1999, 2001).

Work is in progress in our lab to improve selection further through chemical imaging. Analyses have revealed near infrared spectral features that are associated with germination competence, and identifiable with known classes of embryo storage compounds (Timmis, 1999). The extent to which images acquired in the appropriate wavelength bands will improve selection (by allowing chemistry to be mapped with morphology) is currently being evaluated.

9. Containerized delivery systems

Conventional containerized systems are those in which seeds are sown into soil or soil-like particulate medium in arrays of small containers, so that germination can take place under the more favorable and controllable environment of a greenhouse, rather than an outdoor nursery. There are a number of such systems in forestry use, and most are adaptable for use with somatic embryos. We have used miniplug trays because these have the additional advantage of being machine transplantable into forest nursery beds after a relatively short period for somatic embryo germination and acclimation indoors.

In this system (Timmis et al., 1992), the Miniplug trays are filled by machine with a soil mix such as vermiculite and sand, and a precise volume of sucrose-containing nutrient medium is then added. An adhesive is applied to the upper surface of the tray, which is then autoclaved. Embryos are sown ~5mm beneath the soil surface, and then a sterile transparent polyethylene film is placed in contact with the adhesive on the upper surface of the cells to provide protection from airborne contamination. A similar film can be placed on the undersurface. In this way the cells are isolated microbiologically from one another, and the trays can be transferred to a shaded greenhouse for germination.

The somatic embryos germinate into the headspace in about 2 weeks. The polypropylene film, which is more permeable to CO_2 than to water vapor, allows photosynthesis while slowing water loss, so that the plants gradually acclimate to the *ex vitro* environment. The type of adhesive selected allows the top and bottom films to be easily peeled off when the epicotyls reach the top of the headspace. The plants are then watered thoroughly to wash out any remaining sucrose, and thereafter tended much as ordinary containerized seedlings. An alginate gel treatment was found to improve the soil plug's integrity for earlier nursery transplanting (Pullman and Yancey, 1991).

In general, we found that direct-to-soil sowing of somatic embryos was superior to gelled medium because it produced an upright plant with well-branched root system and root hairs. Tests of variations in this method showed its potential for use with other container systems (e.g., Styroblocks, using dry sterile surface powders or particulates for isolation), or even with uncovered containers in sufficiently clean environments.

10. Manufactured seed delivery system

Agricultural processes are well developed for the use of natural seed. Artificial seed has the potential of delivering somatic embryos to the bareroot nursery or farm field using methods that require minimum alteration of current sowing equipment. Ideally, once artificial seeds are sown, the performance of the crop and the nursery or field cultural practices required should not differ from those for current nursery seedlings. No additional care should be necessary to culture the crop. Redenbaugh (1986) defined a synthetic seed as a somatic embryo inside a coating, and as being directly analogous to natural seed. Several names have been used for such "seed", including artificial seed, seed analog and somatic embryo seed. At Weyerhaeuser Company, we use the term "manufactured" seed, to reflect the nature of the construct more accurately. Several patents and papers have been published on synthetic seed design exhibiting low percentage germination. Calcium alginate and polyethylene oxide have both been widely publicized as coatings. Most of these studies have been done under laboratory conditions. Indication from the literature suggests that oxygen is limiting in the calcium alginate bead design. Neither alginate nor polyethylene oxide systems have been very successful under normal agricultural conditions due to their low resistance to desiccation, low nutrient and oxygen availability, failure of the geminating embryo to extract itself from the capsule, low tolerance of mechanical damage and seed bed temperature fluctuation.

Carlson et al. (1993) published a seed design, which overcomes many of these problems. This was accomplished by providing oxygen carrier emulsions, and a hard seed coat (wax impregnated paper) to prevent mechanical damage and desiccation, and a cotyledon restraint system to ensure emergence of the germinating embryo without trapping in the gel. Weyerhaeuser Company has several patents on this technology (Carlson and Hartle, 1995). Manufactured seed research is moving rapidly and will make it possible to deliver somatic embryos to the field at low cost.

11. Implementation

The implementation of somatic embryogenesis technology at Weyerhaeuser and several other forestry organizations worldwide has already begun. Clonal field tests have been established and corresponding clones have been cryostored. Performance of the individual genotypes in these tests will be carefully measured to identify outstanding genotypes as soon as possible. During this evaluation period (5+ years) continuing R&D is expected to bring the costs of somatic embryogenesis and manufactured seed technologies within acceptable limits. The selected genetic material will be retrieved from the clone bank, then multiplied and treated to produce embryos in larger vessels or bioreactors. Mature embryos will be harvested with the aid of a machine vision system, and fed to a manufactured seed assembly line to produce the quantities needed for reforestation. Manufactured seed will be storable, and sown through standard container facilities or nurseries.

12. Conclusion

Conifer embryonal cultures can be grown and multiplied in liquid culture in a bioreactor. However, embryo maturation has not been achieved in submerged liquid culture bioreactor. Perfusion bioreactors using absorbent pads have been used for embryo development and maturation. In conclusion, liquid culture combined with bioreactor, automation technology and manufactured seed delivery are the solution for low-cost mass propagation of conifer species.

References

Attree SM, Pomoroy MK & Fowke LC (1994) Production of vigorous, desiccation tolerant white spruce synthetic seeds in bioreactor. Plant Cell Rep.12: 601-606

Becwar MR (1994) Method for regeneration of conifer plantlets by somatic embryogenesis. US Patent No. 4,13930

Carlson WC, Hartle JE & Bower BK (1993) Oxygenated analogs of botanic seed, U.S. Patent No. 5,236,469

Carlson WC & Hartle JE (1995) Manufactured seeds of woody plants. In: Jain M, Gupta PK, Newton RJ (eds) Somatic Embryogenesis in Woody Plant Vol 1 (pp. 253-264). Kluwer Academic Publishers, Dordrecht

Durzan DJ & Gupta PK (1987) Somatic embryogenesis and polyembryogenesis in Douglas-fir cell suspension cultures. Plant Sci. 52: 229-235

Gupta PK & Pullman GS (1990) Methods for reproducing coniferous plants by somatic embryogenesis U.S. Patent No. 4,957,866

Gupta PK & Pullman GS (1991) Methods for reproducing coniferous plants by somatic embryogenesis using abscisic acid and osmotic potential variation U.S. Patent No. 5,036,007

Gupta PK, Pullman G & Timmis R (1993) Forestry in the 21st Century: The biotechnology of somatic embryogenesis. Bio/Technology 11: 254–459

Gupta PK & Grob JA (1995) Somatic embryogenesis in conifers. In: Jain SM, Gupta P K & Newton RJ (eds) Somatic Embryogenesis in Woody Plant Vol.1 (pp. 81-89). Kluwer Academic Publishers, Dordrecht

Gupta PK (1996) Methods for reproducing coniferous plants by somatic embryogenesis using maltose enriched maintenance medium, 1996. U.S. Patent. No. 5,563,061

Gupta PK & Timmis R (1999) Conifer somatic embryo production from liquid culture. In: Altman A, Ziv M, & Izhar S (eds) Plant Biotechnology and *In Vitro* Biology in the 21st Century (pp. 49-52). Kluwer Academic Publishers, Dordrecht

Gorbatenko O & Hakman I (2002) Desiccation-tolerant somatic embryos of Norway Spruce can be produced in liquid culture and regeneration into plantlets. Intl. J. Pl. Sci. 162: 2111-2118

Ibaraki Y & Kurata K (2001) Automation of somatic embryo production. Plant Cell, Tiss. Org. Cult. 65: 179-199

Ingram B & Mavituna F (2000) Effect of bioreactor configuration on the growth and maturation of *Picea sitchensis* somatic embryo cultures. Plant Cell, Tiss. Org. Cult. 61: 87-96

Ingram B & Mavituna F (1998) *Picea sitchensis* somatic embryogenesis: Proliferation and maturation in bioreactor. In: Proceedings Abstracts of IX International Congress on Plant Tissue and Cell Culture (p 83) Jerusalem, Israel

Moorhouse SD, Wilson G, Hennerty MJ & Selby CM (1996) A plant cell bioreactor with medium-perfusion for control of somatic embryogenesis in liquid cell suspensions. Plant Growth Reg. 20: 53-56

Paques M, Bercetche J & Palada M (1995) Prospects and limits of somatic embryogenesis of *Picea abies*. In: Jain M, Gupta PK, Newton RJ (eds) Somatic Embryogenesis in Woody Plant Vol 1 (pp. 399-414). Kluwer Academic Publishers, Dordrecht

Pullman GS & Gupta PK (1994) Method for reproducing coniferous plants by somatic embryogenesis using mixed growth hormones for embryo culture, U.S. Patent No. 5,294,549

Pullman GS & Yancey MJ (1991) Method of Transplanting a Container Growth Plant US Patent No. 5,060,418

Redenbaugh K, Viss P & Slade D (1986) Scale-up: Artificial seeds. In: Green CE, Sommers DA, Hackett WP (eds) Plant Tissue and Cell Culture (pp. 473-494). New York. Alan R. Liss, Inc

Smith DR (1994) Somatic embryogenesis joins the plantation forestry revolution in New Zealand. TAPPI Proceedings 94: 19-29

Tautorus TE, Lulsdrof SL & Kikcio SI (1994) Nutrient utilization during bioreactor cultures, and maturation of somatic embryos of black spruce and interior spruce somatic embryos. *In Vitro* Cell. Dev. Bio. Plant 308: 58-63

Timmis R (1998) Bioprocessing for tree production in the forest industry: conifer somatic embryogenesis. Biotechnology Progress 14: 156-166

Timmis R, Surerus-Lopez HA & Barton BM (1998) Preliminary experiments on Douglas-fir somatic embryo yield and quality from stirred bioreactors. Abstract. *In Vitro* Cell. Dev. Bio. Plant, 34: 21 A

Timmis R, Kreitinger ME & Yancey MJ (1992) Method and apparatus for culturing autotrophic plants from hetrotrophic plant material. US Patent No. 119588

Timmis R, Toland MR, Gheermay T, Carlson W & Grob J (1999) Methods for classification of somatic embryos. US Patent PCT W/0 99/63057

Timmis R, Toland MR, Ghermay T & Sureus-lopez HA (2001) Automated selection of germination-competent conifer somatic embryos by multi-viewpoint color image analysis. Abst. In Intl. Conf. on Breeding, Wood, Biotechnology (p. 60). Bordeaux

Chapter 31

Potentials for cost reduction in a new model of commercial micropropagation

V.A. Savangikar*, Chitra Savangikar*, R.S. Daga** & Sunil Pathak**

** Consultants in Commercial Plant Tissue Culture, A-6, Patil Presidency, Patil Estate, Vise Mala, College Road, Nashik-422 005. India. Tel: 0091 253 2314896, 2329415; Fax: 0091 253 2571064, E-mail: info@tissuecon.com; woodcon_nsk@sancharnet.in. (Corresponding Author*)*

*** Saubhagya Kalpataru Pvt. Ltd., Sarafa Lane, Hinganghat 442 301, Dist: Wardha, India.*

Abstract: A commercial micropropagation facility, using semi-solid medium in jam bottles as culture containers and a conventional growthroom (with filtration of the air condition system and artificial lighting), was adapted to a new system which uses proprietary liquid medium, improved handling methods to speed up inoculations in polybags and biologically clean stock cultures. Diffused sunlight in the greenhouse is used as light source for the multiplication and rooting of cultures. The temperature of the greenhouse was only controlled by fan-and pad cooling when exceeding 37°C in the afternoons of hot days, allowing ambient temperature of the greenhouse as sufficient control for the cultures. A representative comparison from the daily commercial production records of both the systems was conducted. In the new system, number of propagules produced per workstation per day increased 5 times, and rate of multiplication of propagules per culture in a multiplication cycle increased 3.5 times. Thus, net improvement of performance of the new system is product of these, i.e. 17 times. This is without infusion of automation and with simultaneous reduction in capital costs, running costs as well as interest cost. Further, losses from contamination and during hardening stage is also reduced in the new system. Cost data has been presented on a comparative basis to avoid disclosing commercially sensitive absolute figures. Use of natural light and ambient temperature for culture incubation achieved about a 75% reduction in capital cost as well as in the cost of electricity. Production of cultures per worker, on account of new proprietary handling methods, was 70% more in the new model, a factor relevant to developed economies as this feature can be utilized in their present systems too, whether rest of the features of new system are adapted or not. Potential reduction, extrapolated from actually achieved results, in cost of production is 92 - 98% in Indian context and is estimated to offer about 55-87% for developed countries, depending upon efficiency of hardening. Thus this model is relevant to developing as well as developed countries. Further scope exists in improvement in cost reduction in tropical and semi-tropical climate. In temperate climate, scope is lesser for cost reduction, but the same may yet be practically worthwhile when compared to costs in conventional model. Over and above the advantages mentioned above, sturdiness and survival of plants in hardening was better in plants coming from new system.

A.K. Hvoslef-Eide and W. Preil (eds.), Liquid Culture Systems for in vitro Plant Propagation, 403–414.

Key words: cost effective technology, cost of production, low cost technology, micropropagation

1. Introduction

Commercial micropropagation is the method of choice for multiplication of elite plants in ornamentals, fruit crops and woody species, further for improvement in productivity of agricultural and field crops such as sugarcane, banana, potato and for multiplying selected mutants and genetically engineered plants (http1; http2; http3; http4; http5; DaSilva 1998a; DaSilva 1998b). It has been widely acknowledged, however, that high cost of production is a limited factor in practical application of this technology. Solving this problem is a priority area for many R&D programmes.

Several cost reduction strategies have been adopted by several groups, such as use of jam bottles as culture containers, use of liquid media instead of semi-solid medium to save cost of agar and to get better growth rates, use of alternative gelling agents (Pierik, 1989; Kodym and Zapata, 2001; Nagamori and Kobayashi, 2001; http6), cheaper grades of chemicals, use of natural light for culture growth (Kodym and Zapata, 2001; Savangikar and Savangikar; 2001), securing concessional tariffs for electricity consumption, use of bioreactors or robots. Some of these measures have indeed resulted in some reduction in cost. However, a further reduction is desired.

In developed countries, use of bioreactors and robots are considered as potentially most useful strategies for reduction in cost, since high wages is the dominant factor. However, the bioreactor model, based mostly on somatic embryogenesis, has encountered new problems related to contamination control, loss of embryogenic potential and doubts about genetic stability of the product. Benefits of reduction in requirement of human labour by introduction of bioreactors and robots being offset by increase in capital cost. The use of robots, has yet to come to a technically satisfactory stage, as robotic has yet to adapt the vision and image analysis required for tissue culture operations.

A new comprehensive approach, which combines several cost reduction strategies, based on the genetically conservative technique of meristem and shoot tip culture followed by enhanced axillary proliferation, has been perceived to have high potential of cost reduction (Savangikar and Savangikar, 2001). This approach was adopted by a commercial enterprise in large-scale production to evaluate its commercial utility over the old (conventional) model. Production data maintained by this enterprise after this adaptation and data kept for the old model for production operations carried out before adaptation, has been used as practical tool for comparative evaluation in this work. Special significance of this paper lies in the fact that this is commercial validation of a set of cost effective measures under real commercial conditions, and an analysis of contribution of factors leading to cost reduction. Being a product of real commercial activity, its reproducibility and validity under commercial conditions is already established.

2. Materials and methods

Sugarcane was the crop used for micropropagation. Shoot tip culture was used to initiate the cultures and multiplication was done by enhanced axillary proliferation.

In old model, bottles containing semi-solid medium were used as culture vessels. Class 10,000 air areas (contamination less than 10,000 particles per cubic foot, 0.5 μm and larger) were used for media storage, inoculation rooms and culture incubation rooms. Cultures were illuminated by artificial light from cool fluorescent lights giving 1000 lux at the level of cultures and temperature in growth room was controlled by centralized air conditioning to regulate temperature at 25 ± 2°C.

In the new model, polybags with liquid medium were used as culture containers. Culture inoculation operations were carried out in laminar clean air flow units which were housed in ordinary rooms not provided with class 10,000 air area. Cultures were incubated while hung on ropes, in a fan-and-pad greenhouse built up directly in field with no artificial flooring and under natural light illumination. The fan-and-pad arrangement was capable of giving about 5°C drop in temperature of air and was used whenever temperature inside greenhouse exceeded 37 °C. Day-length in the period in which above work was done ranged around 11.5 hours.

Polybags of polypropylene of 80 to 100 gauge thickness and 20 cm length and 15 cm width were sterilized at 15 pounds pressure and at 121°C for 15 minutes. Propagules were inoculated in sterile bags containing about 30 ml of sterile medium. Allowing enough air inside to keep the bags in inflated condition, the mouth of the bags was folded, first horizontally two times and then vertically several times from both sides, each fold being about 1 cm in width and the final bunch of folds of each bag held together at place by heat sealing (or any alternative means, such as use of rubber band, tying thread etc). At this stage, the bag looks like an inflated triangular balloon with several folds of polypropylene at the top and with medium and propagule inoculated in it. Photograph of inoculated polybag is available on web-site http7. Thread was tied to the sealed bags. Folds can be held together even by applying a rubber band around it or by applying a clip. Length and width of the bags and volume of the medium used can be varied depending upon the size of the inoculum.

For the dimensions of the bags given above, single plants were inoculated, which transformed into a bunch of plants in four to six weeks depending on the ambient temperature. The lower the ambient temperature, the more the time required to come to full development.

Inoculated bags were tied in a group of 18 bags in each taking care to avoid overlap, on a thick rope in the greenhouse. While hanging, the bags may remain horizontal, or usually, medium accumulates in one of the corners of the bag. In a greenhouse of 28 m length and 25 m width, 6 meter height at center, 100,000 bags were accommodated. Mutual shading of the bags was avoided as far as practically possible.

Mother cultures to be used in the new model were adapted to a proprietary medium with better fine-tuning of the ingredients and new conditions of incubation for three passages and were subjected to visual indexing. As a matter of policy, antibiotics, high levels of cytokinin and 2,4-D are not used at any stage in the medium. Multiplication cultures were also initiated using medium of the old model in liquid form (by omitting agar) and cultures incubated in old model as well as in the new model.

The location of the laboratory is near 20° North and 80° East in central India. Climate of the location of the facility is semi-tropical, hot and humid. In hot days in May, ambient temperature reaches up to 43° to 46°C for few hours in the day. In winter months, minimum temperature reaches 6° C approximately. Fan-and-pad cooling is needed for about four months per year, in October, March, April and May, when maximum ambient temperature crosses 36°C in the afternoon. For most of the days in monsoon months of June to September, sky is overcast with clouds. In rest of the days in a year, the sky is clear with full sunshine. Ambient humidity in monsoon remains between 70 to 90%, in October it remains between 50 to 65%, in winter months of November to January it remains between 50 to 80% and in other months, it comes down sharply and remains between 10 to 30%. Primary hardening is usually carried out between August to December, when humidity inside the polyhouse is maintained between 70 to 90% by intermittent misting.

The production data maintained in regular course for operations in old and new model was used to compare the practical efficiencies of both models. The data selected for comparison was based on the steady state period of large-scale operations with the use of 6 laminar clean air flow stations in the new model in a one shift operation and 10 laminar clean air flow stations in old model in a two shift operation, taking the average over the period. From the old system data, an 11 week period was selected for culture production as well as greenhouse hardening. For new system, such data is available from 8 weeks of production and 4 weeks of hardening. These periods are representative enough of practically feasible large scale sustainable process efficiency.

Considering that with the same process efficiencies, economics changes with various production options, such as year round or seasonal production and number of shifts operated per day, extrapolations were made to match some representative combinations of production options. The data is given in a generic form to make it universally relevant, free from currency calculations and inflation effects. To enable generic presentation, after calculations were made in terms of Indian rupees, the cost of production of the Production Option 6 (old system in three shifts operation) is considered to be one currency unit and costs calculated for other models and other options were expressed in proportion to this option. The following six Production Options are considered:

1. One day one shift practically achieved capacity - New and old model: This contains data on average achieved performance. This forms the basic data from which all other extrapolations are made.
2. Extrapolation: One day one shift 10 laminar clean air flow workstations – New and old model: This forms the equal inoculation capacity basis for further extrapolations.
3. Extrapolation: 120 calendar days, one shift 10 laminar clean air flow workstations – New and old model: This is the practical option in which many micropropagation units operate for agricultural crops of seasonal nature.
4. Extrapolation: 120 calendar days, three shifts, 10 laminar clean air flow workstations – New and old model: This is the next practical option in which some micropropagation units operate for agricultural crops of seasonal nature. Some may operate at a two shift operations level, in which case, an extrapolation between (4) and (3) can be made.
5. Extrapolation: 365 calendar days, one shift 10 laminar clean air flow workstations – New and old model: Number of labs operating at this level is few. This option requires a product range, which can cover all parts of the year almost uniformly.
6. Extrapolation: 365 calendar days, three shifts 10 laminar clean air flow workstations – New and old model: This is an ideal system, which is seldom followed in practice. Two-shift operation is more a prevalent practice. In such cases, an extrapolation between (5) and (6) can be made.

In making cost calculations for the new and old model, following factors are considered:

a) Repayment on capital assets considered at 20% each year. Capital cost in new model being 25% of the capital cost for old model. It is presumed that capital costs shall be similar in all countries.
b) Interest (per annum) on capital cost considered at 18% for developing countries and at 4% for developed countries.
c) Depreciation on capital assets considered at 10% per annum.
d) Electrical consumption in new model, 33% of the old model.
e) Labour cost in developing countries is considered to be US $ 0.73 per day (8 hour shift). Same in developed countries is taken to be US $ 40 per day (8 hour shift).
f) Costs in primary and secondary hardening are not included in cost calculations as the location-specific differences, purpose specific requirements/imperatives and species specific imperatives differ widely and should be separately considered. Further, they are not a part of the micropropagation model and should be assessed entirely separately.

3. Results and discussion

Production from the new model was more reliable than the old model in terms of general vigor, contamination control and productivity. Multiplication and rooting were also achieved simultaneously. The multiplication culture lines initiated with use of old model medium in liquid form (by omitting agar) lost vigor slowly with every subculture and despite the application of visual indexing, were subject to high rates of contamination and were discontinued eventually. The new model cultures initiated with the new liquid medium and visual indexing were also subject to fluctuating contamination incidents, but continued to be vigorous and ultimately settled to a steady production rate. These stabilized lines are the inputs for the work which has given the reported and analysed data. It was also seen that success in primary hardening in plants derived from the new model was better than the plants derived from the old model. This was mainly due to the acclimatization effect, which the plants get already in culture developmental stage in the new model i.e. development of hardiness in leaves, stem and profuse rooting in culture itself as response to high intensity of natural illumination of the cultures while they are undergoing differentiation and multiplication while hanging in the greenhouse. Ziv (1986) has also maintained that acclimatization *in vitro* is the best possible strategy to ensure excellent survival rates when the plants are transferred to soil.

These observations show the importance of proper medium, biologically clean cultures having reduced risk of to expression of endophytic contamination, and incubation of cultures in high intensity of diffused natural sunlight and ambient temperature and day-length normally available in most sub-tropical and tropical climates for success of the new model. It is the combined interaction of all above mentioned factors that leads to success of the new model. The logistic feasibility of the new model in the successful use of natural light for culture illumination is mainly due to introduction of polybags as culture containers. The results of analysis of the available data on large-scale production in old and new model on sugarcane are given in table 1. In practice, exact conditions may vary with every facility. However, most of the units will find themselves close to one or other of the Production Options given and relevant cost calculations will be a useful guide for them to assess usefulness of new model for themselves. Quantitative capability during multiplication phase increased almost 17 times due to 5 times increase in number of propagules inoculated through the workstations (as a result of improved handling technique-logistics) in new model as compared to old model, and due to 3.5 times better rates of multiplication in cultures in new model (as a result of improved medium composition and culture health).

The achievement of very high multiplication rates bring in several practical advantages in commercial production planning. The sharp drop in cost of production makes it feasible to consider planning for a far excess production every day than the target, at little extra cost. This excess production very easily remedies or covers situations, as the case may be, such as a tissue culture manager's nightmare of unexpected loss of plants in a certain day's batch by contamination or unforeseen causes, an opportunity arising from a demand exceeding targeted production within a reasonable limit but registered after production has taken place, having at hand widest possible choice of sturdiest lot of plants to be chosen for hardening to ensure uniformity and limiting hardening losses etc. It may be readily appreciated when production costs are high, luxury of planning production in excess of what can be sold can not be opted for.

Table1: Comparative data on cost of production in conventional (old) and new system

Attribute/Production Option	Operative model	Volume of medium autoclaved x 10 liters	Number of workstations used	Number of mother cultures/propagules used for subculture x 1000	Number of propagule/cultures produced x 1000	Ratio of culture production in new model/ conventional (old) model i.e. D (new)/D (old)	Multiplication rate of the cultures	Ratio of multiplication rates in new model/conventional (old) i.e. F (new)/F (old)	Number of workers in media group
		A	B	C	D	E	F	G	H
1. One Day one shift achieved capacity	New	32.5	6	0.94	10.30	3	11.0	3.5	7
	Old	8.5	10	1.10	3.40		3.1		4
2. Extrapolation: One Day one shift 10 laminar clean air flow workstations	New	54.2	10	1.57	17.17	5	11.0	3.5	12
	Old	8.5	10	1.10	3.40		3.1		4
3. Extrapolation: 120 calender days, one shift 10 laminar clean air flow workstations	New	5400.0	10	188.0	2060.0	5	11.0	3.5	1400
	Old	850.0	10	132.0	408.0		3.1		480
4. Extrapolation: 120 calender days, three shifts, 10 laminar clean air flow workstations	New	16200.0	10	564.0	6180.0	5	11.0	3.5	4200
	Old	2550.0	10	396.0	1224.0		3.1		1440
5. Extrapolation: 365 calender days, one shift 10 laminar clean air flow workstations	New	16200.0	10	564.0	6180.0	5	11.0	3.5	4200
	Old	2550.0	10	396.0	1224.0		3.1		1440
6. Extrapolation: 365 calender days, three shifts 10 laminar clean air flow workstations	New	48600.0	10	1692.0	18540.0	5	11.0	3.5	12600
	Old	7650.0	10	1188.0	3672.0		3.1		4320

						Projections (Indian context)			Projections (Developed Countries' context)	
Number of operators on workstation	Number of workers in culture handling	Total number of workers used (H+I+J)	Number of cultures produced per worker	Ratio of production of cultures per worker in new/old model i.e. L (new)/L (old)	Production of No. of plants x 1000: No. of cultures in New Model (D) x 1.6; No. of cultures in old model (D) x 1.16	Ratio of production capability of plants x 1000 in new/old model I.e. N (New)/N (Old)	Relative per plant cost of production in Indian context: Projections	Relative per plant cost of production in Indian context: Projections on 100% increase in production of plants in New system	Relative per plant cost of production in context of developed countries: Projections	Relative per plant cost of production in context of developed countries: Projections on further 50% reduction in cost of production
I	J	K	L	M	N	O	P	Q	R	S
18	9	34	302.9	1.7	16.5	4.2				
13	2	19	178.9		3.9					
30	15	57	301.2	1.7	27.5	7.0				
13	2	19	178.9		3.9					
3000	1500	5900	349.2	1.7	3296.0	7.0	0.20	0.10	1.74	0.87
1300	200	1980	206.1		473.3		7.38	7.38	11.40	11.94
9000	4500	17700	349.2	1.7	9888.0	7.0	0.08	0.04	1.25	0.63
3900	600	5940	206.1		1419.8		2.60	2.60	6.63	6.91
9000	4500	17700	349.2	1.7	9888.0	7.0	0.08	0.04	1.25	0.63
3900	600	5940	206.1		1419.8		2.60	2.60	6.63	6.91
27000	13500	53100	349.2	1.7	29664.0	7.0	0.05	0.02	1.09	0.54
11700	1800	17820	206.1		4259.5		1.00	1.00	5.03	5.03

It was very clear that, by introducing more radical indexing procedures based on microbiological and molecular methods, by further fine tuning of the medium and by improving control on primary hardening, it will be feasible and rational to target a further increase of 100% or more of the number of hardened plants recovered from the cultures inoculated in last cycle. Hence, assuming success in achieving 100% increase over already achieved results, extrapolations are made to understand the potential benefits from new system for developing as well as developed countries.

Since the work done on new system was done as a pilot evaluation of the operating techniques, and the cultures were not initiated specifically to suit for the new model i.e. they were shoot tip cultures rather than meristem tip cultures, with a lesser degree of rejuvenation, there is significant scope for further improvement in performance than is observed in the introductory work.

Efficiency of production per worker used (taking all the workforce of the facility together) increased by 1.7 times. This is a significant improvement for developing as well as developed countries in the context of their major cost factors. For developing countries, this will improve utilization of capital (which attracts interest rate of 18% per year). For developed countries, this means reduction in wage bills.

It is very important to note that the cost of production of plants possible in Production Option 6 in new system, offering 95% reduction in cost of production, is very close to what approximates to the cost farmers in developing countries would find affordable when reasonable profit margin is included in it. Cost of plants in Production Options 2 or 3, which are more practical situations, although costlier, are still very close as they give 80% or higher reduction in cost of production in new system. Thus, new system gives an affordable product in all practical situations. Whereas, even the lowest cost of production in old system, given by Production Option 6, is clearly unaffordable to farmers. Rest of the options are out of question. Thus, in no circumstances will old system be useful for producing plants for agricultural and agroforstry applications, where the plants are not able to command high unit value in the market.

The cost of production in developed countries is already 5 times more than that for developing countries. However, the economic analysis shows that all the options become feasible in their context too when the new model is considered, as the range of reduction in cost extrapolated over the entire range of Production Options works out to 65% to 79%. This potential can further increase in the context of developed countries by mechanising several steps which are carried out manually in developed countries to take strategic advantage of availability of cheap labour.

Gross and Levin (1999) have claimed that by introducing their version of bioreactor technology, for a facility with production of 20 million units per year distributed equally throughout the year (12 cycles per year), capital cost of the conventional model, which stands at US \$ 2,512,950 would come down to US \$ 664,683, a reduction by about 74%, and reduction in cost of production of US \$ 0.16 per unit to 0.07 unit, a drop in production cost by 56%. Their version of bioreactor is applied even for multiplication cultures undergoing enhanced axillary proliferation, involves use of mechanical cutters, reduction in the area under strict environment control and incubation of bioreactors in areas having less rigorous and less expensive environmental controls. There are points of complementation between this and the new model. The propagules multiplied in bio-fermenters need transfer to static cultures for last round of culturing. This can be done in the new model, which will help in further reduction in capital costs and production costs than is achieved so far in the model of Gross and Levin (1999). Above projections of Gross and Levin (1999), however, presuppose contamination rates not exceeding 5%, which is a problem yet to resolve.

The importance of the results lies in the fact that results of commercial micropropagation obtained under uncontrolled ambient conditions are better in quality and at substantially reduced cost than results obtained in sophisticated and highly costly environment controlled production facility. This shows that expensive and highly sophisticated environment controlled facility is not indispensable for commercial micropropagation. Expensive controls on environment have led to several micropropagation units across the globe becoming unviable. This has also led to limiting application of the technology only to plant species capable of commanding high unit value in the market i.e. ornamental plants and denying feasibility of application of this technology to agricultural and plantation crops as high unit value of micropropagated plants in these cases makes the application unaffordable. Interestingly, the main reason for which public funds are sought to be spent on this technology is for applications for increasing productivity of agricultural plantation crops, which will become feasible with a the approach described in this paper.

Although the work is carried out in Indian conditions, the ambient conditions described are applicable to many regions in the world with minor adjustments. They are, incidentally, the type of regions where need of application to agricultural plantation crops is greatest. The results are from a sub-tropical location, and hence, represent an average of conditions available on both the climatic sides i.e. temperate and tropical. Moved to tropical side, the application may become more cost effective, as favourable natural climatic conditions are available throughout the year. Moved to temperate side, more investment may be necessary to make climate of greenhouse congenial in extremes of climate and extent of cost reduction will be lesser, but yet it may cost much less than totally environment controlled facility.

References

DaSilva E (1998a) http://ejb.ucv.cl/content/vol1/issue1/full/1/index.html

DaSilva E (1998b) http://ejb.ucv.cl/content/vol1/issue3/full/6/index.html

Gross A & Levin R (1999) Design Considerations for a Mechanized Micropropagation Laboratory Plant Biotechnology and *In Vitro* Biology in the 21st Century (pp. 637-642) Kluwer Academic Publishers, Dordrecht

http1 - http://www.fao.org/biotech/Conf1.htm

http2 - http://www.fao.org/biotech/Conf2.htm

http3 - http://www.fao.org/forestry/FOR/FORM/FOGENRES/genresbu/125/125e/arte11.stm

http4 - http://www.fao.org/unfao/bodies/COAG/COAG15/X0074E.htm#P95_1296

http5 - http://www.mnd.fh-wiesbaden.de/fag/bio/bo/cost_843/index.html

http6 - http://www.nrdcindia.com/pages/tisuecul.htm

http7 - http://www.tissuecon.com

Kodym A & Zapata-Arias FJ (2001) Plant Cell Tiss. Org. Cult. 66(1): 67-71

Nagamori E & Kobayashi T (2001) Viscous additive improves micropropagation in liquid medium. J. Biosci.Bioeng. 91(3): 283-287

Pierik RLM (1989) *In Vitro* Culture of Higher Plants. Martinus Nijhoff Publishers, The Netherlands

Savangikar VA & Savangikar Chitra (2001) New Perspective in Commercial Micropropagation Technology. In: Sandhya Tewari (ed) Business Opportunities in Biotechnology (pp. 119-121). Confederation of Indian Industry, New Delhi, India

Ziv M (1986) *In vitro* hardening and acclimatization of tissue culture plants. In: Withers L A & Alderson P G (eds.) Plant Tissue Culture and Its Agriculture Application (pp. 187-203). Buttersworths, London

Chapter 32

A new approach for automation: Sorting and sowing dehydrated somatic embryos of *Daucus* and *Coffea* using seed technologies

J.P. Ducos[1], B. Florin[1], J.M. Dupuis[2] & V. Pétiard[1]
[1]*Nestlé Research and Development Center, 101 Avenue Gustave Eiffel, Notre Dame D'Oé BP 9716, 37097 Tours Cedex 2, France; E-mail: jean-paul.ducos@rdto.nestle.com; bruno.florin@rdto.nestle.com;vincent.petiard@rdto.nestle.com.*
[2]*Aventis CropScience, 55 Avenue René Gassin, CP310, 69337 Lyon Cedex 09 France; E-mail: Jean-Marc.Dupuis@aventis.com*

Abstract: Somatic embryos of *Daucus carota* L. and *Coffea canephora* L. (var. Robusta) were dehydrated under a 43 % relative humidity then placed in the hopper of a precision seeding system used in the transplant industry. The seeder was adjusted to distribute the embryos onto horticultural trays, each one containing 240 cells filled with soil. As a preliminary result, 72 % and 88 % of the individual cells received a single embryo, in Daucus and Coffea respectively. The embryo-to-plantlet conversion rate was not affected either by the vibration of the hopper or by the nozzles. In carrot 66 % of the embryos germinated after the use of the seeding system (62% germination for the control). Sorting methods traditionally used for the seeds (e.g. air column, vibrating table) can also be used. Such an approach, based on desiccation as a key step, has the potential for a complete automation of the large-scale handling and delivery of somatic embryos.

Key words: automation, bioreactor, *Coffea canephora* L., *Daucus carota* L., desiccation, embryo sorting, embryo sowing, somatic embyogenesis

1. Introduction

As somatic embryos are relatively small and relatively uniform in size, somatic embryogenesis is more suitable for automation and scale-up than micropropagation *via* organogenesis. Approaches toward automation have been focused on production of "synthetic" seeds involving encapsulation of hydrated somatic embryos, which should allow direct sowing into the soil (Redenbaugh et al., 1987). The need for a sugar supply in the capsules

A.K. Hvoslef-Eide and W. Preil (eds.), Liquid Culture Systems for in vitro Plant Propagation, 415–423.

causes contaminations after transfer to a non-sterile environment (Molle et al., 1993). Mainly due to this difficulty to control the asepsis during the embryo development, this approach is difficult to implement. Consequently, some authors have described another approach where somatic embryos are embedded in sterilized plugs moistened with medium containing sucrose then maintained in an aseptic environment until they become photoautotrophic (Timmis et al., 1991, Gupta et al., 1993; Dupuis et al., 1999). However, automation of embryo distribution from liquid medium onto such plugs causes problems of aggregation and blockage inside the delivery pipes.

As a post-harvest treatment, desiccation has been applied to many species to enhance the embryo-to-plantlet conversion rate and/or the short term storage (Florin et al., 1993; Attree and Fowke, 1993). We investigated whether dehydrated somatic embryos can be handled using seed sowing technologies and equipment. To evaluate this possibility, we used somatic embryos of *Daucus carota* L. and *Coffea canephora* L., two species for which large numbers of somatic embryos can be produced in stirred bioreactors, up to $10^6\ l^{-1}$ in *Daucus* and up to $0.4{\cdot}10^6\ l^{-1}$ in *Coffea* (Ducos et al., 1993 a, b).

2. Materials and methods

2.1 Embryo formation

Daucus carota L.: Cell cultures were initiated from hypocotyl segments of aseptic seedlings of a cytoplasmic male sterile genotype (Clause Semences, Bretigny Sur Orge, France). Embryogenic tissues were developed on semi-solid medium (8.0 g l^{-1} of Bacto-Difco agar) then multiplied in liquid medium of the same composition. This medium contained the macro and micronutrients of Murashige and Skoog (1962), the mixture of organic substances of Halperin (1964), 0.1 mg l^{-1} 2,4-dichlorophenoxyacetic acid (2,4-D) and 20 g l^{-1} sucrose. The medium pH was adjusted to 5.8 prior to autoclaving (1.1 bar, 20 min at 115° C). Cell cultures were maintained by repeated subcultures every 12 days at an inoculation density of 10 g l^{-1}. For embryo production, cultures were successively filtered through two nylon sieves of 100 and 50 µm pore sizes. The cell clumps remaining on the latter filter were washed three times in fresh basal medium lacking 2,4-D and then transferred into the same medium at a density of 0.1 % (v/v) based on packed cell volume. Embryo productions were achieved in 0.5 litre of medium contained in 1-litre flasks. Cultures were grown on a orbital shaker

(110 rpm) at 25°C and under light intensity of 5 µmol m^{-2} s^{-1} with a 16 h photoperiod.

Coffea canephora (var. Robusta) L.: For coffee, young leaves collected from greenhouse-grown trees (clone FRT105) were disinfected for 30 min in a 4 % calcium hypochlorite solution then rinsed 3 times in sterile water. The explants were placed on the callogenesis semi-solid medium described by Yasuda et al. (1985). It was composed of 1/4 strength macronutrients and half strength micronutrients of Murashige and Skoog (1962), B5 vitamins (Gamborg et al., 1968), supplemented with 1.1 mg l^{-1} benzylaminopurine (BAP), 30 g l^{-1} sucrose and 8.0 g l^{-1} Bacto-Difco agar. To establish embryogenic cell suspensions, friable and yellowish calli were selected and transferred into liquid medium of the same composition. The cell suspensions were subcultured every 2 weeks at an inoculation density of 10 g l^{-1}. For the embryo production, 0.1 g of biomass was transferred into 100 ml of liquid production medium contained in 250-ml flasks. The production medium (Ducos et al., 1999) contained macro and micronutrients of Murashige and Skoog (1962), B5 vitamins (Gamborg et al., 1968), 1.1 mg l^{-1} BAP and 30 g l^{-1} sucrose. Adjustment of pH, autoclaving and culture conditions were the same as previously described for carrot.

2.2 *Embryo desiccation*

Somatic embryos were dehydrated as reported by Florin et al. (1993). To induce dehydration tolerance, an osmotic pretreatment was performed by subculturing the embryo suspensions in a liquid medium containing 0.4 mol sucrose during the last week of the embryo production phase. Then the embryos were transferred onto paper disks (Whatman GF/C, diameter 55 mm) and placed into 370-ml glass jars with a 43 % constant relative humidity (R.H.) created by the presence of a saturated solution of K_2CO_3. The jars were placed at 24° C and in darkness.

For the sorting experiments, somatic embryos were bulk collected on 400-µm nylon sieves then placed by layer onto the paper disks for desiccation (approximately 1,500 embryos per disk).

For the sowing experiments, torpedo-shaped somatic embryos were manually selected (0.5-1.5 mm and 1.5 to 3.0 mm, for carrot and coffee respectively) before being placed on the paper disks (150 embryos per disk) for desiccation.

2.3 *Sorting*

Bulk dehydrated embryos were placed at the base of a seed selector (Mag Process, diameter 5 cm, height 50 cm), which divides the seeds according to their weight by an upward airflow. The airflow rate was progressively increased unit by unit (1 minute for each unit). At each step, the embryo population was collected at the exit of the column. The number of embryo per class size was measured using an image analysis system (software Hellix, Microvision Instruments)

2.4 Sowing

Dehydrated embryos were placed in the compartimented hopper of a precision seeding system operating by suction (model Step-O-Mat, Visser). This apparatus uses nozzles to pick up the seeds from the vibrating hopper and deposits them down tubes into individual cells of the horticultural trays. The diameter of the nozzles was 0.25 mm. The seeder was adjusted to distribute the embryos onto polystyrene trays, each one containing 240 cells filled with soil. The experiments were carried out in a non-sterile environment (temperature: 24°C; R.H.: 25-30 %).

2.5 Germination

Carrot embryo germination was achieved according to Dupuis et al. (1999): *i)* somatic embryos were grown on cellulose acetate miniplugs moistened with a sterile liquid germination medium *ii)* a phytoprotection mixture against micro-organisms was added to this medium as the experiment was carried out in non-sterile conditions. This mixture contained a fungicide (iprodione 100 mg l^{-1}) and an antibiotic (cefotaxime 100 mg l^{-1}) (Dupuis et al., 1999). Each miniplug was placed into an individual glass jar. The germination medium was composed of macronutrients of Heller (1953), micronutrients of Murashige and Skoog (1962) and 15 g l^{-1} sucrose. The conversion rate was defined as the percentage of embryos giving rise to normal plantlets after 21 days. The plantlets were characterized by the initiation of at least two leaves and growth of the root.

3. Results and discussion

3.1 Sorting

Bulk populations of dehydrated somatic embryos were collected by gently tapping the paper disks held vertically then placed into a seed selector consisting of an air column. The higher the airflow rate, the longer the embryos collected at the exit of the air column. Starting from a bulk population, it is easy to harvest embryo populations of relatively uniform size: for instance, 95% of the carrot embryos had a size range between 0.4 and 1.6 mm in the population collected when the air flow rate reached the unit 3 (Figure 1). In the case of coffee, 97% of the embryos measured between 0.8- 2.2 mm at the airflow rate unit 8 (Figure 2).

As embryo size is an important characteristic for embryo development, embryos can be roughly sorted according to their developmental stage with this method. Filtration through nylon meshes of various pore size is a very simple and useful method for embryo sorting but presents some difficulties for large volumes of embryo suspensions: the filters should be thoroughly rinsed and blockage problems are difficult to avoid (Rodriguez et al., 1990; Ducos et al., 1993a). Sorting dehydrated embryos does not seem to be more efficient than filtration in liquid medium for size uniformity but it offers the advantage to be more easily scaled-up. Moreover, other sorting methods traditionally used for the seeds can be also assessed (e.g. vibrating table, sifters, alveolus sorter).

3.2 *Sowing*

Dehydrated embryos did not stick together and bounced in the vibrating hopper of the seeder like seeds. In the case of carrot, six trays were tested: 72 % of the cells received one embryo and 5 % two embryos (Table 1). In the case of coffee, the seeder was tested with only one 240-cell tray: as a preliminary result, the filling rate was 88 % with one embryo and only 1 % of cells containing two embryos. Coffee embryos could therefore be more easily sown than carrot embryos probably due to their higher weight. These filling rates can probably be improved by reducing static electricity inside the tubes and/or using another type of seeding system in which the seeds are directly transferred from the nozzles to the trays.

3.3 *Germination*

To evaluate the embryo capacity to develop into plantlets after automatic sowing, some carrot embryos were collected on paper at the exit of the tubes and then sown onto miniplugs. As carrot embryos grow very fast, they developed into plantlets within 21 days despite contamination (Table 2). The potential of bare carrot embryos to develop into plantlets was not affected either by the vibration of the hopper or by the nozzles. Even without phytoprotection, the conversion rate was not lower than in the control.

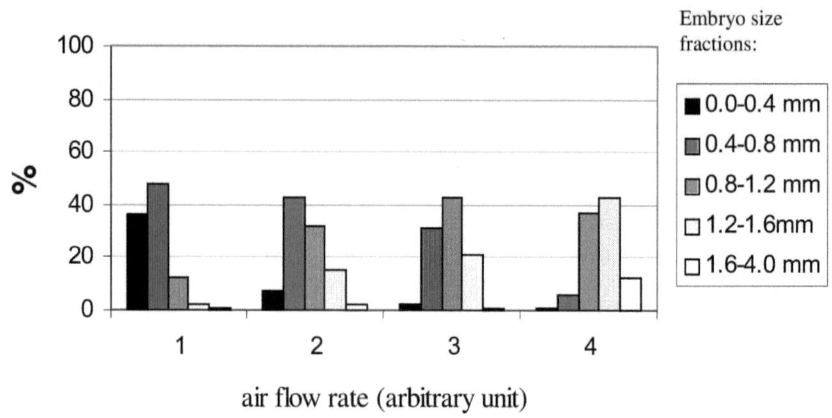

Figure 1: Fractions of carrot somatic embryos after sorting by use of a seed selector (air column).

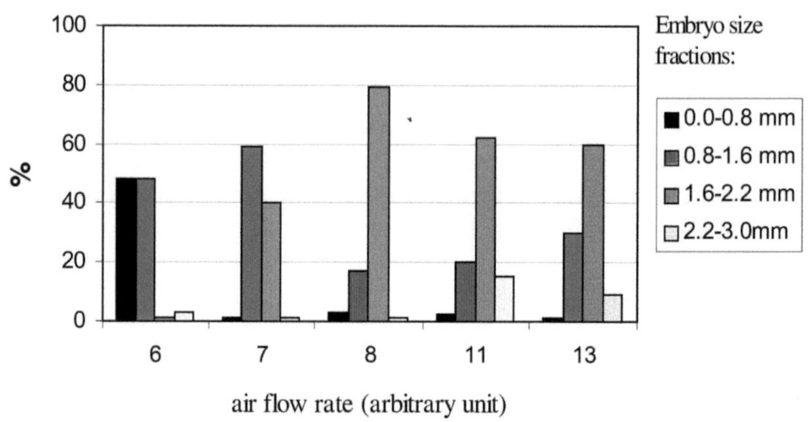

Figure 2: Fractions of coffee somatic embryos after sorting by use of a seed selector (air column).

Table 1: Sowing of dehydrated somatic embryos onto 240-cell trays using a precision seeding system (percentage of individual cells containing 0, 1 or 2 embryos)

Number of embryos per cell	0	1	2
Carrot	23±6	72±6	5±4
Coffee	11	88	1

Data are the mean percentage of six measures for carrot (i.e. six 240-cell trays). Only one measure for coffee (i.e. one 240-cell tray).

Table 2: Effect of the seeding precision system on the embryo-to-plantlet conversion rate of carrot somatic embryos

	Embryos (n)	Plantlets (n)	Conversion (%)
Control	100	62	62
After seeder	50	33	66
After seeder with phytoprotection[a]	80	50	63

[a] according Dupuis et al. (1999)

4. Conclusion

Desiccation of somatic embryos may not be effective for all species and/or must be adapted to their specific requirements. For example, the desiccation procedure described in this work did not maintain the survival of coffee embryos: they can be successfully dehydrated under 75 % R.H. but they regrow through a secondary embryogenesis process under 43 % R.H. conditions (Florin et al., 1995). Consequently, the possibility to handle coffee embryos using seed technologies must be assessed after a desiccation under a 75 % instead of a 43 % R.H. regime. Another key point for the implementation of this method is the scaling-up of desiccation *i.e.* to increase the volumes of bulk dehydrated embryos preventing them sticking together and/or to paper disks or other types of supports.

In the case of species like carrot, with a very fast growth rate, a chemical protection against micro-organisms for two or three weeks could be sufficient for the somatic embryos to reach the photoautotrophic status (Dupuis et al., 1999). For other species like coffee, for which the embryo-to-plantlet conversion is longer than two months, it will be difficult to maintain phytoprotection and, consequently, embryo sorting and sowing might have to be performed under aseptic conditions.

Such an approach (Ducos et al., 1998), based on desiccation, suggests a new strategy for a complete automation of a large-scale handling delivery process (up to 14,000 embryos sown per hour). Indeed, embryos can be delivered onto sterile miniplugs wetted with sucrose–containing media as containerized delivery systems (Timmis et al., 1991), pharmaceutical type capsules (Dupuis et al., 1994) or cellulose acetate mini-plugs (Dupuis et al., 1999). This would make the mass propagation *via* somatic embryogenesis very close to the horticultural transplant industry and economically viable for a large number of plant species.

Acknowledgements

The authors would like to thank D. Chenevotot (Clause Semences) who has directly participated on the sowing experiments.

References

Attree SM & Fowke LC (1993) Embryogeny of gymnosperms: advances in synthetic seed technology of conifers. Plant Cell, Tiss. Org. Cult. 35: 1-35

Ducos JP, Bollon H & Pétiard V (1993a) Production of carrot somatic embryos in a bioreactor. Appl. Microbiol. Biotechnol. 39: 465-470

Ducos JP, Zamarripa A & Pétiard V (1993b) Production of somatic embryos of coffee in a bioreactor. In: Association Scientifique Internationale du Café (ed) 15th Int. Sci. Colloquium Coffee (pp. 89-96). A.S.I.C, Paris

Ducos JP, Florin B, Chenevotot D & Dupuis JM (1998) Method for sowing meristematic tissues. Rhone Poulenc Agrochimie. World patent. Publication Number: WO 98/21932

Ducos JP, Gianforcaro M, Florin B, Pétiard V & Deshayes A (1999) A technically and economically attractive way to propagate elite *Coffea canephora* (Robusta) clones: *in vitro* somatic embryogenesis. In: Association Scientifique Internationale du Café (ed) 18th Int. Sci. Colloquium Coffee (pp. 295-301). A.S.I.C, Paris

Dupuis JM, Roffat C, Derose R & Molle F (1994) Pharmaceutical capsules as a coating system for artificials seeds. Bio/Technology. 12: 385-389

Dupuis JM, Millot C, Teufel E, Arnault A & Freyssinet G (1999) Germination of synthetic seeds: new substrates and phytoprotection for carrot somatic embryos to plant. In: Soh WY & Bhojwani SS (eds) Morphogenesis in Plant Tissue Cultures (pp. 418-441). Kluwer Academic Publishers, Dordrecht

Florin B, Tessereau H, Lecouteux C, Courtois D & Pétiard V (1993) Long-term preservation of somatic embryos. In: Redenbaugh K (ed) Synseeds. Applications of Synthetic Seeds to Crop Improvement (pp. 133-162). CRC Press, Boca Raton

Florin B, Ducos JP, Firmin L, Meshine MC, Thierry C, Pétiard V & Deshayes A (1995) Preservation of coffee somatic embryos through desiccation and cryopreservation. In: Association Scientifique Internationale du Café (ed) 16th Int. Sci. Colloquium Coffee (pp. 542-547). A.S.I.C, Paris

Gamborg OL, Miller RA & Ojima K (1968) Nutrients requirements of suspension cultures of soybean root cells. Exp. Cell. Res. 50: 151-158

Gupta PK, Pullman G, Timmis R, Kreitinger M, Carlson WC, Grob J & Welty E (1993) Forestry in the 21st century. The biotechnology of somatic embryogenesis. Bio/Technology 11: 454-459

Halperin W (1964) Morphogenetic studies with partially synchronized cultures of carrot embryos. Science 146: 408-410

Heller R (1953), Research on the mineral nutrition of plant tissues. Ann. Sci. Nat. Bot. Biol. Veg. 14: 1-223

Molle F, Dupuis JM, Ducos JP, Anselm A, Crolus-Savidan I & Pétiard V (1993) Carrot somatic embryogenesis and its application to synthetic seeds. In: Redenbaugh K (ed) Synseeds. Applications of Synthetic Seeds to Crop Improvement (pp. 257-287). CRC Press, Boca Raton

Murashige T & Skoog F (1962) A revised medium for rapid growth and bioassay with tobacco tissue cultures. Physiol. Plant. 15: 473-497

Redenbaugh K, Viss P, Slade D & Fujii JA (1987) Scale-up: artificial seeds. In: Green CE, Somers DA, Hackett WP & Biesboer DD (eds) Plant Tissue and Cell Cultures (pp. 473-493). A.R. Liss. New York

Rodriguez DL, Kitto SL & Lomax KM (1990) Mechanical purification of torpedo stage somatic embryos of *Daucus carota* L. Plant Cell, Tiss. Org. Cult. 23: 9-14

Timmis R, Kreitinger M & Yancey MJ (1991) Methods and apparatus for culturing autotrophic plants from heterotrophic plant material. Weyerhaeuser. World patent. Publication Number: WO 91/04655

Yasuda T, Fujii Y & Yamaguchi T (1985) Embryogenic callus induction from *Coffea arabica* leaf explants by benzyladenine. Plant Cell Physiol. 26: 595-597

Chapter 33

Use of the temporary immersion bioreactor system (RITA®) for production of commercial *Eucalyptus* clones in Mondi Forests (SA)

B. McAlister[1], J. Finnie[2], M.P. Watt[3] & F. Blakeway[1]
[1]*Mondi Forests, P.O. Box 12, Hilton, 3245, South Africa.*
[2]*School of Botany and Zoology, University of Natal Pietermaritzburg, Private Bag X01, Scottsville, 3209, South Africa.*
[3]*School of Life and Environmental Sciences, University of Natal Durban, King George Avenue, Durban, 4001, South Africa.*

Abstract: In order to optimise tissue culture systems and to meet production targets, Mondi Forests' biotechnology programme has in the last two years concentrated efforts on the use of the RITA® temporary immersion bioreactor system. Protocols have been established for six *Eucalyptus* clones. Results indicate a four- to six-fold increase in yield, in half the time, with the RITA® system when compared with axillary bud proliferation on semi-solid media. Furthermore, plants produced from the RITA® system are hardier and acclimatize better, giving higher yields of hardened-off plants. The establishment of aseptic axillary shoots into the RITA® system is from shoots in the semi-solid system. Highest multiplication was achieved using 30-second flushes of medium every 10 minutes, starting with 50 shoots per vessel. The multiplication cycles in RITA® are between 14 and 18 days, compared with 25 to 28 days in a semi-solid system. There is minimal callus evident on the leaves and bases of the stems of plants in the RITA® system and, in addition, cold-tolerant plants have a greater rooting competence when compared with plants coming from the semi-solid system. *Ex vitro* rooting of RITA®-derived plantlets is substantially better than the plants from the semi-solid media.

Key words: costs, forest tree, liquid vs. semi-solid tissue culture, rooting, woody plants

Abbreviations: BAP - 6-benzylaminopurine; EC - electrical conductivity; IBA - indole-3-butyric acid; MS - Murashige & Skoog (1962) medium; NAA - α-naphthaleneacetic acid

A.K. Hvoslef-Eide and W. Preil (eds.), Liquid Culture Systems for in vitro Plant Propagation, 425–442.

1. Introduction

Eucalyptus species and hybrids are important plantation trees throughout the world, including South Africa, as they are used for a wide variety of products. Due to diverse climatic conditions in South Africa, a variety of *Eucalyptus* species and clones is needed in order to produce appropriately site-matched planting stock in as short a time as possible. In Mondi Forests' *Eucalyptus* clonal programme, there is increasing focus on selected hybrids (viz. *E. grandis* x *E. urophylla* and *E. grandis* x *E. nitens* hybrids) which are disease-resistant, have more homogeneous wood density and withstand stress and climatic conditions (Denison and Kietzka, 1993). At present in the Company, seedlings and macro-cuttings are used to produce commercial planting clones. Additionally, *in vitro* propagation of *Eucalyptus* is used to provide stock and to replace macro-hedges and hydroponic hedges.

The latter approach is not new, as micropropagation has been used commercially for a large number of plant species, including trees, as multiplication of shoots is more rapid than other vegetative methods of multiplication (George, 1993). To date, the most common method of micropropagation of *Eucalyptus* involves the proliferation of shoots *via* a semi-solid system (see review by Le Roux and Van Staden, 1991). While such semi-solid systems have been moderately to highly successful in terms of multiplication yields, it has become increasingly important to improve productivity and reduce the time taken to multiply commercially-important material. In the last few years, reports in the literature have shown that temporary immersion bioreactor systems, such as RITA®, have numerous advantages compared to the semi-solid methods. Temporary immersion systems combine the advantages of gelled and liquid medium, in particular having intermittent total availability of nutrients, but still allowing the plants to grow in an air space. Using RITA®, Escalona et al. (1999) found that immersion increased the multiplication rates for *in vitro* shoots of pineapple. Akula et al. (2000) reported that the immersion frequency and immersion time impacted on multiplication rates of tea. The RITA® system has been found to increase root development in *Hevea brasiliensis* (Etienne et al., 1997). Examples of other advantages listed by various authors include improved micropropagule quality, reduced consumables costs, reduced labour costs (Etienne et al., 1997; Borroto, 1998), better leaf development, reduced hyperhydricity and minimized asphyxiation of tissue (Aitken-Christie et al., 1995). Further, plants from the temporary immersion system have been found to be more suitable for acclimatization and development towards photoautotrophy (Aitken-Christie et al., 1995).

This report describes how the RITA® system has been identified recently as a potentially useful method to increase multiplication yields and rooting of *Eucalyptus* clones, at lower costs. The value of this system is discussed in terms of yields, costs and application to the *Eucalyptus* plantation component of Mondi Forests' (South Africa) tree improvement and nurseries programmes.

2. Materials and methods

2.1 Establishment of axillary buds into RITA® vessels

The following establishment treatments were tested using six clones (GN107, GN108, NH58, GU175, GU180, TAG31) per treatment. After each treatment, explants were placed into RITA® vessels and any contamination was recorded.

Treatment a: Potted parent plants were sprayed with Sporgon® and Bravo® prior to harvesting. Single nodal explants with reduced leaf area were prepared, submerged and aerated for 3 h in 1 g l^{-1} Benlate® and 1 g l^{-1} boric acid. The explants were surface-sterilized with 2 g l^{-1} calcium hypochlorite for 5 minutes and then placed in 0.1 g l^{-1} mercuric chloride, plus two drops of Tween® 20 for two minutes. After which the explants were washed three times with sterilized water and rinsed with 1 ml l^{-1} Bravo®.

Treatment b: Secondary leaders from rooted cuttings were surface-sterilized with 0.5, 1 or 2 g l^{-1} calcium hypochlorite for 5, 10 or 15 min, and rinsed three times with sterile distilled water.

Treatment c: Nodal explants surface-sterilized as in treatment a), were placed onto semi-solid MS medium (Murashige and Skoog, 1962) (pH 5.8) containing kinetin (0.2μmol), NAA (0.2μmol) and BAP (0.4μmol) to enhance axillary shoot growth, and 25 g l^{-1} sucrose and gelled with 2.3 g l^{-1} Gelrite. Contaminant-free shoots were visually selected after two weeks.

Treatment d: Visually contaminant-free, established, multiplying *in vitro* (5 months old) shoots were selected.

Treatment e: Explants were treated as in c), after which contaminant-free shoots were selected visually and placed in 0.1 g l^{-1} rifampicin® supplemented medium on a shaker at 70 rpm for seven days. Contaminant-free shoots were selected.

Treatment f: Treatment f: Visually contaminant-free multiplying *in vitro* shoots were placed in MS medium (pH of 5.8) containing 0.1 g l^{-1} Rifampicin®, 25 g l^{-1} sucrose shaking at 70 rpm for seven days, and contaminant-free shoots were selected thereafter.

2.2 Multiplication

a) Thirty shoots were placed into RITA® vessels. Different flush times (where the plants are submerged in the media for 30 s, 1, 5 and 10 min) and rest times – (where the plants were not covered with media for 5, 10 and 20 min) were used. Multiplication was recorded after 14 days to determine which times gave the highest multiplication.
b) To test the effect, different numbers of shoots per vessels, 50, 100 and 150 shoots were placed into RITA® vessels and flushed for 30 seconds. Rest periods of 10, 20 and 30 min were used for the different numbers of shoots per vessel. Shoot multiplication rates were recorded after 21 days. MS medium (pH 5.8) with 0.8 μmol BAP, 0.01 μmol NAA, and 25 g l^{-1} sucrose was used.
c) Using 50 shoots per vessel, multiplication rates and shoot size of plants, as well as electrical conductivity (EC) of media were recorded at 0, 7, 14 and 21 days. Multiple analysis of variance was undertaken.
d) Multiplication rates of semi-solid and liquid system were compared for five clones during a 21-day period. Exposure time to nutrients in the liquid system was constant throughout, a 30 seconds flush time and a 10 minute rest time was used.

2.3 Rooting

Shoots from multiplication media (semi-solid and RITA®) were placed onto half-strength MS medium (pH 5.8) containing 4.9 μmol IBA and 10 g l^{-1} sucrose for rooting (6 g l^{-1} agar was used for the semi-solid media). After seven days in rooting medium, plants (with and without roots) were placed in the greenhouse. Rooting and survival of acclimatized plants from both systems were recorded.

3. Results and discussion

Establishment of nodal cuttings of six different clones directly into the RITA® vessels was the first attempt at obtaining contaminant-free cultures (Treatment a). Although this material came from pre-treated parent plants, this method of initiation into the RITA® system was unsuccessful. There was 100 % contamination in all the vessels for the six clones used (Table 1). Contamination percentages on the semi-solid medium usually ranged from 20 % to 80 % dependant on the clone and whether material was taken from pre-treated parent plants. Ikemori (1987) found the average contamination rate to be 60 % if nodal sections from 58 *E. grandis* mother trees were used.

This author found 37 % contamination rate if apical buds were used, but necrosis of the buds occurred. The use of different explant material was needed to initiate shoots into the RITA® vessels.

Secondary leaders were taken as explants from rooted cuttings in the greenhouse (Treatment b). They were surface-sterilized using different concentrations of calcium hypochlorite for different periods of time. The use of the secondary leaders as explants for establishment was, however, unsuccessful (Table 1). Contamination (100 %) occurred in all the clones at 0.5 $g.l^{-1}$ of calcium hypochlorite, suggesting that this concentration was too low. However, when 2 g l^{-1} calcium hypochlorite was used for 10 and 15 minutes, death of the shoots occurred. This was due to the fact that the secondary material is young and cannot withstand a harsh surface-sterilization regime. When using soft young material, it is difficult to obtain surface- sterilization regimes that are sufficiently rigorous to destroy the surface microbes without becoming toxic to the young shoots. Ikemori (1987), using *Eucalyptus grandis* epicormic shoots, also found that contaminant-free explants were difficult to obtain without killing the plant tissue when too high a concentration of disinfectant was used.

The use of the semi-solid media (Treatment c) facilitated the removal of fungal contamination, which was the main cause of contamination in the previous initiation treatments (a and b). After placement of the visually contaminant-free shoots into the vessels, bacterial contamination (average of 57 %) occurred across the clones (Table 1). When treatment d was used, where visually contaminant-free, multiplying plantlets from *in vitro* culture (for five months) were placed directly into the RITA® vessels, an average of 32 % bacterial contamination was obtained with the different clones used (Table 1).

Table 1: Percentage of contamination occurring when different explants were sterilized by various methods and initiated into the RITA system (shoots of 6 clones were tested)

Sterilization treatment and explant type*)	Average % contamination (6 clones) of shoots placed into the RITA system
a	100
b	100
c	57
d	32
e	0
f	0

*) see chapter 2.1

Table 2: Multiplication of shoots (from 100 starting shoots) in the semi-solid system (28 days) and RITA® system (14 days) of different *Eucalyptus* clones and average multiplication for three sub-tropical and two cold-tolerant clones (SD - represents a mean SD)

Clone	No. of shoots after 28 days culture on semi-solid medium	No. of shoots after 14 days culture in RITA® vessel
GU177	497	845
GU178	376	722
TAG31	526	637
Average multiplication for the subtropical clones	4.7 times (SD 0.78)	7.3 times (SD 1.05)
GN107	187	237
GN108	294	744
Average multiplication for the cold-tolerant clones	2.4 times (SD 0.54)	4.9 times (SD 2.45)

Table 3: Acclimatization success of shoots transfered to the greenhouse with and without roots from the RITA® and the semi-solid systems (expressed as % of total plants transferred from laboratory to greenhouse)

	Acclimatisation success (%)			
	Cultured on semi-solid medium		Cultured in RITA® vessels	
Clone	Shoots with roots	Shoots without roots	Shoots with roots	Shoots without roots
GU175	43	30	32	9
GU177	47	23	52	33
GU178	50	29	53	15
GU180	39	28	36	18
Average rooting percent for the sub-tropical clones	36		35	
GN108	20	1	63	37
NH58	5	0	67	43
Average rooting percent for the cold-tolerant clones	6.5		53	

The use of an antibiotic in the media was undertaken as part of a pre-treatment to overcome the bacterial problem, which occurred in treatment c and d. According to Phillips, et al. (1981) and Cornu and Michel (1987), Rifampicin was found to be an effective antibiotic with no phytotoxic effects to the plants. The use of 0.1 g l^{-1} Rifampicin resulted in contaminant-free explants and had little effect on the shoots (Treatment e and f). All shoots without visible signs of contamination were then placed into the RITA® vessels, after which no bacterial contamination occurred (Table 1). Obtaining contaminant-free shoots in RITA® by using the semi-solid medium and Rifampicin pre-treatment with visual selection of contaminant-free plants is thus appropriate for the six *Eucalyptus* clones tested. Escalona et al. (1999) and Preil and Hempfling (2002) used established shoots from an agar base as inoculum for the bioreactors, as indicated by most other researchers. Similarly, with *Eucalyptus* clones it was important that elimination of contamination was undertaken in the semi-solid phase after which the shoots were then used for the liquid systems. Unless disease indexing of the parent plant or screening takes place, as described by Cassells (1997) and Holdgate and Zandvoort (1997), it is not possible to place shoots directly into the RITA® vessels without obtaining high losses.

The temporary immersion system provides a highly aerobic system for plant growth, as there is forced ventilation through the vessel lid. However the immersion times, i.e. duration or frequency, is the most decisive parameter for system efficiency (Alvard et al., 1993; Berthouly and Etienne, 2002). The flush time interval of 30 min resulted in the lowest multiplication rate. It was found that the five minute rest period gave significantly lower multiplication i.e. from 2.1 times (30 s flush) to 1.5 times at 10 minutes flush (Figure 1). At the 10 minute flush time with a five minute rest period the *Eucalyptus* shoots became brittle and hyperhydricity occurred. This was probably due to the flush intervals being too long or too frequent. Jackson (2002) stated that an aqueous cover interferes strongly with gas exchange to the outer tissue or cell surface since gas diffusion rates are approximately 10000 times slower in water than in air. This impact is increased with the depth of the aqueous cover or the inclusion of gel matrices such as agar. Thus, by total submersion, or submersion of the plants too frequently for long periods, gaseous exchange for photosynthesis and respiration was reduced even if there was dissolved oxygen and carbon dioxide in the liquid. Fujiwara and Kozai (1995) found that increasing the number of air exchanges avoided shoots becoming hyperhydric with long-term continuous liquid cultures. The 10 minute flush time caused a reduction in the multiplication. However, with the increased rest time between the flushes (10 and 20 minutes rest time) there was an increase in multiplication. These results indicated that there was a relationship between the length of flush and rest time – with an increase in flush time, the shoots required an increase in the rest period. One minute flush time at 10 and 20 minute rest time, and five minute flush time at 20 minute rest time, gave high (3.2x, 3.2x and 3x) rates of multiplication. Different plants require different flush and rest times for optimal multiplication and many researchers found that the immersion time affected the plant growth rate. Preil and Hempfling (2002) found with *Phalaenopsis* that the effect of immersion frequency affected the plant growth rates and that eight immersions for 10 minutes per day were applied gave maximal multiplication. Alvard et al. (1993) found that 20 minutes every two hours was optimal for bananas. By controlling the immersion cycles, Akula et al. (2000) achieved a more consistent and synchronized multiplication and embryo development of tea. They used one minute immersions every six hours to obtain a 24-fold increase. Matre et al. (2001) reported that the immersed stage induced a substantial oxidative stress on *Hevea brasiliensis* callus. This oxidative stress could explain the time variations of the multiplication at the different immersion times. The immersion time intervals play a decisive role in influencing the multiplication rates for different species as this factor affects nutrient supply and composition of the internal atmosphere in the culture vessel (Jimenez et al., 1999). For the *Eucalyptus* shoots, a flush time of 30 seconds with a rest period of 10 minutes gave the highest multiplication rate (3.8x).

It became evident from the study on the interval and submersion times that the number of shoots in the vessels had an effect on the time required between submersions. The results showed that there was a significant difference in multiplication of plants as a result of the number of starting shoots, with 50 shoots per vessel giving the greatest multiplication for all three rest times tested. Starting with 50 shoots per vessel, the multiplication of the three clones tested were significantly greater using a 30-second flush every 10 or 20 minutes (2.74, 2.66x respectively) than with 100 or 150 starting shoots. The rest time of 30 minutes gave the lowest multiplication (1.3x) using 100 and 150 shoots per vessel (Figure 2). With more shoots per vessel a decrease in the length of time between flushes was required as more shoots per vessel led to a decrease in the availability of nutrients. More shoots led to the depletion of nutrients at a faster rate.

With multiplication from 100 starting explants in both systems (RITA® - 50 per vessel and semi-solid – eight per jar), the RITA® system far exceeded the semi-solid system in multiplication. Shoot numbers (axillary buds larger than 0.5 cm) in the RITA® system increased from 423 to 744 between day seven and day 14, and from 744 to 888 between day 14 and 21 (Figure 3). Between day 14 and 21, shoot elongation increased considerably, thus making it feasible to culture the shoots in RITA® for 21 days (Figure 4).

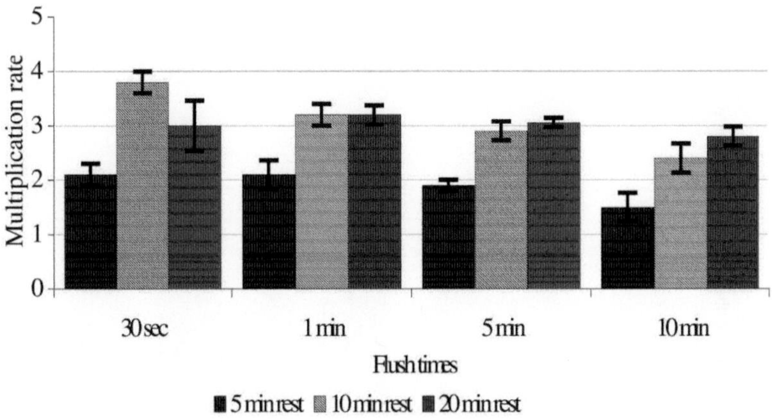

Figure 1: Multiplication rates of the shoots placed in the RITA® vessels at different flush vs. rest times. Each SD value represents a mean SD.

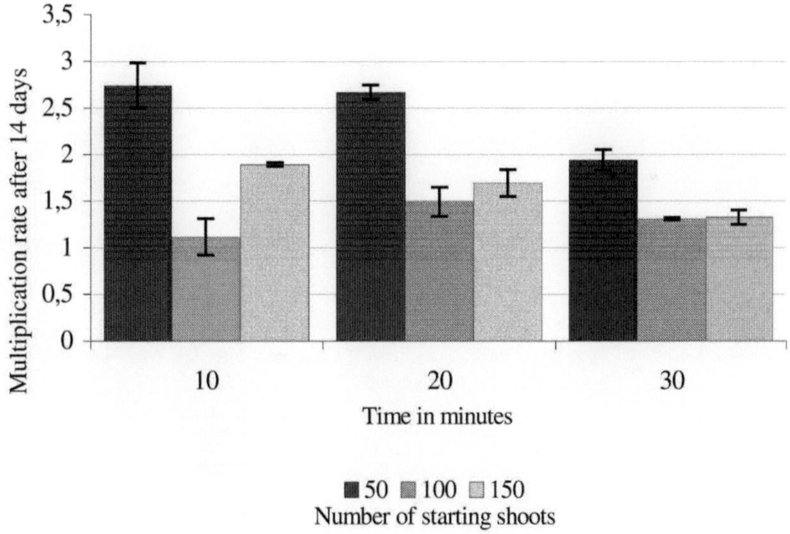

Figure 2: Effect of different flush intervals (10, 20 and 30 min) and shoot numbers per vessel on multiplication rates after 14 days for three clones (exposure time per flush is 30 seconds). Each SD value represents a mean SD.

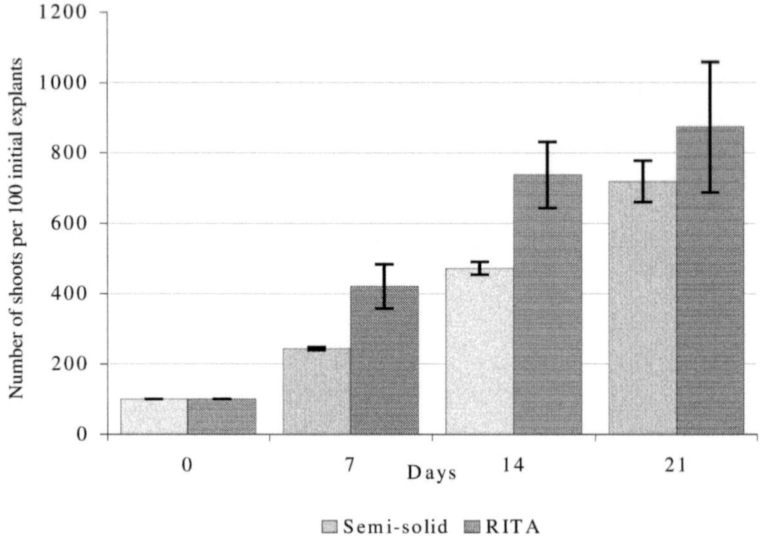

Figure 3: Multiplication in the RITA® and semi-solid systems (per 100 starting shoots). Each SD value represents a mean SD.

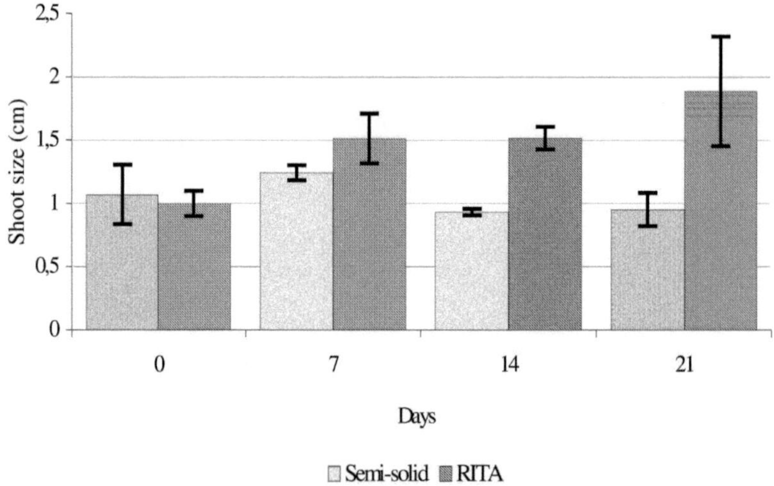

Figure 4: Shoot length (cm) in the RITA® and semi-solid systems (minimum of 100 shoots per system per time period). Each SD value represents a mean SD.

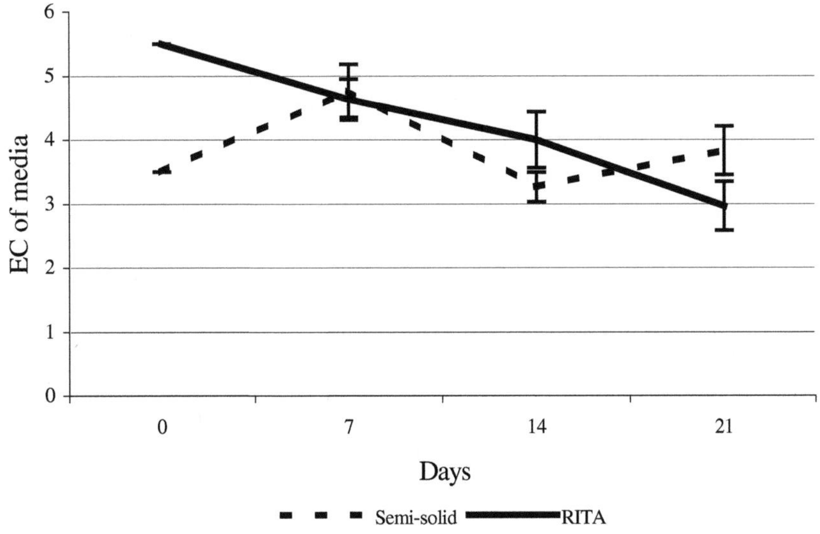

Figure 5: EC (µS) of media from RITA® and semi-solid systems.

The multiplication slows and elongation occurs at this time. The semi-solid system gave smaller plants and multiplication was less and plant numbers were achieved more slowly. There was a decrease in the shoot length of the plants at day 14 in the semi-solid system, which could be due to the manner in which the shoots are excised from the main stem. The EC in the RITA® system was high at the beginning (5.5 µS) (Figure 5). It decreased steadily during the 21-day period, with the greatest decrease between day seven and 14. Between day 14 and 21, there was only a small decrease to 2.9µS. This indicated that there was a rapid uptake of nutrients at the start of the cycle when the plants were multiplying rapidly and by day 21 multiplication had decreased, as had the uptake of nutrients. With the semi-solid system the EC was initially low (3.2µS) and at day seven it increased to 4.5 µS and thereafter dropped again to 3.5 µS - which was higher than the original value. To begin with, the gel in the semi-solid system appeared to be binding the nutrients, and the nutrients only became available to the plants on day seven, allowing uptake. However the uptake of the nutrients is not as great when compared with that in the RITA® system. In RITA®, nutrients were immediately available. After 28 days the plants in RITA® system deteriorated which could be due to a lack of nutrients. In the semi-solid system, plantlet numbers increased slowly between days 21 and 28; multiplication had slowed, but the length of the shoots increased.

Average multiplication rates for three sub-tropical clones and two cold-tolerant clones were calculated for the semi-solid and RITA® systems, and all clones had different multiplication rates (Table 2). All clones multiplied faster in the RITA® system compared with plants in the semi-solid system. After 28 days in the semi-solid system, subtropical clones achieved a multiplication of 4.7x while in the RITA® system the same clones achieved 7.3x during a 14 day period. The shoots of cold-tolerant clones multiplied 2.4x in the semi-solid during 28 days, and 4.9x in the RITA® system during 14 days (Figure 6). The optimum multiplication cycles in RITA® were between 14 and 21 days, whereas in the semi-solid system they were 25 to 28 days. The shoots in the RITA® system began to deteriorate quickly and started to die if they were left longer than 21 days in the system. Preil and Hempfling (2002) found that with *Phalaenopsis* the media had to be changed at two-week intervals as the four-week intervals of media exchange resulted in a distinct reduction of propagation efficiency.

Figure 6: At the beginning of the cycle 8 shoots per jar (top left) and 50 shoots per RITA® (bottom left) are used, and after 28 days in the semi-solid (top right) and 14 days in the RITA® (bottom right) the multiplication which occurred is shown.

The vessel closure regulates the degree to which the physiochemical factors in the growth room impact on the micro-environment, as the type of closure forms the interface between the inside and outside environments of the vessels (Smith and Spomer, 1995). With the semi-solid system, a major barrier to tissue aeration is the enclosing of a vessel to prevent drying out and contamination. Zoybayed et al. (2001) and (Jackson, 2002) reported that sealing of culture vessels could seriously inhibit growth, development, induce hyperhydricity, reduce the leaf chlorophyll content and result in asphyxiation. According to Jackson (2002) forced ventilation allows the plants to become more photoautotrophic which enhances growth; the enhanced oxygen and carbon dioxide availability in the temporary immersion system allows aerobic respiration and photosynthesis to occur without the build-up of ethylene in the vessels. The exchange of gases in the RITA® system could be one of the factors leading to the increased growth rates observed.

This study indicated that plant quality is important for rooting and it can be seen that the plants produced by the RITA® system were superior to those of the semi-solid system, which prompted trials to improve rooting and acclimatization (Figure 7). Minimal callus was evident on the leaves, bases and stems of plants in the RITA® system, with roots developing directly from the base of the stems. This was not the case with the semi-solid system as the plants often formed callus at the base of the stems from which roots grew. This caused problems at the acclimatization stage. The concentrations of O_2 in the vessels affect the root system and where an anaerobic or low O_2 availability condition occurs, rooting is reduced or abnormal roots form (Jackson, 2002). With the semi-solid system a lower concentration of O_2 in the gel may have resulted in poor root development, whereas with the RITA® system there was a continuous supply of O_2 which may have improved rooting.

Plantlets in the RITA® vessel rooted readily *in vitro* using modified MS medium containing IBA. Roots also developed *ex vitro*. Nevertheless, clones were found to have different acclimatization potentials (Table 3). In this regard, percent rooting and acclimatization was determined for four sub-tropical clones known to be 'easy rooters' (GU175, GU177, GU178 and GU180), and for two 'difficult-to-root' cold-tolerant clones (GN108 and NH58). The sub-tropical clones showed no difference in percentage rooting between the semi-solid and the RITA® rooting environments. In contrast, rooting of the cold-tolerant clones was 6.5 % and 53 % in semi-solid and RITA® systems respectively. It seems, therefore, that one of the greatest values of the RITA® system is to facilitate the rooting steps in recalcitrant clones.

Table 4: Costs to produce 10,000 plants (from 100 starting plants) in the semi-solid and RITA® system. Data based on average rooting percentage (cold- tolerant and sub-tropical clones). Costs in US $

Materials or activity	Semi-solid (6 months)	RITA (3 months)
Media	616	104
Transfer	518	155
Media preparation (labour)	518	104
Autoclaving	21	7
Washing	518	26
TOTAL	2 191	396
Cost of vessels	649	5 909
FINAL TOTAL	2 840	6 305

With *Eucalyptus,* acclimatization was improved in the plants that came from the RITA® system as the plants produced were of a better quality. The air exchange that occurred in the RITA® vessels could have led to better stomatal and outer epidermal layer development which may have given the plants an improved chance of survival. The improved acclimatization results obtained in this study were similar to those found by Berthouly and Etienne (2002), in that plant material propagated by temporary immersion performed better during the acclimatization phase than material obtained on semi-solid or liquid media.

A cost analysis was done (Table 4) using the average yields for all the clones. Calculations are based on data obtained to date which indicate that with 100 initial explants for both systems, 10,000 plants can be obtained with the RITA® system in three months, while in the semi-solid system it took six months to achieve that number. Using the RITA® system the costs of the disposable items and running expenses are far lower than that of the semi-solid system. The costs of media and media preparation are reduced substantially by the elimination of a gelling agent in the liquid media and the dispensing time of the media. With the semi-solid media, each aliquot of 25 ml has to be dispensed into each jar. The reduction in autoclaving is due to lower quantities of vessels and media to be autoclaved at each transfer. With the RITA® system the transfer time is considerably shortened as the shoots can be cut and 50 shoots are dropped into the vessel. However, with the semi-solid system, each jar must be opened and seven shoots per jar placed, with care, so that each stem is at a suitable depth in the semi-solid medium.

If new nutrients are required during a cycle, the middle unit of the RITA® vessel may be lifted out and placed into a clean vessel with new nutrients. In contrast, in the semi-solid system, each individual shoot has to be handled. Using the RITA® system, fewer vessels are used and therefore the washing costs are reduced. Less space was required for the production of plants in the RITA® vessels compared to those in the semi-solid system. This increased multiplication in the RITA® system was achieved in a smaller production space compared with that of the semi-solid system. Approximately 1,792 and 3,200 plants per m^2 at the onset of multiplication, for the semi-solid and RITA® systems respectively. In addition to the multiplication rates that were achieved in a smaller space with the RITA® system, the final acclimatized yields (i.e. after greenhouse establishment and ready for planting out) were the most important in terms of evaluating the success of the method.

The initial outlay for the RITA® vessels is high, but the vessels are re-used and this high cost is soon offset by the multiplication rates and turnover of the plants produced. The average yields (cold-tolerant together with sub-tropical) obtained from the RITA® system are greater than those in the semi-solid system. The costs involved in producing plants in the temporary immersion system are lower, as more plants are produced in a shorter time from the medium. Further, for a commercial laboratory, the RITA® system offers flexibility in that newly approved commercial clones can readily replace the commercial clone being produced. The RITA® system is more efficient in producing higher numbers of difficult-to-root clones than the semi-solid system.

The reduction in costs parallels the findings of other researchers of large cost savings using different plants. Etienne (2000) found that the use of the temporary immersion system combined with direct sowing of somatic embryos of coffee, eliminated a labour-intensive stage in tissue culture. They found that the production time was reduced by three months and the handling time was reduced by 6.3 % compared with the standard micropropagation system. The shelving requirements were also reduced by 13 %. Etienne (2000) states that it is reasonable to expect major economic gains since labour and shelving represent 70 % and 10 % respectively of micropropagation costs. Lorenzo et al. (1998) calculated a cost reduction of 46 % for sugarcane propagation in a temporary immersion system compared with that on the agar medium, while Escalona et al. (1999) saved 20 % of production costs per pineapple plant at multiplication stage in a temporary immersion system in comparison with conventional cultures.

Figure 7: Rooted plants grown on semisolid medium and in RITA® vessels (right) after 4 days in the greenhouse.

4. Conclusion

The results indicate that for *in vitro* culture of *Eucalyptus*, the RITA® system results in benefits not yet obtained with the more commonly-used semi-solid protocols for axillary bud propagation. However, with *Eucalyptus*, an initial short-term semi-solid stage is recommended as a quick and economical means of establishing microbial-free plants. In RITA®, multiplication increases with the use of the appropriate numbers of starting shoots in the vessels, as well as with appropriate exposure to media at correct intervals. Plant quality (hardiness and size) for clones tested to date is better in RITA®-produced plants than for plants grown in semi-solid media. In addition, cold-tolerant *Eucalyptus* clones which have extremely difficult to multiply, root, and subsequently acclimatize using semi-solid protocols, have been shown to respond favourably to the RITA® environment. Costs per 10,000 plants produced using RITA® are less than those for the semi-solid system. The RITA® system thus has great potential for *in vitro* production of *Eucalyptus* plants commercially, provided that contaminant-free explants can be obtained *via* a semi-solid system.

Acknowledgements

Mondi Forests for financial support and development opportunities. Dean da Costa, Jacqui Wallis, Khanyie Zitha, Isabel Sokhela, Beatrice Maphamolo, Nelisiwe Dube, Nomusa Nxumalo and Sara-Jane Zuma for their commitment to the team and their dependable assistance.

References

Aitken-Christie J, Kozai T & Takayama S (1995) Automation in plant tissue culture. General introduction and overview. In: Aitken-Christie J, Kozai T & Smith MAL (eds) Automation and Environmental Control in Plant Tissue Culture (pp. 1-18). Kluwer Academic Publishers, Dordrecht

Akula A, Becker D & Bateson M (2000) High-yielding repetitive somatic embryogenesis and plant recovery in a selected tea clone, 'TRI-2025', by temporary immersion. Plant Cell Rep. 19: 1140-1145

Alvard D, Cote F & Teisson C (1993) Comparison of methods of liquid medium culture for banana micropropagation. Effects of temporary immersion of explants. Plant Cell, Tiss. Org. Cult. 32: 55-60

Berthouly M & Etienne H (2002) Temporary immersion system: A new concept for use of liquid medium in mass propagation. First International Symposium on Liquid Systems for *In Vitro* Mass Propagation of Plants (pp. 37-38). Cost 843 Working Group. Aas, Norway

Borroto CG (1998) Temporary immersion bioreactor systems reduced micropropagation costs. Agricell Rep. 30(1): 2

Cassells AC (1997) Pathogen and microbial contamination management in micropropagation – an overview. In: Cassells AC (ed) Pathogen and Microbial Contamination Management in Micropropagation (pp. 1-13). Kluwer Academic Publishers, Dordrecht

Cornu D & Michel MF (1987) Bacterial contaminants in shoot cultures of *Prunus avium* L. choice and phytotoxicity of antibiotics. Acta Hort. 212: 83-86

Denison NP & Kietzka JE (1993) The use and importance of hybrid intensive forestry in South Africa. SA For. J. 165: 55-60

Escalona M, Lorenzo JC, González B, Daquinta M, González JL, Desjardins Y & Borroto CG (1999) Pineapple (*Ananas comosus* L. Merr) micropropagation in temporary immersion systems. Plant Cell Rep. 18: 743-748

Etienne H (2000) Direct sowing of temporary immersion-produced somatic embryos. Agricell Rep. 34(3): 17-18

Etienne H, Lartaud M, Michaux-Ferrière N, Carron MP, Berthouly M & Teisson C (1997) Improvement of somatic embryogenesis in *Hevea brasiliensis* (Müll.Arg) using the temporary immersion techniques. *In Vitro* Cell. Biol. 33: 81-87

Fujiwara K & Kozai T (1995) Physical micro-environment and its effects. In: Aitken-Christie J, Kozai T & Smith MAL (eds) Automation and Environmental Control in Plant Tissue Culture, (pp. 319-369). Kluwer Academic Publishers, Dordrecht

George EF (1993) Plant Propagation by Tissue Culture. The Technology, Vol 1 (pp. 612-635). Exegetics Ltd: Edington

Holdgate DP & Zandvoort EA (1997) Strategic considerations for the establishment of micro-organism-free tissue cultures for commercial ornamental micropropagation. In: Cassells AC (ed) Pathogen and Microbial Contamination Management in Micropropagation (pp. 15-22). Kluwer Academic Publishers, Dordrecht

Ikemori YK (1987) Epicormic shoots from the branches of *Eucalyptus grandis* as an explant source for *in vitro* culture. Comm. For. Rev. 44: 351-356

Jackson MB (2002) Ventilation of plant tissue cultures. First International Symposium on Liquid Systems for *In Vitro* Mass Propagation of Plants (pp. 56-57). Cost 843 Working Group. Aas, Norway

Jiménez E, Pérez N, de Feria M, Barbón R, Capote A, Chávez M, Quiala E & Pérez JC (1999) Improved production of potato microtubers using a temporary immersion system. Plant Cell, Tiss .Org.Cult.. 59: 19-23

Le Roux JJ & Van Staden J (1991) Micropropagation and tissue culture of *Eucalyptus*: a review. Tree Physiol. 9: 435-477

Lorenzo JC, González LB, Escalona M, Teisson C, Espinosa P & Borroto C (1998) Sugarcane shoot formation in an improved temporary immersion system. Plant Cell, Tiss. Org. Cult.54: 197-200

Martre P, Dominique L, Just D & Teisson C (2001) Physiological effects of temporary immersion on *Hevea brasiliensis* callus. Plant Cell, Tiss.Org.Cult. 67: 25-35

Murashige T & Skoog F (1962) A revised medium for rapid growth and bioassays with tobacco tissue cultures. Physiol. Plant. 15: 473-496

Phillips R, Arnott SM & Kaplan SE (1981) Antibiotics in plant tissue culture: Rifampicin effectively controls bacterial contaminants without affecting the growth of short-term explants cultures of *Helianthus tuberosus*. Plant Sci. Lett. 21: 235-240

Preil W & Hempfling T (2002) Application of temporary immersion system in propagation of *Phalaenopsis*. First International Symposium on Liquid Systems for *In Vitro* Mass Propagation of Plants (pp. 47-48). Cost 843 Working Group. Aas, Norway

Smith MAL & Spomer LA (1995) Vessels, gels, liquid media and support systems. In: Aitken-Christie J, Kozai T & Smith MAL (eds) Automation and Environmental Control in Plant Tissue Culture (pp. 371-404). Kluwer Academic Publishers, Dordrecht

Zobayed SMA, Armstrong J & Armstrong W (2001) Micropropagation of potato: Evaluation of closed, diffusive and forced ventilation on growth and tuberization. Ann. Bot. 87: 53-59

Chapter 34

Efficiency in thin-film liquid system for *Hosta* micropropagation

Jeffrey Adelberg
Department of Horticulture, Clemson University, Clemson SC 29634, USA.
E-mail: jadlbrg@clemson.edu

Abstract: Three varieties of *Hosta* ('Striptease', 'Minuteman' and 'Stiletto') at four densities (40, 80, 120 and 200 explants per litre) were micropropagated on semi-solid agar and a thin-film liquid system with intermittent wetting of plant tissue. The mechanics of wetting by a small wave front required a larger rectangular vessel (11 x 27 = 297 cm^2) compared to the common cylindrical baby food jar (18 cm^2). Plants multiplied more rapidly in the agitated thin-film system than on agar. Lower plant densities increased rates of multiplication in liquid, but had little or no effect on multiplication rate on agar. Increasing plant density lowered the overall multiplication rate, but yielded greater numbers of plants per vessel. Yield, tabulated for utilization of shelf-space in growth room, was greater at all densities in rectangular vessels of liquid than conventional jars of agar media. Increased plant density lowered the sugar residual in media following the culture cycle and liquid media had less residual sugar than agar media. A liquid medium with 50g l^{-1} sucrose was concentrated enough so that sugar depletion did not limit growth, even at the highest densities. The liquid system allows the technician to skip the step of manually spacing and orienting the freshly cut bud tissue at the transfer station. Harvesting 75 - 100 plants per vessel from the large rectangular vessels resulted in most efficient use of technician time. Plants from liquid and agar acclimatized to greenhouse. Increased multiplication, space utilization, sugar availability and worker efficiency was demonstrated to be greater in thin-film liquid than more conventional agar-based system.

Key words: bioreactor, liquid culture, mass flow, propagation

Abbreviations: PPF-photosynthetic photon flux

1. Introduction

Hostas are among the most valuable tissue cultured plants in North America (Zimmerman, 1996). Virtually all commercial micropropagtion,

A.K. Hvoslef-Eide and W. Preil (eds.), Liquid Culture Systems for in vitro Plant Propagation, 443–457.

including *Hosta*, uses semi-solid agar-based media. *Hosta* micropropagated on liquid media in shaker flasks had greater dry weights than from agar medium (Adelberg et. al., 2000). The plants from liquid media also acclimatized and grew at a faster rate in the mist bed and outdoor nursery than plants from agar. Acclimatization requires functional changes in leaf anatomy - cuticle, stomata, and mesophyll, and conductive functions in vessel and root systems, before water relations and photosynthetic competence are balanced for growth (Ziv, 1995). Carbohydrates accumulated *in vitro*, are necessary for plant metabolism *ex vitro* until photosynthetic growth is established (Williams, 1995). It is possible that *Hosta* from liquid culture have greater carbohydrate reserves, in addition to greater dry weights.

Labour is the largest cost component in micropropagation (Chu, 1995). Liquid culture allows innovative methods in workstation mechanization for the cutting and transfer during subculture. Further savings in media preparation (labour and materials), dishwashing and the manual removal of sugar-containing agar from plants prior to planting-out, is realized in liquid systems. Mechanical cutters and large liquid culture vessels allows cost reductions greater than 50% to be predicted (Gross and Levin, 1998). A man-motion study of technician activities showed 'sorting and placing' buds in fresh media required 45% of the time at the transfer station (Alper et al., 1994). A 'cut and dump process' was demonstrated for shoot bud clusters, where labour was reduced by non-oriented cutting and bulk transfer of buds into a vessel of soft agar. With 'cut and dump', yield per vessel was reduced to 41%, but overall system efficiency increased by a factor of 4.8 (Alper et al., 1994). *Hosta* buds grown in agitated shaker flasks of liquid were obviously never oriented or spaced by the technician and grew, at least as well as, those planted carefully in semi-solid agar (Adelberg et al., 2000). Liquid micropropagation systems may simultaneously reduce inputs and increase quality of the propagated plants, but will only be implemented when a system is deemed easily adapted and cost-effective for an industry.

Mechanized systems for temporary immersion in liquid media have been designed and allow vigorous growth of high quality plants while increasing the efficiency of labour and laboratory shelf space (Etienne and Berthouly, 2002). The first such system, a tilting rocker for flasks, was described twenty years ago (Harris and Mason, 1983) with no further reports since that time. A simplified, scaled-up rocker was designed, fabricated and tested (Figure 1; Adelberg and Simpson, 2002) to intermittently wet tissue by slowly moving wave fronts in large rectangular vessels (e.g. Nalgene Biosafe, Nalge Nunc Intl., Inc.). High vessel costs, the large number of mechanical parts in each Biosafe vessel, and the failure of this vessel to maintain asepsis during multiple uses, prompted work with Southern Sun

Figure 1: Agitated thin-film rocker system is a simple, temporary-immersion liquid culture.

Figure 2: Rectangular vessels facilitated a slow moving wave of media to migrate the length of the vessel with the motion of the shelf.

BioSystems (Hodges, SC, USA) to devise a less costly vessel with fewer parts for plant micropropagation.

This current study evaluates plant responses during micropropagation in a device developed to facilitate non-oriented bulk transfer process in a simple, scaled-up liquid culture system. Three varieties of *Hosta* were compared in the thin-film culture and semi-solid agar media at different planting densities. Efficiency was described in terms of (1) multiplication rate, (2) facility utilization, (3) sugar availability, and (4) technician labour. Plant quality was verified in greenhouse-finished transplant liners.

2. Materials and methods

Three varieties of *Hosta* ('Striptease', 'Minuteman' and 'Stilletto') from commercial stage II cultures were prepared in liquid media modified from Murashige and Skoog (MS) media (Murashige and Skoog, 1962) that included (1) addition of 170 mg l^{-1} sodium phosphate, (2) increased $CuSO_4 \cdot 5 H_2O$ concentration to 25 mg l^{-1}, and (3) organic constituents of the medium included 0.5 ml l^{-1} of MS vitamin solution (2.0 g l^{-1} glycine, 100.0 g l^{-1} myo-inositol, 0.50 g l^{-1} nicotinic acid, 0.50 g l^{-1} pyridoxine hydrochloride, 0.10 g l^{-1} thiamine hydrochloride), adenine hemisulfate (0.92 mg l^{-1}), sucrose (50 g l^{-1}), benzyladenine (1 μmol). The pH of the media was adjusted to 5.7 before being dispensed. Liquid culture vessels were maintained on a thin-film rocker system (Adelberg and Simpson, 2002) that produced a 1-rpm pitch every 15 minutes. Plant buds were carefully trimmed free of leaves and roots. Crowns were divided into single bud units and agar vessels were planted with vertical and spatial orientations carefully maintained. For each variety, 1, 2, 3 and 5 buds were placed in each of 8 babyfood jars (180 ml) containing 25 ml of media gelled with 0.07% (w/v) agar (PhytoTechnology Plant Tissue Culture Grade Agar A111, Shawnee Mission, KS). Buds for liquid treatments were cut, trimmed and collected in units of 8, 16, 24, or 40 buds per litre empty jar, and dumped with one motion without regard for orientation or spacing. For each variety, 4 rocker vessels (Figure 2; Southern Sun BioSystems, Hodges SC) were prepared at each density, containing 200 ml of liquid media. Densities of 40, 80, 120 and 200 buds per litre media were directly comparable between agar and liquid media. The 180 ml cylindrical babyfood jar containing 25 ml of media typified a standard practice in agar and was compared with a dissimilarly proportioned rectangular vessel of liquid. One rectangular rocker vessel occupies the same shelf space as 8 jars and so a scale-up factor of 8 was applicable to both media volume and shelf space. The relationship for surface area within vessels was approximately 16 : 1 for jar to rocker vessels (18 cm^2 per jar,

297 cm^2 per vessel). Plants were grown for six weeks with 16 h photoperiod at 25 to 35 $\mu mol\ m^{-2}\ s^{-1}$ PPF provided by cool white fluorescent lamps and temperature was maintained at 24 $\pm$ 2°C. After six weeks, shoots were divided into single bud units and returned for a second culture cycle.

Residual sugar concentration was measured from vessels of expended media with a hand-held refractometer. Sugar was assayed from semi-solid agar by disrupting the matrix with repeated forceful shearing of the medium through a pipette tip. After several pump actions, a bead of liquid slurry was expressed on the lens of the refractometer. The refractometer is an inexpensive tool to measure sugar content rapidly with minimal sample preparation. Using the refractometer requires understanding an inherent limitation - the reading of % BRIX in media is a molar summation of sucrose, and its monomers, glucose and fructose. For tissue culture application: (1) a fraction of sucrose hydrolyzes to glucose and fructose in the autoclave, (2) plant cell wall invertase further hydrolyzes some sucrose to glucose and fructose and returns monomers to the media, and (3) water uptake by the growing plant will increase the molar concentration of the solution by decreasing the volume of diluent. These three factors will cause underestimates of sugar uptake when reading % BRIX in residual medium in comparison to initial sucrose concentrations.

A completely randomized design compared semi-solid agar and liquid media with the three varieties, at four different densities and two subculture cycles, for bud multiplication, plant yield and residual sugar concentrations. Sixteen jars, or eight rocker vessels per treatment factor level combination were pooled from two subculture cycles. Analysis was made on JMP version 3.2.6 (SAS Inst., Cary NC).

Following the second culture cycle, 72 plants from liquid and agar, for each variety, were planted in a greenhouse mist bed. Two weeks later plants were moved to a conventional greenhouse bench and observations of survival were made after a total of four weeks in greenhouse culture.

A data archive from the commercial micropropagation process at the beta-site was assembled. The technicians performing subcultures on 40 varieties of *Hosta* in Nalgene Biosafe vessels had logged, during 6 months, time spent in the production hood hours, and yield data. The technicians were not aware of any experimental design or bias.

3. Results and discussion

Plants multiplied equally well during both cycles of subculture and observations were pooled for analysis (data not presented). The three varieties of *Hosta* had multiplied at similar rates. All three varieties were more prolific in the agitated, thin-films of liquid media than on agar (Table 1). Better growth in liquid systems has been reported on numerous occasions and may be attributed to the lack of impurities from agar, better water availability, better nutrient availability and larger vessels (Smith and Spomer, 1995; Berthouly and Etienne, 2002).

Plant density had a great effect on multiplication rate for the liquid system. On agar, plant density did not affect multiplication rate of two varieties, and had a relatively small effect on the third (Table 1). With a density of 200 explants per litre multiplication rates were on average 1.6 and 2.3, respectively, for agar and liquid media. This equates to 5 buds in a babyfood jar with 25 ml media or 40 buds in a rocker box with 200 ml media. When bud density was decreased to 40 per litre (1 bud in 25 ml agar or 8 in the liquid culture vessel), multiplication rates increased, to an average of 2.0 on agar or 3.4 in liquid. If efficient micropropagation were simply to be a comparison of the predicted numbers of plants propagated from a hypothetical bud in an annual crop cycle, and 8, six-week subcultures were performed in that year, than the liquid system with 40 per litre would produce 60,717 plants (3.4^8). Similarly, the agar system at the same density would only produce 256 plants (2^8) in the same year. It can be seen that for a valuable new plant where the overriding need may be the speedy increase in plant numbers, then the efficient use of a liquid system such as the large rectangular rocker box with a low density of tissue (8 plants per vessel with 200 ml media) could be most effective.

In practice, many considerations besides multiplication of buds impact efficiency of a micropropagation laboratory. The total number of plants generated by that facility is critical since the bench space available for plant growth puts seasonal limitations on production rates. When larger numbers of buds were used to initiate a culture, more plants were harvested six weeks later (Table 1). Reducing plant density increased multiplication rate, but the highest yields were still from highest density planting. The significant linear increase in yield through the ranges of planting density tested, implies that higher yields are possible with greater explant density. In all cases, yields were higher in liquid than on agar, per volume media. Eight jars placed in 4 x 2 arrangement create the same 'footprint' on the culture room shelf as one rectangular thin film vessel. Within the parameters established for this experiment, media volume and area shelf space were directly related by the same factor (8 jars : 1 vessel) and comparing plant yield between vessels and jars for shelf utilization in the growth room is identical to yield comparisons by media volume. Fewer, large rectangular vessels create less void space between vessels on a culture room shelf than larger numbers of smaller cylindrical vessels. Eight cylindrical jars of agar (8 x 18 = 144 cm^2) has less than half the interior surface for the growth of plants than one rectangular

Table 1: Multiplication rate, yield and sugar residual in media for 3 varieties of *Hosta* after 6 weeks of culture in semi-solid agar and thin-film liquid rocker system

	'Minuteman'					'Stiletto'					'Striptease'				
	Density (explant per litre)					Density (explant per litre)					Density (explant per litre)				
	40	80	120	200	Prob.>F	40	80	120	200	Prob.>F	40	80	120	200	Prob.>F
System	Multiplication[a]														
Agar	1.9	1.6	1.7	1.4	L*	2.3	2.0	2.1	1.8	n.s.	2.0	1.4	1.4	1.9	n.s.
Liquid	2.8	2.3	2.2	1.7	L***	3.9	2.5	3.1	2.9	L*Q*	3.4	3.1	2.8	2.2	L***
	Yield[b]														
Agar	76	128	200	276	L***	92	156	252	356	L***	80	116	196	372	L***
Liquid	112	184	264	336	L***	160	204	368	584	L***	136	244	336	448	L***
	% Sugar[c]														
Agar	6.1	6.1	5.3	4.7	L***	6.2	6.3	6.1	5.9	L*	6.8	6.3	5.2	4.1	L***
Liquid	5.3	3.8	3.2	2.3	L*Q**	5.6	4.7	3.9	3.2	L*Q*	5.6	3.5	3.9	3.0	L**

ANOVA	Multiplication Rate	Yield	% Sugar
Variety (V)	n.s.	n.s.	n.s.
Agar vs. Liquid (S)	**	*	***
Density (D)	***	***	***
V * S	n.s.	n.s.	*
V * D	n.s.	***	**
S * D	**	**	**
V * S * D	n.s.	**	**

L and Q indicate fit of Linear and Quadratic terms, respectively, with ***, **, * indicating significance at P<0.001, 0.01, 0.05, respectively, as determined by ANOVA, and n.s. was not significant.

[a]Multiplication rate was the ratio of buds harvested: buds initiated, for the six weeks of culture.

[b]Yield was recorded as number of plants harvested per liter of media.

[c]Sugar residual was determined from samples of expended media analyzed with refractometer (BRIX).

vessel of liquid (11 x 27 = 297 cm^2). The increased yield in the liquid system was partially due to the greater area of growth surface inside the large, rectangular vessels. When compared to agar, the liquid allowed plants to multiply more vigorously and grow more densely in the same volumes of medium. This raised concern as to whether the liquid media contained adequate nutrients to support the increased growth. The residual sugar in semi-solid agar media following six weeks of culture was generally higher than the 5% (w/v) sucrose used at the onset of cultures (Table 1). With 3% (w/v) sucrose as an initial concentration in strawberry and apple, the hydrolysis of sucrose proceeded to near completion during the culture cycle (Kozai et al., 1991; Kahru, 1997), but with kiwi fruit, residual sugar remained mostly as sucrose (Monaclean et al., 2003). In our current work with 5% (w/v) sucrose as at initiation, finding residual BRIX values greater than 5% (w/v) at the end of culture did not mean sugar was not used, but indicated hydrolysis and water uptake were relatively large compared to sugar uptake. Sugar residual in agar at the highest plant density had decreased slightly below 5% (w/v) in two of the three varieties (Table 1). It is clear that the 5% (w/v) sugar supplied was ample to support plant growth in agar for the density ranges tested.

There was more sugar used during plant growth on liquid media than agar (Table 1). At the higher plant densities in liquid media there was a further reduction in concentrations of residual sugar. At the highest densities in this experiment, the 2 - 3% (w/v) sugar residual (BRIX) present made it unlikely that sugar depletion had limited growth. Even if sugars were completely in monomer form, significant concentrations of sugars were present throughout the culture cycles. However sugar may become growth-limiting if an optimization plan further increased plant density. Similarly, other nutrients that were not monitored may become growth-limiting sooner than sugar, with high-density thin-film cultures of *Hosta*.

Sugars are known to have regulatory, nutritive and osmotic effects on *in vitro* plant growth. HPLC analyses of mono and di-saccharides and/or colorimetric procedures measure sugars more specifically than the refractometer. Analyzing endogenous concentrations of soluble sugars in plant tissue would yield more critical information to optimize plant quality. However, the refractometer assay of media, with zero reagent cost and real-time feedback, is so readily adaptable to practical situations, the author suggests use of this tool be considered preferable to the likely alternative of proceeding without any information about sugar concentrations.

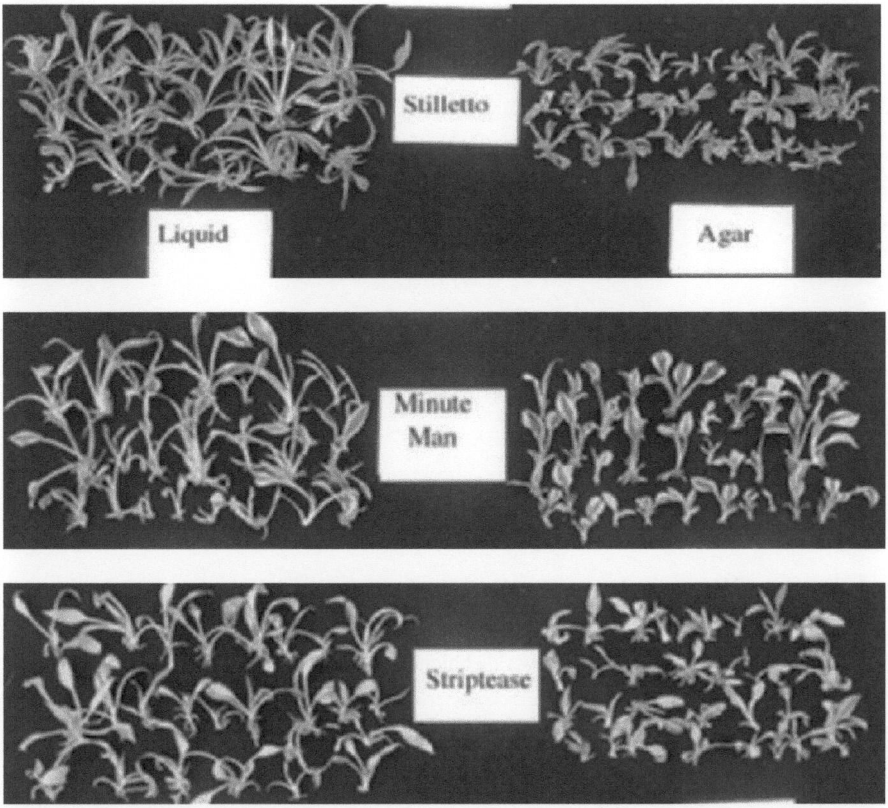

Figure 3: The three hosta varieties after 6 weeks in culture on liquid, thin-film culture (left) and agar media (right), supplemented with 1 μmol BA.

Practical alternative micropropagation methods should not increase the amount of man-hours spent in the transfer hood. A bulk dump process eliminated the significant time component when the technician individually oriented and spaced each bud on agar. Spacing of plants in a thin-film rocker is a mechanized-passive process and the buds orient themselves with respect to light during the culture cycle. *Hosta* grew well in these conditions, and plants were larger with longer petioles than when grown on agar (Figure 3). Passive spacing and orientation in the large vessels allowed high yields despite the larger size of the plants.

Labour efficiency is of critical importance for commercial application. While designing a vessel for a thin-film bulk dump process in the Nalgene

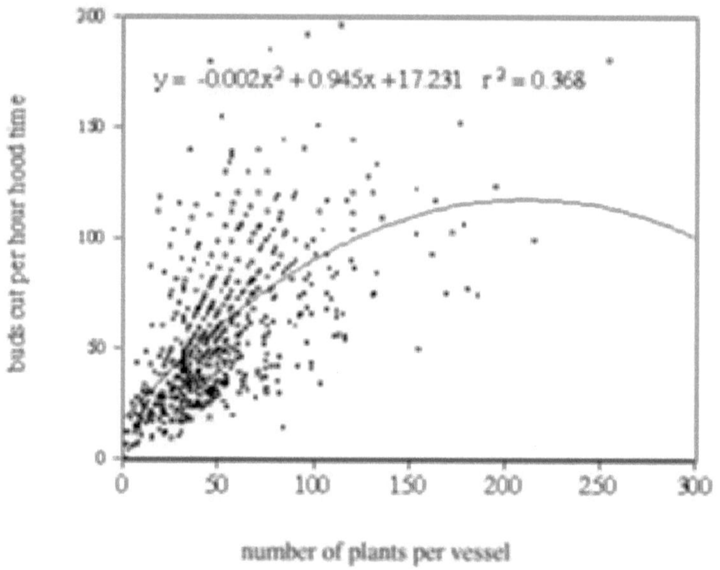

Figure 4: Labour input for twenty-two technicians propagating 40 varieties of *Hosta* from more than six months of observations.

Biosafe, technicians logged their time spent at transfer station and output was quantified as buds cut per hour. Variance in cuts per hour was partitioned for main effects of treatment factors including: the individual technician, plant variety, media formulation, time of day, day of week, weeks in culture, volume of media, number of explants in, and number of plants harvested per vessel (data not presented). The most significant factor that impacted technician efficiency was the number of plants harvested from the vessel (Figure 4). Observations during six months of production data in Nalgene Biosafe with 22 technicians cutting 40 varieties of *Hosta*, indicated that worker efficiency at the transfer station increased dramatically as numbers of plants harvested reached about 100 per vessel. In this current work, labour efficiency reached the optimal range when a minimum of 40 buds were dumped into 200 ml of liquid media and plant yield was 77-103 per vessel (95% confidence).

There were two causes for the quantitative relationship between buds harvested and cuts per hour dedicated at the hood station. First, a certain fixed portion of off-task time per vessel included setting up tools, opening the vessel, recording data and sealing the vessel. This remained relatively constant regardless of the number of plants per vessel and obviously

Figure 5: The three *Hosta* varieties from liquid thin-film and agar media, after 4 weeks acclimatization and growth in greenhouse.

constituted a lower proportion of hood time as the worker processed more buds per vessel. Secondly and more subtle, there was a likely period of concentrated effort when cutting proceeded at a brisk pace and was not interrupted as frequently by the off-task functions.

These observations were not from a time-motion study designed for costing this system. Information was logged about a vessel, the Nalgene Biosafe, which was eliminated as being too awkward. The numbers of cuts per hour also involve technicians cutting every plant into single bud divisions and counting each division for quantification. The purpose of the data was to establish factors related to improved worker performance in bulk dump process. The closure system for the rocker vessel, presentation of the rocker vessel from the autoclave to the hood, and securing vessel closure in the hood were unfinished at time of reporting this research. Time-motion studies and cost analyses need follow completion of the mechanical vessel system and could not be abstracted from the information presented.

Hundreds of thousands of *Hosta* plants from the commercial beta-site have been acclimatized to greenhouse and nursery. With retained sub-samples, plants from liquid were compared to plants from agar culture in the greenhouse during acclimatization and plant quality was good (Figure 5). All plants from liquid and agar survived and grew well in the greenhouse. Liquid-cultured *Hosta* plants tend to grow more quickly than agar-cultured plants in the greenhouse and nursery (Adelberg et al., 2000) and it can be surmised that *Hosta* from thin film systems would convey the same advantage.

4. Conclusion

Liquid culture in a thin-film rocker system presents opportunities for more efficient micropropagation than agar. The magnitude of these efficiencies depended largely on selecting appropriate plant density. Low-density culture results in the most rapid multiplication and has the largest incremental advantage compared with agar-based systems if high-value plants are in scarce supply. High-density culture results in the most efficient use of hood labour, least cost, and the highest yield from an optimized process (Figure 6). It was observed that as density increased, so did sugar utilization, and nutrient formulation may warrant further consideration, if this system is to be further optimized for high-density culture. In other reports on micropropagation in liquid media, dependent on the crop species - sugar, nitrogen salts, phosphate or water may be most limiting to growth (Hale et al., 1992; Desamero et al., 1993; Moncalean et al., 2003; Adelberg et al., 1997).

Figure 6: Thin film rocker system scaled-up for efficient space utilization in culture room.

The thin-film system and rocker vessels offered several improvements in plant growth and system efficiencies in comparison to agar-based methods. Plants were enabled to perform functions of spacing and orientation without direct human intervention. Solute transfer from medium to plant is greater and allows more rapid growth. Technicians are more focused on the germane tasks of carefully dividing the plant buds. The ergonomic and biological benefits may help to justify up-front inputs required to implement a new system.

Acknowledgements

This work is contribution no. 4788 of the SC Agriculture Research and Forestry Service (SCAFRS). The author acknowledges the cooperation of Southern Sun BioSystems Inc. Special acknowledgement is given to Jacqueline Naylor-Adelberg - manager at the Norris test site, and her staff, who were most helpful in during this experiment.

References

Adelberg J & Simpson EP (2002) Intermittent immersion vessel apparatus and process for plant propagation. Intl. S/N: PCT/US01/06586

Adelberg J, Kroggel M & Toler J (2000) Greenhouse and nursery growth of micropropagated Hostas from liquid culture. HortTech 10: 754-757

Adelberg J, Desamero N, Hale SA & Young (1997) Long-term nutrient and water use during micropropagation of *Cattleya* orchid on liquid/membrane system. Plant Cell, Tiss. Org. Cult. 48: 1-7

Alper Y, Young R, Adelberg J & Rhodes B (1994) Mass handling of watermelon microcuttings. Trans. Amer. Soc. Ag. Eng. 37: 1337-1343

Chu I (1995) Economic analysis of automated micropropagation. In: Aitken-Christie J, Kozai T & Smith MAL (eds) Automation and Environmental Control in Plant Tissue Culture (pp. 19-26). Kluwer Academic Publishers, Dordrecht

Desamero N, Adelberg J, Hale SA, Young R & Rhodes B (1993) Nutrient utilization in liquid/membrane system for watermelon micropropagation. Plant Cell, Tiss. Org. Cult. 33: 265-271

Etienne H & Berthouly M (2002) Temporary immersion systems in plant micropropagation. Plant Cell, Tiss. Org. Cult. 69: 215-231

Gross A & Levin R (1988) Design considerations for a mechanized micropropagation laboratory. In: Altman A, Ziv M & Izhar S (eds) Plant Biotechnology and *In Vitro* Biology in 21st Century (pp. 637-642), Kluwer Academic Publishers, Dordrecht

Hale A, Young R, Adelberg J, Keese R & Camper D (1992) Bioreactor development for continual-flow, liquid plant tissue culture. Acta Hort. 319: 107-112

Harris R & Mason E (1983) Two machines for *in vitro* propagation of plants in liquid media. Can. J. Plant Sci. 63: 311-316

Karhu ST (1997) Sugar use in relation to shoot induction by sorbitol and cytokinin in apple. J. Am. Soc. Hort. Sci. 122: 476-480

Kozai T, Iwabuchi K, Watanabe K & Watanabe I (1991) Photoautotrophic and photomixotrophic growth of strawberry plants *in vitro* and changes in nutrient composition of the medium. Plant Cell, Tiss. Org. Cult. 25: 107-115

Moncalean P, Jesus Canal M, Fernandez H, Fernandez B & Rodriguez A (2003) Nutritional and gibberellic acid requirements in kiwifruit vitroponic cultures. *In Vitro* Cell. Dev. Biol. - Plant 39: 49-55

Murashige T & Skoog F (1962) A revised media for rapid growth and bioassays with tobacco cultures. Physiol. Plant. 15: 473-497

Smith MAL & Spomer LA (1995) Vessels, gels, liquid media and support systems. In: Aitken-Christie J, Kozai T & Smith MAL (eds) Automation and Environmental Control in Plant Tissue Culture (pp. 371-404). Kluwer Academic Publishers, Dordrecht

Williams RR (1995) The chemical microenvironment. In: Aitken-Christie J, Kozai T & Smith MAL (eds) Automation and Environmental Control in Plant Tissue Culture (pp. 405-439). Kluwer Academic Publishers, Dordrecht

Zimmerman R (1996) Commercial application of tissue culture to horticultural crops in the United States. J. Kor. Soc. Hort. Sci. 37: 486-490

Ziv M (1995) *In vitro* acclimatization. In: Aitken-Christie J, Kozai T & Smith MAL (eds) Automation and Environmental Control in Plant Tissue Culture (pp. 493-516). Kluwer Academic Publishers, Dordrecht

Chapter 35

Aeration stress in plant tissue cultures

Michael B. Jackson
School of Biological Sciences, University of Bristol, Woodland Road, Bristol BS8 1UG, U.K.; and Plant Ecophysiology, Faculty of Biology, University of Utrecht, Sorbonnelaan 16, 3584 CA, Utrecht, The Netherlands. E-mail: mike.jackson@bristol.ac.uk

Abstract: The requirement for both sterility and the avoidance of dehydration in plant tissue cultures can impose sealing requirements that severely limit the rate of gas exchange in and out of the culture vessel. Conditions within the culture vessel, such as the depth of any water cover, the presence of gelling agents, the bulk and porosity of the tissue and the temperature, also strongly influence *in vitro* rates of gas exchange, primarily driven by diffusion. This article uses elements of Fick's Law of Diffusion to identify key factors limiting gas exchange between a culture and its immediate surroundings. In particular, it identifies static liquid media, gelling agents, large tissue mass and warm temperatures as imposing severe limits on diffusive flux for gases such as O_2, CO_2 and ethylene. The principal barrier to diffusive exchange of gases between the *in vitro* and *ex vitro* atmospheres is the wall of the enclosing vessel. This is invariably made of glass or plastic that is gas-impermeable and well-sealed against evaporative drying or entry of micro-organisms. Cultures enclosed in this way will, inevitably, asphyxiate unless a compensating pathway for diffusive gas exchange is contrived or replaced by some system of convective flow that carries gases to and from the tissue. Supplementing diffusive aeration with convective flow is the basis of most successful hydroponics systems for whole plants and may be a prerequisite for securing levels of aeration suitable for autotrophic cultures. The paramount consideration is the extent to which the total rate of consumption or production of a particular gas by the cultured tissues is matched by the maximum rates of gas transport imposed by the culture itself, its immediate surroundings and the ventilation and sealing system of the culture enclosure.

Key words: plant aeration, gas diffusion, ethylene, oxygen deficiency

Abbreviations: A - cross sectional area of diffusion pathway; C_{in} - concentration at start of diffusion pathway; C_{out} - concentration at end of diffusion pathway; D - diffusion coefficient of the medium through which the gas is moving; K_m - Michaelis-Menten constant; Q/t - amount of gas moved per second (flux density); Th - length of diffusion pathway

A.K. Hvoslef-Eide and W. Preil (eds.), Liquid Culture Systems for in vitro Plant Propagation, 459–473.

1. Introduction

Techniques of plant tissue culture are a stimulus for basic research and also a tool with potential and proven applications in agriculture, horticulture and forestry. Plant tissue culture has wide application, encompassing vector and vector-free genetic transformation of cells, embryo rescue, somatic embryogenesis and commercial clonal propagation. Much plant biotechnology requires at least one stage where plant tissue culture is obligatory (Thorpe, 2000). For success, the conditions under which cultures are housed and treated must be strictly adhered to, and many papers record how minor changes to the conditions, especially chemical components of the culture medium, enable hitherto recalcitrant cultures to perform in the desired manner.

Ventilation is an issue because the technology of tissue culture is inextricably bound up with a need for sterility and preventing dehydration. Such methods almost always involve some form of sealing of culture vessels. Inevitably, this interferes with the free exchange of gases between the cultured material and the outside atmosphere. The problems can be made worse by use of liquid cultures where cells, callus or tissues can be immersed in liquid medium. Much of what is known about the consequences of the resulting poor aeration for healthy plant growth or survival can be inferred from studies of flooding or submergence experienced by plants in the outside world (Vartapetian and Jackson, 1987). Here too, gas exchange is severely compromised by virtue of a dramatic slowing of gas diffusion in and out of the inundated tissue imposed by the water. Well-sealed plastic or glass culture vessels and the nutrient medium they contain, impose much the same aeration stress since they also impede gas exchange severely. One major difference between the well-sealed tissue culture vessel and plant submergence (the most extreme form of flooding stress) is that in the later case, well-adapted species such as *Callitriche platyarpa*, rice (*Oryza sativa*) and *Rumex palustris* can escape by means of an invigorated underwater elongation that renews contact between shoot and aerial environment (Jackson 1990; Voesenek and Blom, 1999). This is obviously not an option for plant tissues and cells growing in culture vessels.

Despite the self-evident importance of the gaseous environment for tissue cultures, it is an often neglected component of their technical specification. This shortcoming can have unwelcome consequences for culture performance because of the strong physiological impact of the gases involved, notably O_2, CO_2 and C_2H_4 (Jackson et al., 1994). This article summarises the processes of tissue culture aeration and highlights some recent developments intended to ensure its adequacy.

2. Normal plant aeration

The gases of greatest concern are O_2, CO_2 and ethylene. Oxygen and CO_2 are principal substrates or products of aerobic respiration and photosynthesis and thus intrinsic to the most basic life-sustaining metabolic pathways of plant cells. By contrast, ethylene is a plant hormone capable of influencing developmental processes such as cell expansion, senescence and differentiation at relatively small concentrations [0.01 - 10 ppm (v/v)]. The gas is correctly considered as a growth regulator rather than a toxin. The lack of metabolic breakdown and the ease with which it can be measured at small concentrations make it a suitable marker of aeration status (Jackson et al., 1987). Other gaseous growth regulators such as methyl jasmonate and nitric oxide may also influence tissue culture development but these have received relatively little research attention in this context.

All living tissues of higher plants, not just the photosynthetic parts, require unimpeded gas exchange to function normally. Figure 1 provides some approximate figures illustrating the scale of gas influx and efflux that underpins normal plant growth and development. Severe impediment to gas movement for more than a few hours is usually fatal to growing cells, while partial interference can influence growth rate and the pattern of development. As the term 'gas exchange' implies, both inward and outward efflux of gases are slowed or arrested. This leads not only to a shortage of gases normally provided by the environment, but also to an internal accumulation of gases made by the plant and normally excreted at a similar rate by outward diffusion. Thus, when gas exchange is slowed by external conditions, plant cultures will experience shortages of oxygen and (in the light) carbon dioxide while accumulating ethylene, a plant hormone that can have a range of responses depending on tissue sensitivities. Other plant-made volatiles such as ethanol and acetaldehyde (Righetti et al., 1990), the free radical nitric oxide and the hormone-like molecule methyl jasmonate, may also build up; effects of these have not been explored to any great extent but deserve investigation. Nitric oxide could be generated from nitrite produced by nitrate reductase activity. It can be harmful by causing oxidative damage *via* the production of oxygen superoxide (O_2^-) as well as being a possible signalling molecule for certain developmental processes (Sakihama et al., 2002). There is a Clarke-type NO electrode and a fluorescence method for NO detection: NO can also be measured by laser photoacoustic devices.

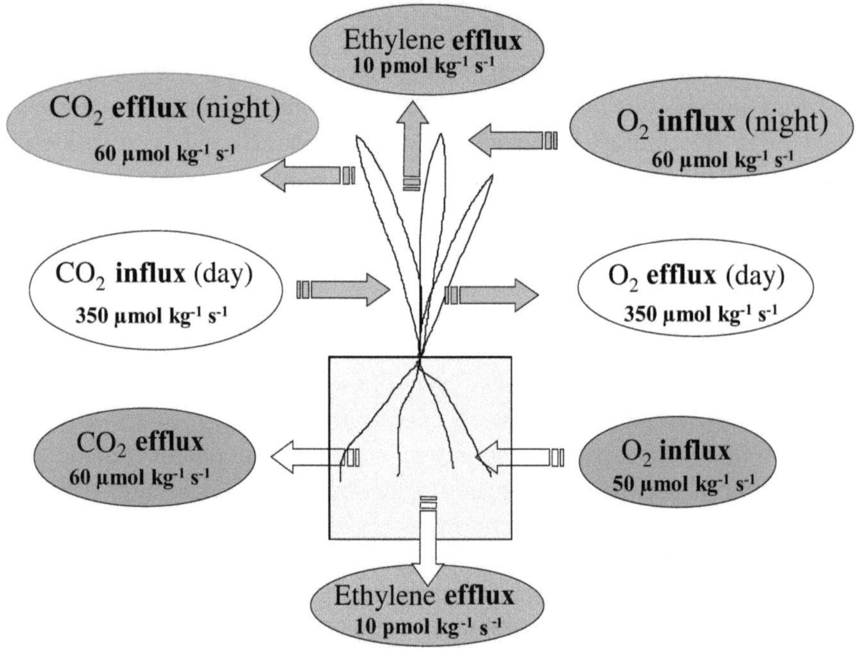

Figure 1: Schematic illustration of the pattern and approximate scale of gas efflux and influx that underlies normal plant growth and development. Sources for gas flux data include: Haupt-Herting and Fock (2002), Jackson (1980), Armstrong et al. (1991) and Poorter et al. (1991).

3. Aeration by diffusion

The principal mechanism that propels gas exchange in and out of a plant tissue is net diffusion down concentration gradients. The process is summed up by Fick's Law of Diffusion that can be expressed as $Q/t = D \cdot A\ (C_{in} - C_{out})/Th$, where Q/t = amount of gas moved per second (flux density); D = diffusion coefficient of the medium through which the gas is moving; A = cross sectional area of diffusion pathway; C_{in} = concentration at start of pathway; C_{out} = concentration at end of pathway; Th = length of diffusion pathway. Factors that influence the components of this equation and thus impact on the rate of gas transport (flux density) are now examined.

3.1 Flux density

Flux density (Q/t) is the first term in the Fick's Law equation ($Q/t = D \cdot A (C_{in} - C_{out})/Th$). Its value is affected by the physical environment of the tissue and the morphology of the tissue itself. Whatever the maximum possible flux of a gas to the surface of the tissue or cell turns out to be in a given set of circumstances, it must be judged against what the tissue or cell actually requires to sustain normal rates of metabolism or to keep internally generated gases below some activity threshold. Oxygen flux density in relation to root growth will be used as an example, in a situation where the maximum oxygen flux rate to a root is depressed by the high diffusion resistance imposed by submergence in water. Cereal root respiration rate in the elongating apex can be in the order of 1.56 µmol kg^{-1} s^{-1} or 50 ng O_2 cm^{-2} of root surface min^{-1} at 10 °C. The expression using surface area is especially useful because it highlights the importance of area available for gas capture or release. Using platinum electrodes, Blackwell and Wells (1983) showed that when the equilibrium concentration of dissolved oxygen in the flooded soil fell from 21 % to 15 % in response to consumption by micro-organisms, the maximum possible inward flux density of O_2 toward the root decreased from > 80 ng O_2 cm^{-2} of root surface min^{-1} (the value for freshly flooded soil) to below 30 ng O_2 cm^{-2} min^{-1}, a value close to the respiratory demand of the root. At this point, root elongation became retarded. They also showed that, for elongation to be stopped completely, the flux density must fall almost to zero. From this study we learn that (i) some loss of gas diffusion capability for O_2 can be sustained without damage (from >80 ng O_2 cm^{-2} min^{-1}down to < 30 ng O_2 cm^{-2} min^{-1}, in this case), (ii) to depress growth, flux densities must fall to below those potentially generated by rate of respiration and (iii) an almost total arrest of O_2 influx is needed to stop growth and possibly kill the root.

The seeming tolerance of quite substantial decreases in O_2 flux density is partly attributable to the large amount of O_2 present in the air (21 % v/v). This ensures a maximum flux density in water that can often exceed the needs of respiring tissue (see below for qualifications). This can be true, in well-stirred water, despite the limited solubility of O_2 in water that reduces the concentration from 8.31 mol m^{-2} in air to 0.25 mol m^{-3} in water (at 25° C). An exception may be roots in tissue culture growing at optimal temperatures and supplied *in vitro* with external sugar (Asplund and Curtis, 2001). Here, local root-tip respiration is so intense that fluxes through well-stirred water in equilibrium with air are too slow to sustain maximum respiration rates. A second explanation for the insensitivity of tissue to modest depressions in O_2 availability is that the affinity for oxygen shown by cytochrome oxidase, the terminal oxidase in mitochondrial electron transport, is extraordinarily strong

(K_m approx. 0.14 mmol m^{-3}). It is instructive to compare these figures with those for CO_2, the photosynthetic progenitor of environmental O_2. CO_2 flux densities required to support photosynthesis are similar or higher than those required to support respiration with external O_2 (Figure 1). Although CO_2 is also much more soluble in water than O_2; at 25 °C (the ratio of dissolved CO_2/gas-phase being approximately 0.76 at neutral pH, while the ratio for oxygen is only 0.03), the small amount of CO_2 in air (approximately 360 ppm, v/v) is insufficient to generate dissolved concentrations that match those of dissolved O_2. At 25 °C, CO_2 concentrations may attain only 11.4 mmol m^{-3} compared with the concentration of O_2 in aerated water of 0.266 mol m^{-3}, a difference of almost 2000. Since the inward flux densities of CO_2 needed for photosynthesis are similar or greater than those of O_2 needed for respiration, it is clear that autotrophic plants are much more at risk from CO_2 starvation than from O_2 starvation. A further contributing factor to this greater sensitivity to decreased availability of CO_2 is the low affinity for CO_2 of Rubisco, the enzyme largely responsible for capturing photosynthetic CO_2 (K_m for CO_2 - 10 mmol m^{-3} compared to 0.14 mmol m^{-3} for O_2). This high K_m for CO_2 by Rubisco is almost the same as that for the concentration of CO_2 dissolved in water in equilibrium with air, indicating the impossibility of submerged leaves from photosynthesising normally. Data for O_2 and CO_2 used in these comparisons can be found in Armstrong (1979).

This analysis makes it apparent that while plants can sustain considerable interference in ventilation before mainstream respiration is strongly inhibited, almost no interference in aeration is possible in autotrophic plants before photosynthesis is depressed. This means that diffusion-driven CO_2 supply is rarely adequate for photosynthesis in tissue cultures, imposing the necessity of the sugar supplements almost universally used in tissue cultures. Even for oxygen, the leeway in diffusion losses can be sustained before damage sets-in, is strongly eroded by other factors such as warmer temperatures and decreasing surface to volume ratios. Thus, although the physical processes of diffusion is largely unresponsive to temperature *per se*, the flux required by the tissue increases dramatically with warming, typically doubling every 10 °C (Figure 2). At about 30 °C, demand by a cereal root matches the maximum flux that it possible from water in equilibrium with air. Clearly, increasing temperature markedly increases the required flux, placing a much larger demand on the gas exchange capacity of the surroundings. The flux required is also raised very considerably by radial growth of the tissue. This is because, as tissue bulk increases, the surface area available to service the gas exchange needs of each unit of tissue volume, decreases. In a tissue culture, the leeway for how much aeration can be impaired before a tissue culture will suffer from O_2 shortage is abolished as size of tissue mass increases. The effect can only be ameliorated if the tissue mass is highly porous and permeated by gas-filled intercellular spaces, which is uncommon. It may be concluded that any margin that exists before restricted aeration damages tissue cultures through O_2 shortage can be lost to warmer temperatures and to increasing tissue volume.

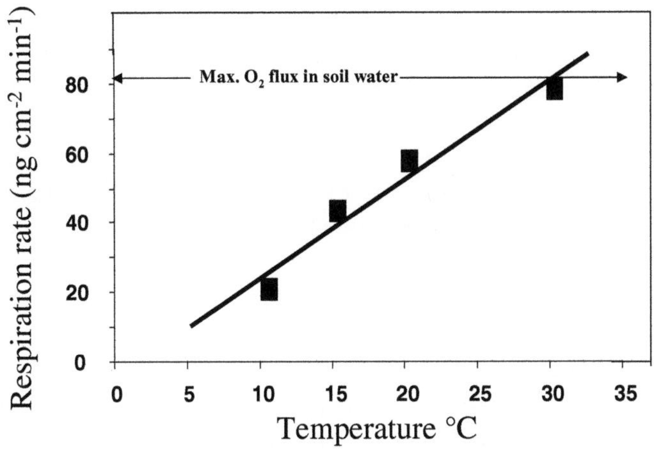

Figure 2: Relationship between temperature and the rate of respiration expressed in terms of an inward flux density of O_2 required to service respiratory demand of plant tissue such as a cereal root axis. While at cool temperatures (e.g., 15 °C) the requirement is smaller than the maximum flux density possible in water equilibrated with air (approx. 80 ng cm^{-2} min^{-1}), it is barely adequate to service the complete needs of roots at 30 °C.

3.2 *Diffusion coefficient of the diffusion medium*

The diffusion coefficient (D) of the medium, or barrier surrounding the tissue, is the second component of the Fick's Law equation ($Q/t = D \cdot A (C_{in} - C_{out})/Th$). This value quantifies the relative ease with which the gas diffuses through a given medium (e.g., air, water, tissue) along its path to the tissue or cell. The very small coefficient of gas diffusion in water compared to that in air makes water a principal enemy of tissue aeration. For O_2, ethylene and CO_2, the diffusion coefficient in water is almost 10,000 times smaller than in air (in air: 0.201 cm^2 s^{-1}; in water: 2.1×10^{-5} cm^2 s^{-1}). Adding a gelling matrix such as agar or similar setting agent, increases the barrier to gas diffusion even more. It also imposes a large unstirred layer that, in effect, strongly decreases the concentration of externally derived gases at the tissue surface thus slowing the rate of gaseous diffusion into the cells. This effect also enhances the trapping effect of water on metabolically generated gases such as ethylene. The work of Barrett-Lennard and Dracup (1988) and Verslues, Ober and Sharp (1998) illustrates the highly damaging impact of gelling agents on aeration and growth.

In most tissue culture systems, the strongest diffusion barrier is the container itself, which is usually fabricated from material that is totally impermeable to gases. For cultures to do well in such containers they must either be very large in relation to the size of the culture or be made to leak gas in some way. Obvious ways to achieve leakage include loosening the closure or inserting a gas-permeable membrane (e.g., polypropylene). Cultures of *Ficus elastica* in 305 ml plastic Magenta boxes were found to benefit from leakage rates that gave a half-time for gas replacement of about 8 h. The benefit was attributed to a decrease in accumulated ethylene, resulting in enhanced leaf expansion (Jackson et al., 1991).

3.3 Cross sectional area of the diffusion pathway

The cross sectional area of the diffusion pathway (A) of the medium or barrier is the third component of the Fick's Law equation ($Q/t = D \cdot A (C_{in} - C_{out})/Th$). Obviously, the larger the area across which gases can diffuse, the more gas can be moved in total within a given time. One important aspect of this effect is the influence of the size and format of the tissue mass; a larger mass creating a smaller area for diffusion per unit volume of tissue. The impact of cross section area also means that the amount of *internal* surface area of tissue that is in contact with a gas phase (e.g., that created by aerenchyma) will also affect the gaseous flux in proportion to the diffusive surface area that this creates. These effects have already been considered above.

3.4 Length of the diffusion pathway

The length of the diffusion pathway (Th) is the final term in the Fick's Law equation ($Q/t = D \cdot A (C_{in} - C_{out})/Th$). The longer the path, the greater the total resistance to gaseous diffusion. Tissue itself impedes its own aeration. After all, cells are mostly composed of water. If tissues enlarge radially, the total path length to the centre of the mass also increases. For this reason, a large tissue mass may result in a poorly aerated interior leading, possibly, to anaerobic or hypoxic tissue cores, while more peripheral cells remain better aerated. There is an additional problem connected with the thickness of the tissue barrier. For oxygen, it is that more and more of it is consumed as the pathway lengthens. This adds to the probability of deficiency deeper into the tissue because the concentration gradient driving further inward flow, reduces as more and more oxygen is used in respiration.

These effects are additional to the small surface–to-volume ratios that typify larger masses of tissue and raise, detrimentally, the requirements for gas exchange at the tissue external surface (as discussed above). Diffusion, within tissue can also be impeded more strongly by certain kinds of differentiation (e.g., lignified parts). These can impose sharply internally delineated radial-diffusion barriers (Armstrong et al., 2000).

Outside the tissue, the thickness (depth) of any water covering also impacts on gas diffusion. The effect is illustrated in calculations for ethylene accumulating at the surface of a root producing ethylene at a known rate (Jackson, 1979). When the radius of a water covering over the root of radius 0.025 cm is increased from 0.25 cm to 2.0 cm, ethylene build-up at the root surface is enhanced 10-fold. A similar effect in reverse is seen for O_2. Thus, there is a benefit to aeration from maintaining water cover that is as shallow as possible or is intermittent to allow periods of relief from gas entrapment or exclusion.

3.5 Driving force for diffusion

The driving force for directional net diffusion is the fourth bracketed term in the Fick's Law equation ($Q/t = D \cdot A\ (C_{in} - C_{out})/Th$). It comprises the difference in concentration between source and sink. For O_2 and CO_2 it is appropriate to consider C_{out} as the atmospheric concentrations and C_{in} usually lying somewhere between atmospheric and zero. For ethylene we take C_{out} to be zero and C_{in} to be up to tens of parts per million and thus possibly above physiologically active levels. Clearly, C_{in} is the physiologically active component and is affected by all the factors considered above that influence gaseous flux density. Thus, managing tissue culture aeration is about managing C_{in} to avoid the development of physiologically damaging concentrations of gases.

4. Approaches towards improving tissue culture aeration

The key to aerating, adequately, the cultured tissue itself, is to maximise the concentration gradient for the gas between the interior and the immediate exterior of the tissue and to minimise the flux density per unit of surface area need to sustain normal growth, respiration and where appropriate, photosynthesis. Our analysis of components of the Fick's Law equation has indicated this can be achieved by increasing diffusive ventilation of culture volume, minimising resistances to the movement of gas down that gradient, keeping cultures cool, minimising thickness of any water cover and avoiding gelled / semi-solid media such as agar. A further important consideration is minimizing tissue volumes. This both decreases the flux density per unit of surface area of tissue required to aerate the tissue, and shortens internal diffusion pathways. Actual success is determined by achieving the flux density of gas needed to support the desired growth rate or pattern of development. This is especially demanding if the cultures are to be autotrophic, because CO_2 is almost always in short supply. The aim is to optimise C_{in} for key gases such as CO_2, O_2 and ethylene. A simple approach to achieving this aim is to provide culture vessels that are large in relation to the amount of tissue, because this provides greater reserves of O_2 or CO_2 and a more effective dilution of metabolically generated gases such as ethylene.

However, in practice, satisfying the aeration needs of tissue cultures by facilitating diffusive aeration and maximising vessel volumes alone is almost impossible, especially if autotrophic cultures are needed. Thus, additional measures are required. One approach is to side-step some of the problems. The most widespread such approach is to overcome the need for adequate flux of external CO_2 by supplying respirable sugars such as sucrose. This is widely practised and does not warrant further discussion here except to emphasise the absolute requirement for sterility that the use of sugar demands, and to note the potential problems of culture adaptation to autotrophic metabolism on final transfer *ex vitro.* The absence of photosynthetic CO_2 fixation also deprives the cultures of the O_2 that would be generated by the photosynthesis. A second example of side-stepping is to minimise C_{in} of ethylene by reducing C_{out} by absorption, by the use of such compounds as alkaline potassium permanganate or mercuric perchlorate. Enhanced leaf expansion rates in *Ficus lyrata* have been reported using this approach (Jackson et al., 1987; 1991). A closely related measure is to incorporate an inhibitor of ethylene action, such as silver nitrate (e.g., Armstrong et al., 1997) or 1-methylcyclopropene, a recently-developed and very effective gaseous inhibitor of ethylene action (Sisler et al., 1996)

More satisfactory than side-stepping, is to introduce an element of convection flow (mass flow) of the air (or water where liquid cultures are uses) to augment the contribution from diffusion alone. With this mechanism of aeration, it is useful to consider the surrounding water or air as a physical carrier of gases to and from the tissue rather than a diffusion medium or barrier. For convection to be effective, the air or water must move across the cultured material. Various methods have been adopted in attempts to achieve this simply and effectively.

5. Harnessing convective flow

An elegant non-mechanized solution to the problem of how to ventilate tissue culture vessels with convection is provided by Armstrong and colleagues (Armstrong et al., 1997; Zobayed et al., 2001). It is based on a mechanism operating in whole plants of wetland species such as *Phragmites australis*. Here, mass flow of gases is driven initially by diffusion of N_2 and O_2 through stomata into the humid interior of the leaf down a concentration gradient. This gradient arises as a result of extra dilution of N_2 and O_2 in the plant's interior by abundant water vapour. The pore size of stomata is such that the resistance they offer to inward diffusion is less than outward mass flow of gas. Thus, inward diffusion of N_2 and O_2 (and other more minor components of air) results in an internal pressurization of the plant's gas spaces. This force is then capable of driving mass flow along interconnected gas-filled spaces. This mechanism of humidity-induced convective throughflow can be harnessed to ventilate tissue culture vessels with a sterile flow of air. Pressurization is achieved by connecting the culture vessel to a port with an overlying microporous membrane with a small pore size (0.03 µm) overlying but no quite touching water held in a small reservoir. This creates a humidity-driven inflow and pressurizes the culture vessel. A second port protected by a membrane of larger pore size (0.2 µm) creates an exhaust of low resistance to mass flow (Figure 3). The outcome is a convective throughflow of air that is fast enough to clear half any accumulated ethylene within 30 min and to sustain greater rates of dry matter accumulation through enhanced CO_2 supply (Zobayed et al., 2001).

For tissue cultures where liquid medium is used and especially where mass production of somatic embryos in bioreactors is required, highly mechanised systems for introducing convective aeration have been devised. Here, three new elements are introduced to supplement diffusive aeration with mass flows. Firstly, air is pumped over the cultures to carry O_2 and CO_2 to the plants and substrate and aid removal of metabolically produced gases from the tissues. Secondly, the air flow is supplemented with CO_2 to encourage further photosynthetic assimilation if convention flow of air does not provide a CO_2 flux density for a desired rate of photosynthesis Thirdly, the roots are grown in a porous medium to promote their aeration by diffusion. This is also periodically flushed with aerated nutrient solution to bring these resources to the roots by mass flow. A set-up such as this is described by Afreen et al. (2002). The plants produced by such a system were notable for their rapid adaptation to conditions *ex vitro*.

A quite different approach to overcoming aeration problems was taken by Barry-Etienne et al. (2002): they recognised that the *in vitro* nature of culture systems is the principle problem for aeration. Thus, they proposed keeping any *in vitro* phase in micropropagation as short as possible and returning the tissues to an open non-aseptic environment at the earliest opportunity. This philosophy was anticipated several years earlier by Firn et al. (1994). In the system of Barry-Etienne, autotrophic somatic embryos, generated in a sophisticated illuminated bioreactor were transferred at an early stage (the cotyledon stage) to high density cells of sterilised soil-containing compost in a glasshouse under shade and carefully watered. The 60 % that survived grew on quite quickly to form well-rooted 'normal' plants without requiring troublesome rehabilitation conditions.

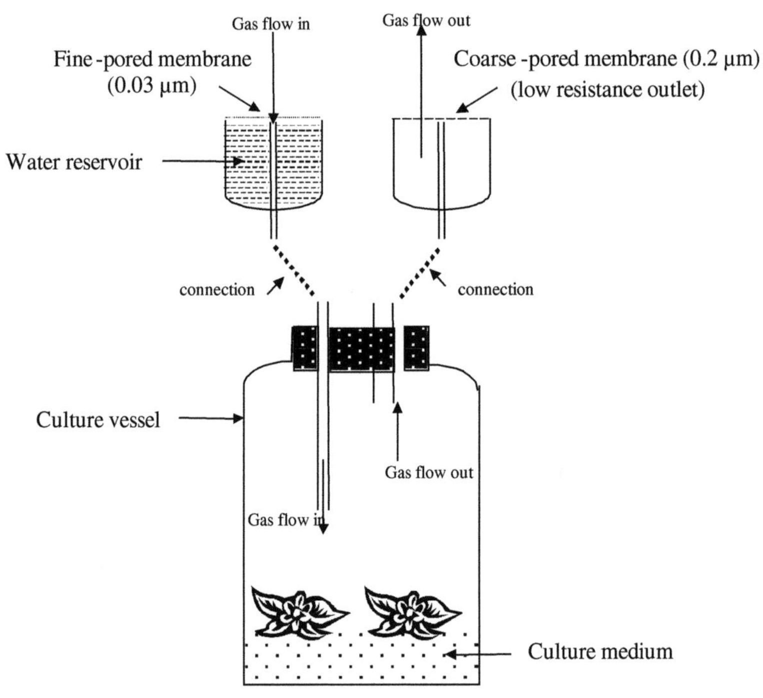

Figure 3: Diagram illustrating how humidity-driven inward diffusion of air across a membrane of fine pores can create a pressure-driven flow of sterile air through the culture vessel and out through a second port protected by a membrane of larger pore size. Taken from Armstrong et al. 1997.

6. Conclusion

Gas exchange in tissue culture can involve two different although not entirely unconnected processes. These are simple diffusion down concentration gradients and/or convective (mass) flow. In the latter, gases are delivered to plant tissue by a flow of solvent (water) or carrier gas (air) that is propelled over the surface of cells or tissue by an external force, usually a pressure gradient. However, within the tissue mass itself, gas movement is effected almost exclusively by diffusion. Only in whole plants growing in the natural environment where extensive interconnected porous tissue (aerenchyma) is present, can internal convective flow play a significant role in aeration (e.g., in *Phragmites*, Armstrong et al., 1992). However, convective flow within culture vessels can readily be harnessed to increase rates of tissue culture gas exchange by contriving slow pressure-driven gas flows (e.g., Zobayed et al., 2001) or by flowing or stirring aerated medium over the cultured tissue. Opportunities for enhancing the diffusive aeration of tissue cultures are many and include temporary immersion set-ups, fitting gas permeable ports to culture vessels, and adopting loose-fitting lids. The demands on the ventilation system can be minimised by restricting the depth of culture medium to thin films, keeping tissue bulk to a minimum and by adopting cool temperatures. The biological effectiveness of any particular system will be affected by many factors. Paramount is balancing the total rate of consumption or production of a particular gas by the culture with the influx and efflux rates imposed by the enclosing environment and ventilation system.

References

Afreen F, Zobayed SMA & Kozai T (2002) Photoautotrophic culture of *Coffea arabusta* somatic embryos: development of a bioreactor for large scale plantlet conversion from cotyledonary embryos. Annals of Botany 90, 21-29

Armstrong W (1979) Aeration in higher plants. Advances in Botanical Research 7: 225-332

Armstrong J, Armstrong W & Beckett PM (1992) *Phragmites australis*: Venturi-induced and humidity-induced pressure flows enhance rhizome aeration and rhizosphere oxidation. The New Phytologist: 120: 197-207

Armstrong W, Beckett PM, Justin SHFW & Lythe S (1991) Modelling, and other aspects of root aeration by diffusion. In: Jackson MB, Davies DD & Lambers H (eds) Plant Life under Oxygen Deprivation: Ecology, Physiology and Biochemistry (pp. 267-282). SPB Academic, The Hague

Armstrong J, Lemos EEP, Zobayed SMA, Justin SHFW & Armstrong W (1997) A humidity-induced convective throughflow and ventilation system benefits *Annona sqamosa* L. explants and coconut calloid. Annals of Botany 79: 31-40

Armstrong W, Cousins D, Armstrong J, Turner DW & Beckett PM (2000) Oxygen distribution in wetland plant roots and permeability barriers to gas exchange with the rhizosphere: a microelectrode and modelling study with *Phragmites australis*. Annals of Botany 86: 687-703

Asplund PT & Curtis WR (2001) Intrinsic oxygen use kinetics of transformed plant root culture. Biotechnology Progress 17: 481-489

Barry-Etienne D, Bertrand B, Vasquez N & Etienne H (2002) Comparison of somatic embryogenesis-derived coffee (*Coffea arabica* L.) plantlets regenerated *in vitro* and *ex vitro*: morphological, mineral and water characteristics. Annals of Botany 90: 77-85

Blackwell PS & Wells EA (1983) Limiting oxygen flux densities for oat root extension. Plant Soil 73: 129-139

Barrett-Lennard EG & Dracup MA (1988) A porous agar medium for improving the growth of plants under sterile conditions. Plant and Soil 108: 294-298

Firn RD, Sharma N & Digby J (1994) Physiology, growth and development of plants and cells in culture – the way ahead. In: Lumsden PJ, Nicholas JR & Davies WJ (eds) Physiology, Growth and Developments of Plants in Culture (pp. 409-421). Kluwer Academic Publishers, Dordrecht

Haupt-Herting S & Fock HP (2002) Oxygen exchange in relation to carbon assimilation in water-stressed leaves during photosynthesis. Annals of Botany 89: 851-859

Jackson MB (1980) Aeration in the nutrient film technique of glasshouse crop production and the importance of oxygen, ethylene and carbon dioxide. Acta Hortic. 98: 61-78

Jackson MB (1990) Hormones and developmental change in plants subjected to submergence or soil waterlogging. Aquatic Botany 39: 49-72

Jackson MB, Blackwell PS, Chrimes JR & Sims TV (1984) Poor aeration in NFT and a means for its improvement. J. Hort. Sci. 59: 439-448

Jackson MB, Abbott AJ, Belcher AR & Hall KC (1987) Gas exchange in plant tissue cultures. In: Jackson MB, Mantell SH & Blake J (eds) Advances in the Chemical Manipulation of Plant Tissue Cultures Monograph No. 16 (pp. 57-72). British Plant Growth Regulator Group, Bristol

Jackson MB, Abbott A, Belcher AR, Hall KC, Butler R & Cameron J (1991) Ventilation in plant tissue cultures and effects of poor aeration on ethylene and carbon dioxide accumulation, oxygen depletion and explant development. Annals of Botany 67: 229-237

Jackson MB, Belcher AR & Brain P (1994) Measuring shortcomings in tissue culture aeration and their consequences for explant development. In: Lumsden PJ, Nicholas JR & Davies WJ (eds) Physiology, Growth and Developments of Plants in Culture (pp. 191-203). Kluwer Academic Publishers, Dordrecht

Poorter H, Vanderwerf A, Atkin OK & Lambers H (1991) Respiratory energy-requirements of roots vary with the potential growth-rate of a plant-species. Physiologia Plantarum 83: 469-475

Righetti B, Magnanini E, Infante R & Predieri S (1990) Ethylene, ethanol, acetaldehyde and carbon dioxide released by *Prunus avium* shoot cultures. Physiologia Plantarum 78: 507-510

Sakihama Y, Nakamura S & Yamasaki H (2002) Nitric oxide production mediated by nitrate reductase in the green alga *Chlamydomonas reinhardtii*: an alternative NO pathway in photosynthetic organisms. Plant and Cell Physiology 43: 290-297

Sisler EC, Serek M & Dupille E (1996) Comparison of cyclopropene, 1-methylcyclopropene and 3.3-dimethylcyclopropene as ethylene antagonists in plants. Plant Growth Regulation 18: 169-174

Thorpe TA (2000) History of plant cell culture In: Smith RH (ed) Plant Cell Tiss. Cult. Techniques and Experiments (pp. 1-32). Academic Press, London

Vartapetian BB & Jackson MB (1997) Plant adaptation to anaerobic stress. Annals of Botany 79 (Supplement A): 3-20

Verslues PE, Ober ES & Sharp RE (1998) Root growth and oxygen relations at low water potentials. Impact of oxygen availability in polyethylene glycol solutions. Plant Physiology 116: 403-1412

Voesenek LACJ & Blom CWPM (1999) Stimulated shoot elongation: a mechanism of semi-aquatic plants to avoid submergence stress. In: Lerner HR (ed) Plant Responses to Environmental Stresses: from Phytohormones to Genome Reorganization (pp. 431-448). Marcel Dekker, New York

Zobayed SMA, Armstrong J & Armstrong W (2001) Micropropagation of potato: evaluation of closed, diffusive and forced ventilation on growth and tuberization. Annals of Botany 87: 53-59

Chapter 36

Macro- and micronutrient nutrition of plants in greenhouses, hydroponic systems, and *in vitro* culture on gelled media

Hans R. Gislerød, Ravichandran Selliah, Kwadwo Owusu Ayeh & Anne Kathrine Hvoslef-Eide

Department of Horticulture and Crop Sciences, Agricultural University of Norway, P.O.Box 5022, 1432 Aas, Norway. E-mail: hans.gislerod@ipm.nlh.no

Abstract: Nutrition for *in vitro* and greenhouse production systems is reviewed and found to be broadly similar. The optimal pH (5.0-6.0) is independent of the growing system used, with some lower or higher pH values required for special crops. The pH of nutrient solutions is regulated either with acid or via the NH_4:NO_3 ratio. For *in vitro* solutions, it is possible to use buffers (MES or TRIS), while in greenhouse systems the pH buffering capacity relies mostly on the quantity of colloids in the growing medium.

The normal nutrient solution conductivity for vegetative propagation is 0.8-1.2 mS cm^{-1} and 1.5-2.0 mS cm^{-1} for growth in greenhouse production. In micropropagation on gelled media the conductivity will be between 3.0-6.5 mS cm^{-1}. When spraying nutrient directly on the leaves in greenhouse production either to increase uptake of a particular element or to prevent a deficiency, the same concentration as for micropropagation should be used.

The air humidity will usually be greater in micropropagation systems than in greenhouse production. In *in vitro* culture of plant parts, uptake of the different nutrient elements will be mainly by diffusion. When comparing the proportion between *in vivo* and *in vitro* growing systems, there is a surprisingly low content of Ca and P for micropropagation. For micronutrients, micropragation media are low in Fe and Cu content and high in Mn and Zn content, compared to that required for greenhouse production. Light has been found to have an effect on the stability of iron-chelates and thus on the quantity of iron available in the media for uptake. We present results to show the importance of the right nutrient content of the medium and how it affects growth and development *in vitro*, but mostly the increase in development rate to flowering.

Key words: air humidity, conductivity, light, pH, proportion of the nutrient element

Abbreviations: BA(P)–6-benzylaminopurine; EC–electrical conductivity; IAA–indoleacetic acid; MES–2-(N-morpholino)ethanesulfonic acid = $C_6H_{13}NO_4S{\cdot}H_2O$; M. et al.–Murashige et al. (1972) medium; MS–Murashige and Skoog (1962) medium; NFT–nutrient film technique; SSE–soil saturated extract; TRIS–Tris (hydroxymethyl) aminomethane = $C_4H_{11}NO_3$, 2-amino-2-hydroxy-methyl 1, 3,propanodiol

A.K. Hvoslef-Eide and W. Preil (eds.), Liquid Culture Systems for in vitro Plant Propagation, 475–492.

1. Introduction

The composition of the nutrient solution plays a key role in plant nutrition. In modern greenhouses, the quantity of nutrient available at any moment in a growing system is limited, and usually covers only a small proportion of the total crop requirements. Therefore, to ensure maximum production, usually a nutrient solution is supplied *via* hydroponics several times each day, having a composition that is well suited to the needs of the crop and the growing conditions. Incorrect supply of nutrients easily unbalance the composition of the 'soil solution' in the root environment, and can lead to deficiency or toxicity. Also nutrient solutions can also be sprayed onto greenhouse plants as a foliar feed, especially to cure a nutrient deficiency or to prevent a deficiency developing. In a foliar feed, the concentration of the element required can be 2-10 times greater than in a nutrient solution supplied to the plant roots.

The compositions of nutrient media for *in vitro* cultures follow the same principles as for highly productive greenhouse production of whole plants under high humidity. It may be that *in vitro* shoots without roots have the same nutrient uptake mechanisms *via* the stem tissues as foliar nutrition. However, there has been little exchange of knowledge about uptake mechanisms between the greenhouse and micropropagation fields. Recently there has been some interest in optimising the nutrient supply *in vitro* to obtain improved proliferation and quality of propagules (e.g. Leifert et al., 1995; Ayeh, 1998; Bouman and Siekstra, 2002). Essentially, we have been doing the same type of experiments as those published by Murashige and Skoog in 1962; analysing the plant mineral content and providing a medium composition most appropriate for that particular plant. In liquid cultures, such adjustments of media composition are even more important than soil-based systems.

This chapter aims at link knowledge of greenhouse production of whole plants with that of *in vitro* culture of shoots, and underlines the importance of the composition of the *in vitro* nutrient media. In (whole) plant nutrition, three facets are especially important:

1. The pH of the solution - which will determine how easily available are the nutrients.
2. The element composition of the nutrient solution - which will determine to a great extent the nutrient uptake.
3. The root environment and the atmosphere – which influence the uptake of nutrients and, in turn, determines plant growth and quality.

2. Effects and adjustments of pH

Because the pH of a nutrient solution will determine how accessible are the nutrient elements, it is of vital importance to be able to regulate the pH accurately. Optimum pH for greenhouse production would be 5.2 – 6.0 for growth, with 4.5 – 7.0 as the outer limits. For vegetative propagation, the pH optimum has a slightly narrower range; with 5.0 – 5.5 as optimal and 4.5 – 6.5 as the outer range which is acceptable (Bævre and Gislerød, 1999). Optimum pH for *in vitro* propagation would be in the range 5.0 – 6.0 (4.5 – 7.0), with an initial pH set to 5.2 for woody species (Lloyd and McCown, 1980) and 5.8 for herbaceous species (Murashige and Skoog, 1962). The optimal pH is therefore comparable between hydroponics systems in greenhouses and in liquid cultures systems *in vitro*.

The pH of a medium is influenced by:

a) the balance between negative and positive ions in the nutrient solution including NH_4^+ and NO_3^-
b) the amount of bicarbonate in the tap water
c) the different nutrient uptake pattern of plants (Stensvand and Gislerød, 1992)
d) the type of growing system (buffering, ‘closed’ or ‘open’)

To reduce the pH in greenhouse cultures, HNO_3, H_3PO_4 and/or ammonium-nitrogen can be used. The quantity of ammonium ions added to the nutrient solution calculated on the basis of the amount of bicarbonate in the tap water used to make up the solution. Normally, ammonium ions account for 8-12 % of the total nitrogen in the nutrient solution when the bicarbonate content is approximately 30 mg per litre. This amount of ammonium can in some situations improve growth (Stensvand and Gislerød, 1992; Sonneveld, 2002). Addition of more than 25 % of the total nitrogen as ammonium ions can, for some crops (e.g., tomato, lettuce) and situations, reduce growth and cause toxicity and/or deficiency (Ingestad, 1972; Ikeda and Osawa, 1983; Feigin et al., 1984).

Other ways to reduce the pH are either by adding acids using a fertilizer injector, or by adding CO_2 directly into the nutrient solutions. To increase the pH of the growth medium in greenhouse cultures, K_2CO_3 may be used. In addition, all the ammonium may be removed from the fertilization programme. With nutrient film technique (NFT)-systems, it is possible to use NaOH or KOH to increase the pH, but this is rarely needed as the pH in most systems gradually increases during culture.

When preparing media for micropropagation, NaOH and/or HCl are commonly used to adjust pH to optimum value. For liquid cultures in bioreactors, where monitoring the pH is easy, we have noticed that the pH often changes a little during autoclaving. With the same medium composition, it will always be in a consistent manner. In bioreactor cultures, pH can be adjusted continuously using sterile filtered HCl or NaOH/KOH, similar to NFT-systems in the greenhouse. Mostly, the pH in the bioreactors is not adjusted, but is free to change with ion uptake from or excretion into, the culture medium. To stabilise the pH during the culture on gelled media, pH buffers can be used (MES or TRIS).

The contents and the formulation of nitrogen in the medium are important because the ratio of NO_3^- to NH_4^+ will influence the pH of the medium. Uptake of NO_3^- ions by plant cells leads to the extrusion of anions into the substrate, which increases the pH. On the other hand, NH_4^+- uptake results in the excretion of H^+ ions from the roots, with a consequent decrease of pH in the media. Uptake of other anions and cations also influences the pH in the medium. Working with embryogenic callus cultures, Niedz (1994) showed that the pH of the gelled medium depends on the amount of nitrogen and the proportion between NH_4^+ and NO_3^- added to the agar-gelled medium. We have seen the same in the Plant Cell Laboratory in NLH (Hvoslef-Eide et al., this volume). In bioreactors where we constantly monitor the pH, the pH of the liquid medium first declines (until the ammonium is used), and then rises a little again (when the nitrate is being taken up). How reduced is the pH value and how much it rises thereafter is, again, dependant upon the ratio between NH_4^+ and NO_3^-.

The ion uptake pattern of different plants has a subsequent effect on pH (Stensvand and Gislerød, 1992). With about 5 % of nitrogen as ammonium, the pH in hydroponics was constant for *Chrysanthemum*, while it decreased for *Cordyline* and increased for *Passiflora*.

3. Nutritional responses to changes in the root environment

3.1 Salinity

The electrical conductivity (EC) is an important variable, and has to be controlled according to the sensitivity of the crop. Since the EC is the sum of both the nutrients added and the ions already in the water supply, considerable attention has been given to the accumulation of impurities from the water, particularly in closed greenhouse systems. Deionised water is used for preparation of media in microhydroponics systems. In greenhouse production, conductivity is in the range of 0.8 – 5.0 mS cm^{-1}. This depends on the plant species; *Rhododendron* will have a soil conductivity of 0.8 – 1.2 mS $cm^{-1}$and *Saintpaulia* of 1.2-1.5 mS cm^{-1}. However, tomato can be grown at 4.0 – 5.0 mS cm^{-1}, when changing tomatoes from vegetative to generative growth, or when a higher dry matter content and a better taste are required for the tomato fruit (Adams, 2002).

During vegetative plant propagation, the EC is kept at 0.8-1.2 mS cm^{-1} in the growing medium, while for growth of established plants the conductivity should be raised to 1.5 – 2.0 mS cm^{-1}, measured by the SSE-method (Soil Saturated Extract). By increasing the conductivity and the concentration of the nutrient elements, the uptake will increase as long as the growth increases, but when the growth start to decrease because of an increasing conductivity, the total nutrient uptake also decrease (Table 1). How quickly a decrease in growth will commence because of increased conductivity depends of the humidity of the growing medium. Plants in dry medium give quicker response to conductivity than plants in a wet medium.

Uptake of Ca^+ increases with increasing concentration of Ca^+, when the amount of the other elements are constant. By increasing conductivity, the uptake of Ca^+ usually decreases (Table 2). If, on the other hand, only Ca^+ is increased in the nutrient solution, the uptake of Ca^+ will also increase (Gislerød, 1997).

Table 1: Effect of salinity (EC) on the uptake of water (l) and nutrients (g) by tomato plants in NFT from April till August (from Adams, 1994)

	EC of nutrient solution (mS cm^{-1})				
	3.2	4.8	6.6	7.8	9.3
Uptake per plant					
Water (l)	98	109	105	93	82
N (g)	24	26	25	23	21
P (g)	6.0	6.4	5.7	5.1	4.6
K (g)	31	35	30	28	23

Table 2: The effect of salinity (EC) on growth and Ca accumulation by cucumber plants grown in NFT for 7 weeks in autumn (from Adams, 1994)

	EC of nutrient solution (mS cm^{-1})		
	3.0	5.5	8.0
Dry weight per plant (g)	140	113	76
Total Ca in plant (g)	4.71 (3.4 %)	3.44 (3.0 %)	1.62 (2.1 %)

A low EC will normally be advantageous for the growth of the plants, if they are supplied with sufficient quantities of the different nutrient elements: this depends on the growing system. A crop grown in NFT will have a maximum growth at a wide range of EC values, compared to a crop grown in soil and fertilized a few times during a season. Massey and Winsor (1980) showed that there was no difference in yield of tomato in NFT with nitrate-nitrogen concentrations ranging from 10 to 320 mg l^{-1}. The appearance of the plants gave no indication of the difference in nitrogen supply, even at 10 mg l^{-1}. Similar responses were found for potassium (Adams and Grimmett, 1986). This showed that by growing plants in a NFT system ('flowing' solution), it is possible to maintain the macronutrient elements at a concentration of 10 % of the 'normal' concentration, and have good crop growth, but the fruit quality may be poor (Adams and Grimmett, 1986). On the other hand, it is not harmful to grow at a 'normal' concentration or higher, because the growing medium is liquid at all times; in a solid medium, however, there will be considerable variation in water content and hence in conductivity. Gislerød (1993) showed for two cultivars of cut roses in peat/bark compost that a rise in EC from 2.8 mS cm^{-1} to 4.8 mS cm^{-1} increased the yield by 25 % during a winter season with a light input of 130-370 μmol m^{-2} s^{-1}. For some other cultivars, this made no difference or caused a little decrease in yield.

In micropropagation on gelled medium, the conductivity range is from 3.0 – 6.5 mS cm^{-1}. This indicates that the tolerance for high salt uptake through the stem is similar to that of leaves, which is expected because most *in vitro* systems are without roots. Root-inducing media, in general, have reduced salt concentrations, to stimulate root growth; as shown for *Cordyline fruticosa* (Hvoslef-Eide, 1990).

Leaves of intact plants in greenhouses can be sprayed with solutions with an EC equivalent to that of micropropagation media (3.0-6.5 mS cm^{-1}) provided that the air humidity is high at the time of spraying and during the following hours. This also indicates that the air humidity must be close to saturation in the *in vitro* culture vessels, otherwise the plantlets would have become desiccated. If air humidity in the culture vessels is reduced due to water loss through non-airtight lids, the conductivity of the medium has to be reduced to prevent plant tissue from excessive nutrient uptake. Some plants are more susceptible to high salts than others and need to start on a lower concentration to get established. This was the case with *Nephrolepis exaltata*, which otherwise was highly variable in morphology (Borgen and Næss, 1987), and *Begonia x cheimantha,* which otherwise would not proliferate from the initial leaf discs (Borgen, 1983). Later, when the cultures were established, it was possible and sometimes necessary to increase the salt concentration in the medium. In *Begonia* and *Saintpaulia*, the high salt concentration increases both the multiplication rate and the total amount of material produced (Selliah, unpublished results from Ulvik Plantelab AS, Norway). In *Begonia* and *Saintpaulia,* the high salt concentration increased both the multiplication rate and the total amount of material produced.

3.2 Nutrient concentrations

The response to different concentrations of each nutrient has been studied with many crops in soil, and later in organic substrates. Once hydroponics was introduced for commercial production, establishing suitable concentrations of nutrient for the crops became vital. In contrast to crops grown in soil and organic substrates, all the nutrients must be supplied in solution. Thus, provided that the pH is suitable, nutrient availability is no longer a problem. However, the very availability of the nutrients means that rapid depletion of the solution could limit growth and yield: 1) if the root volume is too small, 2) if the rate of solution supply in an open system is insufficient, 3) if nutrient replacement in a closed system is insufficient (Adams, 2002). In greenhouse production one tries to obtain optimal conditions both climatically and for nutrient uptake by the plants. Because of environmental regulations in many countries, more and more of crop production is in closed systems. This has forced growers to be more aware of the proportion of the different nutrient elements in solutions. If the grower is not fully aware, it could easily happen that one or more elements will accumulate and become toxic, or perhaps depleted, causing deficiency.

The same situation is the case for microhydroponics systems. If one changes the medium frequently, it is not necessary to be so accurate with the proportion of the macro- and microelements. This is probably why MS medium has been useful for such a range of species for so long. But if one wants to extend the period between plantlet transfers, and leave the culture for a longer time in the same medium, the nutrient composition has to be fairly accurate and optimised to what the plant tissue in the culture vessels really need.

3.3 Components of the nutrient media

Media for *in vitro* cultures are composed of macronutrients, micronutrients, vitamins, amino acids and amides, sugar and sugar alcohols, organic acids, and plant growth regulators. Buffers can also be included (MES for pH 5.5 to 6.7 and TRIS for pH 10.0 to 11.5). These media can be made from the basic salts, or a "ready to use pack" of most tissue culture media can be purchased with or without added pH buffers.The most

Table 3: Nutrient content and relative relation between these in a conventional greenhouse fertilizer solution and the classical, much used MS medium

Nutrient	mg l^{-1} in typical greenhouse crop fertilizer	mg l^{-1} in MS medium	Rel.amounts* N, Fe resp. = 100 in greenhouse crop ferilizer	Rel.amounts* N, Fe resp. =100 in MS medium
Nitrogen (N)	200**	840	100	100
Phosphorous (P)	40	39	20	5
Potassium (K)	200	784	100	93
Calcium (Ca)	150	119	75	15
Magnesium (Mg)	30	37	15	5
Sulphur (S)	40	48	20	6
Iron (Fe)	2	5	100	100
Manganese (Mn)	0.6	4	30	80
Boron (B)	0.25	1.1	13	22
Copper (Cu)	0.1	0.006	5	0.1
Zinc (Zn)	0.15	2	8	40
Molybdenum (Mo)	0.03	0.1	2	2
Chlorine (Cl)	-	212	≅0	25

* To compare nutrient solution in a typical greenhouse crop fertilizer with that of a typical *in vitro* culture medium, the relative amounts of macronutrients in each case have been compared with the amount of nitrogen (N), while the micronutrients are compared with iron (Fe).

**The nitrogen form will mainly be NO_3-N. The amount of NH_4-N will vary from 4 to 20 % of the total amount of nitrogen.

Table 4: Nutrient element content of three species (Anon., 1992)

Nutrient	Begonia (7 % DM)*	Poinsettia (17 % DM)*	Saintpaulia (4 % DM)*
	mmol kg^{-1} DM	mmol kg^{-1} DM	mmol kg^{-1} DM
K	500-750	400-800	900-1500
Ca	250-500	300—600	300-500
Mg	150-250	200-350	250-350
N-total (N-deficient)	2500-3500 (<2000)	3500-4000 (<3000)	1500-2500
P (P-deficient)	100-200	200-350 (<65)	200-500
Fe	1.0-2.0	2.0-4.0	2.0-4.0
Mn	0.5-2.3	1.4-2.7	0.5-2.0
Zn	0.60-1.50	0.50-2.30	1.00-4.00
B	2.0-5.0	2.8-9.2	4.0-10.0
Cu (μmol kg^{-1} DM)*	100-200	100-200	100-600

*DM – dry matter

Table 5: Media composition of two RS-1997 media tested for *Saintpaulia, Begonia* and *Gypsophila*, compared with Murashige et al. (1972) medium

	RS-1997-1 mg l^{-1}	M.et al. (1972) mg l^{-1}	RS-1997-3 mg l^{-1}
MACRO elements			
$Ca(NO_3)_2.4H_2O$	1300		1200
NH_4NO_3	800	1650	800
KNO_3	1200	1900	800
$CaCl_2.2H_2O$	40	440	
$MgSO_4.7H_2O$	750	370	750
KH_2PO_4	400	170	300
NaH_2PO_4		170	
$(NH4)_2SO_4$	396		264
NaFe- EDTA	40	40	40
MICRO elements			
H_3BO_3	18.6	6.2	12.4
$MnSO_4.H_2O$	16.8	16.8	25.2
$ZnSO_4.7H_2O$	8.6	8.6	1.72
KI	2.5	0.83	1.66
$Na_2MoO_4.2H_2O$	0.75	0.25	0.25
$CuSO_4.5H_2O$	0.25	0.025	0.25
$CoCl2.6H_2O$	0.1	0.025	0.025
ORGANIC compounds			
Vitamins	MS	MS	MS
myo-inositol	100	100	100
Casein hydrolysat	200	200	200
IAA	2.0	2.0	2.0
BA(P)	0.08	0.08	0.08
Sucrose	30g	30g	30g
Agar	6g	6g	6g
pH	5.8	5.8	5.8
EC (SSE) mS cm^{-1}	6.27	6.48	5.30

common *in vitro* culture medium today is Murashige and Skoog's medium (MS), published in 1962 to support rapid growth of tobacco tissue culture. Murashige and Skoog analysed ash from tobacco and composed the nutrient composition based on the mineral contents. The macronutrient composition of Murashige and Skoog has been used in approximately 25% of the published cell and tissue culture methods and is used in more than 50% of all plants produced by micropropagation (Leifert et al., 1991). MS medium contains 40 meq l^{-1} (66%) NO_3^- -nitrogen and 20 meq l^{-1} (34%) NH_4^+ -nitrogen, 4.5 meq l^{-1} PO_4^{3-}, 3 meq l^{-1} SO_4^{2-}, 6 meq l^{-1} Cl^-, 21.5 meq l^{-1} K^+, 6 meq l^{-1} Ca^{++} and 3 meq l^{-1} Mg^{++}. Authors have often tended to investigate growth on only one medium (in many cases MS), and, if satisfactory growth was obtained, have not compared growth rates on other media (Leifert et al., 1995). It is therefore unclear whether published methods describe nutrient compositions that support maximal growth rates of a particular plant species.

When comparing the solutions in table 3, one will see that generally the concentration of the different nutrient elements is approximately four times greater in the solution for tissue culture compared with those for greenhouse production. The ratios between the macro elements and the microelements are different in the two solutions. The amount of Ca is about one fifth, and for P, Mg and S about one-third in MS-medium compared to a nutrient solution for greenhouse production. In addition, there is quite a large amount of Cl in the MS-medium. This is a clear disadvantage for Cl intolerant plants. Reducing the Cl content and increasing the P and Mg content could, potentially, prolong the period of growth and give better quality plantlets.

The proportion of the microelements Fe, Mn, B, Cu, Zn, Mo in the nutrient solutions are also important for many ornamentals and flowering pot plants. By slightly increasing these elements in the *in vitro* media, we assume that better quality plants can be produced from micropropagation. MS has a low content of the microelements compared to N and K. The Fe and Cu content in MS-medium compared to a nutrient solution for greenhouse plants is also low.

When *Begonia x hiemalis* is cultured on MS-medium, plants must be subcultured at three week intervals. Failing to transfer to fresh medium results in plant senescence from the 4th week (Selliah, unpublished results). In *Saintpaulia*, the plantlets show symptoms of phosphorus and boron deficiencies. Prolonging the culture periods and reducing the number of subcultures, while maintaining good quality of the shoots, could save valuable labour time. Based on the nutrient elements in different ornamentals (Table 4), Selliah composed two alternative media (Table 5) for *Begonia x hiemalis* and *Saintpaulia* in 1997 (called RS-1997). This comparison is shown in table 5. The purpose of testing alternative media was to obtain a more optimised, balanced medium for these ornamentals and to prolong

Table 6: The effect of alternative RS-media compared with M. et al. (1972) medium on biomass yield, number of plants and survival after rooting in *Saintpaulia ionantha* (from Ayeh, 1998)

At transplanting (per jar)			
Media	Biomass (g)	No. of plants	Survival after rooting (%)
M. et al. (1972)	11.2 b	42.0 b	50.4 b
RS-1997-1	14.4 ab	49.9 ab	87.5 a
RS-1997-3	15.0 a	62.1 a	89.9 a

* Different letters represent statistical differences at 5% level analysed by SAS Statistical package (ANOVA).

Table 7: The effect of alternative RS-media compared with M. et al. (1972) medium on number of days from potting to first visible bud, the first open flower and plant on size in *Saintpaulia ionantha* (from Ayeh, 1998)

Media	First visible bud (days)	First open flower (days)	Size (cm^2)	
			single leaf	whole plant
M. et al. (1972)	104.2 b	112.2 b	17.4 c	161.1 c
RS-1997-1	85.2 ab	93.4 ab	20.5 b	198.2 b
RS-1997-3	71.6 a	80.0 a	24.5 a	245.5 a

* Different letters represent statistical differences at 5% level analysed by SAS Statistical package (ANOVA).

subculturing periods. Ayeh and Hvoslef-Eide tested these for *Saintpaulia*, *Begonia* and *Gypsophila* at NLH (Ayeh, 1998). The results from *Saintpaulia* are presented here (tables 6 and 7).

Saintpaulia ionantha was propagated on these three media from table 5 and compared for multiplication rate, survival rate at transfer, days-to-flowering and plant size and quality. The experiments were repeated six times. Table 6 shows that the best medium (RS-1997-3) for *Saintpaulia* gave greater biomass per jar, greater number of plants, as well as greater survival rate when transferred to greenhouse conditions. The most dramatic effect of this medium was flowering 32 days earlier (Table 7), than in the medium proposed by Murashige et al. (1972). The plants also produced a greater number of flowers in the best nutrient medium (Table 8). Leifert and his co-workers have also demonstrated that the PO_4^{3-} concentration of MS medium can be too low to sustain optimal growth in *Hemerocallis, Iris* and *Delphinium* plants (Leifert et al., 1991; Pryce et al., 1993, 1994). The same is true for *Gallium verum* cell suspension cultures (Strobel et al., 1990).

The effect of the optimal medium on the number of days-to-flowering in *Saintpaulia,* as well as the size and quality of the plant, are highly valuable in a commercial setting. This affects both the commercial propagator, who can reduce labour costs through less subculturing, and the greenhouse grower who can sell his crop 3-4 weeks earlier and save costly greenhouse space. Optimising the nutrient content of the *in vitro* medium can therefore be cost-effective. Bouman and Tiekstra (2002) compared adapted media (optimised to element analysis of the plants) for *Cymbidium* and *Gerbera* with the medium of Murashige and Skoog (1962) and that of Driver and Kuniyuki (1984). They concluded that the adapted medium was better for multiplication and growth in both species. Normally, concentrations of tissue culture media are between two to ten times greater than that used in hydroponics. For copper (Cu), the situation is the opposite, Murashige and Skoog (1962) has only one fifth of the concentration used in closed hydroponic systems. Increasing the copper content 16 times compared to Murashige and Skoog was highly beneficial to *Gerbera* (Bouman and Tiekstra, 2002).

Leifert et al. (1995) give detailed descriptions on how to analyse the depletion of nutrient media to elucidate which elements are in deficiency concentrations. This is, of course, one way of developing the nutrient medium for optimal plant growth *in vitro.* We think, as did Bouman and Tiekstra (2002), that using Murashige and Skoog's method from 1962 of simply analysing the ash of a particular plant, and composing the medium thereafter, may well be a much simpler and more cost-effective way of optimising the nutrients.

3.4 The quality enhancement of the micropropagation procedure

The growth conditions and health status of the mother plants directly affect the quality of their progeny. A balanced and optimum nutrient supply to the mother plants is necessary to produce high quality *in vitro* materials. Table 9 shows the effect of different nitrogen and calcium treatments on *Begonia x cheimantha* mother plants for one month on subsequent shoot production. The plants receiving the highest concentration (300mg l^{-1} N, 368mg l^{-1} Ca) gave a greater number of shoots per explant, and also a higher percentage of the explants (83.2 %) produced shoots (Borgen, 1983). When setting up this experiment, we were convinced that the highest nitrogen/calcium concentration was too high, but to our surprise, it turned out to be the best. Certainly, we would never have recommended a commercial grower such a high fertilization rate because of the danger of root decay at such high SSE. We must emphasise that with such high rates of fertilization of mother plants, careful attention to the water content of the pots is essential; they must never be allowed to dry out.

Table 8: The effect of alternative RS-media compared with M. et al. (1972) medium on some floral characters of *Saintpaulia ionantha* (from Ayeh, 1998). Registrations were done 3 weeks after appearance of the first flower, i.e. after 17 (M-72), 16 (RS-1997-1) and 14 weeks (RS-1997-3)

Mean number per plant			
Medium	Floral buds	Floral stalks	Flowers
M- 72	20.2 b	3.3 a	8.3 b
RS-1997-1	24.2 ab	3.0 a	10.0 ab
RS-1997-3	32.5 a	4.9 a	19.3 a

* Different letters represent statistical differences at 5% level analysed by SAS Statistical package (ANOVA).

Table 9: The effect of nitrogen fertilization ($Ca(NO_3)_2$ to the mother plants on percentage of shoot forming leaf discs and number of shoots per leaf disc (*Begonia x cheimantha;* Borgen, 1983) as determined after 10 weeks of culture

Fertilizer $Ca(NO_3)_2$ (N=15.5%, Ca = 19.45%)	Nitrogen and calcium fertilisation in mg l^{-1}					
	N 75	Ca 86	N 150	Ca 184	N 300	Ca 368
No. of shoots per leaf disc	48.6		40.0		81.4	
% of leaf discs producing shoots	55.0		66.0		83.2	

4. Uptake of nutrients related to climate

The greenhouse environment consists of many factors that affect both the uptake of nutrients and their distribution within the plant, thus affecting the growth of the crop, the yield and the quality of the product.

4.1 Light intensity

Light intensity is the major factor affecting photosynthesis and greenhouse temperatures during the day. In addition the total hours of artificial lighting per cycle is important. The uptake of water and nutrients generally increase as the light intensity and air temperature increase, while the air humidity decreases. Adams (1994) showed that the uptake of K and N for a tomato crop followed the water uptake, which was closely, related to the solar radiation, but was about an hour delayed. Increasing light intensity causes increased growth, and that will demand an increased rate of nutrient supply. This may be particularly important with use of artificial light in winter, when the air humidity usually is high.

In micropropagation carbohydrate is used as an energy source, because of low rates of photosynthesis. In the phase of *in vitro* to *in vivo* the acclimatisation is easier when the carbohydrate content of the medium has been reduced the last weeks before transplanting (Selliah, unpublished results).

Giving more light than optimal can, in some situations, cause iron deficiency. This is shown for both greenhouse crops and in micropropagation in figure 1, and may happen when using EDTA as a chelate for iron (Papathanasiou et al., 1996). Organic chelates such as EDTA, are sensitive to light, and Fe deficiency may occur in some species under high light conditions. To eliminate this problem, NaFe-EDTA can be replaced by Fe-EDDHA; alternatively, the light intensity can be reduced. Fe-EDDHA is a more light-stable chelated Fe source than Fe-EDTA. It is now possible to purchase the pre-mixed tissue culture media with Fe-EDDHA.

4.2 Humidity

Recent measures to conserve energy in greenhouses have resulted in higher humidity regimes, which reduce transpiration, and hence Ca movement into the leaves, as Ca moves only in the xylem in most greenhouse crops. Conversely, high humidity allows more Ca to move into the plant organs that have a low rate of transpiration, such as tomato fruit. Humidity, therefore, is a key factor in controlling the distribution of some nutrients within the plant.

Increasing air humidity from 55-60 % RH to 90-95 % RH reduced the concentration of the nutrient element in the leaves of ornamental greenhouse plants (Gislerød et al., 1987). Therefore the concentration of the nutrient solution should be increased when the plants are grown at a high RH. Adams (1994) showed for tomato grown at high air humidity that the margin of the leaves had a lower content of Ca and K than for plants grown at a "normal" humidity. Also that in short days the air humidity during the night had a more pronounced effect than that during the day. Experiments with *Eustoma* show that high air humidity is the main reason for tip burn and a lower Ca in the tips of the leaves (Islam et al., 2004).

The air humidity in a micropropagation system is close to saturation and therefore reduces the Ca uptake. Therefore there is a need to give a higher Ca concentration, and consequently, a higher conductivity in a micropropagation media. Roche and Cassells (1996) found high humidity to cause shoot-tip and leaf-tip necrosis in nodal cultures of *Helianthus tuberosus*. They suggest that an alternative to increasing the Ca content of the medium could be to reduce the humidity by using more permeable plastic film to cover the cultures or using bottom-cooling of the incubation room shelves.

Figure 1: *Populus* cultivated *in vitro* with 3000 lux (45 μmol $m^{-2}s^{-1}$) indicating Fe-deficiency (left) and cultivated with 1000 lux (15 μmol $m^{-2}s^{-1}$): no Fe-deficiency symptoms (right).

4.3 Carbon dioxide (CO_2)

Increasing CO_2 to 700-800 ppm increased the growth by 15-30 % and reduced the transpiration by about 15-20 % (Mortensen and Gislerød, 1989). This suggests that it may be necessary to increase the conductivity of the nutrient solution by about 30 %, compared to growing without supplementary CO_2.

5. Conclusions

Well-balanced nutrient concentrations in the growing media combined with the optimum growth environment, are beneficial not only to the growth in the greenhouse and the culture room for *in vitro* cultures, but the after-effects are also of high value. In addition to this, optimum nutrient concentrations have to be supplied for the mother plants before propagation by conventional methods as well as for the *in vitro* culture to achieve maximum growth. Growing plants in liquid cultures, whether in greenhouses or *in vitro*, needs even more careful attention to plant nutrition than growing on solid or gelled media, and more knowledge of the climatic conditions that interact with nutrition.

Comparing the most used media for *in vitro* cultures (Murashige and Skoog, 1962; Murashige et al., 1972), with nutrient solutions used for NFT in greenhouses, the relative concentrations between the nutrients are far from ideal. We strongly suggest that more attention should be paid to the natural mineral content of the plant being propagated, and the relative nutrient concentrations. These analyses should be used when designing optimised media for *in vitro* cultures. This should improve plant quality from *in vitro* cultures by enhancing proliferation *in vitro*, as well as having a positive after-effect on the subsequent growth and quality *ex vitro*.

References

Adams P (1994) Nutrition of greenhouse vegetables in NFT and hydroponics systems. Acta Hortic. 361: 145-257

Adams P (2002) Nutritional control in hydroponic. In: Saavas D & Passam H (eds) Hydroponic production of vegetables and ornamentals. Embryo publications, Athens, Greece. ISBN 960-8002-12-5

Adams P & Grimmett MM (1986) Some responses of tomatoes to the concentration of potassium in recirculating solution. Acta Hortic. 178: 29-35

Ayeh KO (1998) Optimisation by mass propagation through tissue culture of *Saintpaulia, Begonia, Gypsophila* and cocoa. Master Thesis at University of Oslo/Agricultural University of Norway, 121 pages

Borgen, Hvoslef-Eide AK (1983) *In vitro* culture of *Begonia x cheimantha.* Master thesis at the Agricultural University of Norway, 27 pages

Borgen AK & Næss SK (1987) Homogeneity and plant quality of *in vitro* propagated *Nephrolepis exaltata* 'Bostoniensis'. Acta Hortic. 212: 433-438

Bouman H & Tiekstra A (2002) Adaptations of tissue culture media derived from composition of hydroponic substrates and plant elemental analysis. In: Hvoslef-Eide AK (ed) Abstract Book from the 1st Symposium on Liquid Systems for *in vitro* Mass Propagation of Plants, Aas NLH, Norway 29 May-2 June 2002 (pp. 58-59). Agricultural University of Norway, ISBN 82-996332-0-6

Bævre OA & Gislerød HR (1999) Plantedyrking i regulert klima. Landbruksforlaget, Oslo, Norway

Driver JA & Kuniyuki AH (1984) *In vitro* propagation of Paradox walnut rootstock. Hort. Sci. 19: 507-509

Feigin A, Ginzburg C, Ackerman A & Gilead (1984) Response of roses growing in a volcanic rock substrate to different NH_4/NO_3 ratios in the nutrient solution. In: Proc.6th Intern. Congr. Soilless Culture (pp. 207-214). Lunteren

Gislerød HR (1993) The effect of supplementary light and electrical conductivity on growth and quality of cut flowers. Acta Hortic. 342: 51-60

Gislerød HR (1997) The role of calcium on several aspects of plant and flower quality from a floricultural perspective. Acta Hortic. 481: 345-352.

Gislerød HR, Selmer-Olsen AR & Mortensen LM (1987) The effect of air humidity on nutrient uptake of some greenhouse plants. Plant and Soil 102: 193-196

Hvoslef-Eide AK (1990) The effect of irradiance and temperature on *in vitro* culture of *Nephrolepis exaltata* and *Cordyline fruticosa.* Gartenbauwissenschaft 55(6): 259-264

Ingestad T (1972) Mineral nutrition requirements of cucumber seedlings. Plant Physiol. 52: 332-338

Ikeda H & Osawa T (1983) Effect of ratios of NO_3 to NH_4 and concentration of each N source in the nutrient solution on growth and leaf N constituents of vegetable crops and solution pH. J.Japan. Soc. Hort. Sci. 52: 150-166

Islam N, Grindal Patil G, Torre S & Gislerød HR (2004) Effect of RH, light and calcium fertilization on tipburn and calcium content in the leaves of *Eustoma grandiflorum* (Raf.) Shinners. Europ.J. Hort. Sci. 69(1): 29-36.

Leifert C, Lumsden PJ, Pryce S & Murphy K (1991) Effects of mineral nutrition on growth of tissue cultured plants. In: Goulding KH (ed) Horticultural exploitation of recent biological developments (pp. 43-57). Lancashire Polytechnic Publication Service, Preston, UK

Leifert C, Murphy KP & Lumsden PJ (1995) Mineral and carbohydrate nutrition of plant cell and tissue cultures. In: Critical Reviews in Plant Sciences 14(2): 83-109

Lloyd G & McCown B (1980) Commercially feasible micropropagation of mountain laurel *Kalmia latifolia,* by use of shoot-tip culture. Proc. Int. Plant. Prop. Soc. 30: 421-427

Murashige T & Skoog F (1962) A. revised medium for rapid growth and bio assays with tobacco tissue cultures. Physiol. Plant.15: 473-497

Murashige T, Shaba MN, Hasagawa PA, Taktori FH & Jones JB (1972) Propagation of Asparagus through shoot apex culture 1. Nutrient for medium for formation of plantlets. *J. Amer.* Soc.Hort.Sci. 97: 158-161

Mortensen LM & Gislerød HR (1989) Effect of CO_2, air humidity, and nutrient solution concentration on growth and transpiration of *Begonia x hiemalis* Fotsch. Gartenbauwissenschaft 54 (4): 184-189

Niedz RP (1994) Growth of embryogenic sweet orange callus on media varying in the ratio of nitrate to ammonium nitrogen. Plant Cell, Tiss. Org. Cult. 39: 1-5

Papathanasiou F, Selby C & Harvey BMR (1996) Soluble iron lost from MS medium pre-exposed to light but growth of potato plantlets is not inhibited. Plant Cell, Tiss. Org. Cult. 46: 117-121

Pryce S, Lumsden PJ, Berger F & Leifert C (1993) Effect of macronutrient nutrition on *Delphinium* shoot cultures. J. Hort. Sci. 68: 807-813

Pryce S, Lumsden PJ, Berger F & Leifert C (1994) Effect of macronutrient nutrition on *Iris* shoot cultures. In: Lumsden PJ, Nicholas JR & Davies WJ (eds) Physiology, growth and development of plants in culture (pp. 67-71). Kluwer Academic Publishers, Dordrecht

Roche TD & Cassells AC (1996) Gaseous and media-related factors influencing *in vitro* morphogenesis of Jerusalem artichoke (*Helianthus tuberosus*) 'Nahodka' node cultures. Acta Hort. 440: 588-593

Sonneveld C (2002) Composition of nutrient solutions. In: Saavas D & Passam H (eds) Hydroponic production of vegetables and ornamentals. Embryo publications, Athens, Greece. ISBN 960-8002-12-5

Stensvand A & Gislerød HR (1992) The effect of the NO_3/NH_4 ratio on the nutrient solution on growth and mineral uptake in *Chrysanthemum x morifolium, Passiflora caerulea*, and *Cordyline fruticosa*. Gartenbauwissenschaft 57 (4): 193-198

Strobel J, Hieke M, Gebauer E, Wind E & Gröger D (1990) The influence of organic and inorganic chemical factors on cell growth and anthraquinone formation in suspension culture of *Gallium vernum*. Biochem. Physiol. Pflanzen 186: 117-124

Chapter 37

Adaptions of the mineral composition of tissue culture media on the basis of plant elemental analysis and composition of hydroponic substrates

Han Bouman* & Annemiek Tiekstra
Applied Plant Research, Wageningen UR, Group Tissue Culture and Biotechnology, P.O.Box 85, 2160 AB LISSE, The Netherlands.
**Requests for offprints: E-mail: Han.Bouman@wur.nl.*

Abstract: For the improvement of *in vitro* propagation protocols of *Gerbera* and *Cymbidium*, we adapted the macronutrients according to the elemental composition of adult leaves. In comparison with normally used media, in particular, the Ca and P concentrations were much higher. Using the adapted media, for both crops growth was much improved. The relative concentrations of macronutrients in these modified media formulations is much more in accordance with the mineral composition of nutrient solutions used in hydroponic cultures. For *Gerbera*, we also made adaptations to the micronutrients according to the elemental composition of nutrient media in hydroponics (*viz*., increased Cu-content and decreased Mn-content). This resulted in additional improvement of growth; by this adaptation, the plants became larger and also greener (only with the Cu adaptation).

Key words: *Cymbidium*, *Gerbera*, hydroponics, mineral nutrients

Abbreviations: CAM – *Cymbidium* adapted medium; DKW – Driver and Kuniyuki walnut medium (1984); DW-dry weight; FW – fresh weight; GAM – *Gerbera* adapted medium; MS – Murashige and Skoog medium (1962)

1. Introduction

For the development of *in vitro* propagation protocols, most usually, research is carried out to obtain optimal auxin and cytokinin concentrations. Modifying the mineral composition of the medium is an alternative or additional strategy to improve propagation, but examining dose-response curves for the various elements and their interactions to identify optimal concentrations is excessively time-consuming. Therefore, usually, either a

A.K. Hvoslef-Eide and W. Preil (eds.), Liquid Culture Systems for in vitro Plant Propagation, 493–505.

small range of commercial available media is examined, or various concentrations of a conventional medium, in particular MS (developed by Murashige and Skoog in 1962 for optimal growth of tobacco callus), are tested.

In this paper, two other strategies for the modification of mineral nutrient composition are investigated with *Cymbidium* and *Gerbera*: one is based on the elemental analysis of the plant species, and the other on the mineral composition of hydroponic substrates used by growers. Growth on adapted media was compared with growth on normal-used MS (Murashige and Skoog, 1962) and DKW media (Driver and Kuniyuki, 1984). The latter medium was included because the macro-elemental composition of DKW-medium is more similar to the composition of plants than MS-medium (see for examples tables 1 and 3). Henry et al. (1999) used a similar approach (adaptation according to analysis of plant tissues) applying, however, only simple adaptations. They found a strong increase of growth and of the production of secondary metabolites. Recently, Monteiro et al. (2000) and Cos Terrer and Frutos Tomas (2001) also used this approach with success. Earlier, Rugini (1984) cultured olive cultivars using analytical data from developing shoots and embryos to formulate the basal medium. In preliminary experiments based on elemental analysis of adult leaves, we also found improved growth and propagation for a number of plant species, including *Gerbera*, *Cymbidium*, apple and rose (Bouman et al., 2001; Bouman and Tiekstra, 2001).

2. Material and methods

2.1 Plant material

Cultures of three *Cymbidium* cultivars and two *Gerbera* cultivars were kindly provided, respectively by P+S Plantlab, Assendelft, Netherlands and by SBW International, Roelofarendsveen, Netherlands.

2.2 Growth conditions

Cymbidium was grown in 100 ml Erlenmeyer flasks with 20 ml medium on a gyratory shaker at 120 rpm at 22 °C, and 16 h light (30 μmol m^{-2} s^{-1}). For each cycle, at least 15 Erlenmeyer flasks with 1.5 g protocorm material were used per medium. The plant material was weighed and visually evaluated after 3 weeks. Normally, these *Cymbidium* cultivars are grown in half–strength MS macro- and micronutrients, 2 % (w/v) sucrose, MS vitamins, pH 6.0 (according to the tissue culture laboratory that had supplied the cultures). For the experiments, two other mineral media were tested: half-strength DKW medium and CAM medium (macronutrients adapted according to the elemental analysis of *Cymbidium*-, see 'Results and Discussion' section; plus ½ MS micronutrients). All compounds, and ready-made MS and DKW macro- and micronutrients were obtained from Duchefa (Haarlem, Netherlands).

Gerbera plantlets were grown in glass jars (10 propagules per jar) on 150 ml medium with MS macro- and micronutrients with extra 50 µmol NaFeEDTA, 3 % (w/v) sucrose, vitamins (300 mg l^{-1} thiamine HCl, 100 mg l^{-1} nicotinic acid, 10 mg l^{-1} pyridoxine HCl and 1 g l^{-1} m-inositol), and 12 µmol kinetin, pH 5.8, 0.6 % (w/v) BBL Granulated (Becton Dickinson, Aalst, Belgium) or Daishin (Brunschwig Chemie, Amsterdam) agar at 25 °C and 16 h light (30 µmol m^{-2} s^{-1}). Every 4 – 5 weeks, the clusters of plantlets were weighed, separated into individual plantlets, counted and divided into size classes (arbitrary units: 1: small, 2: medium and 3: large). Plants of approximately the same size (mostly 3, occasionally 2) were used for the next cycle. For experiments, the same composition was used except for the minerals. The adapted media are presented in the 'Results and Discussion' section. For experiments with hydroponic adaptations in *Gerbera*, the concentrations of two microelements were changed: extra Cu was added (1.8 µmol), whereas for Mn, a lower concentration was used (10 µmol).

2.3 Elemental analysis of plant tissues and of hydroponic solutions

The elemental compositions of adult leaves of *Gerbera* and *Cymbidium* are from Bergmann (1992). The concentrations of elements used for hydroponic cultures are from data and advice from the IKC organisation (Anonymous, 1999). For selected media and plantlets, elemental analysis was performed by contracted service laboratories.

3. Results and discussion

3.1 Formulation of media

For the formulation of the adapted media, some initial choices had to be made. First, the type of tissue used 'for reference' had to be decided upon: whether for the whole plant or just a specific organ (e.g., leaf, embryo, root), the age, the growing nutrient conditions of the reference-plants and their source (field, greenhouse or tissue culture). We used the analyses of young-adult leaves of healthy growing plants as reference material. This is the same material used for adaptations and improvements of hydroponic media for many plants (Anonymous, 1999). For olive, Rugini (1984) used the composition of developing shoots and embryos, which led to rather high P and much lower Ca concentrations. In preliminary experiments, tissue-cultured and young greenhouse-grown *Gerbera* reacted unfavourably to mineral adaptations derived from analyses of young plants, which had similar P and Ca concentrations as in Rugini's media.

A second choice concerns how to obtain the mineral concentrations for the essential macronutrients without adding Na and Cl. The very low concentrations of these two elements, necessary for growth, are invariably present in other salts: high concentrations of Na and Cl can be detrimental for growth (George, 1993). For the final modified formulations, usually compromises had to be made to obtain a suitable composition for satisfactory plant growth.

The third choice concerned the data obtained from the tissue analyses: results were in mmol per g DW or FW for the various elements, but for the media, the basis of the concentrations (mmol per litre) had to be decided upon. We decided to start from a nitrogen concentration similar to that used in conventional protocols and then to calculate the concentrations of the other elements in relation to the concentration value for nitrogen.

Fourthly, it had to be decided in which molecular form the nitrogen would be supplied, *viz.*, how much ammonium and how much nitrate. Whilst it is recognised that the ammonium-nitrate ratio is important for growth and development (Avila et al., 1998), this ratio is not dealt with in the research reported here. Experiments showed that the ratio had some influence, especially for *Gerbera*, but provided that the ratio was not extreme, its effect was not as profound as the mineral adaptations.

A general problem, which often occurred during the preparation of the modified media for *Gerbera* (but also for other species) was the precipitation of white insoluble salts mostly composed of Ca and P. When mixing the salts, no precipitate was noticed, but the endothermic reaction was started at high temperatures during autoclaving, resulting in precipitates after autoclaving. Because of this precipitation, infections became more difficult to detect, and a loss of specific minerals occurred and the pH was reduced because H^+ was released from both HPO_4^{2-} and $H_2PO_4^-$. The analysis of the medium after such precipitations showed that the amount of Ca and P that had precipitated was low and reduced the medium-concentration of P by less than 10%. However, because of the pH reduction, there was softening of the agar gel: in some cases the agar did not gel sufficiently. DKW medium was opaque and showed a yellow precipitation before autoclaving. This precipitate contained Fe, probably linked to phosphate, with some $Ca_3(PO_4)_2$. Schenk et al. (1991) also reported these medium-precipitates after autoclaving, and suggested that Fe salts and P salts should be autoclaved separately. Dalton et al. (1983) reported Fe precipitates with losses up to 50 % of Fe and 25 % of Zn; thus this phenomenon can be a severe problem.

Interestingly, when liquid media were used, the precipitates redissolved during the culture period. This was probably due to the decrease of pH in the medium, and/or because of plant uptake of dissolved constituents in the medium and the resultant shift in the dissociation equilibrium, leading to solubilisation of the precipitates. In agar media there was only a much smaller decrease in pH and the precipitates did not redissolve. It was observed for *Dahlia* in liquid culture (on DKW with a precipitate at the start) that the first new leaves were yellowish but that the later ones were green. In agar cultures, yellow leaves developed on DKW, but not on the adapted medium made according to elemental analysis.

Table 1: Mean dry weight content of adult, healthy *Cymbidium* leaves, and concentrations of macroelements in calculated 'ideal' medium, media used in experiments and used in hydroponics

Plant content		Medium content	Calculated medium	CAM	½ MS	½ DKW	Hydro-ponics
mmol g^{-1} DW		mmol l^{-1}					
N	1.8	NH_4^+/NO_3^-	30	11.6/18.4	10/20	8.5/16.5	1.0/4.0
P	0.09	PO_4^{3-}	1.6	1.6	0.61	0.98	0.8
K	0.65	K^+	10.8	10.0	10	9	2.8
Ca	0.41	Ca^{2+}	6.7	6.7	1.5	4.7	1.0
Mg	0.18	Mg^{2+}	2.8	2.8	0.75	1.5	0.75
S	n.a.	SO_4^{2-}	n.a.	7.5	0.75	6.0	1.25

n.a. = not available

Table 2: Growth of *Cymbidium* pseudoprotocorm cultures on three different media: weight increase factor per 3 weeks

Medium	Cultivar 1	Cultivar 2	Cultivar 3
½ MS	2.82 ± 0.08	1.98 ± 0.05	2.78 ± 0.06
½ DKW	3.56 ± 0.07	2.99 ± 0.06	3.19 ± 0.07
CAM	3.87 ± 0.14	3.33 ± 0.12	3.70 ± 0.15

3.2 Cymbidium

The elemental analysis of adult leaves of *Cymbidium* is shown in table 1 and originates from Bergmann (1992). From these values, the composition of *Cymbidium* adapted medium (CAM) was calculated (Table 1). As orchids are normally grown on low-salt concentration media (in this case ½ MS, personal communication P+S Plantlab), a N-content of 30 mmol was adopted and on the basis of this concentration, the other elemental concentrations were calculated. In CAM, P, Ca and Mg contents are much higher in comparison with ½ MS. Almost no precipitation occurred, even after autoclaving, probably because of the relatively low concentrations of Ca and P and the relatively high concentration of sulphate. During other adaptations, it was often observed that, while other concentrations were kept the same, increasing salt concentration with potassium sulphate diminished the precipitation of salts after autoclaving, possibly because sulphate has some buffering and complex-forming capacity.

For all three cultivars, growth increased significantly by culture with CAM compared to ½ MS and ½DKW (Table 2). For shoot formation, rooting and acclimatisation there were no differences in mineral media and the performance of the propagules was the same irrespective of the original propagation medium. It is possible that further improvements could be achieved by mineral adaptations in these steps. The nutrient concentration ratios used in ½MS compared with those applied in hydroponics (Table 1) are 1.5 for Ca, 1.0 for Mg and 0.76 for P. In ½DKW these ratios are 4.7, 2.0 and 1.2, respectively and in CAM 6.7, 3.7 and 2.0, respectively. This indicates that the concentrations of Ca, Mg and P are too low in MS and -to a lesser extent- in DKW. The CAM medium supplies these three elements at a more appropriate concentration. For K and N in all 3 media, these ratios are similar viz., ca. 3.5 and 5.5, respectively.

After the 3-week cycle, the media were analysed for elemental content (data not shown). None of the major minerals was exhausted with the notable exception of P (8 % of the original concentration remained in the medium). Phosphorous, particularly, is known to be consumed in a 'luxury' way, i.e., the plant takes up more than needed for immediate growth (George, 1993). Concerning nitrogen, it was found that the final amount of ammonium was rather low (ca 10 % remained), but that ca 50 % of the nitrate was still present. Especially in the CAM medium with the faster growing plant material, sucrose concentration became low, but was not exhausted (approximately 10 % remained).

Table 3: Mean dry weight content of adult, healthy *Gerbera* leaves, and concentrations of macrominerals in calculated 'ideal' medium, in media used in experiments and used in hydroponics

Plant content		Calculated medium	GAM	DKW	MS	Hydro-ponics
mmol g^{-1} DW		mmol l^{-1}				
N total	2.1	40				
NH_4^+	n.a.	n.a.	7	17	20	0.75
NO_3^-	n.a.	n.a.	33	33	40	7.25
K	1.1	20.7	11.7	18	20	4.5
Ca	0.35	7.7	8	9.3	3	1.6
Mg	0.12	2.4	2.5	3	1.5	0.4
SO_4^{-2}	n.a.	n.a.	2.5	12	1.5	0.7
PO_4^{-3}	0.10	1.9	1.7	1.95	1.25	0.6
µmol g^{-1} DW		µmol l^{-1}				
Fe	n.a.	n.a.	100	120	100	25
Mn	1.2	23	100	200	100	5
Zn	0.8	15	30	72	30	3
B	3.2	61	100	78	100	20
Cu	0.13	2.5	0.1	1	0.1	0.5
Mo	0.004	0.075	1	1.6	1	0.5

n.a. = not available

3.3 *Gerbera*

Table 3 shows the elemental composition of *Gerbera* leaves according to Bergmann (1992). The adapted medium was a compromise to approximate to this formulation using the salts available in the lab, but minimizing the content of Na and Cl. Adapted media for *Gerbera* always showed some white precipitate. The pH of the adapted media was reduced by 0.4 pH units after autoclaving with BBL agar, whereas for MS this was 0.2 pH units, and for DKW 0.3 pH units. For Daishin agar this pH reduction was greater, and agar stayed also more gelled after autoclaving. It is known that physical and chemical properties of different agar brands can vary considerably (Scholten and Pierik, 1998b). The problem of precipitation could be overcome by autoclaving P separately and mixing it with the rest of the medium after autoclaving (Dalton et al., 1983; Schenk et al., 1991). In practice, such procedure is undesirable as it involves much additional labour. Analysis of fresh adapted media showed that the precipitation had no significant effect on the available concentrations of Ca and P. Table 5 shows that propagation was much improved by GAM.

Comparison of mineral concentrations in adapted media with concentrations used in hydroponics showed that the former concentrations are mostly 2 to 10 times greater than the latter. So, in a few experiments, a medium was included based on mineral concentrations 5 times greater than used in hydroponic cultures. This '5x'-medium also gave good results, comparable with the adapted medium.

Analyses of *Gerbera* media after culture of plant material did not show exhaustion for any specific macro-element (results not shown). For ammonium, the lowest residue was found (15 % remained), but there was still more than 50 % of the nitrate present. This means that the differences in growth on the media investigated were unlikely due to exhaustion of one of the elements but likely due to their relative concentrations in the tissue. Analysis of plant tissues (results not shown) indicated that the main differences in elemental content were in S, P (MS only), Mg, Ca and Fe. This was not only caused by the higher concentrations of these elements in the adapted medium; for example, although K was rather low in GAM, the plants from GAM had the same K- content as plants from MS and DKW media with higher K-concentration. The higher uptake and therefore values found in plants from adapted media especially for Ca, Mg and Fe could be the main reason for better growth on these media. Because the Fe concentration in the three media is the same, greater Fe uptake should then be a result of the differences in relative ion concentrations, whereas a greater Ca concentration in the adapted medium is directly responsible for its greater Ca content.

In our experiments, initially we changed only the relative concentrations of the macronutrients. From the composition of hydroponic solutions, and also from analysis of the elemental content of plant tissues, it was evident that with respect to micronutrients, Mn is present at a high concentration in MS and DKW, whereas Cu is present at a low concentration, particularly in MS, if compared with hydroponic solutions (see ratios in Table 4). Table 5 shows that growth on GAM was much improved by decreasing the Mn and by increasing the Cu concentration. *Gerbera* plantlets grown on low Mn, however, were chlorotic. This was reflected by the elemental analysis, which

Table 4: Ratio between concentrations of minerals in tissue culture media and closed system hydroponics for Gerbera; * **bold:** Cu and Mn values after 'adaptation' to hydroponic concentrations

Mineral		'Ideally' adapted med.		GAM		MS		DKW	
N total	NH_4^+ NO_3^-	5.0	not known	4.8	8.0 4.4	7.5	26.7 5.5	6.3	22.6 4.6
K		4.6		2.6		4.4		4.0	
Ca		4.8		5.0		1.7		5.8	
Mg		6.0		6.3		3.8		7.5	
$S(O_4^{-2})$		n.a.		3.6		2.1		17.1	
$P(O_4^{-3})$		3.2		2.8		1.8		2.8	
Fe		n.a.		4.0		4.0		4.8	
Mn		4.6		20 > **2** *		20		40	
Zn		5.0		10		10		24	
B		3.1		5.0		5.0		3.9	
Cu		5.0		0.2 > **3.2** *		0.2		2.0	
Mo		0.15		2.0		2.0		3.2	

n.a. = not available

Table 5: Propagation results for *Gerbera* after 4 weeks of growth, for a representative propagation cycle on six media

Medium	Weight of cluster in mg	Number of plantlets	Average size of plantlets (a.u.)	Colour of plantlets, habit
MS	166 ± 11	3.8 ± 0.2	1.01 ± 0.01	Green, normal
DKW	277 ± 19	4.9 ± 0.2	1.38 ± 0.04	Green, normal
GAM	411 ± 24	6.4 ± 0.3	1.33 ± 0.04	Green, more adult leaflets*
GAM–Mn	453 ± 22	6.1 ± 0.3	1.55 ± 0.04	Often chlorotic
GAM+ Cu	590 ± 27	6.3 ± 0.3	1.56 ± 0.05	Dark green
GAM–Mn +Cu	631 ± 25	5.8 ± 0.3	1.74 ± 0.06	Sometimes chlorotic

*all'GAM' plantlets had a more adult habit than plantlets from MS and DKW
a.u. = arbitrary units

showed that these plantlets had a much lower Fe content (ca 40 %) than MS- and Cu-plantlets. So maybe lowering of Mn-concentration on its own - although growth is better- should be applied carefully. In contrast, plantlets grown with increased Cu were much greener than MS-plantlets, but their Fe-content was not increased.

4. General discussion

Mineral adaptations for tissue culture media consist usually of lowering the concentration of the MS macroelements (e.g., half concentration of MS or half concentration of nitrogen in rooting media), or screening a number of ready-made media (MS, B5, DKW etc.). Suggestions to increase some minerals, among others, Ca and Mg (Singha et al., 1987; George, 1993) were made years ago, but have been generally ignored.

In our research, healthy young-adult leaf material was chosen for elemental analysis (see above). Rugini's medium (1984) was based on data from young plant parts, i.e., shoots and embryos: in his medium, P was much greater and Ca (much) less. Preliminary experiments with gerbera showed that such composition was less beneficial for growth than the adapted medium used here. Monteiro et al. (2000) used mineral analysis of adult leaves of passion fruit for total mineral adaptation to define a specific medium. Their successful medium had a greater concentration of Ca, P, Mg, S, and especially Cu, than MS, confirming our findings. For Mn they used a greater concentration than in MS, which contrasts our results where we had an improvement of growth by reducing the Mn concentration, similar to Kintzios (2001; see below). Cos Terrer and Frutos Tomas (2001) also used adult leaf analyses for developing media for peach-almond hybrids. They only adapted the macrominerals, but did make variations for each cultivar. They compared growth on their media with growth on MS, but not DKW.

Higher mineral concentrations in the medium resulted in higher concentrations in the plant for several elements, e.g., S and Ca (Singha et al., 1990), but not for K. *Gerbera* plantlets from GAM had the same or even greater K content as plantlets grown on MS or DKW in spite of the much lower K concentration in GAM. On the other hand, plantlets from DKW had a high S content. "Luxury consumption" can lead to depletion of the medium of some elements, but because of redistribution within the plant, this does not mean, necessarily, that at the same time there would be a deficiency in the plant. With *Cymbidium,* P was almost exhausted in CAM (<10 % remaining), after the 3- week cycle, but growth was not retarded until about 2 weeks later.

In tissue culture all ingredients of the media are provided at the start of the culture at high (possibly supra-optimal) concentrations, whereas at the end of culture the concentrations will be much less (possibly sub-optimal). Additionally to this, Amiri (2001) describes local deficiencies around explants for slow-diffusing ions such as Ca in banana cultures. In liquid media local deficiencies cannot occur because of convection, diffusion and mixing through agitation; even precipitates -as in DKW media - disappear as we recorded in relation to the Fe- precipitate that redissolved into solution during culture. Although not further dealt with here, proof was found for this phenomenon with *Dahlia* during liquid culture on DKW: as the precipitate is still present the first appearing leaves after subculture were yellowish, but the later developing leaves had a normal, green colour. On gelled DKW media, the plants did not grow well and plants became yellowish, because the 'Fe'-precipitate did not dissolve.

Our second approach was in adapting the mineral contents of tissue culture media to the formulations of hydroponic nutrient solutions. In hydroponics, minerals are generally supplied at concentrations one fifth of the concentrations used in tissue culture (Table 1 and 3). This reflects the uptake by the full-accessible root system and the continuous supply of all nutrients, which are monitor and re-adjusted frequently. There are, however, minerals for which the concentration ratio is very different (see Table 4), in particular for Cu and Mn. We adapted Cu and Mn concentrations in MS media for *Gerbera,* increasing Cu from 0.1 μmol to 1.6 μmol and decreasing Mn from 100 μmol to 10 μmol. There are various reports of the beneficial effects of greater concentrations of Cu. In barley, Dahleen (1995) found with 5 μmol Cu (50 x MS) much better regeneration and he also emphasised that in the original paper of Murashige and Skoog (1962) no clear optima were obtained for micronutrients. For chilli pepper, Kintzios (2001) found a greater callus growth rate and much increased somatic embryogenesis with Cu concentration at x10 that in MS; 0.1 x Mn (compared to MS) resulted in better growth too, confirming the result we recorded for *Gerbera*. Although we used the Zn concentration of MS, table 4 shows that in MS and especially in DKW the Zn concentration is rather great compared to that in the hydroponic solution, so it may be beneficial to use lower Zn concentrations for extra improvements.

The initial hypothesis in this research is that growth in tissue culture is improved when the elementary composition in the medium is more similar to the nutrient composition in well-growing plants. It might be that by future additional changes, e.g. increasing P. that additional improvements could be obtained.

5. Conclusion

The advantages of using media adapted for specific crops instead of general media such as MS are clear: using media adapted according to the elemental composition of plant tissue or of nutrient solutions in hydroponics, substantial improvements for *in vitro* propagation of two species have been made. Using this approach, improvements were also found for *in vitro* propagation of other species including rose, apple and *Dahlia.* Often the effect was more profound after the *in vitro* phase; e.g., rose plantlets grew much better after micro-elemental adaptations, although no significant differences were found during the *in vitro* phase.

We conclude that it can be valuable for improvement of *in vitro* propagation to apply the strategy of adapting the mineral content of tissue culture media to the elemental analysis of the plant. If the species is grown in hydroponics, adaptations derived from hydroponic substrate composition can have very beneficial effects too.

Acknowledgements

PT Trade Board for Horticulture and The Ministry for Agriculture, Nature Management and Fisheries supported this research. We thank B. Morris for technical assistance, and G.J de Klerk for reading the manuscript.

References

Amiri ME (2001) Mineral uptake by banana (*Musa acuminata* L.) explant *in vitro.* Acta Hortic. 560: 387-389

Anonymous (1999) Bemestingsadviesbasis Glastuinbouw IKC Akker- en Tuinbouw, afd. Glasgroente, Aalsmeer & Naaldwijk, Netherlands

Avila A de L, Pereyra SM & Argüello JA (1998) Nitrogen concentration and proportion of NH_4^+-N affect potato cultivar response in solid and liquid media. Hort. Science 33: 336-338

Bergmann W (1992) Nutritional disorders of plants. Fischer, Jena

Bouman H, Morris B & Tiekstra A (2001) Development of new tissue culture media, using the relation between mineral composition of plant and medium. Acta Hortic. 560: 373-376

Bouman H & Tiekstra A (2001) Mineral nutrition in tissue culture: Influence on propagation and quality of the plantlets. In: Horst WJ et al. (eds) Plant Nutrition - Food Security and Sustainability of Agro-Ecosystems (pp. 316-317). Kluwer Academic Publisher, Dordrecht

Cos Terrer J & Frutos Tomas D (2001) Determination of macronutrients to be included in *vitro* culture media according to leaf concentrations. J.Horticult. Sci. Biotech. 76: 484-488

Dahleen LS (1995) Improved plant regeneration from barley callus cultures by increased Cu levels. Plant Cell, Tiss. Org. Cult. 43: 267-269

Dalton CC, Iqbal K & Turner DA (1983) Fe P precipitation in Murashige & Skoog media. Physiol. Plant. 57: 472-476

Driver JA & Kuniyuki AH (1984) *In vitro* propagation of Paradox walnut rootstock. Hort. Science 19: 507-509

George EF (1993) Plant propagation by tissue culture. Part 1. The Technology (pp. 292-295 and 344-361), and Part 2. In Practice (pp. 579-581). Exegetics Ltd., Edington, Great Britain

Henry M, Fulcheri C & Morard P (1999) Development of new plant tissue culture media. In: Altman A et al. (eds) Plant Biotechnology and *In Vitro* Biotechnology in the 21st Century (pp. 647-650). Kluwer Academic Publisher, Dordrecht

Kintzios S, Drossopoulos, JB & Lympropoulos C (2001) Effects of vitamins and inorganic nutrients on callus growth and somatic embryogenesis from leaves of chilli pepper. Plant Cell, Tiss. Org. Cult. 67: 55-62

Monteiro ACB de A, Higashi EN, Goncalves AN & Rodriguez APM (2000) A novel approach for the definition of the inorganic medium components for micropropagation of yellow passion fruit (*Passiflora edulis* Sims. f. *flavicarpa* Deg.) *In Vitro* Cell. Dev. Biol.-Plant 36: 527-531

Murashige T & Skoog FL (1962) A revised medium for rapid growth and bio assays with tobacco tissue cultures. Physiol. Plant. 15: 473-497

Rugini E (1984) *In vitro* propagation of some olive (*Olea europea sativa* L.) cultivars with different rootability, and medium development using analytical data from developing shoots and embryos. Scientia Hortic. 24: 123-124

Schenk N, Hsiao KC & Bornman CH (1991) Avoidance of precipitation and carbohydrate breakdown in autoclaved plant tissue culture media. Plant Cell Rep. 10: 115-119

Scholten HJ & Pierik RLM (1998a) Agar as a gelling agent: differential biological effects *in vitro*. Scientia Hortic. 77: 109-116

Scholten HJ & Pierik RLM (1998b) Agar as a gelling agent: chemical and physical analysis. Plant Cell Rep. 17: 230-235

Singha S, Oberl GH & Townsend EC (1987) Changes in nutrient composition and pH of the culture medium during *in vitro* shoot proliferation of crab apple and pear. Plant Cell, Tiss. Org. Cult. 11: 209-220

V. Biomass for Secondary Metabolite Production

Chapter 38

Development and validation of an efficient low cost bioreactor for furanocoumarin production with *Ruta graveolens* shoot cultures

E. Gontier[(1)], S. Piutti[(1)], A. Gravot[(1)], S. Milesi[(1)], A. Grabner[(1)], B. Massot[(1)], K. Lievre[(1)], M. Tran[(1)], J.L. Goergen[(2)] & F. Bourgaud[(1)]
[(1)]UMR 1121 INPL-ENSAIA-INRA, [(2)] UPR CNRS 6811
Ecole Nationale Supérieure d'Agronomie et des Industries Alimentaires, Institut National Polytechnique de Lorraine, 2 avenue de la forêt de Haye, 54505 Vandoeuvre les Nancy, France. Correspondence: Eric.Gontier@ensaia.inpl-nancy.fr; Tel: (+33)3 83 59 57 81
Fax: (+33)3 83 59 57 99

Abstract: Despite efforts made to produce plant secondary metabolites from cell suspensions, only a few industrial applications have been successful. Generally, higher yields are obtained when cultivating organs (roots or leafy stems) instead of undifferentiated cells. In this case, new problems arise because of the structure of the plant material, and special bioreactors have to be built for such applications. Furthermore, the high cost of commercial bioreactors may limit the number available for the researcher to carry out many experiments in parallel. Because of this, we developed a very low cost system (i.e; bioreactors) that allows good growth of *Ruta graveolens* L. shoots and production of secondary metabolites (*i.e.* furanocoumarins). The development of a very simple auto-priming siphon allows the use of common jars ranging from 3 to 20 litres for temporary immersion cultures. The very low cost of such a home-made bioreactor allows scientists to run many different experiments at the same time. It thus saves time in optimising the culture medium parameters and in replicating trials before reaching the step of final culture system development with highly equipped (costly) bioreactors.

Key words: psoralen, temporary immersion, Tween20, permeabilisation, secondary metabolites, simplified bioreactor system

Abbreviations: FC - furanocoumarin; FW - fresh weight; DW - dry weight; TIS - temporary immersion system; 5Mop - 5-methoxypsoralen; 8Mop - 8-methoxypsoralen; 5,8Mop -5,8-methoxypsoralen; Pso - psoralen; vvm - volume of gas per volume of liquid per minute

A.K. Hvoslef-Eide and W. Preil (eds.), Liquid Culture Systems for in vitro Plant Propagation, 509–524.

1. Introduction

Plant secondary metabolites are of wide use for pharmaceutical applications. For more than twenty years, many scientists have studied the potentials of cultivated plant cells *in vitro* to produce such compounds (Dornenburg et al., 1995; Bourgaud et al., 2001). Most often, despite efforts to improve cell lines, culture media and process engineering, very few examples of industrial development have been successful. These poor results are linked to low productivities obtained with undifferentiated cells and sometimes to genetic instability. At the end of the eighties, some reports showed that, as compared to cell suspension, organ cultures could be a better source of secondary metabolites (Bourgaud et al., 2001). For example, root cultures and more specifically transformed roots (hairy root cultures) often grow quite well *in vitro* and produce amounts of secondary metabolites equivalent – sometime higher – than in the entire plant. Thus, another research area was opened up, which consisted of evaluating the potential of organ culture for secondary metabolites production. The first difficulty encountered was due to the physical properties of organs. As compared to isolated cells, it is difficult to inoculate a bioreactor by pumping the organs (roots or shoots) in the classical tubing usually employed for plant cells, bacteria or yeasts. Moreover, the "classical" fermentors employed for microbiological applications are not very well-adapted to the plant organ's needs. For instance, because of the presence of probes (pH, pO_2, temperature etc.) and of possible other metallic parts that plunge into the tank, plant organs tend to be trapped in some parts of the reactor when agitation is applied (Boitel-Conti et al., 1995a). This phenomenon leads to heterogeneity of the culture conditions and sometime, to the death of a part of the biomass. The best solution to avoid such problems is to eliminate all the vertical (and, if necessary, horizontal) parts that plunge into the tank. Then, partial oxygen pressure, pH, temperature can not be easily controlled, but the organ culture may be drastically improved in this way (Boitel-Conti et al., 1995a).

Then, starting from this analysis, another strategy can be developed. Firstly, a classical commercial bioreactor is an expensive investment (20-30,000 Euros or more). Secondly, if it is necessary to dismantle at least a part of it (all the metallic matrix), why should one pay for such a part? Thus it seems easier to custom build a bioreactor. This can be realized with glass or autoclavable plastic (polypropylene) jars, using Hepa Vent filters and silicone tubing. Then the most important difficulty that arises is to built water and gas tight connexions within the cap (screw closure) of the flask. This can be performed by gluing, welding the connector (fittings) to the cap or preferably by using disposable polypropylene or inox screw fittings.

Based on this analysis, and because we were interested in furanocoumarin production from *Ruta graveolens* shoot cultures (Massot et al., 2000), we initiated such an approach. *Ruta graveolens* is a Mediterranean plant from the lemon tree family that produces linear furanocoumarins (FCs), including psoralen, 5-methoxypsoralen (5Mop), 8-methoxypsoralen (8Mop) and 5,8-methoxypsoralen (5,8Mop). These molecules (Figure 1) are of pharmaceutical interest for the treatment of vitiligo and psoriasis (Pathak and Fitzpatrick, 1992) and also show interesting properties for multiple sclerosis treatment (Koppenhofer et al., 1995).

Reference cultures in 250 ml Erlenmeyer flasks and in a classical industrial bioreactor were established. Then, home-made permanent or temporary immersion system were developed and tested in terms of *Ruta graveolens* shoot growth and furanocoumarin production.

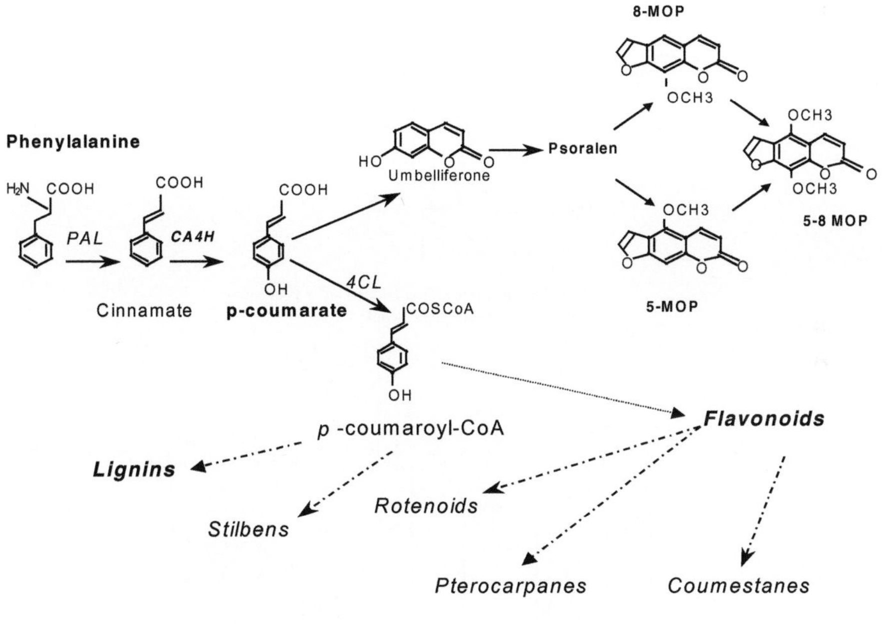

Figure 1: Phenylpropanoids (furocoumarins) pathway.

2. Materials and methods

2.1 Plant material and culture media

Ruta graveolens shoot cultures were initiated from sterile seedling (seeds provided by Ets Bertrand Frères SA, Orléans, France) as previously described (Massot et al., 2000). The plant material stock was cultivated in 250 ml Erlenmeyer flasks containing 100 ml of B5 (Gamborg et al., 1968) medium containing 9 µmol (2 mg l^{-1}) of 2,4-D and kinetin and 30 g l^{-1} of sucrose. This medium – called B5(30)K2D2 – was used for all experiments, except in the case of sucrose trials where sucrose was lowered to 10 g l^{-1} (then: medium called B5(10)K2D2).

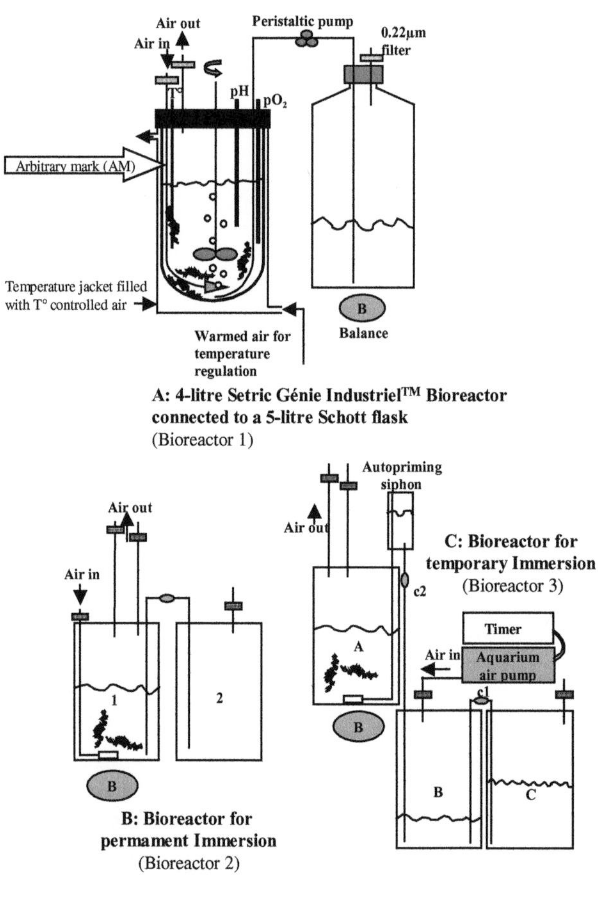

Figure 2: Scheme of the different bioreactors tested in these experiments.

2.2 *Bioreactor experiments*

A 4-litre fermentor provided by Setric Genie Industriel (France), equipped with a temperature, an oxygen and a pH probe was coupled with a 5-litre Schott flask that allows the weighting of shoots drained weight in the fermentor after transfer of the medium into the flask (Figure 2, Bioreactor 1). The culture was performed in 4.8 litres (2 litres in the bioreactor and 2.8 litres remaining in the 5-litre Schott flask) of B5(30)K2D2 at 24°C (pO_2 set at 50% of air saturation). The impeller agitation rate was set at 10 rpm. Measurement of shoot biomass was performed on a weight-volumetric method as follow: before culture experiments, the mass of medium (M_o) necessary to fill the empty reactor up to an arbitrary mark (AM on Figure 2, Bioreactor 1) was measured. Afterwards, for each measurement, all the medium contained in the bioreactor was transferred into the Schott flask which was weighted. Then, the medium was transferred back into the reactor up to the marked level (AM) and the Schott flask was weighted again. The difference of mass between the two weightings corresponds to the mass of medium (M_t) necessary to complete the volume (in the fermentor which contains the shoots) up to the marked level. Both densities of the medium and of the shoots are very close to 1 (± 1%). Thus, for each measurement, the difference between M_o and M_t corresponds to the weight of the shoots, which can be estimated with a final measured error of 2%.

2.3 *Bubble flask system*

The bubble flask consists of a 3-litre glass vessel, containing various volumes of medium (classically 1.5 litre) in which air is bubbled (0.01 vvm) through a sparger. Inlets and outlets are protected by 0.22 μm Hepa-vent filters. The culture flask is connected to another 3-litre flask in which medium is transferred for weighting the biomass (Figure 2, Bioreactor 2).

2.4 *Temporary immersion system*

A temporary immersion system (Figure 2, Bioreactor 3) has been developed on the basis of the previous bubble flask (Bioreactor 2). The culture medium (classical volume for experiments: 1.5 litre) is contained in a flask (flask C in Figure 2, Bioreactor 3) and partially transferred into another one (flask B in Figure 2, Bioreactor 3) at inoculation time. Further transfers are carried out during the culture depending on the shoot growth (see results). During the culture, the medium is pulsed from flask B to flask A (in which the biomass is inoculated and cultivated) using an air-pump controlled by a timer. When the air-pump stops, an auto-priming siphon that we developed drains the medium back to flask B, allowing shoot emergence. The typical immersion cycle used in these studies is 15 minutes each hour.

2.5 Chemical analysis

Psoralen (Pso), 5-methoxy psoralen (5Mop), 8-methoxy psoralen (8Mop) and 5,8-dimethoxy psoralen (5,8Mop) purchased from Extrasynthese (Lyon, France) are analysed by HPLC as previously described (Milesi et al., 2000). Sugar was analysed using an enzymatic method from Boehringer Mannheim (ref. 716260) as previously described (Massot et al., 2000).

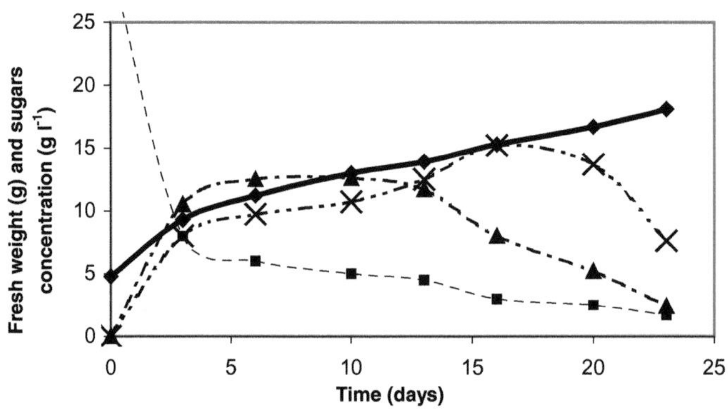

Figure 3: Time course of growth and sugar consumption for *Ruta* shoots cultivated in 250 ml Erlenmeyer flasks.

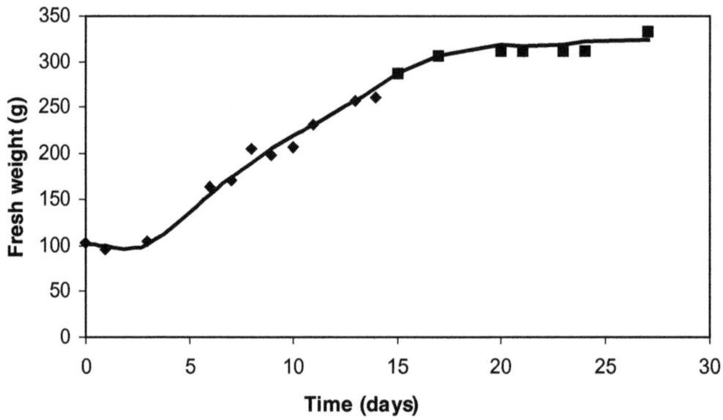

Figure 4: *Ruta* shoot growth in the 4-litre Setric Génie Industriel bioreactor (Figure 2, Bioreactor 2).

3. Results and discussion

3.1 Control culture in 250 ml Erlenmeyer flasks

At first, shoot growth (Figure 3) was characterized in 250 ml Erlenmeyer flasks (Figure 12 A). From a 5 g FW inoculum, after a lag period of 3 days, linear growth followed, reaching a final biomass fresh weight of 18 g after 23 days. As a function of time, sucrose is rapidly hydrolysed and glucose and fructose progressively consumed. In such conditions, the total FCs level is typically of 1.2 % DW (Massot et al., 2000). Shoot growth occurred at a mean rate of 0,06 g DW per day, corresponding to 1.36 mg of FCs produced per day of culture and per gram DW of inoculum over 23 days (6.8 mg FCs day^{-1} l^{-1}).

3.2 Establishment of a reference bioreactor culture

Ruta shoot culture was performed in a 4-litre SGI bioreactor coupled to a 5 litre Schott (Bioreactor 1: Figure 2A and Figure 12 B), the whole containing 5.8 litres of B5(30)K2D2 medium with a 115 g FW inoculum. As presented in figure 4, growth occurred with a maximum rate of 13 g FW day^{-1} for 16 days and thereafter, only 2 g of fresh weight were produced each day for the further 11 days of the experiment. Starting from 115 g FW, the total biomass reached 332 g FW within 27 days. Sucrose was completely hydrolysed in 6-7 days whereas glucose and fructose were progressively consumed during the culture (data not shown). At day 27, only 1 g of glucose and fructose still remained in the culture medium. The furanocoumarin concentration in the medium oscillated between 1 and 6 mg l^{-1} and the relative proportion of the four furanocoumarins is relatively constant during the culture period. Shoot growth occurred at a mean rate of 0.8 g DW per day corresponding to a calculated FC production of 0.98 mg of FCs per day of culture and per gram DW of inoculum in 4.8 l of medium (2 mg FCs day^{-1} g^{-1} l^{-1}). The productivity per volume unit is three times lower than the results obtained with the Erlenmeyer flask cultures, but the total daily FC production is much higher (9.78 *vs* 0.68 mg FC per day).

In this bioreactor system, the biomass spatial distribution exhibited notable heterogeneity. Indeed, the shoots were rapidly pinned between bioreactor glass wall and vertical parts (tubes and probes). This heterogeneity of shoot distribution in this bioreactor led us to think that it could be, in the future, a source of variability in terms of growth and FC production. Furthermore, the biomass cannot be inoculated *via* the tubing of the reactor, meaning that a further scaling up is hardly conceivable.

3.3 Establishment of bubble flask cultures

The above results were promising but the heterogeneity was quite important and it was not possible to carry out many experiments in parallel with a single bioreactor. It was then decided to build a simplified bioreactor in which oxygen pressure and pH were not controlled. This system consisted of a 3-litre flask in which the shoot culture was performed (Figure 12 C and Figure 2, Bioreactor 2). This system was coupled with a 2.5-litre bottle that allowed draining of the culture medium for weighting the biomass.

The first growth experiments led to good results in terms of growth and reproducibility (Figure 5). In these conditions, furanocoumarin levels were equivalent to those measured in Erlenmeyer flasks or Setric bioreactor experiments. The mean biomass production within 42 days was 0.4 g DW per day of culture corresponding to a calculated productivity of 0.97 mg FC per day of culture and per gram DW of inoculum in 1.5 litres (3.2 mg FCs day^{-1} l^{-1}). On this basis, it was demonstrated that lowering the sucrose concentration in the medium (10 g l^{-1} instead of 30 g l^{-1}) improved the fresh weight to nearly double (Figure 6). This led to values of: 1.8 mg FC per day of culture and per gram DW of inoculum in 1.5 litres. Because it was easy to build up many copies of the bioreactor (low technology and low cost), it was possible to run many experiments (and replicates) in parallel (Figure 12 D).

3.4 Effect of temporary immersion

The permanent immersion regime (in bubble flasks, Figure 2, Bioreactor 2) and temporary immersion one (Figure 2, Bioreactor 3) were compared (Figure 7). Our results show that, over a 44 day culture, temporary immersion is beneficial for growth (a 17% increase). It seems also to be positive for shoot quality, as no vitrification occurred whereas it happened very often in the permanent immersion regime. No significant differences appeared in FCs contents. The calculated productivity was 2.6 mg FCs produced per day of culture and per gram DW of inoculum in 1.5 litres (1.7 mg FCs day^{-1} l^{-1}).

In order to try to improve the culture process, we performed some cultures in temporary immersion and replaced the culture medium by a fresh one, each time the growth slowed down (Figure 8). Then, very high densities of biomass were obtained and productivity was steadily high over a long period (140 days). The mean biomass production and calculated FCs productivity were 0.6 g DW per day and 3.8 mg of FCs per day and per gram DW of inoculum (5 mg FCs day^{-1} l^{-1}). Furthermore, when reaching such densities, shoots were found to be deeply coloured green all around the mass, more yellow-green inside and yellow-white in the centre of the culture (Figure 12 E).

We analysed the FCs content of these different qualities of the biomass (Figure 9) and it revealed that the less green was the biomass, the lower its content was (less than 50%).

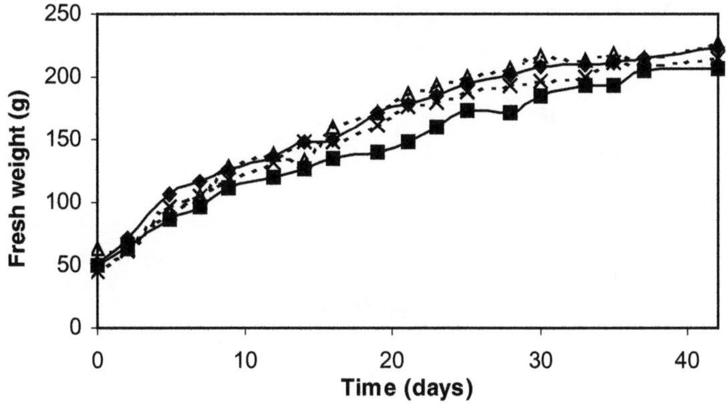

Figure 5: Time course of *Ruta* shoots growth when cultivated in 3-litre bubble flask (Figure 2, Bioreactor 2).

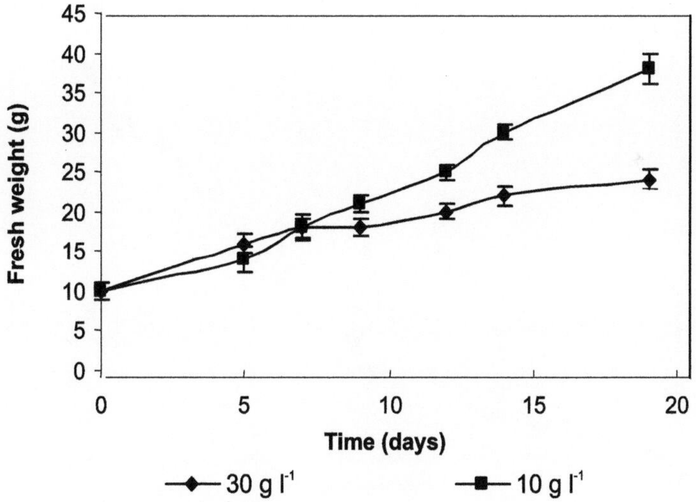

Figure 6: Comparison of *Ruta* shoot growth cultivated in 3-litre bubble flasks (Figure 2, Bioreactor 2) in a medium containing 10 or 30 g l^{-1} of sucrose.

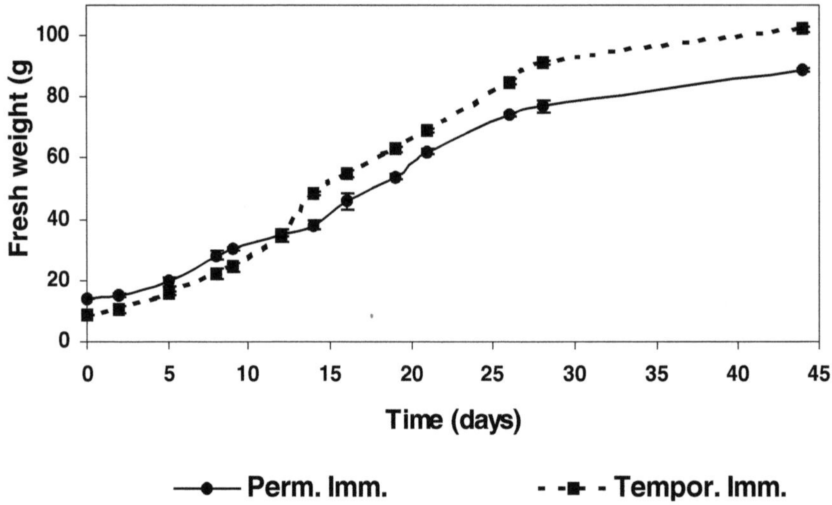

Figure 7: Comparison of *Ruta* shoot growth in permanent (3-litre bubble flask = bioreactor 2) or temporary immersion (Figure 2, 3-litre TIS = Bioreactor 3) system.

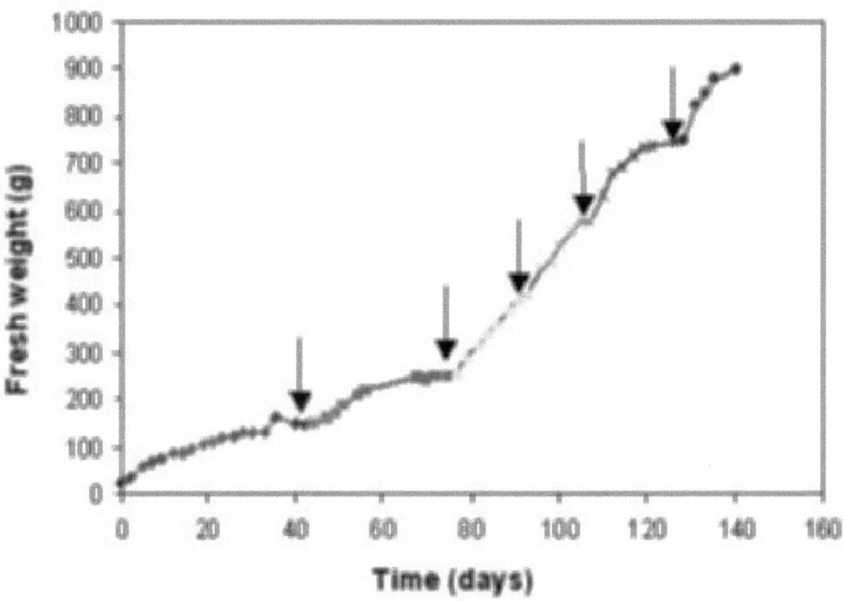

Figure 8: *Ruta* shoot growth in 3-litre bubble flask (Figure 2, Bioreactor 2) when medium is changed (arrows) by the time.

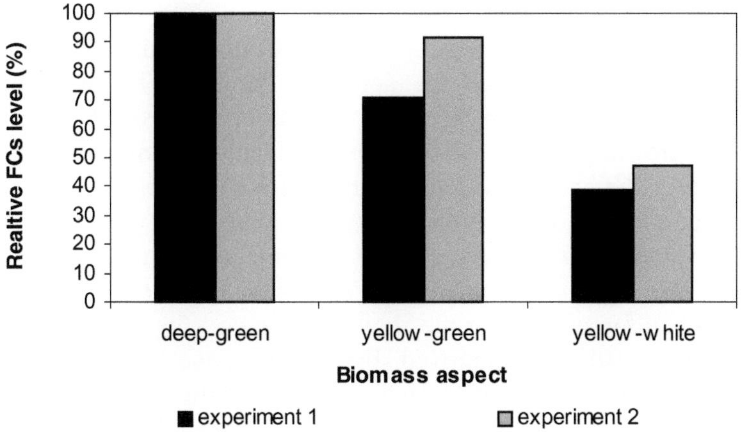

Figure 9: Furanocoumarin levels in the *Ruta* shoots cultivated up to high density in the 3-litre bubble flask bioreactor (Figure 2, Bioreactor 2). Shoot contents are presented as a function of their colour: deep-green (external part), yellow-green (medium part) and yellow-white (centre of the shoot mass).

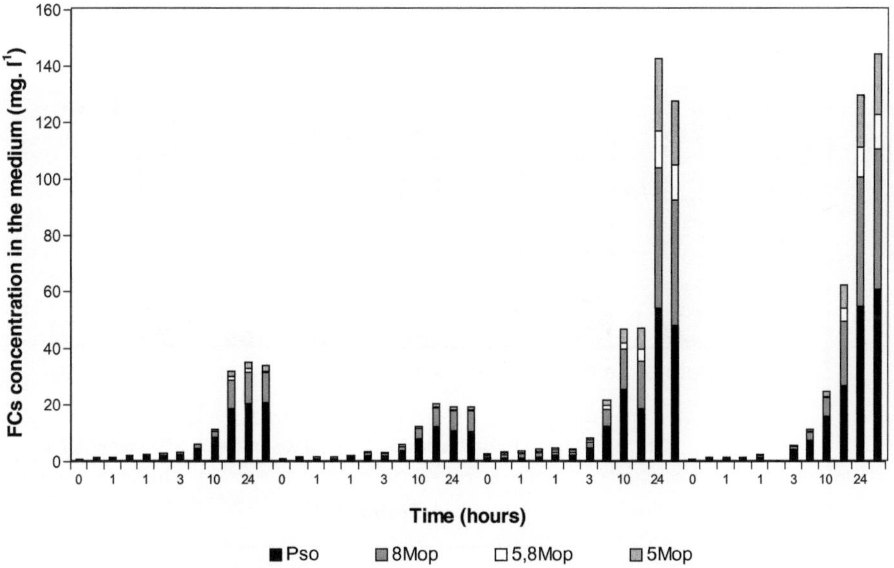

Figure 10: Furanocoumarin release by *Ruta* shoots when the culture medium is replaced by a fresh one (control) containing umbelliferone and or Tween20 as precursor and permeabilizing agent, respectively. Experiments conducted in 3-litre bubble flasks (Figure 2, Bioreactor 2) after 20 weeks of growth.

3.5 *Effect of permeabilization*

Based on our previous experience (Boitel-Conti et al., 1995-1997; Gontier et al., 2002), we tested the potential effects of Tween20 as a membrane permeabilizing agent for FCs release in *Ruta* shoot cultures. Thus, after the growth phase (20 weeks, 4 parallel cultures corresponding to figure 8) some of the shoots were permeabilized with Tween20 3 % (v/v) with or without addition of umbelliferone (an FC precursor). The evolution of FCs in the culture medium was measured as a function of time (Figure 10) and the best results were obtained for the Tween treatment. Following addition of umbelliferone, neither the shoot FC content (Figure 11) nor the FC release (Figure 10) were improved. FCs levels in the biomass (Figure 11), measured on a sample before and after the treatment, significantly decreased only when a high increase was observed for the FCs concentration in the medium (*i.e.* after Tween treatment). Furthermore, the biomass could be rinsed and cultivated again for weeks (at least) after being permeabilized, showing that cell viability was not completely altered.

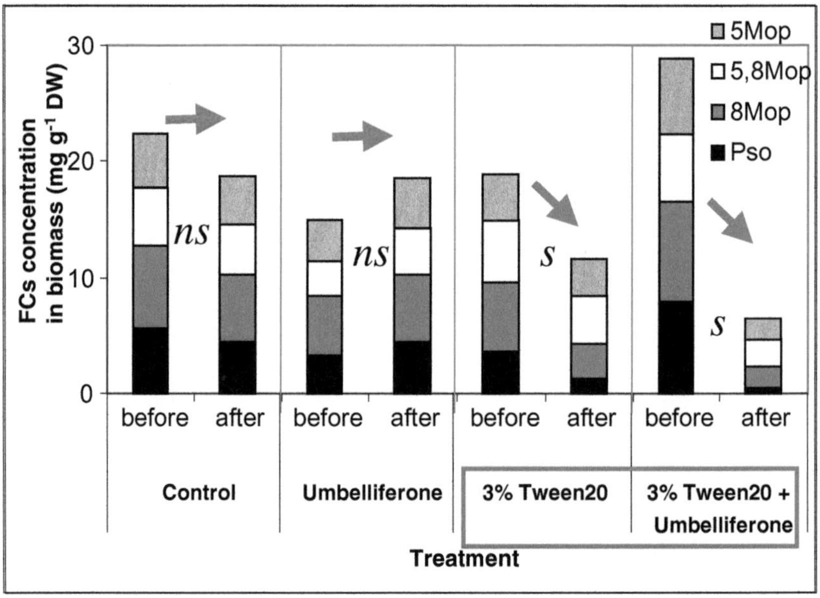

Figure 11: Ruta shoot furanocoumarin levels before and after the culture medium is replaced by a fresh one (control) containing umbelliferone and/or Tween20 as precursor and permeabilizing agent respectively. Experiments conducted in 3-litre bubble flasks (Figure 2, Bioreactor 2) after 20 weeks of growth [ns: non significant difference; s: significant difference; $p<0.05$ Student's t-test).

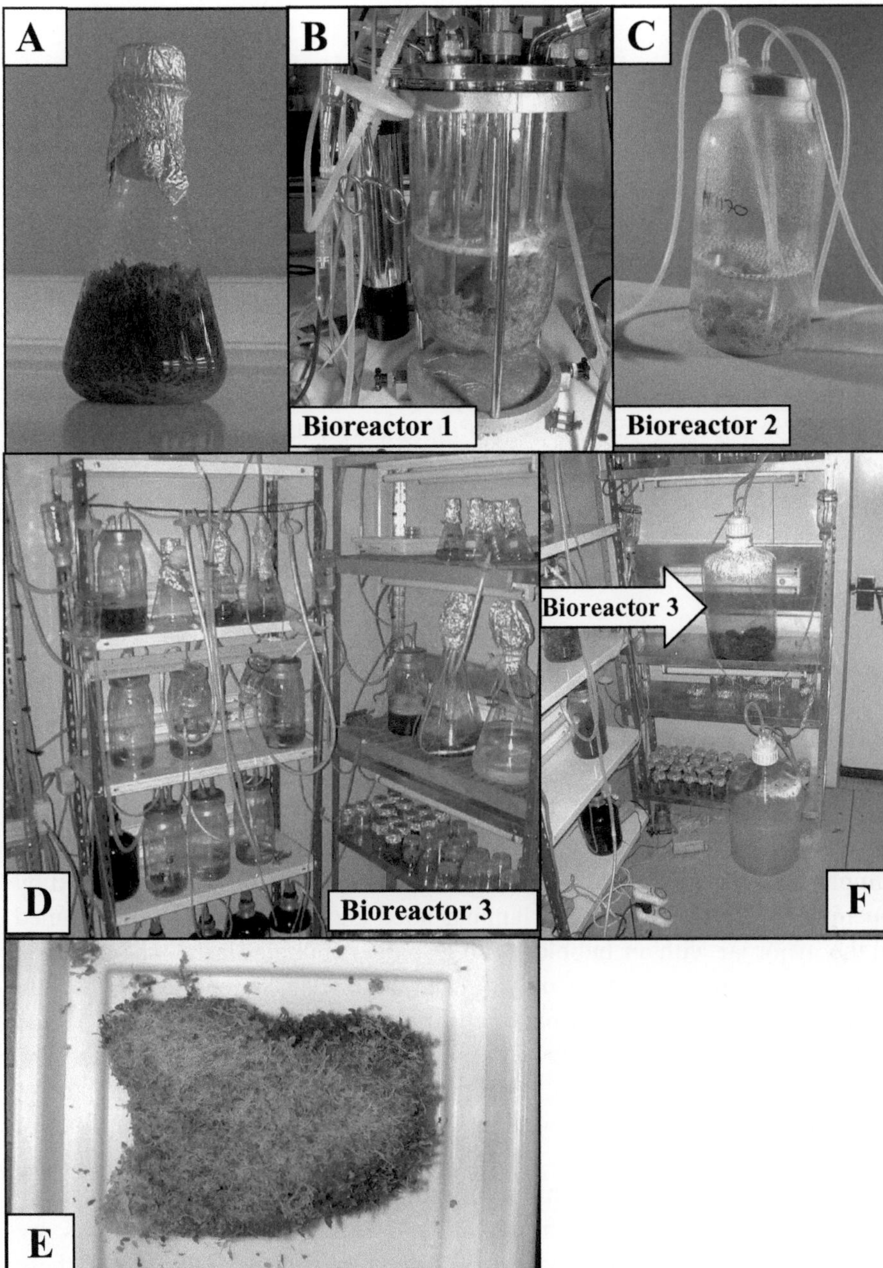

Figure 12: Ruta shoots obtained in 250ml flask (A); in a 4-litre Setric bioreactor (B), in 3-litre bubble flask (C), in 3-litre (D) or 20-litre (F) temporary immersion systems. Aspect of the *Ruta* shoots mass obtained at high density (E).

3.6 Scale up of the system

In order to scale up our temporary immersion system (TIS), we developed a 20-litre vessel composed of three Nalgene polypropylene flasks. This larger TIS allowed us to produce kilograms of biomass (Figure 12 F) with FCs levels comparable to those obtained in the 3-litre TIS (growth: 1.6 g DW day^{-1} and calculated FC productivity: 3.8 mg FCs day^{-1} l^{-1}). This 20-litre size is the maximum that we could reach within the lab because of the size of our autoclave and of our laminar hood (in which the inoculation is performed). As compared to the results obtained in the 3-litre devices, heterogeneity in biomass aspect was similar (low chlorophyll content within the shoot core). Thus, we estimate that 20 l is the maximum vessel size (volume) in which we can cultivate *Ruta* shoots without the need for further light supply (i.e. development of a photobioreactor).

4. Conclusions

Firstly, shoot cultures were initiated in liquid medium in 250 ml Erlenmeyer flasks. After studying growth and FCs production in these conditions, a 4-litre Setric™ bioreactor was inoculated with shoots and the culture parameter were studied. As expected, because of the presence of a metallic matrix in the tank, a part of the plant material was trapped in some parts of the reactor and a high heterogeneity of the biomass could be observed. Thus, we developed a home made bioreactor consisting in a simple glass jar coupled to a second one. The inoculation was performed by opening the screw cap under a laminar hood and the culture was performed in the major jar with air bubbled with a sparger (bubble flask). The coupling of the major jar to an annex one allowed us to transfer the culture medium from the first to the second one and to weigh this containing the biomass. This allowed us to measure growth in real time. This simple bioreactor also allowed us to carry out many experiments and repetitions in parallel and to show, for example, that a lower sucrose concentration (10 g l^{-1} instead of 30 g l^{-1}) gave best results in terms of growth and FCs production. Furthermore, because temporary immersion has been described as a possible way to obtain a higher growth rate and less vitrification of the biomass, we built a very simple (and low cost) system to perform cultures in these conditions. The trials showed that temporary immersion lead to results equal or superior to those of permanent immersion regime. Shoot growth was improved and its quality (no vitrification, high FCs content) was better also.

Then, based on our in-house technology (bubble flasks for permanent or temporary immersion regime) we evaluated the maximum biomass and FCs production that could be reached with such bioreactors. We showed that very high densities of *Ruta* shoots can be obtained and that a scale up step at 20 l can be performed successfully. This plant material contains large amounts of FCs that can be released when permeabilizing the biomass with detergent (Tween20). This process can be performed without killing the tissues, allowing further culture and permeabilization cycles.

In conclusion, based on the experience we gained in the past on hairy root cultures (Boitel-Conti et al., 1995 a, b, 1996, 1997), we developed a home-made, simple and low costs bioreactor system that allowed a good biomass and FCs production. These results are of interest for potential applications of plant organ cultures for secondary metabolite production. They are also very promising for developing useful bioreactor systems for plant micropropagation (Gontier et al., 1993; Lorenzo et al., 1998, Hsiao et al., 1999).

References

Boitel-Conti M, Laberche JC, Gontier E & Sangwan-Norreel BS (1995a) Culture de racines transformées de *Datura innoxia* Mill. en bioréacteur: comparaison de 3 types de procédés. C.R. Acad. Sci. Paris 318: 593-598

Boitel-Conti M, Gontier E, Laberche JC, Ducrocq C and Sangwan-Norreel BS (1995b) Permeabilization of *Datura innoxia* hairy roots for release of stored tropane alkaloids. Planta Medica 61: 287-290

Boitel-Conti M, Gontier E, Laberche JC, Ducrocq C & Sangwan-Norreel BS (1996) Inducer effect of Tween 20 permeabilization treatment used for release of stored tropane alkaloids in *Datura innoxia* Mill. hairy root cultures. Plant Cell Reports 16: 241-244

Boitel M, Gontier E, Assaf C, Laberche JC, Ducrocq C & Sangwan-Norreel BS (1997) Procédés de production de métabolites primaires et secondaires à partir d'organes végétaux cultivés *in vitro*, Patent: France n°FR 97 00641

Bourgaud F, Gravot A, Milesi S & Guckert A (2001) Production of plant secondary metabolites: a historical perspective. Plant Science 161: 839-851

Dornenburg H & Knorr D (1995) Strategies for the improvement of secondary metabolite production in plant cell cultures. Enzyme and Microbial Technology 17: 674-684

Gamborg OL, Miller RA & Ojima V (1968) Nutrient requirements of suspension cultures of soybean root cells. Exp. Cell. Res. 50: 151-158

Gontier E, Clement A, Tran TLM, Lievre K, Gravot A, Guckert A & Bourgaud F (2002) Hydroponic combined with natural or forced root permeabilization: a promising technique for plant secondary metabolites production. Plant Science (in press)

Gontier E, Fliniaux MA, Barbotin JN & Sangwan-Norreel BS (1993) Tropane alkaloid levels in the leaves of micropropagated *Datura innoxia* Mill. plants. Planta Medica 59: 432-435

Hsiao TY, Bacani FT, Carvalho EB & Curtis WR (1999) Development of a low capital investment reactor system: application for plant cell suspension culture. Biotechnol. Prog. 15: 114-122

Koppenhofer E (1995) Kaliumkanalblocker in der Neurologie: 5-MOP zur Behandlung MS-bedingter Funktionsdefizite. TW Neurologie Psychiatrie: 585-591

Lorenzo JC, Gonzàlez BL, Escalona M, Teisson C, Espinosa P & Borroto C (1998) Sugarcane shoot formation in an improved temporary immersion system. Plant Cell, Tissue and Organ Culture 54: 197-200

Massot B, Milesi S, Gontier E, Bourgaud F & Guckert A (2000) Optimized culture conditions for the production of furanocoumarins by micropropagated shoots of *Ruta graveolens*. Plant Cell, Tissue and Organ Culture 62: 11-19

Milesi S, Massot B, Gontier E, Bourgaud F & Guckert A (2001) *Ruta graveolens* L.: a promising species for the production of furanocoumarins. Plant Science 161: 189-199

Milesi S, Massot B, Piutti S, Goergen JL, Murano E, Toffanin R, Bourgaud F, Gontier E & Guckert A (2000) *Ruta graveolens* cell cultures for psoralens production: culture in bioreactor, effect of elicitation and calcium alginate immobilization. International Journal of Biochromatography 5: 281-296

Pathak MA & Fitzpatrick TB (1992) The evolution of photochemotherapy with psoralens and UVA (PUVA): 2000 BC to 1992 AD. Journal of Photochemical and Photobiology B: Biology 14: 3-22

Chapter 39

Comparison of secondary plant metabolite production in cell suspension, callus culture and temporary immersion system

Dirk Wilken[1*], Elio Jiménez González, Annette Hohe[1], Miguel Jordan[3], Rafael Gomez Kosky[2], Guillermo Schmeda Hirschmann[4] & André Gerth[1]
[1]*BioPlanta GmbH, Deutscher Platz 5, D-04103 Leipzig, Germany. E-mail: info@bioplanta-leipzig.de;*
[2]*Instituto de Biotecnología de las Plantas, Carretera a Camajuaní Km. 5 ½, Santa Clara, Cuba. E-mail: ejimenez@uclv.edu.cu; rgkosky@uclv.edu.cu;mailto:rgkosky@uclv.edu.cu*
[3]*Departamento de Ecología, Pontificia Universidad Católica de Chile, Alameda 340, Santiago, Chile. E-mail: mjordan@genes.bio.puc.cl;*
[4]*Instituto de Quimica de Recursos Naturales, Universidad de Talca, Casilla 747, Talca, Chile. E-mail: schmeda@pehuenche.utalca.cl*

Abstract: Cell and organ cultures of *Lavandula officinalis*, *Hypericum perforatum*, *Cymbopogon citratus* and *Fabiana imbricata* were established for the production of secondary metabolites *in vitro*. Shoot multiplication was performed by conventional micropropagation on agar-solidified medium as well as in temporary immersion systems (TIS), the latter resulted in higher multiplication rates compared to the culture in microcontainers for all plant species tested. The concentration of bioactive compounds was determined in different *in vitro* cell and organ cultures and was compared to field grown plants. For *Lavandula* the highest content of rosmarinic acid was found in cell cultures, for the other three species in field grown plants. Concentrations of bioactive compounds were always higher in plant material grown in TIS compared to cell suspension and callus cultures.

Key words: bioactive compounds, bioreactor culture, *Cymbopogon, Fabiana, Hypericum, in vitro* culture, *Lavandula,* pharmaceuticals, suspension culture

Abbreviations: BA - 6-benzylaminopurine; 2,4-D - 2,4-dichlorophenoxyacetic acid; GC-MS - mass spectroscopy combined with gas chromatography; HPLC - high performance liquid chromatography; NAA - α-naphtalene acetic acid; TIS - temporary immersion system

A.K. Hvoslef-Eide and W. Preil (eds.), Liquid Culture Systems for in vitro Plant Propagation, 525–537.

1. Introduction

Bioactive compounds obtained from plants are widely used as pharmaceuticals and ongoing research is still revealing new plants and metabolites of interest. However, the quality and quantity of compounds from plants harvested in nature or cultivated in fields fluctuates and the composition varies depending on environmental conditions such as climate, season, weather and soil (Bhojwani and Razdan, 1996). *In vitro* culture of plants can overcome these problems, since the environmental conditions that affect plant metabolism can be strictly controlled. Thus, growth conditions e.g. pH (Payne et al., 1988; Jardin et al., 1991), light quality and quantity (Berlin, 1986; Drapeau et al., 1987), gas composition (Schlatmann et al., 1994; Tisserat et al., 2001; Zobayed et al., 2003) as well as the medium composition (Fujita et al., 1982; Moreno et al., 1995; Schlatmann et al., 1992) can be manipulated in order to optimise the production of the desired secondary metabolites. Extensive work on this topic has been done mainly on callus and suspension cultures, as the latter can be scaled up to industrial production in bioreactors (e.g. *Lithospermum erythrorhizon*, Fujita et al., 1982 or *Catharanthus roseus*, Schiel and Berlin, 1987). However, the yield of secondary metabolites from cell cultures is often instable and decreases throughout the time. Moreover, highly sophisticated laboratory techniques for plant cell cultivation are of commercial value only for very high value compounds or for chemicals for which market demand is very high (Bhojwani and Razdan, 1996).

Temporary immersion systems (TIS) are low-cost semi-automated plant tissue culture systems, that are used for a wide variety of cultures e.g. shoots (Lorenzo et al., 1998), somatic embryos (Akula et al., 2000) and microtubers (Jiménez et al., 1999). Since TIS allows the industrial scale-up of organ cultures, this technique represent an alternative also for secondary metabolite production avoiding the disadvantages of cell cultures stated above. However, up to now there have been no reports on the use of TIS for *in vitro* production of secondary plant metabolites. We, therefore, established TIS cultures of *Lavandula officinalis*, *Hypericum perforatum*, *Cymbopogon citratus* and *Fabiana imbricata* and compared the yield of secondary metabolites with that obtained from cell suspensions, callus cultures and field grown plants.

2. Materials and methods

2.1 General in vitro culture conditions

All *in vitro* cultures were performed on MS basal medium containing macroelements, microelements and vitamins according to Murashige and Skoog (1962) in microcontainers (250 ml polypropylene vessels). For callus and suspension cultures of *Fabiana*, *Hypericum* and *Lavandula* the MS medium was additionally supplemented with 1 mg l^{-1} nicotinic acid, 1 mg l^{-1} pyridoxine HCl, 10 mg l^{-1} thiamine HCl, 0.25 mg l^{-1} NAA, 0.25 mg l^{-1} 2,4-D, 0.25 mg l^{-1} kinetin and 30 g l^{-1} sucrose. For shoot multiplication of *Hypericum* and *Lavandula* the medium was supplemented with 0.5 mg l^{-1} nicotinic acid, 0.5 mg l^{-1} pyridoxine HCl, 9.9 mg l^{-1} thiamine HCl, 1 mg l^{-1} BA and 30 g l^{-1} sucrose. For shoot multiplication of *Fabiana* half concentrated MS macro- and microelements were used supplemented with 100 mg l^{-1} myo-inositol, 1 g l^{-1} peptone, 2 mg l^{-1} glycine, 0.5 mg l^{-1} nicotinic acid, 0.5 mg l^{-1} pyridoxine HCl, 0.1 mg l^{-1} thiamine HCl, 1 mg l^{-1} NAA, 1 mg l^{-1} gibberellic acid A_3 and 20 g l^{-1} sucrose. For *Cymbopogon* also MS macro- and microelements as well as vitamins were used. For callus and suspension cultures the medium was supplemented with 0.9 mg l^{-1} thiamine HCl, 5 mg l^{-1} 2,4-D, 0.8 mg l^{-1} kinetin and 20 g l^{-1} sucrose. For shoot multiplication of *Cymbopogon* the additional medium supplements were 0.3 mg l^{-1} BAP and 30 g l^{-1} sucrose. Solid medium was prepared by addition of 2.5 g l^{-1} Phytagel, the pH was adjusted to 5.7 before autoclaving. The culture temperature was 26°C, the light intensity was 50 µmol $s^{-1}m^{-2}$ for 16 hours per day. Callus cultures were kept in the dark.

2.2 Suspension and bioreactor culture

Cell suspensions were initiated by culturing 3 g callus in 20 ml liquid medium in 100 ml Erlenmeyer flasks. The cultures were kept in the dark on an orbital shaker (100 rpm) at 24°C. Obtained cell suspensions were subcultured every week by inoculating 60 ml medium in 300 ml Erlenmeyer flasks with 8 g cell fresh weight.

The bioreactor culture of *Hypericum* was performed in a 2-litre stirred tank bioreactor (Chemap A.S., Switzerland). The bioreactor was equipped with a bubble free aeration system using silicone tubes, the setpoint of dissolved oxygen was 80 % of air saturation. Stirring was performed by using a large blade stirrer running with 40 rpm. For *Lavandula* a 5-litre bioreactor (Braun Biotech, Germany) was equipped with a bubble aeration (setpoint of dissolved oxygen concentration: 60 %) and a marine impeller running at 60 rpm. The culture temperature was 26 °C. Both bioreactors were inoculated with 20 g l^{-1} cell fresh weight and harvested after 19 days.

2.3 *Temporary immersion system (TIS)*

A temporary immersion system was constructed using 5-litre glass twin vessels. Under standard conditions the system was run with 2 litres of medium that was changed every 2 or 3 weeks. The TIS vessels were inoculated with 30 g of *in vitro*-grown shoots. Shoots were immersed every 4 h for 5 min. The temperature and the light intensity were as given above. The cultures were harvested after 8 weeks.

2.4 *Field grown plants*

Fabiana imbricata was collected in the Western Andean slopes near Altos de Chillan, Las Trancas, Chile. *Lavandula officinalis* and *Hypericum perforatum* plants were obtained from the Botanical Garden of the University of Leipzig. For analysis of the leaf oil composition of *Lavandula officinalis* commercially dealed material originated from Corsica. *Cymbopogon citratus* was collected in Bàyamo, Cuba.

2.5 *Analysis of secondary plant metabolites*

Contents of bioactive compounds were determined in powdered tissue samples that were oven-dried at 40 °C. The metabolites of interest were: rosmarinic acid and essential oil in *Lavandula*, hypericin in *Hypericum*, α- and β-citral in *Cymbopogon* and oleanolic acid in *Fabiana*.

2.5.1 *Sample preparation and determination of rosmarinic acid*

According De-Eknamkul (1984), Gracza und Ruff (1984) and Lopez et al. (1994) a method for determination of the content of rosmarinic acid in cultures of *Lavandula* was established. Samples of 20 mg of dried and powdered plant material were extracted two times with 3 ml methanol in an ultrasonic bath at room temperature for 10 minutes. The filtrates were collected, filtrated and analysed by HPLC. A Gynkotek HPLC system (Dionex) equipped with a diode array detector (UVD 340S) and a LiChrocart RP18 Supersphere Colomn (250mm x 4 mm; 4 µm) was used. A mixture of methanol/water/phosphoric acid (50/59.7/0.3; v/v) was used as eluant at a flow rate of 0.5 ml min^{-1}. All measurements were carried out at 30°C under isocratic conditions. Rosmarinic acid was detected at 333 nm and showed a retention time of 9.8 minutes. A solution of 1 mg ml^{-1} rosmarinic acid was used as standard.

2.5.2 *Determination of the composition of essential oil from Lavandula leaves*

To obtain the essential oil an apparatus described in the European Pharmacopeia was used (European Pharmacopeia 2002). 20 g of dried *in vitro* plants or dried leaves from *in vivo* or field plants were heated with 500 ml water and steam distilled for 3 hours. Mass spectra of the essential oil obtained were recorded on a Finnigan MAT ITD-800 gas chromatography – mass spectroscopy instrument (injection volume: 0.2 µl, Split 1:100). The GC column used was DB1 60 m x 0.25 mm ID x 0.25 µm film thickness. It was heated from 60°C to 220°C with a speed of 2°C min^{-1}. Helium flow-rate was 1 ml min^{-1}. The ionization mode used was electron impact ionization with an ion trap temperature of 200°C.

2.5.3 *Sample preparation and determination of hypericin*

The content of hypericin in the plant biomass was analysed using a modification of a method developed by Kartnig et al. (1996). Samples of 100 - 400 mg of dried and powdered plant material were extracted three times with 3 ml methanol in an ultrasonic bath at room temperature. The filtrates were collected, filtered and analysed by HPLC. A Gynkotek HPLC system (Dionex) equipped with a diode array detector (UVD 340S) and a LiChrocart RP18 Supersphere Colomn (250 mm x 4 mm; 4 µm) were used. A mixture of methanol/ethylacetate/phosphate (3.6/1/1.2; v/v) was used as eluant with a flow of 0.9 ml min^{-1}. All measurements were carried out at 30°C under isocratic conditions. Hypericin was detected at 595 nm. A solution of 0.53 mg hypericin in 10 ml methanol was used as standard. Under the conditions described a retention time of 18 min was observed.

2.5.4 *Sample preparation and determination of α- and β-citral*

Samples of 10-20 mg dried and powdered plant material were extracted each with 1.0 ml methanol 10 min in an ultrasonic bath. The extracts were filtered (0.4 µm) and analysed by HPLC. A Nucleosil 100-5 column (C18, 250 x 3 mm) produced by Knauer, Berlin was used. The mobile phase consisted of methanol and water (70/30; v/v) with a flow rate of 0.8 ml min^{-1}. The column was heated to 50 °C. All measurements were carried out at isocratic conditions. Citral was detected at 240 nm. Under the conditions described a retention time of 3.8 min (α-citral) and 4.2 min (β-citral) was observed.

2.5.5 Sample preparation and determination of oleanolic acid

Samples of approximately 190-210 mg of dry material were extracted with 90 ml dichloromethane in a Soxhlet apparatus for 30 min. The organic phase was filtered and taken to dryness under reduced pressure. Percentage recovery was estimated by spiking an inert support with a known amount of oleanolic acid. The oleanolic acid content of the samples was determined by HPLC using calibration curves as reported by Halkes (1998). A Merck-Hitachi (Darmstadt, F.R.G.) HPLC equipment with UV detector (Model Series L-4000, Column: LiChrocart 5 μm RP 18 Select B, 250 mm) and a data processor were used. The mobile phase consisted of acetonitrile, water and acetic acid (70:30:0.5; v/v), the flow rate was 1.0 ml min^{-1}. The signal was detected at 220 nm. The identity of oleanolic acid was checked by co-injection of a reference sample isolated from *Fabiana*. Under the working conditions described, the retention time for oleanolic acid was 14.1 – 14.5 min.

3. Results

The multiplication rates after four weeks for shoot and callus cultures in microcontainers as well as for shoots in TIS are given in table 1. Although the increase of biomass differed considerably between the four species investigated, shoot growth was always highest in the TIS compared to culture in microcontainers.

Multiplication rates of the callus cultures ranged from 3.1 within four weeks for *Cymbopogon* to 8.6 for *Lavandula* (Table 1). These callus cultures were used to establish cell suspensions which in turn were used as inoculum for *Lavandula* and *Hypericum* bioreactor cultures. The growth curves are given in figure 1.

Contents of bioactive compounds for all *in vitro* cultures in comparison with plants grown in the field are shown in table 2. For *Lavandula* the content of rosmarinic acid was higher for shoots and cells grown in any tissue culture system compared to field grown plants (containing 4.8 mg rosmarinic acid per g dry weight). The highest concentration of rosmarinic acid (108 mg g^{-1} dry weight) was found in callus cultures. For *Lavandula* also the composition of the essential oil from field grown as well as from *in vitro* grown shoots was investigated (Figure 2). This analysis revealed that

Table 1: Multiplication rates (increase of biomass) of four plant species producing secondary metabolites in different tissue culture systems within a period of 4 weeks

	Lavandula	*Hypericum*	*Cymbopogon*	*Fabiana*
Shoots from microcontainers	2.8	3.5	9.0	3.2
Shoots from TIS	16.5	7.7	18.0	5.0
Callus culture	8.6	5.0	3.1	8.1

Table 2: Contents of bioactive compounds for *in vitro* cultures of four plant species in comparison to plants grown in the field (concentrations are given in mg g^{-1} dry weight)

Species	*Lavandula*	*Hypericum*	*Cymbopogon*		*Fabiana*
Compound	rosmarinic acid	hypericin	α-citral	β-citral	oleanolic acid
Shoots from microcontainers	55.3	0.15	0.27	0.46	0.01
Shoots from TIS	5.7	0.18	0.35	0.54	0.01
Callus culture	108	0.00	0.00	0.00	0.005
Suspension culture	50	0.00	0.00	0.00	0.004
Bioreactor culture	60	0.00			
Field plant	4.8	0.52	3.95	10.25	0.28

Table 3: Concentration of bioactive compounds in plant material of *Lavandula* and *Hypericum* from TIS vessels. Material was fractionated according to the colour of the material: Depending on the position of the material within the vessel, the biomass was green (top and side layers), yellow (inner part of the vessel) or brown (bottom layer)

Colour of plant material fractions	Content of rosmarinic acid in *Lavandula* (mg g^{-1}dry weight)	Content of hypericin in *Hypericum* (mg g^{-1} dry weight)
green (top layer)	7.5	0.12
green (side layer)	4.0	
yellow	1.9	0.11
brown	0.0	0.23

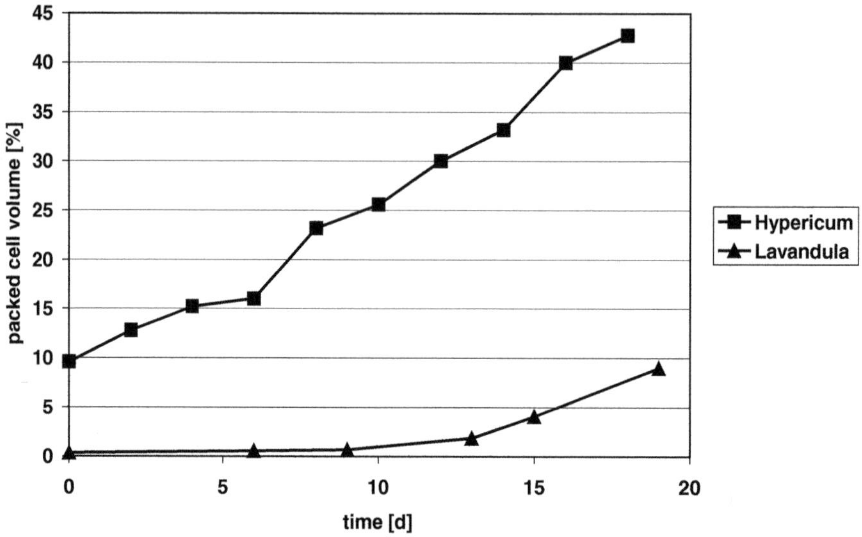

Figure 1: Growth of *Hypericum* and *Lavandula* suspension cultures in bioreactors. Growth curves were obtained by determining the packed cell volume (percentage of suspension volume occupied by settled cells).

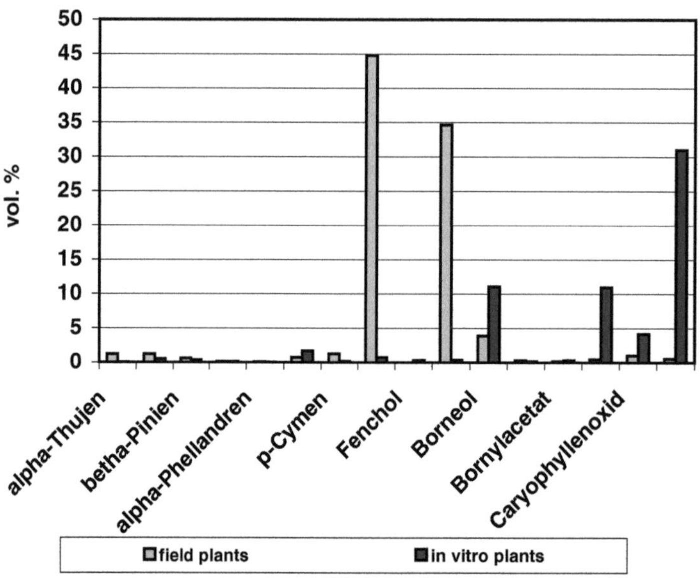

Figure 2: Composition of the essential oil, determined by GC-MS, from field-grown and *in vitro*-grown *Lavandula* shoots.

the composition varied considerably. Some of the compounds were more concentrated in the *in vitro* material compared to the field grown plant.

For the other three species tested the content of bioactive compounds was always lower in the *in vitro* cultures compared to field grown plants. However, comparing the different *in vitro* culture systems, the content was higher in shoot cultures compared to cell cultures. Moreover, the biomass of *Lavandula* and *Hypericum* from TIS vessels was fractionated according to the colour of the material: depending on the position of the material within the vessel, the biomass was green (top and side layers), yellow (inner part of the vessel) or brown (bottom layer). The result of the analysis of the different fractions is shown in table 3. For *Lavandula* the highest concentration of rosmarinic acid (7.5 mg g^{-1} dry weight) was found in the green top layer and nothing in the brown biomass from the vessel centre. In contrast, the highest content of hypericin in the biomass of *Hypericum* was found in brown material (0.23 mg g^{-1} dry weight) and lower concentrations in the yellow (0.11 mg g^{-1} dry weight) and green (0.12 mg g^{-1} dry weight) biomass.

4. Discussion

The *in vitro* production of plant secondary metabolites as bioactive compounds for the pharmaceutical and cosmetic industry presents several advantages compared to the extraction of these compounds from field grown plants: *in vitro* cultures are independent of season, climate and weather and there is no risk of crop failure due to natural hazards. Moreover, the danger of extinction of species due to over collection from natural populations is avoided (Bhojwani and Razdan, 1996). Starting in the nineteen seventies, much work has been carried out on the use of plant cell cultures for the production of secondary metabolites (for reviews see e.g Charlwood and Rhodes, 1990; Fowler, 1992; Bhojwani and Razdan, 1996), however, only few products from cell cultures have been commercialised so far, e.g. shikonin from *Lithospermum erythrorhizon* (Fujita, 1988, 1990). In most cases, cell culture has not become a cost-effective technology, since scale-up in bioreactors is expensive and the yield of metabolites from cell cultures is often low and unstable (Bhojwani and Razdan, 1996). Therefore, we cultivated plant shoots in temporary immersion systems to compare with sophisticated bioreactor technology, TIS are low-cost culture systems, moreover shoots are genetically and physiologically more stable compared to cell cultures.

Working with the model species *Lavandula*, *Hypericum*, *Cymbopogon* and *Fabiana*, there were considerable differences between the content of bioactive compounds of plant material grown in different *in vitro* culture systems and the field grown plants. For *Lavandula*, the content of rosmarinic acid was higher in *in vitro* grown plant material, especially in callus and cell cultures. This corresponds to results of Ulbrich et al. (1985) who produced rosmarinic acid in cell cultures of *Coleus blumei* in a concentration that was 7 times higher compared to whole plants. The same is also true for other compounds e.g. as early as 1977 Zenk et al. reported higher production of ajmalicine in cell cultures of *Catharanthus roseus* compared to whole plants and also shikonin is more concentrated in cell cultures of *Lithospermum erythrorhizon* (Fujita, 1988). For *Hypericum*, *Cymbopogon* and *Fabiana* the concentration of the desired compounds was lower compared to field grown plants. This is a well known phenomenon and attributed to the fact that for many compounds plant growth and the production of secondary metabolites are inversly related (Bhojwani and Radzan, 1996). In the case of *Hypericum* this hypothesis is confirmed by our data analysing different biomass fractions from temporary immersion vessels: the highest content of hypericin was found in yellow and brown biomass indicating that during the transition from active plant growth to the stationary phase the corresponding enzymes involved are induced. For *Lavandula*, in the same experiment the highest content of rosmarinic acid was found in actively growing green biomass. This corresponds well with the data discussed above that for this plant species, fast growing cell cultures produce more rosmarinic acid compared to whole plants. Thus, rosmarinic acid can be obtained from *Lavandula* cultures in the exponential growth phase and, therefore, an optimised continuous suspension culture system can be developed. However, for *Hypericum* it might be useful to establish a two-stage-culture system with a first stage of excessive plant growth and a second stage for secondary product formation induced by a change of the physical or chemical growth conditions.

An important result of our investigations is that in the case of plants with low product yields in *in vitro* cultures (*Hypericum, Cymbopogon, Fabiana*) the concentration of the desired compounds was always higher in shoot cultures compared to cell or callus cultures. All of our model plants could be grown with high multiplication rates in TIS, these systems represent an attractive alternative to cell cultures, especially since environmental growth parameters still remain to be optimised, which will probably further increase the product yield. For example, Tisserat and Vaughn (2001) have recently shown, that the concentration of thymol in *in vitro* grown thyme plants could be increased 317-fold by growing the plants under high CO_2-levels. Zobayed et al. (2003) demonstrated that the concentration of hypericin-pseudohypericin and hyperforin in bioreactor culture of *Hypericum perforatum* can be altered selectively by elevated carbon supply. A higher concentration of sucrose 45 g l^{-1} in the culture medium resulted in an increased content of hyperforin but a decreased content of hypericin/pseudohypericin. Thus it is possible to create specific spectra of bioactive compounds by choosing appropriate cultivation conditions.

For many compounds the economic aspect of the production in plant cells grown in bioreactors has been discussed (e.g. Scragg, 1986; Lambie, 1990; Moreno et al., 1995) concluding that it is of commercial value only for very costly compounds. A main reason for this are the high costs of bioreactor equipment which is around 20,000 Euro for a fully equipped 5-litre standard stirred tank bioreactor. In contrast, the material costs for a 5-litre temporary immersion vessel as used in our laboratory including all technical equipment for automated operation is only 50 Euro. Thus, our data show that the production of plant secondary metabolites in TIS represent an attractive alternative both compared to biomass production in the field or collected from nature as well as in comparison to the costly and often non economic production in bioreactor cell cultures.

Acknowledgements

The authors wish to thank the German Ministry for Education and Research and the Free State Saxony and the European Community for financial support. Performance of analyses and bioreactor cultures by SIAB (Saxon Institute of Applied Biotechnology) is also gratefully acknowledged.

References

Akula A, Becker D & Bateson M (2000) High-yielding somatic embryogenesis and plant recovery in a selected tea clone, 'TRI-2025', by temporary immersion. Plant Cell Rep. 19: 1140-1145

Berlin J, Sieg S, Strack D, Bokern M & Harms H (1986) Production of betalains by suspension cultures of *Chenopodium rubrum*. Plant Cell, Tiss. Org. Cult. 5: 163-174

Bhojwani SS & Razdan MK (1996) Plant Tissue Culture: Theory and Practice, a revised edition. Elsevier, Amsterdam, Lausanne, New York, Oxford, Shannon, Tokyo

Charlwood BV & Rhodes MJC (1990) Secondary Products from Plant Tissue Culture. Clarendon Press, Oxford

De-Eknamkul W & Ellis B (1984) Rosmarinic acid production and growth characterization of *Anchusa officinalis* cell suspension cultures. Planta Med. 50: 346-350

Drapeau D, Blanch HW & Wilke CR (1987) Ajmalicine, serpentine, and catharanthine accumulation in *Catharanthus roseus* bioreactor cultures. Planta Med. 53: 373-376

European Pharmacopeia (2002) 4th Edition, Council of Europe, 67075 Strasbourg Cedex, France 2001, ISBN: 92-871-4587-3: 183-184

Fowler MW (1992) Plant cell culture, process systems and product synthesis. In: Fowler MW, Warren GS & Moo-Young M (eds) Plant Biotechnology: Comprehensive Biotechnology, Second Supplement (pp. 79-88)

Fujita Y (1988) Shikonin: production by plant (Lithospermum erythrorizon) cell cultures. In: Bajaj YPS (ed) Biotechnology in Agriculture and Forestry Vol. 4. Medicinal and Aromatic Plants 1 (pp. 225-236). Springer, Berlin

Fujita Y (1990) The production of industrial compounds. In: Bhojwani SS (ed) Plant Tissue Culture: Applications and Limitations (pp. 259-275). Elsevier, Amsterdam

Fujita Y, Tabata M, Nishi A & Yamada Y (1982) New medium and production of secondary compounds with the two-staged culture method. In: Fujiwara A (ed) Plant Tissue Culture (pp. 399-400) Jpn. Assoc. Plant Tissue Cult. Tokyo

Gracza L & Ruff P (1984) Rosmarinsäure in Arzneibuchdrogen und ihre HPLC-Bestimmung. Arch. Pharm. 317: 339-345

Halkes SBA (1998) *Filipendula ulmaria*: a study on the immunomodulatory activity of extracts and constituents. PhD Thesis, Chapter 5 (pp. 91-92). University Utrecht, The Netherlands

Jardin B, Tom R, Chavarie C, Rho D & Archambault J (1991) Stimulated indole alkaloid release from *Catharanthus roseus* immobilized cultures. Initial studies. J. Biotechnol. 21: 43-62

Jiménez E, Pérez N, de Feria M, Barbón R, Capote A, Chávez M, Quiala E & Pérez JC (1999) Improved production of potato microtubers using a temporary immersion system. Plant Cell, Tiss. Org. Cult. 59: 19-23

Kartnig T, Gröbel I & Heydel B (1996) Production of hypericin, pseudohypericin and flavanoids in cell cultures of various *Hypericum* species and their chemotypes. Planta Med. 62: 51-53

Lambie AJ (1990) Commercial aspects of the production of secondary compounds by immobilized plant cells. In: Charlwood BV & Rhodes MJC (eds) Secondary Products from Plant Tissue Culture (pp. 265-278). Clarendon Press, Oxford

Lopez-Arnaldos T, Lopez-Serrano M, Ros Barcelo A, Calderon AA & Zapata JM (1994) Tentative evidence of a rosmarinic acid peroxidase in cell cultures from lavandin (*Lavandula intermediata*) flowers. Biochem. Molecular Biol. Internat. 34: 809-816

Lorenzo JC, González BL, Escalona M, Teisson C, Espinosa P & Borroto C (1998) Sugarcane shoot formation in an improved temporary immersion system. Plant Cell, Tiss. Org. Cult. 54: 197-200

Moreno PRH, van der Heijden R & Verpoorte R (1995) Cell and tissue culture of *Catharanthus roseus*: a literature survey. II. Updating from 1988 to 1993. Plant Cell, Tiss. Org. Cult. 42: 1-25

Murashige T & Skoog F (1962) A revised medium for rapid growth and bioassays with tobacco tissue cultures. Physiol. Plant 15: 473-497

Payne GF, Payne NN, Shuler ML & Asada M (1988) *In situ* adsorption for enhanced alkaloid production by *Catharanthus roseus*. Biotechnol. Lett. 10: 187-192

Schiel O & Berlin J (1987) Large scale fermentation and alkaloid production of cell suspension cultures of *Catharanthus roseus*. Plant Cell, Tiss. Org. Cult. 8: 153-161

Schlatmann JE, Moreno PRH, Ten Hoopen HJG, Verpoorte R & Heijnen JJ (1994) Effect of oxygen and nutrient limitation on ajmalicine production and related enzyme activities in high density cultures of *Catharanthus roseus*. Biotechnol. Bioeng. 44: 461-468

Schlatmann JE, Ten Hoopen HJG & Heijnen JJ (1992) Optimization of the medium composition for alkaloid production of *Catharanthus roseus* using statistical experimental designs. Med. Fac. Landbouw., Univ. Gent 57: 1567-1569

Scragg AH (1986) The economics of mass cell culture. In: Morris P (ed) Secondary Metabolism in Plant Cell Cultures (pp. 202-207). Cambridge University Press, Cambridge

Tisserat B & Vaughn SF (2001) Essential oils enhanced by ultra-high carbon dioxide levels from *Lamiaceae* species grown *in vitro* and *in vivo*. Plant Cell Rep. 20: 361-368

Ulbrich B, Weisner W & Arens H (1985) Large scale production of rosmarinic acid from plant cell culture of *Coleus blumei*. In: Neumann KH (ed) Primary and Secondary Metabolism of Plant Cell Cultures (pp. 293-303). Springer, Berlin

Zenk MH, El-Shagi H, Arens H, Stockigt J, Weiler EW & Deus D (1977) Formation of the indole alkaloids serpentine and ajmalicine in cell suspension cultures of *Catharanthus roseus*. In: Barz WRE Reinhard E & Zenk MH (ed) Plant Tissue Culture and its Biotechnological Application (pp. 27-44). Springer, Berlin

Zobayed SMA, Murch SJ, Rupasinghe HPV & Saxena PK (2003) Elevated carbon supply altered hypericin and hyperforin contents of St. John's wort (*Hypericum perforatum*) grown in bioreactors. Plant Cell, Tiss. Org. Cult. 75: 143-149

Chapter 40

Cultivation of root cultures of *Panax ginseng* in different bioreactors and in temporary immersion - Comparison of growth and saponin production

Tomáš Vaněk, Lenka Langhansová* & Petr Maršík
Institute of Organic Chemistry and Biochemistry, AS CR, Flemingovo nám. 2, 166 10, Praha 6, Czech Republic.
**Author for correspondence; Fax: +420-220183574; E-mail: lenka@uochb.cas.cz*

Abstract: Different systems of large-scale cultivation of multiple adventitious roots of Panax ginseng C. A. Meyer were compared to cultivation in Erlenmeyer flasks. Adventitious roots were isolated from plantlets regenerated from somatic embryos and cultivated separately in liquid media. Multiplication of adventitious roots was performed in liquid Schenk and Hildebrandt (1972) medium containing 3 % sucrose, and 24.6 μmol indole-3-butyric acid. The highest saponin content of 28.51 mg g^{-1} of dry weight was found in adventitious roots cultivated in the RITA® temporary immersion system (TIS). The best production of biomass was achieved in RITA® vessels and standard Erlenmeyer flasks placed on rotary shaker, followed by the Applikon 3-litre bioreactor and a simple airlift reactor. Saponin production in Erlenmeyer flasks was 10.07 mg g^{-1} of the dry weight while the production in the Applikon 3-litre bioreactor was only 3.60 mg g^{-1}. Other bioreactor systems tested showed neither significant saponin production nor high biomass production.

Key words: bioreactor, saponin, *Panax ginseng*, adventitious roots

Abbreviations: IBA – indole-3-butyric acid; MS – Murashige and Skoog (1962); NAA – naphthalene acetic acid; SH – Schenk and Hildebrandt (1972); TIS – temporary immersion system

1. Introduction

Panax ginseng C. A. Meyer (*Araliaceae*) is an herbaceous plant, which in Oriental medicine has a strong reputation since ancient times for being a tonic, regenerating, and rejuvenating, even though its pharmacological activity has not been fully elucidated. Ginseng grows wild in mountain areas, from Nepal to Manchuria, and from eastern Siberia to Korea but has been

A.K. Hvoslef-Eide and W. Preil (eds.), Liquid Culture Systems for in vitro Plant Propagation, 539–546.

overexploited. Recently it was reported that ginsenosides and polyacetylenes isolated from ginseng roots have cytotoxic activity (Newal et al., 1996).

The current supply of ginseng mainly depends on field cultivation, which is a long and laborious process. Native ginseng plants need 5 – 7 years growth prior to harvest and the content of ginsenosides is low. *In vitro* mass production in large-scale systems seems to be potentially a more efficient alternative for the production of the bioactive components of ginseng.

In our previous study we compared saponin production in different ginseng tissue cultures namely callus, suspension culture and adventitious roots. We found a full range of ginsenosides distributed analogous as in the roots of native plants in adventitious roots cultivated in liquid Schenk and Hildebrandt medium supplemented with 24.6 μmol IBA while the saponin content in callus as well as in cell suspension cultures was limited to major ginsenosides, Rb1 or Rg1 (Langhansová et al., 2002; Langhansova et al., 2003b).

The aim of this study was to establish *in vitro* cultures producing ginsenosides from adventitious roots. We compared different cultivation systems to find a way for more efficient large-scale production.

2. Material and methods

2.1 Adventitious root induction

Adventitious roots of *Panax ginseng* were isolated from plantlets regenerated from somatic embryos and rooted on 1/3-strength MS medium (Langhansova et al., 2003a). Separated roots were transferred to liquid MS medium supplemented with 1 μmol NAA and cultivation was carried out in 500 ml Erlenmeyer flasks on a rotary shaker (125 rpm). Root proliferation occurred on liquid SH medium supplemented with 24.6 μmol IBA (Choi et al., 2000).

2.2 Bioreactor cultures

Adventitious roots cultures were cultivated in different bioreactors and temporary immersion systems (TIS) and in standard cultivation system in Erlenmeyer flasks:

1. Rocking temporary immersion system of our own construction (Figure 1b)
2. Simple airlift bioreactor (Figure 2b)
3. "Mafe" ½ l Bioreactor (Figure 3b)
4. RITA® – temporary immersion system (Figure 4b)
5. "Applikon" Bioreactor (Applikon, Netherlands) with total volume of 3 litre (Figure 5b)
6. Control: 250 ml Erlenmeyer flasks on a rotary shaker (125 rpm) (Figure 6b)

The roots were inoculated in culture vessels according to vessel volume (approx. 1 g per 100 ml) and cultivated for period of two months with the exception of simple airlift reactor where the roots were cultivated for three months. Cultivation was in dark at 24 ± 1 °C.

The Rocking TIS of our own construction is made of steel vessel (width 150 mm, length 240 mm, high 50 mm). Roots were placed on polyurethane foam floating loose in liquid medium in the reactor. Mixing was provided by rocking the reactor (Figure 1b).

The simple airlift bubble reactor made from a glass separation funnel was of a simple conical shape with gas-sparged mixing. The aeration was provided by air entering by a glass pipe from the top opening through a sparger at the bottom and as the air bubbles rise the biomass is lifted and the oxygen required is provided (Figure 2b).

"Mafe" (New Brunswick Scientific Co., INC., USA), a simple ½-litre bioreactor (i.d. 90 mm, height 160 mm) was assembled from an agitator, agitated by magnetic mixer, and a semipermeable silicone tubing system providing aeration. The culture was placed on the top of the supporting plate, which was perforated (10 holes (Ø 1 mm) per cm^2) and mounted 10 mm under medium level. The agitation ensured regular distribution of nutrients and dissolved gasses in the medium (Nepovim and Vanek, 1998) (Figure 3b).

RITA® (Vitropic s.a., France), temporary immersion system. Roots were placed on a polyurethane foam disc fixed in the upper container. The upper basket is mounted above a bell immersed in liquid medium. Flooding was set to 5 min h^{-1} and is provided by pressure applied to sterile air in the lower container which pushes the liquid medium into the upper container holding the roots for 5 min (Figure 4b).

In the "Applikon®" (Applikon, The Netherlands) glass autoclavable bioreactor with a stirred tank, the aeration was provided by a steel pipe ending in a sparger at the bottom of a glass tank. Mixing was set to 60 rpm (Figure 5b).

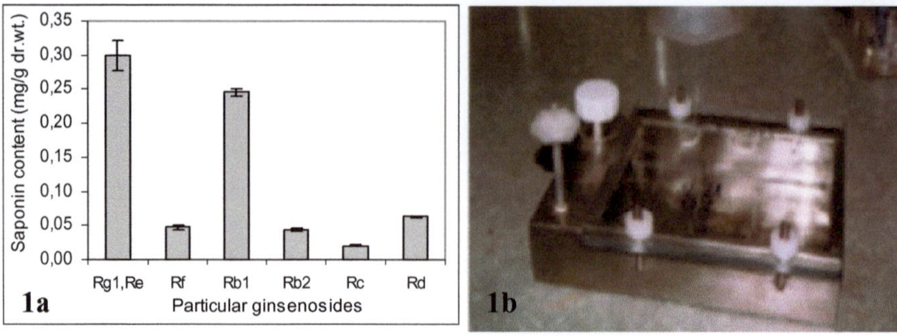

Figure 1: Results from (la) rocking TIS of our own construction (1b)

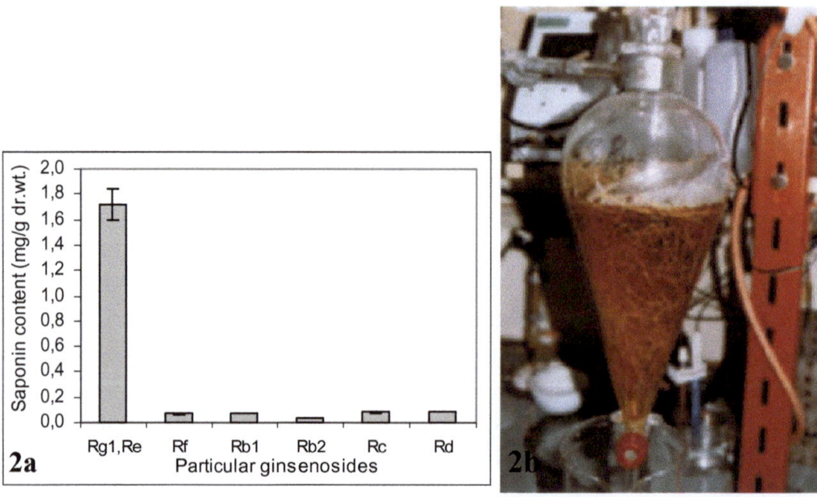

Figure 2: *Results from (2a) s*imple airlift bioreactor of conical shape made of separation funnel (2b)

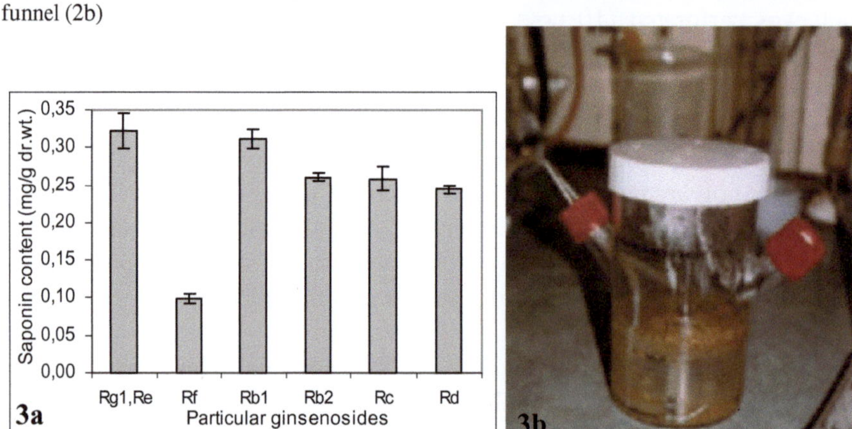

Figure 3: Results from (3a) MAFE, ½ litre Bioreactor (3b)

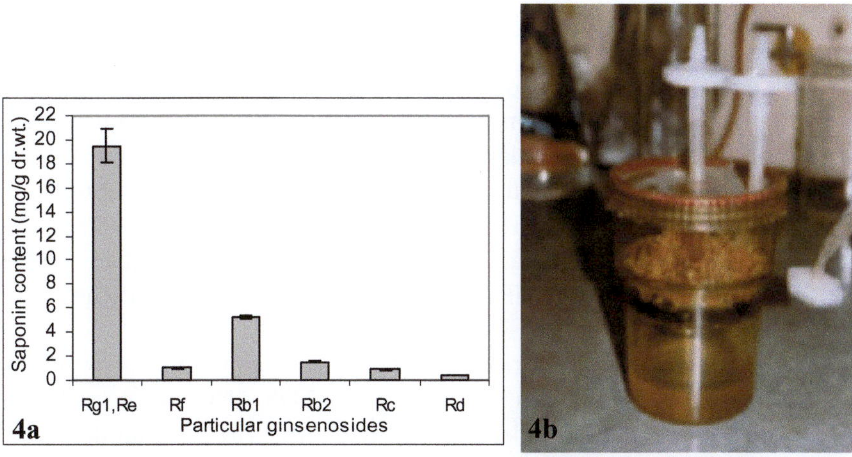

Figure 4: Results from (4a) RITA® – Temporary immersion system (4b)

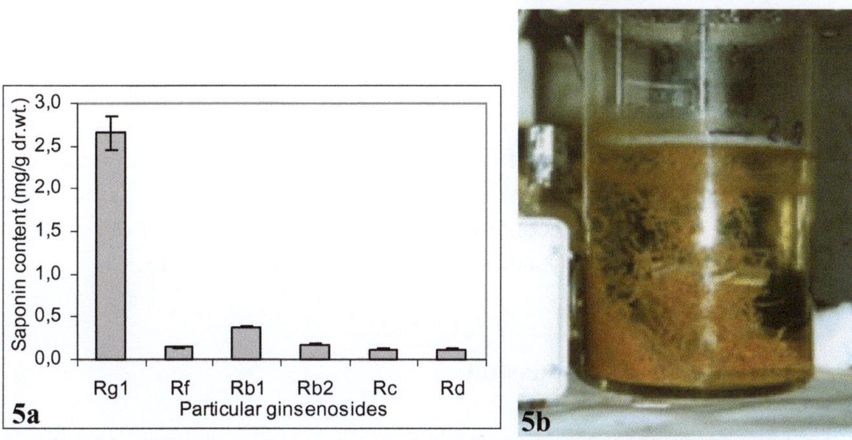

Figure 5: Results from (5a) the “Applikon” Bioreactor (Applikon, Netherlands) with total volume of 3-litre (5b)

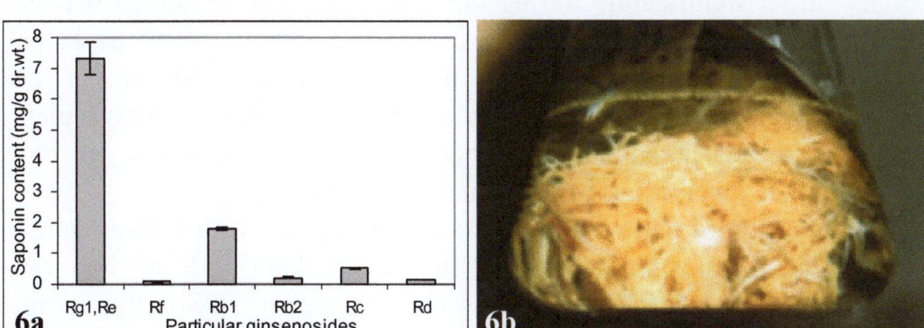

Figure 6: Results from (6a) 250 ml Erlenmeyer flasks on rotary shaker (125 rpm) (6b)

2.3 Detection of ginsenosides by HPLC

Adventitious roots were homogenized and extracted with methanol (7 ml g^{-1} fresh weight) for 5 days at room temperature. The sample was then filtered, and evaporated to dryness under vacuum. The residue of extract was re-dissolved in distilled water and partitioned with dietyl-ether, and twice in n-BuOH saturated with water. The n-BuOH layer was concentrated *in vacuo* to obtain the crude saponin fraction (Sanada, 1974). The n-BuOH soluble fraction was analyzed by HPLC for detection and quantification of ginsenosides Rb_1, Rb_2, Rc, Rd, Re, Rf and Rg_1.

HPLC analyses were performed using a system consisting of two high pressure pumps (DeltaChrom, SDS 020 a SDS 030) with a mixer (SunChrom GmbH) and PDA detector (JASCO, MD 1510); a stainless steel column (250 x 4 mm) packed by reverse phase Si-C18, 7 µm (Biospher); flow-rate 1 ml min^{-1}. The injection volume was set up at 20 µl in the autosampler (TSP, AS300). Eluents: (A) 15 % acetonitrile and water, (B) 100 % acetonitrile; Gradient elution profile: 0 – 40 min, 0 – 35 % B; 40 to 45 min, 35 % B. The peaks were monitored by UV detection at 203 nm (Soldati and Sticher, 1980; Pietta et al., 1986; Petersen and Palmqvist, 1990).

Each ginsenoside was identified by comparison of retention time and UV spectra with authentic ginsenosides purchased from Carl Roth GmbH and Co., Germany. Ginsenoside content was expressed in mg g^{-1} of dry weight. Presence of ginsenosides was additionally confirmed by LC-MS.

3. Results and discussion

We have monitored ginsenoside production in callus cultures previously (Langhansova et al., 2002). We found callus cultures to have a high proliferation rate and also high yielding. However, the production of saponins varied significantly during the year. The content ranged between 0 – 2 % dry weight and the saponin content in callus as well as in cell suspension cultures was limited to two major ginsenosides, Rb_1 or Rg_1. The full range of ginsenosides distribution analogous to roots of native plants was only detected in adventitious roots. Hence, in this study we have concentrated on adventitious root production *in vitro*.

The proportion of particular ginsenosides compared with native material is almost identical after cultivation in the rocking TIS of our own construction (Figure 1a). This suggests that the rocking TIS is the optimal way of producing these adventitious roots *in vitro*. The range in protopanaxadiol (ginsenosides: Rb_1, Rb_2, Rc and Rd) and protopanaxatriol (ginsenosides: Re and Rg_1) groups were similar also in RITA® TIS

(Figure 4a) and in Erlenmeyer flasks (Figure 6a). In roots cultivated in the simple airlift bioreactor (Figure 2a) and in Applikon bioreactor (Figure 5a), we observed inhibition of the protopanaxadiol group while in ½-litre "Mafe" bioreactor the biosynthesis of the protopanaxadiol group was significantly stimulated (Figure 3a).

Total saponin content in the root of nature Panax ginseng is commonly in the range of 1 - 3 % dry weight (Bruneton, 1995). The highest content of 28.5 mg g^{-1} dry weight (2.85 %) of ginsenosides (Rb_1, Rb_2, Rc, Rd, Re, Rf and Rg_1) in adventitious roots was achieved in RITA® TIS (Figure 4a). However, in order to reach effective production we have to combine two factors,; both the saponin content and the biomass growth. The biomass growth in RITA® TIS was high and comparable to cultivation in the maintenance system of 500 ml Erlenmeyer flasks. High growth was achieved also in roots cultivated in the Applikon 3-litre bioreactor and in the simple airlift reactor. Growth values in the rocking TIS and in the "Mafe" bioreactor were considerably lower (Figure 7). We considered these systems except RITA® TIS not suitable for ginseng adventitious roots biomass production because of the distribution of the different saponins and because the growth rate was much too low to be economical.

In conclusion, here we found root culture in the RITA® TIS optimal for ginsenoside production. It is possible to use Erlenmeyer flasks system as a proliferating step since this also gives a high growth rate, but sadly not the right profile of the saponins. Our next investigation will lead toward developing a better production system which should be based on establishment of

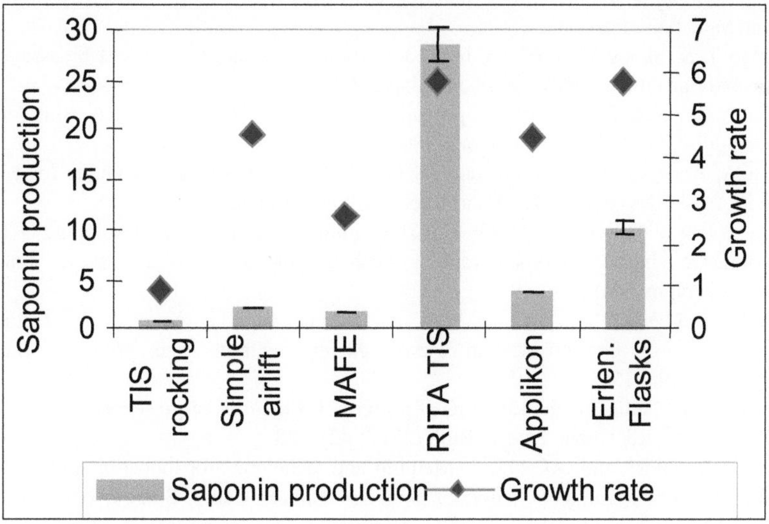

Figure 7: Total production of saponins and biomass growth in different bioreactor systems.

media composition in order to increase a biomass growth using the Erlenmeyer flask system in a proliferation step and to enhance ginsenoside production using RITA® TIS in following production step.

Acknowledgement

This work was supported by COST 843.10 and 521/02/P064 grant.

References

Bruneton J (1995) Pharmacognosy, Phytochemistry, Medicinal Plants (p. 915) Lavoisier Publishing

Choi SM, Son SH, Yun SR, Kwon OW, Seon JH & Paek KY (2000) Pilot-scale culture of adventitious roots of ginseng in a bioreactor system. Plant Cell, Tiss. Org. Cult. 62 (3): 187-193

Inomata S, Yokoyama M, Gozu Y, Shimizu T & Yanagi M (1993) Growth pattern and ginsenoside production of Agrobacterium - transformed Panax ginseng roots. Plant Cell Rep. 12: 681-686

Langhansová L, Nepovím A, Maršík P & Vaněk T (2002) Saponin production from *in vitro* cell and root cultures of *Panax ginseng* C. A. Meyer. The 10th IAPTC&B Congres. Plant Biotechnology: 2002 and Beyond. (82-A) Orlando, USA

Langhansova L, Konradova H & Vanek T (2003a) Polyethylene glycol and abscisic acid improves maturation and regeneration of *Panax ginseng* somatic embryos. Plant Cell Rep. (in press)

Langhansova L, Vanek T & Marsik P (2003b) Production of saponins from *Panax ginseng* suspension and adventitious roots cultures and its „scale up" to laboratory 3-l bioreactor. Planta Med. (submitted)

Murashige T & Skoog F (1962) A revised medium for rapid growth and bioassays with tobacco tissue cultures. Physiol. Plant. 15:473-497

Nepovim A & Vanek T 1998. *In vitro* propagation of *Stevia rebaudiana* plants using multiple shoot culture. Planta Med. 64 (8): 683-780

Newall CA, Anderson LA & Phillipson JD (1996) Herbal Medicines. A Guide for Health–care Professionals (p. 296) The Pharmaceutical Press. London

Petersen TG & Palmqvist B (1990) Utilizing column selectivity in developing a high - performance liquid chromatographic method for ginsenoside assay. Jour. of Chromatography 504: 139-149

Pietta P & Rava PM (1986) Improved high - performance liquid chromatographic method for the analysis of ginsenosides in *Panax ginseng* extracts and products. Jour. of Chromatography 356: 212-219

Sanada S (1974) Studies on the saponins of ginseng. I Structures of ginsenoside - Ro,- Rb_1, - Rb_2, - Rc and - Rd. Chem. Pharm. Bull. (22) 2: 421-428

Schenk RU & Hildebrandt AC (1972) Medium and techniques for induction and growth of monocotyledonous and dicotyledonous plant cell cultures. Can. J. Bot. 50: 199-204

Soldati F & Sticher O (1980) HPLC Separation and quantitative determination of ginsenosides from *Panax ginseng*, *Panax quinquefolium* and from ginseng Drug Preparations. Planta Medica 38: 348-357

Chapter 41

Optimisation of *Panax ginseng* liquid cell cultures for biomass accumulation and ginsenoside production

C. Kevers[1*], G. Bare[2], T. Gaspar[3], P. Thonart[4] & J. Dommes[5]
[1, 3, 5]*Plant Molecular Biology and Hormonology, Institute of Botany B 22, University of Liège, Sart Tilman, BE-4000 Liège, Belgium.*
**Corresponding author: Fax: 32-4-366 38 72; E-mail: c.kevers@ulg.ac.be;*
[2, 4]*CWBI, B 40, University of Liège, Sart Tilman, BE-4000 Liège, Belgium*

Abstract: Solid calli and derived liquid cell cultures were initiated from one-year-old roots of *Panax ginseng* CA Meyer. Half-strength Murashige and Skoog medium supplemented classically with an auxin and a cytokinin did not appear favourable for biomass accumulation nor for a high ginsenoside content. Changes in the levels of mineral nutrients, sucrose and growth regulators were preliminary investigated here to improve growth and ginsenoside production in liquid cultures. The hypothesis that ginseng cells released growth inhibitors in the medium was not supported by the results obtained in experiments involving frequent transfers to fresh growth medium.

Key words: auxin, benzo[b]selenienyl acetic acid, biomass, cytokinin, ginseng, inoculum size, medium composition

Abbreviations: BSAA – benzo[b]selenienyl acetic acid; 2,4-D – 2,4-dichlorophenoxyacetic acid; DW – dry weight; HPTLC – high performance thin layer chromatography; Kin – kinetin; MS – Murashige and Skoog medium (1962); ZR - zeatin riboside

1. Introduction

Panax root has been used in Oriental medicine since ancient times. The crude root extract is known to have tonic, stimulatory and adaptogenic properties (Hu, 1976) due to the presence of a wide range of saponins and sapogenins (Li, 1995). Recently, ginseng has become a popular tonic and health food complement in Western countries. Therefore, the demand for the plant has increased dramatically worldwide. Ginseng is expensive because of its long-term conventional (5-7 years) and troublesome production cycle. As

A.K. Hvoslef-Eide and W. Preil (eds.), Liquid Culture Systems for in vitro Plant Propagation, 547–555.

a result, propagation methods of ginseng by plant tissue culture and particularly by somatic embryogenesis have been investigated. Somatic embryogenesis has been successfully induced on solid media, from derived calli of root (Chang and Hsing, 1980, Asaka et al., 1992, 1993a, Tirajoh et al., 1998), leaf (Tirajoh et al., 1998) or flower bud (Shoyama et al., 1997), or directly from zygotic embryos (Choi and Soh, 1996) or cotyledons (Choi and Soh, 1994). Culture of ginseng tissues in bioreactors was developed in order to produce fresh material containing ginsenoside saponins (Asaka et al., 1993b).

With the same purposes, we already have developed techniques for plantlet multiplication through somatic embryogenesis (Kevers et al., 2000, Monteiro et al., 2002) and for biomass production through *in vitro* hairy root cultures (Kevers et al., 1999). In the present work, we investigated liquid cell cultures as a tool for biomass accumulation and ginsenoside production.

2. Material and methods

One-year-old roots of *Panax ginseng* CA Meyer (from a Belgian field culture; provided by ORTIS Laboratories, Elsenborn, Belgium) were surface sterilised by successive incubations in ethanol (70%) for 3 min and in sodium hypochlorite (3%) for 20 min, and rinsed three times with sterile distilled water. Calluses were initiated from 3 mm long root sections that were cultured on a solid MS (Murashige and Skoog, 1962) basal medium supplemented with 2,4-dichlorophenoxyacetic acid (2,4-D 0.3 mg l^{-1}) and kinetin (Kin 0.1 mg l^{-1}). The cultures were incubated in darkness at 25 ± 2 °C in 9-cm diameter Petri dishes for 6 weeks. The callus was subcultured every four weeks on a solid MS basal medium, supplemented with 2,4-D (1 mg l^{-1}) and Kin (0.1 mg l^{-1}) in darkness.

Finely minced callus (0.3 to 10 g) was transferred to liquid medium in 100 ml conical flasks containing 50 ml of half strength MS (MS/2) medium supplemented with auxin 3-benzo(b)selenienyl acetic acid (BSAA) or 2,4-D, 0.3 or 1 mg l^{-1} and cytokinin (kinetin or zeatin riboside, ZR, 0.2 mg l^{-1}) for one month with shaking (80 rpm) at 25 ± 2 °C in darkness. To evaluate the effect of varying concentration of some mineral nutrients (KNO_3, KH_2PO_4, NH_4NO_3, FeNaEDTA) and sucrose in the medium on the growth and the ginsenoside content of the callus, an experiment based on the statistical design of Plackett-Burman (Plackett and Burman, 1946) was performed. The method allows to screen a large number of factors to identify those that may be important, with the least number of observations. One way to design such

Table 1: Lower, medium and higher values of concentration (in g l^{-1}) used for the different parameters studied in the statistical design of Plackett-Burman

	Lower values	Medium values	Higher values
KH_2PO_4	0.34	1.7	3.4
KNO_3	4	19	38
NH_4NO_3	3.3	16.5	33
FeNaEDTA	0.81	4	20
Sucrose	6	30	90

experiments is to confront all interactions with main effects. Table 1 indicates the values (lower and higher than normal) used for the experiments. This method allows the evaluation of the random error variability (confident limits), and the statistical significance of the measured growth rates and ginsenoside contents (m_i).

To evaluate the weight of the biomass at the end of the assay, the culture was filtered through inox filter (2.5 μm pore diameter) and the biomass was lyophilised.

Ginsenosides were analysed in the dry lyophilised plant material and in the culture medium. Dry biomass was dissolved in methanol (70%). Culture medium was mixed (1:1) with methanol. All samples were boiled during one hour. After centrifugation, the samples were applied to a sep-pak C18 cartridge. The ginsenosides were eluted with methanol (4 ml). The solvent was evaporated to dryness under vacuum and the residue was dissolved in 250μl methanol. The samples were analysed by HPTLC (Camag automatic TLC sampler and a Desaga CD60 scanner). 5 μl of sample or standard (the standards, ginsenosides Rb1, Rb2, Rc, Rd, Re, Rg1 and Rf, were provided by the company Extrasynthese, France) were loaded on HPTLC silica gel 60 (Merck 13748) plates. The plates were developed in a mixture of chloroform/ isobutanol/ethanol/H_2O (20/40/15/20).

All results (except those of the Plackett-Burman design) are the means of measurements in at least three independent experiments.

3. Results

The accumulation of biomass was monitored during four weeks after inoculation of 0.3 to 10 g of minced callus in liquid MS/2 medium. A 200% increase in four weeks was observed with the lowest inoculum size (0.3 g). The rate of biomass accumulation was lower and lower with increasing volume of inoculum. Inoculation of 10 g of callus led to an increase in biomass of only 4% (Figure 1).

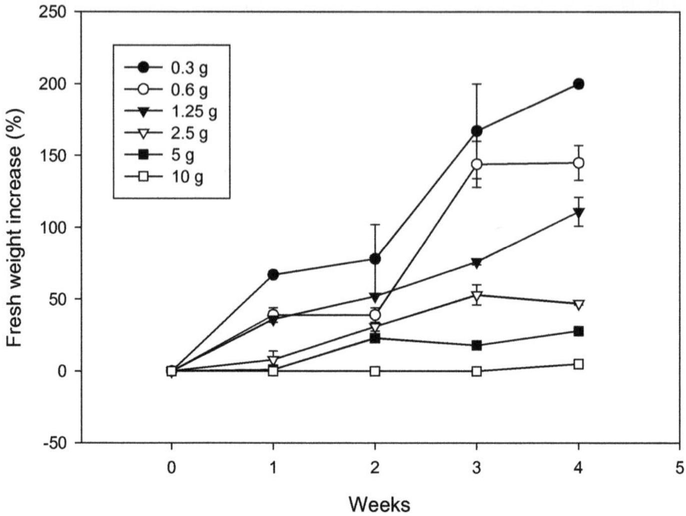

Figure 1: Fresh weight increase of ginseng cells in liquid cultures inoculated with different amounts of callus (0.3 to 10 g in 50 ml of medium).

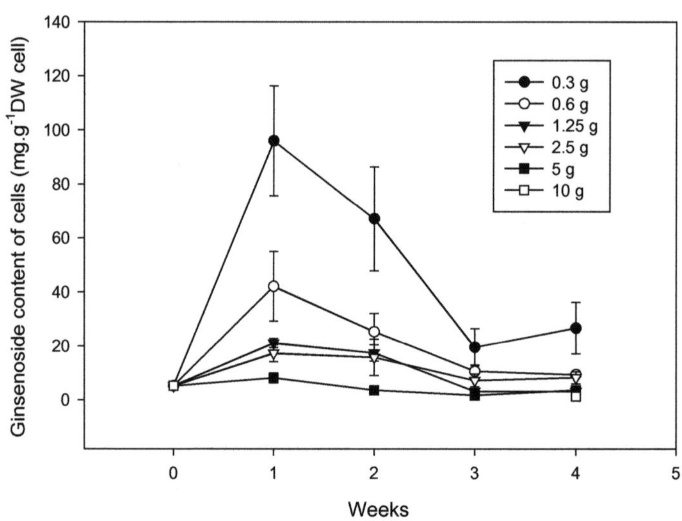

Figure 2: Changes in the ginsenoside content of ginseng cells in liquid cultures inoculated with different amounts of callus (0.3 to 10 g in 50 ml of medium).

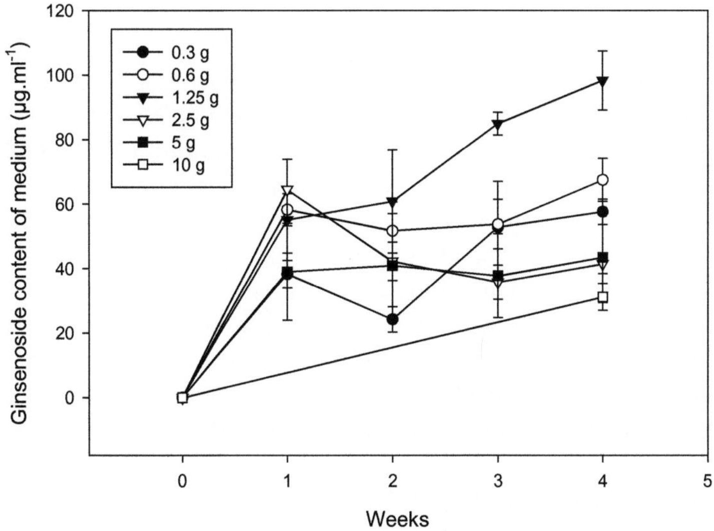

Figure 3: Changes in the ginsenoside content released in the liquid media (50 ml) used for the culture of ginseng cells from different initial inoculum sizes (mg g^{-1} cell DW).

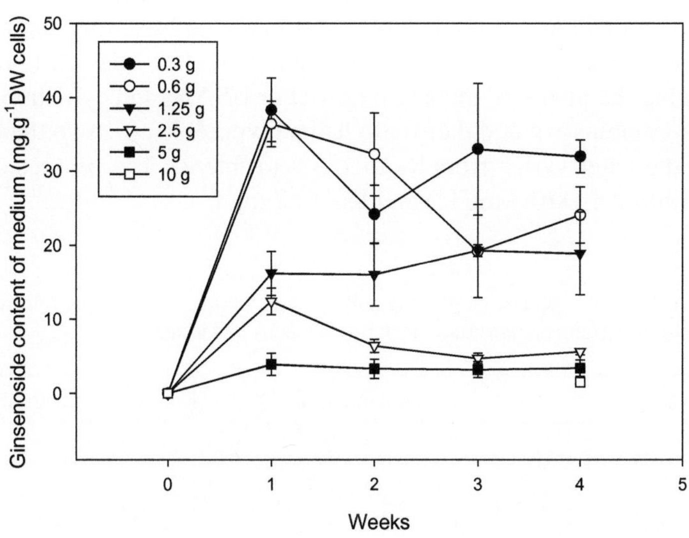

Figure 4: Changes in the ginsenoside content in the liquid culture media (µg ml^{-1}) of ginseng cells from different initial inoculum sizes.

With the smallest inoculum, in the same culture conditions, the ginsenoside level of the biomass passed from 5.2 to 96 mg g^{-1} DW within the first week and decreased progressively down to 26.7 mg g^{-1} DW after four weeks (Figure 2). The same trend was observed with the other sizes of inocula but with a far more lower production after the first week. The levels of ginsenoside released in the media were proportionally the highest for the two smaller (0.3 and 0.6 g) inocula (± 30 mg g^{-1}DW) but remained quite stable during the four-week culture period (Figure 3). When the relative concentrations of ginsenosides in the liquid culture media were considered, an increase was observed with the culture time with an inoculum size varying from 0.3 to 1.25 g (Figure 4). With an inoculum higher than 1.25g, the ginsenoside levels remained quite stable after one week. Moreover, the best production in the culture media was obtained after four weeks with an inoculum size of 1.25g.

To optimise the concentrations of KNO_3, KH_2PO_4, NH_4NO_3, FeNaEDTA and sucrose, a series of modified culture media were tested in a Plackett-Burman statistical design. Lower concentrations of KNO_3 and FeNaEDTA had a favourable effect on biomass accumulation but and no effect on ginsenoside content of the cultures using an inoculum of 2.5 g (Figure 5).

Keeping the same inoculum size (2.5 g), the regular transfer to fresh MS/2 culture medium (the first four days) did not change the biomass accumulation in 7 days. Two transfers of the cells to fresh medium at days 2 and 4 improved slightly the biomass accumulation after 7 days (Figure 6).

Changing the nature of the auxin (2,4-D or BSAA) and cytokinin (Kin or ZR), their combination and their ratio had a favourable effect on the biomass increase: the latter varied from 37 to 183% (Table 2). The best results were obtained with 2,4-D (0.3 mg l^{-1}) and Kin (0.2 mg l^{-1}).

Table 2: Fresh weight increase of ginseng cells in liquid cultures containing different auxins and cytokinins at different concentrations during 4 weeks in darkness

Auxin (mg l^{-1})	Cytokinin (mg l^{-1})	Weight increase (%)
BSAA 0.3	ZR 0.2	65 ± 5
BSAA 1	ZR 0.2	40 ± 4
BSAA 0.3	Kin 0.2	37 ± 4
BSAA 1	Kin 0.2	145 ± 12
2,4-D 0.3	ZR 0.2	118 ± 2
2,4-D 1	ZR 0.2	152 ± 10
2,4-D 0.3	Kin 0.2	183 ± 4
2,4-D 1	Kin 0.2	38 ± 5

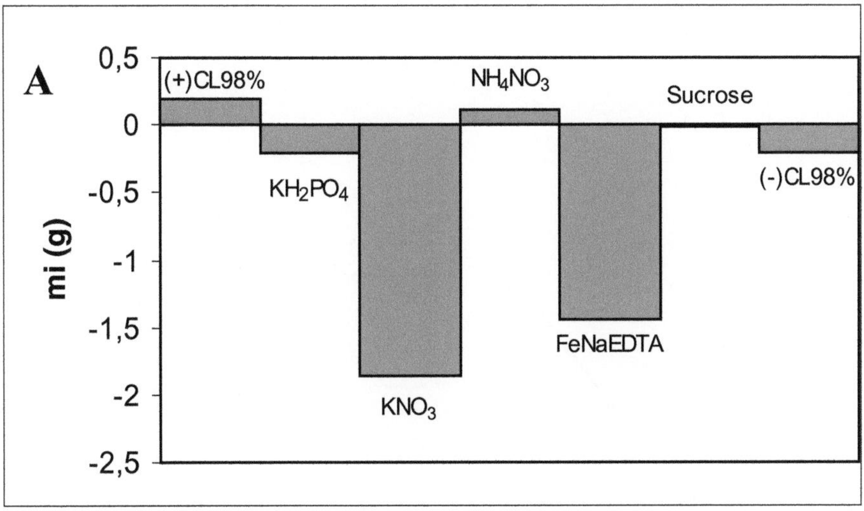

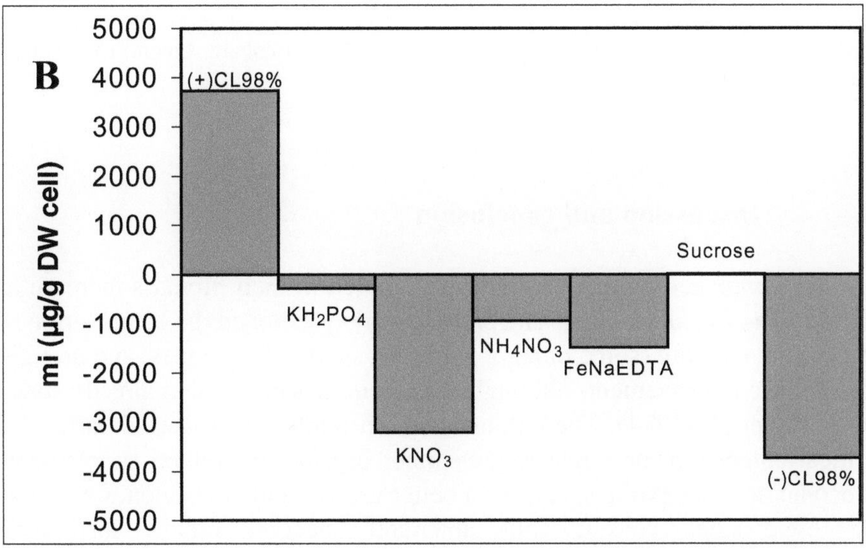

Figure 5: Effect of varying concentrations of KNO_3, KH_2PO_4, NH_4NO_3, FeNaEDTA and sucrose (Table1) on the increase in fresh weight (A) and the ginsenoside content (mi) (B). The mi values and confident limits (CL) were calculated from the statistical design of Plackett-Burman.

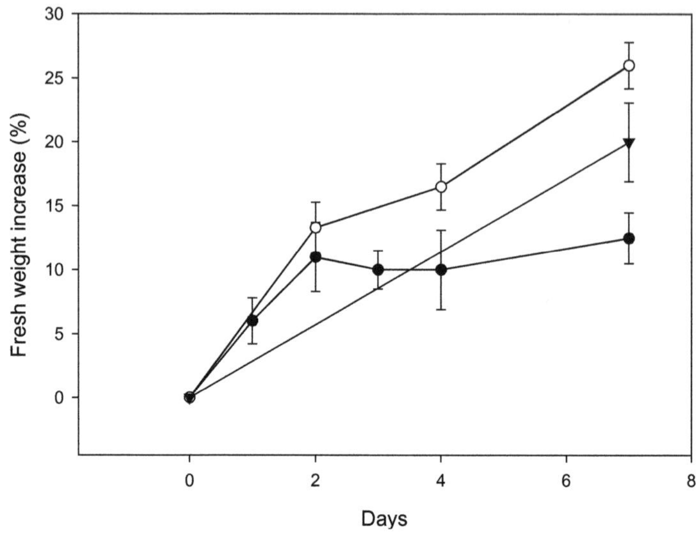

Figure 6: Fresh weight increase of ginseng cells in liquid media (50 ml) transferred to fresh medium every 1 (●) or 2 (○) days during the first 4 days of culture. Control (▼) without change.

4. Discussion and conclusion

Our *Panax* cell cultures are actually limited in their biomass increase as in their ginsenoside content. The present work questioned the composition of the culture medium (some mineral nutrients, sucrose, hormones) and brought about some improvements although still not quite satisfying as already stated by Jhang et al. (1974). The culture medium might be exhausted early of a critical component or might require other organic or mineral supplements (coconut or yeast extracts, etc) for a better growth. Other hormones will also be tested since ginseng cells apparently become rapidly insensitive (independent) to an auxin such as 2,4-D and it was shown that the capacity of saponin production was significantly lower in habituated callus than in normal callus (Furuya et al., 1983). It was also suggested that growth inhibitors were released in the medium (Kevers et al., 1999) but the results of regular transfers to fresh medium do not clearly support this hypothesis. The explant source and the conditions used for the initiation of the cell cultures will be also further investigated to overcome the observed limitations.

Acknowledgements

This research was financed by ORTIS Laboratories (Elsenborn, Belgium) and the Wallonian Ministry for Technology.

References

Asaka I, Ii I, Yoshikawa T, Hirotami M & Furuya T (1992) Embryoid formation by high temperature treatment from multiple shoots of *Panax ginseng*. Planta Med. 59: 345-346

Asaka I, Ii I, Hirotami M, Asada Y, Yoshikawa T & Furuya T (1993a) Mass production of ginseng (*Panax ginseng*) embryoids on media containing high concentrations of sugar. Planta Med. 60: 146-148

Asaka I, Ii I, Hirotami M, Asada Y & Furuya T (1993b) Production of ginsenoside saponins by culturing ginseng (*Panax ginseng*) embryogenic tissues in bioreactors. Biotech. Lett. 15: 1259-1264

Chang WC & Hsing YI (1980) Plant regeneration through somatic embryogenesis in root-derived callus of ginseng (*Panax ginseng* C.A. Meyer). Theor. Appl. Genet. 57: 133-135

Choi YE & Soh WY (1994) Origin of somatic embryo induced from cotyledons of zygotic embryos at various developmental stages of ginseng. J. Plant Biol. 37: 365-370

Choi YE & Soh WY (1996) Effect of plumule and radicule on somatic embryogenesis in the cultures of ginseng zygotic embryos. Plant Cell, Tiss. Org. Cult. 45: 137-143

Furuya T, Yoshikawa T, Ishii T & Kajii K (1983) Effects of auxins on growth and saponin production in callus cultures of *Panax ginseng*. Planta Med. 47: 183-187

Hu SY (1976) The genus *Panax* (ginseng) in Chinese medicine. Econ. Bot. 30: 11-28

Jhang JJ, Staba EJ & Kim JY (1974) American and Korean ginseng tissue cultures: growth, chemical analyses, and plantlet production. *In Vitro* 9: 253-259

Kevers C, Jacques Ph, Thonart Ph & Gaspar Th (1999) *In vitro* root cultures of *Panax ginseng* and *P. quinquefolium*. Plant Growth Regul. 27 : 173-178

Kevers C, Le Gal N, Monteiro M, Dommes J & Gaspar Th (2000) Somatic embryogenesis of *Panax ginseng* in liquid cultures: a role for polyamines and their metabolic pathways. Plant Growth Regul. 31: 209-214

Li TSC (1995) Asian and American ginseng: a review. Hort. Technology 5: 27-34

Monteiro M, Kevers C, Dommes J & Gaspar Th (2002) A specific role for spermidine in the initiation phase of somatic embryogenesis in *Panax ginseng* CA Meyer. Plant Cell, Tiss. Org. Cult. 68: 225-232

Murashige T & Skoog F (1962) A revised medium for rapid growth and bioassays with tobacco tissue culture. Physiol. Plant. 15: 473-497

Plackett RL & Burman JP (1946) The design of optimum multifactorial experiments. Biometrika 33: 305-325

Shoyama Y, Zhu XX, Nakai R, Shiraishi S & Kohda H (1997) Micropropagation of *Panax notoginseng* by somatic embryogenesis and RAPD analysis of regenerated plantlets. Plant Cell Rep. 16: 450-453

Tirajoh A, Kyung TS & Punja ZK (1998) Somatic embryogenesis and plantlet regeneration in american ginseng (*Panax quinquefolium* L.). *In Vitro* Cell. Dev. Biol. 34: 203-211

Chapter 42

Tissue culture of *Corydalis yanhusuo* (*Fumariaceae*) and its medicinal compound production from somatic embryo-derived plants

S. M. Nalawade[1], A. P. Sagare[2], C. L. Kuo[3], S. F. Lo[4], C. Y. Lee[1], Y. L. Lee[5] & H. S. Tsay[1*]

[1]*Institute of Biotechnology; Chaoyang University of Technology; 168, Gifeng E. Road, Wufeng; Taichung 413; Taiwan.*

**Corresp. author: tel: 886-4-2330-4920; fax: 2330-4921; e-mail: hstsay@mail.cyut.edu.tw*

[2]*Division of Neurovascular Biology, Center for Aging and Developmental Biology, University of Rochester Medical Center, Rochester, New York, USA.*

[3]*Institute of Chinese Pharmaceutical Science, China Medical College, Taichung, Taiwan.*

[4]*Department of Agronomy, Chiayi Agricultural Experiment Station, Agricultural Research Institute, Chiayi, Taiwan.*

[5]Department of Food and Nutrition, Fu-Jen University, Taipei, Taiwan.

Abstract: An efficient method for regeneration of entire plants *via* somatic embryogenesis in *Corydalis yanhusuo* using tuber-derived callus has been developed. Primary callus was obtained by culturing tuber explants on Murashige and Skoog's (MS) (1962) medium supplemented with 2.0 mg l^{-1} N^6-benzyladenine (BA) and 0.5 mg l^{-1} α-naphthaleneacetic acid (NAA) in darkness for one month. Somatic embryos were induced by subculturing the primary callus on MS medium supplemented with 0.5-4.0 mg l^{-1} BA, kinetin, or zeatin, within 2 weeks of culture in light. Rooting in the embryos with well developed cotyledonary leaves was achieved by transferring them to half-strength liquid MS medium supplemented with 1.0 mg l^{-1} zeatin riboside for three weeks. Converted somatic embryos were cultured on half-strength MS medium supplemented with 6% (w/v) sucrose, and with 0.5-10.0 mg l^{-1} abscisic acid (ABA), paclobutrazol, or ancymidol, 0.5-5.0 mg l^{-1} gibberellic acid (GA_3) and 15-100 mg l^{-1} polyethylene glycol (PEG) 4000 for further development of plantlets and tuber formation *in vitro*. Recurrent somatic embryogenesis has been observed in the region at the junction of cotyledon and root when the converted somatic embryos were cultured on MS basal medium supplemented with 0-10 mg l^{-1} ABA. The use of liquid MS basal medium with 0.1 mg l^{-1} GA_3 enhanced embryo conversion within fifteen days of culture. Phenotypically normal plants were recovered from converted somatic embryos and the plants showed well-developed tuber formation and *in vitro* flowering after 4 months of culture. Also, the effects of 0.5-5 mg l^{-1} ABA, paclobutrazol and 0.5-2 mg l^{-1} ancymidol, 0.5-5 mg l^{-1} GA_3 and 15-100 mg l^{-1} PEG 4000 supplemented in half-strength MS medium on the production of the two major protoberberine-type alkaloids (D,L-tetrahydropalmatine and D-corydaline) by the tubers of somatic embryo-derived plants of *C. yanhusuo* were examined. The high

A.K. Hvoslef-Eide and W. Preil (eds.), Liquid Culture Systems for in vitro Plant Propagation, 557–565.

performance liquid chromatography (HPLC) analysis revealed that the contents of D,L-tetrahydropalmatine and D-corydaline in the tubers of somatic embryo-derived plants were greater than the marketed crude drug and varied with growth regulator / PEG-4000 treatment and the age of the plant.

Key words: D-corydaline, D,L-tetrahydropalmatine, medicinal plants, protoberberine alkaloids, recurrent somatic embryogenesis

Abbreviations: ABA - abscisic acid; Ancymidol or Anc - α-cyclopropyl-α- (4-methoxy-phenyl)-5-pyrimidine methanol; BA - N^6-benzyladenine; DW - dry weight; GA_3 - gibberellic acid; kinetin - 6-furfuryl amino purine; NAA - α-naphthaleneacetic acid; Paclobutrazol or Pac - (2RS, 3RS)-1- (4-chlorophenyl)-4, 4-dimethyl-2-(1,2,4-triazol-1-yl) pentan-3-ol; PEG-4000 - polyethylene glycol-4000

1. Introduction

The genus *Corydalis* (*Fumariaceae*) comprises about 320 species, widely distributed in the northern-hemisphere, of which about seventy species have been used in traditional herbal remedies in China, Japan, and Korea (Kamigauchi and Iwasa, 1994). The dried and pulverized tubers of *C. yanhusuo*, also called *Rhizoma corydalis* or yan-hu-suo, are a rich source of several pharmacologically important alkaloids (Lee et al., 2001). These are used in traditional Chinese medicine for the treatment of gastric and duodenal ulcers, cardiac arrhythmia disease (Kamigauchi and Iwasa, 1994), rheumatism and dysmenorrhea (Tang and Eisenbrand, 1992). *C. yanhusuo* is a slow-growing herb susceptible to fungal diseases which causes serious crop loss and also affects tuber quality. To achieve high productivity, homogeneity and good quality of the tubers, it is necessary to have pathogen-free planting material (Sagare et al., 2000). Plant regeneration *via in vitro* culture of *C. yanhusuo* would be useful for the rapid mass propagation of this important medicinal plant. From the tissue culture-raised plants, usually a uniform content of the secondary metabolites could be obtained (Hatano et al., 1988). Here, we describe the protocols for complete plant regeneration *via* somatic embryogenesis from tuber-derived callus, recurrent somatic embryogenesis and production of bioactive compounds such as D,L-tetrahydropalmatine and D-corydaline (Figure 1) from the tubers of somatic embryo-derived plants standardized in our laboratory.

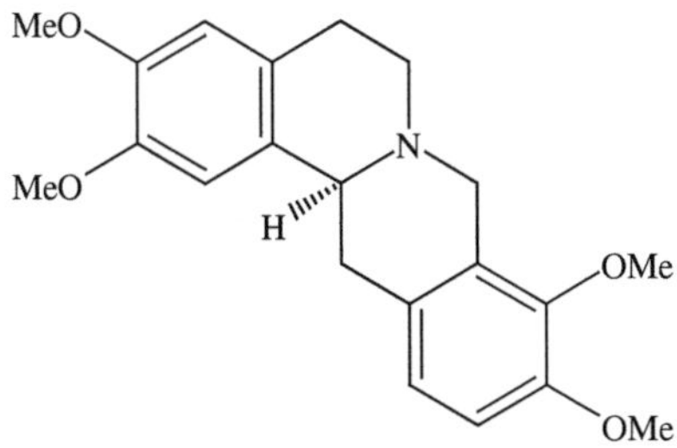

DL-tetrahydropalmatine

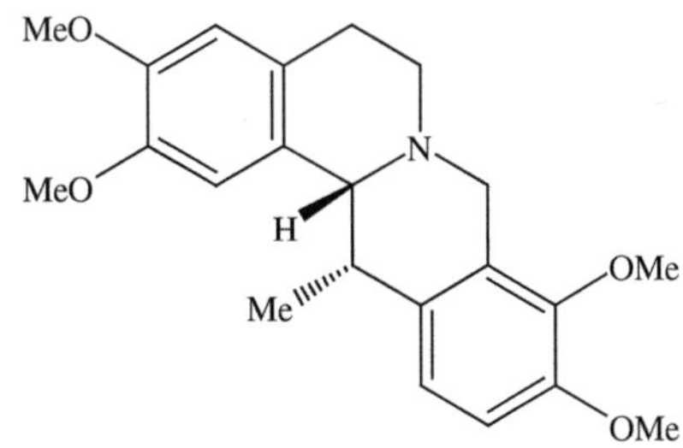

D-corydaline

Figure 1: Chemical structure of D,L-tetrahydropalmatine and D-corydaline.

2. Materials and methods

The surface sterilized mature tubers of *C. yanhusuo* were cut into pieces 5×5×2 mm and used as explants. These explants were cultured on medium containing Murashige and Skoog's (MS) salts and vitamins (Murashige and Skoog, 1962), 3% (w/v) sucrose, 0.9% (w/v) Difco Bacto agar (Difco Laboratories, Detroit, MI) and 2.0 mg l^{-1} BA in combination with 0.5 mg l^{-1} NAA. The cultures were incubated for one month at 25 ± 1°C in darkness. Primary callus (200 mg) was transferred to MS basal medium with 3% (w/v) sucrose, 0.9% (w/v) Difco Bacto agar and supplemented with 0, 0.5, 1.0, 2.0 and 4.0 mg l^{-1} BA, kinetin or zeatin. The cultures were incubated at 25 ± 1°C under cool white fluorescent light at 38 µmol $m^{-2.}$ s^{-1} with a 16-h daily photoperiod for five weeks.

The calli and attached somatic embryos were transferred to 250-ml Erlenmeyer flasks with 20 ml of half-strength MS liquid medium supplemented with 3% (w/v) sucrose and 1.0 mg l^{-1} zeatin riboside. The flasks were rotated (100 rpm) on an orbital shaker under cool white fluorescent light at 38 µmol m^{-2} s^{-1} with a 16-h photoperiod for 2 weeks.

Converted somatic embryos with well-developed shoot and roots were transferred individually to 10 ml of half-strength MS medium supplemented with 6% (w/v) sucrose, 0.9% (w/v) Difco Bacto agar ABA (0.5 - 5 mg l^{-1}) or ancymidol or paclobutrazol (0.5 - 10 mg l^{-1}) or GA_3 (0.5 - 5 mg l^{-1}) or PEG-4000 (MW 4000) (15 - 100 mg l^{-1}) in glass test tubes (22 x 120mm) The experiment was done thrice with five replicates for each treatment. The pH of all media was adjusted to 5.7 ± 0.1 with 1mol NaOH or HCl before autoclaving at 121°C, 105 kPa for 15 min.

Plantlets with well-developed tubers were transferred to half-strength MS medium supplemented with 2% (w/v) sucrose, 0.18% (w/v) Gelrite (Sigma) and 0.1 mg l^{-1} GA_3 in 250-ml Erlenmeyer flasks with 100 ml medium and incubated for 3 weeks. Plantlets with well developed roots, shoots and tubers were washed under tap water and then dipped in 0.1% (w/v) Benlate (Du Pont De Nemours and Co. Inc., Taoyuan, Taiwan) for one minute and transferred to plastic pots containing a mixture of autoclaved sand and peat moss (1:1 v/v). The pots were kept in a growth chamber for 15 days under light (100 μmol m^{-2} s^{-1}) for a 16-h photoperiod and day/night temperatures of 20°C/16°C.

The tubers were harvested after one month of culture (treatments described above), dissected to remove the aerial parts and rootlets and freeze-dried in a lyophilizer (FTS Systems™, New York, USA). The freeze-dried tubers (50 to 200 mg), obtained from different treatments, were individually finely ground at room temperature in 2 ml methanol with mortar and pestle. The extraction and quantitative analysis of D,L-tetrahydropalmatine and D-corydaline was performed as described in our eariler report (Lee et al., 2000). Analysis used a Waters high-performance liquid chromatograph (Waters™, Milford, Massachusetts, USA), connected to an Intersil ODS-3, 5 μm, 4.6 mm x 250 mm HPLC column (GL Sciences Inc., Shinjuku, Tokyo, Japan) fitted with a Guard-Pak™ precolumn (Waters). Authentic, HPLC grade D,L-tetrahydropalmatine and D-corydaline compounds (Purity: >98%) for calibration, were obtained from the Institute of Chinese Pharmaceutical Sciences, China Medical College, Taichung, Taiwan and Yoneyama Pharmaceuticals, Osaka, Japan, respectively.

3. Results and discussion

The induction of primary callus was achieved on MS medium supplemented with 2.0 mg l^{-1} BA and 0.5 mg l^{-1} NAA using mature tuber pieces as explants. The callus proliferated as yellow friable calli (Figure 2 A) when separated from the parent tissue and transferred to fresh medium every 20 days. Among the various cytokinins (BA, kinetin or zeatin) tested to assess the morphogenetic response, the efficiency of phytohormones in inducing somatic embryos varied with the cytokinin type and concentration. The embryo induced on BA- and kinetin-containing media reverted to callus. The development of somatic embryos occurred on the surface of the tuber-derived primary callus. The embryos progressed through the globular, late-globular, heart, early cotyledonary and cotyledonary stages. After five weeks of culture, somatic embryos showed development of cotyledonary

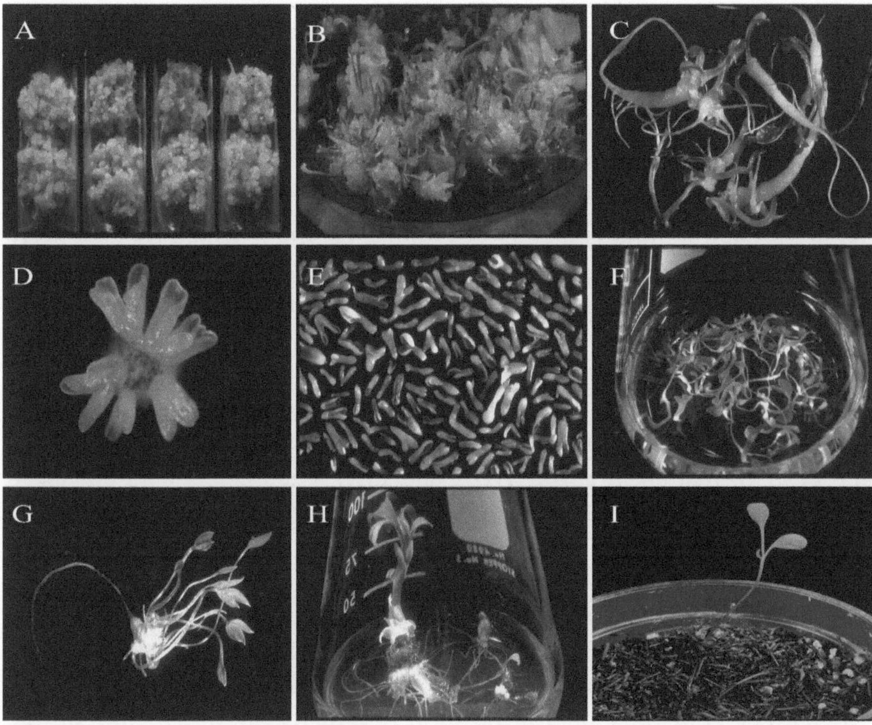

Figure 2: In vitro propagation of *Corydalis yanhusuo via* somatic embryogenesis and recurrent somatic embryogenesis.

(A) Proliferating tuber derived callus on MS basal medium supplemented with 2 mg l^{-1} BA and 0.5 mg l^{-1} NAA.

(B) Somatic embryos showing development of cotyledonary leaves after five weeks of culture on MS basal medium supplemented with 1.0 mg l^{-1} zeatin.

(C) Converted somatic embryo with well-developed roots and cotyledonary leaves on half strength MS basal medium supplemented with 6 % sucrose for one month.

(D) Development of secondary somatic embryos on the surface of converted somatic embryo after two months of culture on MS basal medium supplemented with ABA.

(E) Synchronized development of secondary somatic embryos on converted primary somatic embryos on 2.0 mg l^{-1} ABA containing medium.

(F) Converted secondary somatic embryos after culture in MS liquid medium supplemented with 0.1 mg l^{-1} GA_3.

(G) Converted secondary somatic embryos cultured on medium with 5.0 mg l^{-1} GA_3 showing about eight shoots and roots.

(H) Converted embryos showing developed tuber on MS medium devoid of phytohormones after four months of incubation.

(I) Hardened plant derived from the somatic embryo.

leaves on the medium supplemented with 1.0 mg l^{-1} zeatin (Figure 2B), however, root development was arrested. The development of roots occurred in embryos with well-developed cotyledonary leaves when they were cultured in half-strength liquid MS medium supplemented with 1.0 mg l^{-1} zeatin riboside for three weeks.

Our initial attempts to transfer the tissue culture plant to soil immediately after conversion in liquid medium containing 1.0 mg l^{-1} zeatin riboside were not successful. Therefore, to further develop roots, shoots and tuberization *in vitro*, converted somatic embryos were cultured on medium supplemented with 6% (w/v) sucrose and different levels of either ABA, growth retardants (paclobutrazol, ancymidol), PEG-4000, or GA_3. Similar to our study, higher sucrose concentration (Khuri and Moorby, 1995) and plant growth retardants (Seabrook et al., 1993; Vreugdenhil et al., 1994) have been used for potato microtuber formation in *vitro*. ABA and PEG as an osmoticum enhanced the accumulation of storage reserves in somatic embryos of white spruce (Leal et al., 1995; Roberts et al., 1990).

Of all the treatments tested, the medium with 6% (w/v) sucrose alone promoted tuber, root, and shoot development (Figure 2 C). The plantlets with well-developed roots, shoots and tubers from the converted somatic embryos were hardened for two months in growth chambers.

Recurrent somatic embryos developed directly in the region of the junction of the cotyledonary leaf and root of converted somatic embryos when cultured on ABA- containing media (Figure 2 D). It has been reported that ABA has an important role in early embryonic events (Senger et al., 2001). Among the various ABA treatments tested, the maximum number of secondary embryos were observed on medium containing 8 mg l^{-1} ABA. In the medium supplemented with higher ABA concentrations (8 - 16 mg l^{-1}), a decrease in the number of somatic embryos with browning of the primary somatic embryo was observed. The somatic embryos induced on lower concentrations (0.5 and 1.0 mg l^{-1}) of ABA favored precocious conversion. Embryos induced on 4 - 16 mg l^{-1} ABA showed development of secondary embryos. The development of somatic embryos directly without an intervening callus phase was confirmed by scanning electron microscopy and histology. Histological study revealed that the somatic embryos arose directly from the epidermal layer of cells. Generally, the development of the secondary embryos is not synchronized and the development is also up to the pre-embryonic mass or globular embryo stages. (Raemakers et al., 1995; Fernando and Gamage, 2000).

Table 1: Effect of different concentrations of ABA, paclobutrazol, ancymidoe, GA_3, and PEG-4000 in the medium on the production of of D,L-tetrahydropalmatine and D-corydaline (mg/g^{-1} dry weight) by culturing tubers of somatic embryo-derived plants of *Corydalis yanhusuo* for one month

Treatment ($mg\ l^{-1}$)		THP content ($mg\ l^{-1}$ DW ± standard error)	COR content ($mg\ l^{-1}$ DW ± standard error)
None		1.01 ± 0.14	0.53 ± 0.22
ABA	0.5	1.33 ± 0.55	2.23 ± 0.51
	1.0	1.44 ± 0.35	0.81 ± 0.28
	2.0	1.45 ± 0.15	1.93 ± 0.49
	5.0	2.30 ± 0.98	2.22 ± 1.58
Pac	0.5	1.04 ± 0.13	1.19 ± 0.25
	1.0	1.06 ± 0.28	1.22 ± 0.69
	2.0	0.49± 0.16	0.68 ± 0.28
	5.0	0.86 ± 0.17	0.54 ± 0.14
	10.0	0.50 ± 0.10	0.41 ± 0.29
Anc	0.5	0.13 ± 0.03	0.15 ± 0.04
	1.0	0.82 ± 0.35	1.86 ± 0.78
	2.0	1.24 ± 0.42	1.54 ± 0.48
	5.0	1.10 ± 0.44	1.40 ± 0.50
GA_3	0.5	0.39 ± 0.03	0.19 ± 0.03
	1.0	1.22 ± 0.21	0.24 ± 0.10
	2.0	1.23 ± 0.15	0.69 ± 0.17
	5.0	0.62 ± 0.09	0.32 ± 0.15
PEG	15	0.52 ± 0.36	0.44 ± 0.23
	25	1.06 ± 0.38	1.28 ± 0.33
	50	1.04 ± 0.29	0.84 ± 0.26
	100	0.59 ± 0.18	0.38 ± 0.15
Commercial crude drug		0.35 ± 0.08	0.33 ± 0.10

With our regeneration system, a synchronized development of embryos on converted primary somatic embryos, by using appropriate concentration of ABA, can be achieved. Secondary embryos developed on 2.0 mg l^{-1} ABA were uniform in size (Figure 2 E) and the embryos converted within 15 days of culture in media with different phytohormones. The conversion of embryos was found to be optimum in the GA_3-containing media (Figure 2 F). Converted somatic embryos cultured on medium with 5.0 mg l^{-1} GA_3 and sucrose 6 % (w/v) developed about eight thin shoots and roots (Figure 2 G). Converted embryos developed tubers on MS medium without phytohormones when incubated for four months (Figure 2 H). The plantlets were hardened for two months in growth chambers (Figure 2 I).

HPLC analysis revealed presence of both D,L-tetrahydropalmatine and D-corydaline in the tubers. A marked variation in the content of D,L-tetrahydropalmatine and D-corydaline among different treatments was observed. However, in most of the treatments, D,L-tetrahydropalmatine and D-corydaline content in the tubers was greater than the marketed crude drug and varied depending on the treatment (Table 1). The low amount of the compounds in the crude drug may be due to their partial loss during processing which usually involves either boiling in water, vinegar or wine (Anonymous, 1999).

4. Conclusions

Using the plant regeneration protocol described above, somatic embryos were induced in tuber-derived callus with subsequent recurrent somatic embryos from epidermal cells of the converted primary embryos of *C. yanhusuo.* Tuber formation *in vitro* has been achieved in the converted somatic embryos. By using appropriate growth regulators and prolonged incubation under controlled conditions it is possible to produce good quality tubers with uniform high contents of the two protoberberine-type alkaloids D,L-tetrahydropalmatine and D-corydaline. These protocols can be used for mass propagation, biochemical or genetic transformation studies.

Acknowledgements

This research was supported by a grant from the National Science Council (NSC 88-2317-B-055-001), Taiwan.

References

Anonymous (1999) Yanhusuo. In: Chung-Hua-Ben-Tsao, Vol. 3 (pp. 643-649). Shanghai Science Technology Press, Shanghai (in Chinese)

Fernando SC & Gamage CKA (2000) Abscisic acid induced somatic embryogenesis in immature embryo explants of coconut (*Cocos nucifera* L.). Plant Sci. 151: 193-198

Hatano K, Kamura K, Shoyama Y & Nishioka I (1988) Clonal multiplication of *Conitum armichaeli* by tip tissue culture and alkaloid contents of clonally propagated plant. Planta Med. 54: 152-155

Kamigauchi M & Iwasa K VI (1994) *Corydalis* spp.: *In vitro* culture and the biotransformation of protoberberines. In: Bajaj Y P S (ed) Biotechnology in Agriculture and Forestry, Vol. 26 (pp. 93-105). Springer-Verlag, Berlin

Khuri S & Moorby J (1995) Investigations into the role of sucrose in potato cv. Estima microtuber formation *in vitro*. Ann. Bot. 75: 295-303

Leal I, Misra S, Attree SM & Fowke LC (1995) Effect of abscisic acid, osmoticum and desiccation on 11S storage protein gene expression in somatic embryos of white spruce. Plant Sci. 106: 121-128

Lee YL, Sagare AP, Lee CY, Feng HT, Ko YC, Shaw JF & Tsay HS (2001) Formation of protoberberine-type alkaloids by the tubers of somatic embryo-derived plants of *Corydalis yanhusuo*. Planta Med. 67: 839-842

Murashige T & Skoog F (1962) A revised medium for rapid growth and bioassays with tobacco tissue cultures. Physiol. Plant. 15: 473-497

Raemakers CJJM, Jacobsen E & Visser RGF (1995) Secondary somatic embryogenesis and applications in plant breeding. Euphytica 81: 93-107

Roberts DR, Flinn BS, Webb DT, Webster FB & Sutton BCS (1990) Abscisic acid and indole-3-butyric acid regulation of maturation and accumulation of storage proteins in somatic embryos of interior spruce. Physiol. Plant. 78: 355-360

Sagare AP, Lee YL, Lin TC, Chen CC & Tsay HS (2000) Cytokinin-induced somatic embryogenesis and plant regeneration in *Corydalis yanhusuo* (*Fumariaceae*) - a medicinal plant. Plant Sci. 160: 139-147

Seabrook JEA, Coleman S, & Levy D (1993) Effect of photoperiod on *in vitro* tuberization in potato (*Solanum tuberosum* L.), Plant Cell, Tiss. Org. Cult. 34: 43-51

Senger S, Mock HP, Conrad U & Manteuffel R (2001) Immunomodulation of ABA function affects early events in somatic embryo development. Plant Cell Rep. 20: 112-120

Tang W & Eisenbrand G (1992) *Corydalis turtschaninovii* Bess. f. *yanhusuo* Y.H. Chou et C.C. Hsü. In: Tang W & Eisenbrand G (eds) Chinese Drugs of Plant Origin, Chemistry, Pharmacology, and Use in Traditional and Modern Medicine (pp. 377-393). Springer-Verlag, Berlin

Vreugdenhil D, Bindels P, Reinhoud P, Klocek J & Hendriks T (1994) Use of the growth retardant tetcyclacis for potato tuber formation *in vitro*. Plant Growth Regul. 14: 257-265

Chapter 43

Production of taxanes in callus and suspension cultures of *Taxus baccata* L.

K.Bruňáková, Z. Babincová & E.Čellárová
Institute of Biology and Ecology, Faculty of Science, P. J. Šafárik University, Mánesova 23, 041 54 Košice, Slovakia. E-mail: brunakov@kosice.upjs.sk

Abstract: The selection and cloning of cell lines for the establishment of long-term cell cultures of *Taxus baccata* L. is important for the prospective biotechnological production of taxanes. Callus cultures were established from young stems of adult trees and aseptically grown seedlings. All parts of young seedlings were more suitable for cell proliferation on callus-induction media as compared to young stems of adult trees. The best growing calli successfully achieved an average 8-fold increase in fresh weight after 20 months of culture. The growth characteristics of two seedling-derived cell lines were determined. The best growing calli were used for establishment of suspension cultures. Gamborg's B5 medium supplemented with 3 mg l^{-1} 2,4-dichlorophenoxyacetic acid, 0.5 mg l^{-1} kinetin and 1.5% polyvinylpyrrolidone, a phenolic-binding agent, was used as agar-solidified and liquid medium. A 20-month-old callus culture of *T. baccata* (VI/Ha) produced paclitaxel up to 0.0109 ± 0.0037 % of extracted dry weight basis. The content of taxanes was determined by high performance liquid chromatography or a competitive inhibition enzyme immunoassay system (CIEIA). A kinetic study of callus growth and taxane production was performed.

Key words: Cell culture, kinetics study, taxane, paclitaxel, yew

Abbreviations: E.D.W. – extracted dry weight; V/Kle – callus culture derived from cotyledon of an embryo-derived seedling (genotype V); VI/Ha – callus culture derived form hypocotyl of an embryo-derived seedling (genotype VI); GI – growth rate; PVP - polyvinylpyrrolidone

1. Introduction

Plant cell culture is an alternative method for commercial propagation as well as production of secondary metabolites *in vitro* (George and Sherrington, 1984; Barz and Ellis, 1981). Paclitaxel (TAXOL®; *Bristol-Myers Squibb Co.*) (Figure 1) and docetaxel (TAXOTERE®; *Rhône-Poulenc Rorer*

A.K. Hvoslef-Eide and W. Preil (eds.), Liquid Culture Systems for in vitro Plant Propagation, 567–574.

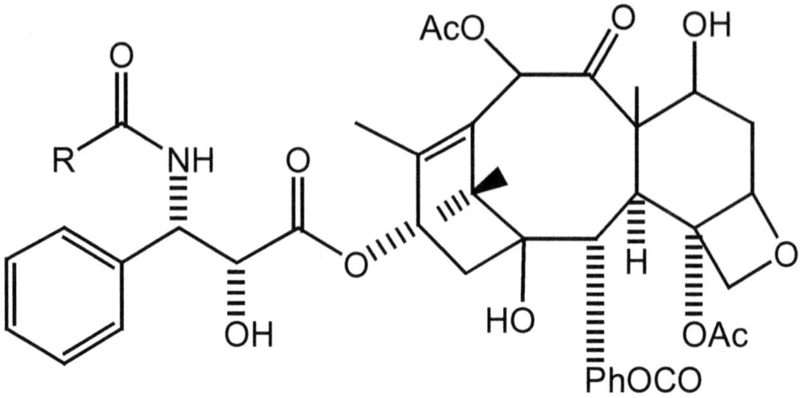

Figure 1: The structure of paclitaxel and cephalomannine; paclitaxel: **R**-C_6H_{10}; cephalomannine: **R**-C_4H_7

Co.) are derived from natural sources. At least 40 % of the phytopharmaceuticals being currently used in Western countries is of natural origin (Rout et al., 2000). With a unique mechanism of action – the inhibition of microtubule depolymerization – taxanes are considered as the most important drug in the treatment of breast carcinoma since antracyclines (Wagnerová and Andrašina, 1996). In 1992, TAXOL® was approved by the United States Food and Drug Administration for treating ovarian and breast cancer. The history of paclitaxel, the chemistry and biosynthesis as well as mechanism of its action and broad antitumour spectra are well documented (Heinstein and Chang, 1994; Jennewein and Croteau, 2001).

The enormous requirements for supplying TAXOL®, originally derived from the bark of the Pacific yew, occurred as a result of TAXOL® effectiveness in clinical trials of cancer treatment and can lead to devastation of natural yew resources (Song and Dumais, 1991). Plant cell cultures of *Taxus* sp. seem to be promising as a potential candidate for commercial TAXOL® production in the near future. While the early work was concentrated on establishing cell lines and taxoid assay (Zhong, 1995; Wickremesinhe and Arteca, 1993), on selection of high-paclitaxel-producing cell and protoplast cultures for efficient paclitaxel production (Aoyagi et al., 2002), recently the emergence of recombinant DNA techniques has opened new possibilities of studying and modifying expression of genes related to taxane biosynthesis (Jennewein and Croteau, 2001). To achieve higher productivity, the selection of cell lines with appropriate genetic, biochemical and physiological properties is necessary.

2. Material and methods

2.1 Culture conditions

Callus cultures were established from embryo-derived seedlings and young stems of mature trees of *T. baccata* L. growing in the Botanical Garden of P. J. Šafárik University, Košice, Slovakia, during 1999 and 2000. Suspension cultures were established using stable growing two years old callus cultures as inoculum. Cells were maintained as callus and suspension cultures on Gamborg's B5 medium (Gamborg et al., 1968) supplemented with an antioxidant 1.5% polyvinylpyrrolidone (PVP) and growth regulators 3 mg l^{-1} 2,4-dichlorophenoxyacetic acid (2,4-D) and 0.5 mg l^{-1} kinetin. Media were prepared with double distilled water, solidified with 0.6% agar, adjusted to pH 5.6 before autoclaving and sterilization at 121 °C for 15 min. The cultures were incubated in the culture room under dark conditions at 22±2 °C and 34 % relative humidity. For the induction of callus the primary explants were placed on 5 ml of medium in test tubes. The best growing calli were used for the establishment of prospective long-term cultures cultivated in 100 ml Erlenmeyer flasks on 20 ml of agar-solidified medium. The transfers to fresh medium were done after 28-35 (callus) and 14 (suspension) days of culture. Cell suspension cultures were incubated in 100 ml Erlenmeyer flasks on a rotatory shaker (90 rpm). The cell viability of suspension cultures was microscopically detected by methyl-blue staining.

2.2 Growth measurements

The results are presented as a growth rate based on fresh weight $GI=(Wt_{Final} - Wt_{Initial})/Wt_{Initial}$. Kinetics curves were determined by weighing the fresh cell mass, each data point represents the mean of 9 replications ± standard error.

2.3 Taxane analysis

Callus samples were harvested during subculture, frozen, lyophilised, and then extracted with 0.1 or 1.0 ml methanol. Determination of taxane content was done by HPLC using a reverse phase (Wickremesinhe and Arteca, 1993). A competitive inhibition enzyme immunoassay system (CIEIA) was used (Grothaus et al., 1995) to monitor level of paclitaxel in some callus cells because of low limit of detection (sensitive above 0.0035 mg l^{-1}). Each data is presented as a mean value from 3 independent flasks ± standard error.

3. Results and Discussion

All parts of young seedlings were more responsive to cell proliferation on callus-induction media as compared with young stems of adult trees. Following the first subculture all calli grew slowly. After 1-2 years of culture some of them improved their growth. Two callus cultures (labelled V/Kle and VI/Ha) were selected for subsequent study.

3.1 Growth characteristics

The fast-growing white-coloured and friable V/Kle callus culture was initiated from cotyledons of embryo-derived seedlings. After 42 days of culture, GI was 7.96±0.49 (n=9; the 24th subculture) with the doubling time estimated in the range of 12-14 days (Figure 2). The majority of callus samples originating from white-coloured calli released a red-coloured compound into the solid or liquid medium, grew at a slow rate and did not survive in subsequent subcultures. According to Wickremesinhe and Arteca (1994) the production of these red-coloured exudates – possibly phenolics – is a totally random and unpredictable process, which eventually leads to cell death. V/Kle suspension culture was pale yellow with the presence of individual oval- or oblong-shaped cells and cell aggregates ranging between 0.5-2.0 mm in diameter. Methyl-blue staining revealed the high viability of cells with relatively large vacuoles and a thin layer of cytoplasm.

The slowly growing, light-brown and more aggregated VI/Ha callus culture was initiated from hypocotyls of embryo-derived seedling. Unlike V/Kle, callus exudates, no red-coloured compound was produced in the presence of a phenolic-binding compound (PVP). GI was 3.45±0.21 (n=9; the 23rd subculture) (Figure 2). The relatively slow growth rate and the doubling time in the range of 14-16 days are comparable with other slow-growing woody species (Fett-Neto and DiCosmo, 1997).

3.2 Taxane yield

The concentrations of paclitaxel in callus cells reached the maximum of 0.0109 ± 0.0037 % of the extracted dry weight in slow-growing VI/Ha callus culture. This amount of paclitaxel is comparable to that found in the bark of the intact plant of various yew species (Fett-Neto et al., 1992; Vance et al., 1994; Jaziri et al., 1991). The highest value obtained from V/Kle callus selected for faster growth represented a paclitaxel yield only of 0.00006 ± 0.00003 % of the extracted dry weight. Both callus cultures have

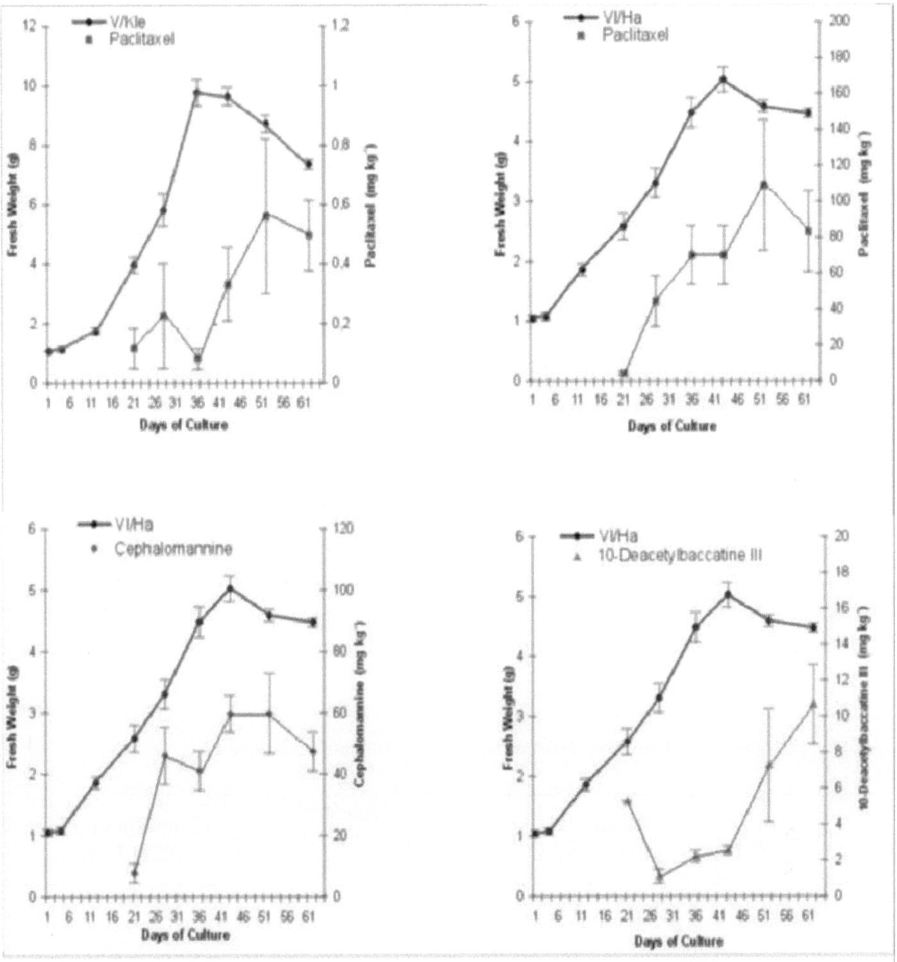

Figure 2: Changes in taxane concentrations of *T. baccata* L. callus cultures (V/Kle and VI/Ha) during a subculture cycle. The accumulation of taxanes shows a biphasic pattern with two peaks. Taxane data are averages of measurements from three independent flasks ± standard error.

been in culture for more than 20 months. The amount of important paclitaxel precursors 10-deacetyl baccatin III and cephalomannine was also investigated during the same subculture (Table 1). The amount of paclitaxel and 10-deacetylbaccatin III peaked after approximately 62 days of culture and cephalomannine after 52 days of culture.

The study on the growth kinetics and taxane production indicated an inverse relationship between growth and taxane accumulation (Sakuta and Komamine, 1987). The pattern of accumulation of all taxanes was similar and mostly had two peaks (Figure 2). The first one appeared in the exponential growth phase of the callus and the second one was typical for the phase of progressive deceleration or stationary phase. As mentioned by Fett-Neto and DiCosmo (1997), the early release of paclitaxel into the medium of cell suspensions of *T. cuspidata* could be a response to the transfer into fresh medium due to absorption of ammonium and a second peak occurring at a stationary phase could result from excretion by the cells that reached their highest level of paclitaxel.

Our observations are consistent with the suggestion that in comparison with the younger pale-yellow callus containing most mitotically active cells, the older brown callus with more non-dividing cells contains a higher amount of paclitaxel (Wickremesinhe and Arteca, 1993). As mentioned by these authors the selection for friable and faster-growing callus may lead to the selection of cells that produce lower amounts of secondary metabolites.

Table 1: Taxane yield from two different *in vitro* cultures of *T. baccata* L.

Callus Culture	Paclitaxel (% E.D.W.)	10-Deacetyl baccatin III (% E.D.W.)	Cephalomannine (% E.D.W.)
V/Kle	0.00006 ± 0.00003	—	—
VI/Ha	0.0109 ± 0.0037	0.0011 ± 0.0002	0.0059 ± 0.0013

4. Conclusion

Stable-growing seedling-derived callus and suspension cultures were established using callus-induction Gamborg's B5 medium. We have demonstrated the ability to produce paclitaxel and its analogues from two stable-growing callus lines with different growth characteristics. The amount of paclitaxel was comparable to that found in the bark of the intact plant of various *Taxus* sp. An inverse relationship between growth and taxane accumulation was confirmed. To achieve higher productivity of taxanes, the optimisation of culture conditions may be necessary.

Acknowledgement

This research was granted by pharmaceutical company Slovakofarma, Hlohovec (Slovakia) and Ministry of Economy of Slovak Rebublic. We wish to thank Drug Research Institute Modra (Slovakia) for chemical analyses of the samples.

References

Aoyagi H, DiCosmo F & Tanaka H (2002) Efficient paclitaxel production using protoplats isolated from cultured cells of *Taxus cuspidata.* Planta Med. 68: 420-424

Barz W & Ellis B (1981) Potential of plant cell cultures for pharmaceutical production. In: Beal JL and Reinhard E (eds) Natural Products as Medicinal Agents (pp. 471-507). Hippokrates, Stuttgart

Fett-Neto AG, DiCosmo F, Reynolds WF & Sakata K (1992) Cell culture of *Taxus* as a source of the neoplastic drug taxol and related taxanes. Biotechnol. 10: 1572-1575

Fett-Neto AG & DiCosmo F (1997) Taxol and taxane production by cell culture. Encyclopedia of Molecular Biology and Molecular Medicine 6 (pp. 10-17). VCH Verlagsgesellschaft mbH, Weinheim

Gamborg OL, Miler RA & Ojima K (1968) Nutrient requirements of suspension cultures of soybean root cells. Exp. Cell Res. 50: 148-151

George EF & Sherrington PD (1984) Plant Propagation by Tissue Culture. Exegetics Ltd., Eversley, England

Grothaus PG, Bignami GS, O´Malley S, Harada KE, Byrnes JB, Waller DF, Raybould TJG, McGuire MT & Alvarado B (1995) Taxane-specific monoclonal antibodies: measurement of taxol, baccatin III, and "total taxanes" in *Taxus brevifolia* extracts by enzyme immunoassay. J. Nat. Prod. 58: 1003-1014

Heinstein PF & Chang C (1994) Taxol. Ann. Rev. Plant Mol. Biol. 45: 663-74.

Jaziri M, Diallo BM, Vanhaelen MH, Vanhaelen-Fastré RJ, Zhiri A, Becu AG & Homes J (1991) Enzyme-linked immunosorbent assay for the detection and the semi-quantitative determination of taxane diterpenoids related to taxol in *Taxus* sp. and tissue cultures. J. Pharm. Belg. 46: 93-99

Jennewein S & Croteau R (2001) Taxol: biosynthesis, molecular genetics and biotechnological applications. Appl. Microbiol. Biot. 57: 13-19

Rout GR, Samantaray S & Das P (2000) *In vitro* manipulation and propagation of medicinal plants. Biotechnol. Adv. 18: 91-120

Sakuta M & Komamine A (1987) Cell growth and accumulation of secondary metabolites. Cell Culture and Somatic Cell Genetics of Plants 4: 97-110

Song JI & Dumais MR (1991) from yew to us: The curious development of taxol. JAMA 266: 1281

Vance NC, Kelsey RG & Sabin TE (1994) Seasonal and tissue variation in taxane concentrations of *Taxus brevifolia.* Phytochemistry 36: 1241-1244

Wagnerová M & Andrašina I (1996) Liečba karcinómov mliečnej žľazy a paclitaxel (TAXOL®). Zborník abstraktov z XXXIII. Onkologických dní (pp. 5-9). Bratislava

Wickremesinhe ERM & Arteca RN (1993) *Taxus* callus cultures. Initiation, growth optimization characterization and taxol production. Plant Cell, Tiss. Org. Cult. 35: 181-193

Wickremesinhe ERM & Arteca RN (1994) *Taxus* cell suspension cultures: optimizing growth and production of taxol. J. Plant. Physiol. 44: 183-188

Zhong JJ (1995) Recent advances in cell cultures in *Taxus* sp. for production of natural anticancer drug taxol. Plant Tiss. Cult. Biotechnol. 1: 75-80

Acknowledgments

The editors thank the Editorial Committee members Dr. Alan Cassells, Dr. Christopher Hunter and Dr. Geert-Jan de Klerk for their valuable contributions throughout. We are also indepted to Mrs. Brita Hvidsten for all her hard work and patience during the final editing of the book.

Plant Index

Subject Index